D0221842

CONTINUOUS GEOMETRY

PRINCETON LANDMARKS IN MATHEMATICS AND PHYSICS

Non-standard Analysis,
by Abraham Robinson

General Theory of Relativity,
by P.A.M. Dirac

Angular Momentum in Quantum Mechanics,
by A. R. Edmonds

Mathematical Foundations of Quantum Mechanics,
by John von Neumann

Introduction to Mathematical Logic,
by Alonzo Church

Convex Analysis,
by R. Tyrrell Rockafellar

Riemannian Geometry,
by Luther Pfahler Eisenhart

The Classical Groups,
by Hermann Weyl

Topology from the Differentiable Viewpoint,
by John W. Milnor

Algebraic Theory of Numbers,
by Hermann Weyl

Continuous Geometry,
by John von Neumann

Linear Programming and Extensions,
by George B. Dantzig

Operator Techniques in Atomic Spectroscopy,
by Brian R. Judd

CONTINUOUS GEOMETRY

BY
John von Neumann

FOREWORD BY
ISRAEL HALPERIN

PRINCETON UNIVERSITY PRESS
PRINCETON, NEW JERSEY

Published by Princeton University Press, 41 William Street,
Princeton, New Jersey 08540
In the United Kingdom: Princeton University Press, Chichester, West Sussex

L.C. Card: 59-11084
ISBN 0-691-05893-8

Princeton University Press books are printed on acid-free paper and meet the guidelines for permanence and durability of the Committee on Production Guidelines for Book Longevity of the Council on Library Resources

First printing, in the Princeton Landmarks in Mathematics and Physics series, 1998

http://pup.princeton.edu

Printed in the United States of America

1 3 5 7 9 10 8 6 4 2

FOREWORD

This book reproduces the notes of lectures on Continuous Geometry given by John von Neumann at Princeton. Part I was given during the academic year 1935—36, and Parts II and III were given during the academic year 1936—37. The notes were prepared, while the lectures were in progress, by L. Roy Wilcox, and multigraphed copies were distributed by the Institute for Advanced Study. The supply was soon exhausted, and the notes have not been reproduced until now.

In the present edition many slips in typing have been corrected, in Part I with the help of Wallace Givens. I have inserted a few editorial remarks and I have made a small number of changes in the text. These changes, together with some comments, are listed at the back of the book. Of these changes, only one is essential and it was authorized explicitly by von Neumann.

Continuous geometry was invented by von Neumann in the fall of 1935. His previous work on rings of operators in Hilbert space, partly in collaboration with F. J. Murray, had led to the discovery of a new mathematical structure which possessed a dimension function (cf. F. J. Murray and J. von Neumann, *On Rings of Operators*, Annals of Mathematics, vol. 37 (1936), page 172, Case (II_1)). The new structure had incidence properties resembling those of the system L_n (L_n denotes the lattice of all linear subsets of an $n-1$ dimensional projective geometry), but its dimension function assumed as values all real numbers in the interval $(0, 1)$.

Von Neumann set out to formulate suitable axioms to characterise the new structure. It happened that just previously, K. Menger and G. Birkhoff had characterised L_n by lattice-type axioms; in particular, Birkhoff had shown the structures L_n could be characterised as the complemented modular irreducible lattices which satisfy a chain condition. Von Neumann dropped the chain condition and replaced it by two of its weak consequences: (i) order completeness of the lattice (Axiom II on page 1 of this book), and (ii) continuity of the lattice operations (Axiom III on page 2 of this book). Lattices which are complemented, modular, irreducible, satisfy (i) and (ii), but do not satisfy a chain condition, were called by von Neumann: *continuous geometries* (reducible continuous geometries were considered later, in Part III). It is easy to see that in a continuous geometry there can be no minimal element — that is, no atomic element or *point*.

The structure previously discovered by Murray and von Neumann in their research on rings of operators was an example, the first, of a continuous geometry.

Von Neumann's first fundamental result was the construction, for an arbitrary continuous geometry, of a dimension function with values ranging over the interval (0, 1). The construction was based on the definition: x and y are to be called equidimensional if x and y are in perspective relation, that is: for some w the lattice join and meet of x with w are identical with those of y with w. The essential difficulty is to prove that the perspective relation is transitive.

Part I of this book discusses the axioms for continuous geometry and gives the construction of the dimension function. These results were summarized in von Neumann's note, *Continuous Geometry*, Proceedings of the National Academy of Sciences, vol. 22 (1936), pages 92—100.

In a second note, *Examples of Continuous Geometries*, Proceedings of the National Academy of Sciences, vol. 22 (1936), pages 101—108, von Neumann used a simple device to construct an extensive class of new continuous geometries. He started with finite dimensional vector spaces over an arbitrary but fixed division ring $\mathscr{D}$ and showed how to imbed an n-dimensional vector space in a kn-dimensional one with $k \geqq 2$; a countable number of repetitions of this imbedding procedure, followed by a metric completion, gives a continuous geometry. This construction uses a countable number of factors k, each $\geqq 2$, but the final result is independent of the particular choice of the k. This was shown by von Neumann in a manuscript (still unpublished): *Independence of F_∞ From The Sequence v* (MS written in 1936—37).

If $\mathscr{D}$ is the ring of all complex numbers, the continuous geometry obtained by the imbedding procedure is non-isomorphic to, and simpler than, the continuous geometry obtained from a ring of operators. This non-isomorphism is established in von Neumann's paper *The Non-Isomorphism Of Certain Continuous Rings*, Annals of Mathematics, vol. 67 (1958), pages 485—496 (MS actually written in 1936—37). The imbedding procedure, the new examples, and the non-isomorphism theorem are not mentioned in the present book.

Von Neumann's next fundamental result was a deep and technically remarkable generalization of the classical Hilbert – Veblen and Young coordinatization theorem. This classical theorem asserts that if $n \geqq 4$, then the points of L_n can be coordinatized with homogeneous coordinates, the coordinates to be taken in some suitable division ring $\mathscr{D}$. Von Neumann first expressed this classical theorem in the following two equivalent forms which do not mention points explicitly and which coordinatize *all* the linear subsets in L_n: L_n, as a *lattice*, is isomorphic to

(i) the lattice of all right ideals in $\mathscr{D}_n$ ($\mathscr{D}_n$ denotes the ring of n-th order matrices with elements in $\mathscr{D}$)

and to

(ii) the lattice of all right submodules of $\mathscr{D}^n$ ($\mathscr{D}^n$ denotes the right module over $\mathscr{D}$ of all n-tuples $(x^1, \cdots, x^n)$ with all x^i in $\mathscr{D}$).

Then he established these coordinatization-isomorphism theorems for every complemented modular lattice of order $n \geqq 4$ (this restriction is a lattice generalization of the condition $n \geqq 4$ for L_n). But in formulating the general coordinatization theorem, von Neumann made some changes: first, the division ring $\mathscr{D}$ was replaced by a more general ring $\mathfrak{S}$ with the properties: $\mathfrak{S}$ has an identity element and for each x in $\mathfrak{S}$, $xyx = x$ for some y in $\mathfrak{S}$ (such rings $\mathfrak{S}$ were called by von Neumann: *regular*); secondly, in the isomorphism with right ideals in $\mathfrak{S}_n$ only *principal* right ideals were to be used; finally, in the isomorphism with right submodules of $\mathfrak{S}^n$ only those right submodules were to be used which are spanned by a *finite* number of vectors.

The properties of regular rings and the proof of the coordinatization-isomorphism theorem are the main content of Part II of this book. This material was summarized in two notes: *On Regular Rings*, Proceedings of the National Academy of Sciences, vol. 22 (1936), pages 707—713; and *Algebraic Theory of Continuous Geometries*, Proceedings of the National Academy of Sciences, vol. 23 (1937), pages 16—22.

The coordinatization is carried out and the ring $\mathfrak{R} = \mathfrak{S}_n$ ($\mathfrak{S}_n$ is a regular ring along with $\mathfrak{S}$) is obtained, without using any of the following axioms: completeness of the lattice, continuity of the lattice operations, irreducibility. However, if these axioms do hold, the ring $\mathfrak{R}$ has special properties: a rank function $R(x)$ can be defined for all x in $\mathfrak{R}$. This rank function leads to a natural metric in $\mathfrak{R}$ and $\mathfrak{R}$ is topologically complete with respect to this metric. Rings with a rank function were called by von Neumann: *rank rings*, and when complete with respect to the rank metric: *continuous rings*. Rank rings and continuous rings are studied in Chapters XVII and XVIII of this book.

Chapter IV of Part II considers complemented modular lattices L of order $n \geqq 3$. If L can be coordinatized by a regular ring $\mathfrak{R}$ (the coordinatization theorem shows that $n \geqq 4$ is *sufficient*), then as von Neumann shows, every lattice isomorphism of L is generated by a ring isomorphism of $\mathfrak{R}$, and every dual-automorphism of L can be generated by an anti-automorphism of $\mathfrak{R}$. If the anti-automorphism of L is an orthogonalization, then the corresponding anti-automorphism of $\mathfrak{R}$ is a Hermitian conjugation: $x \to x^*$, which is non-singular in the sense that $x^*x = 0$ implies $x = 0$.

The continuous geometries obtained from rings of operators do, in fact, possess such orthogonalizations; the Hermitian conjugation in the coordinatizing continuous ring coincides with the operation of taking the Hermitian adjoint operator in Hilbert space.

In a continuation of Part II (not mentioned in this book) von Neumann studied continuous geometries which possess an orthogonalization. In particular, he analysed such geometries which, in addition, possess a transition probability function $P(x, a)$. This means: $P(x, a)$ is defined for all x and all non-zero a in the geometry, $0 \leqq P(x, a) \leqq 1$, and $P(x, a)$ has certain properties characteristic of the transition probability function in quantum mechanics. Von Neumann showed that every such geometry can be obtained from a ring of operators in a suitable Hilbert space (the dimensionality of the Hilbert space must not be restricted). This result is embodied in a manuscript (still unpublished) *Continuous Geometries With A Transition Probability* (MS written in 1936—137) and is important in applications to quantum mechanics. Von Neumann lectured on this topic at Princeton in 1937—38, but detailed lecture notes did not become available.

In a continuation of Part II in another direction, von Neumann developed the theory of arithmetic of continuous rings. In a manuscript (still unpublished (*Arithmetics of Regular Rings Derived From Continuous Geometries* (MS written in 1936—37), von Neumann proved that every element in a continuous ring has a natural decomposition into a set of algebraic parts together with a purely transcendental part; he also gave other arithmetic theorems. These results were summarized in a note *Continuous Rings And Their Arithmetics*, Proceedings of the National Academy of Sciences, vol. 23 (1937), pages 341—349. These results are not mentioned in the present book.

Part III of this book is concerned with lattices which are continuous geometries except that irreducibility is not assumed. The center of such a geometry may be any continuous Boolean algebra. Von Neumann analysed the reducible geometry relative to its center and introduced the basic concept of central envelope of an element a — that is, the least element e in the center which satisfies $e \geqq a$. He used central envelopes to establish the transitivity of perspectivity for reducible continuous geometries. Then he began the construction of the dimension functions.

At this point the lecture notes break off abruptly. In a letter dated November 12, 1936, von Neumann wrote: "I can get the 'central decomposition' of the various dimension functions and in particular their existence, enumeration, etc., in 'reducible' continuous geometries, but I have not yet succeeded in decomposing these lattices themselves as completely as I could do it for bounded operator rings in Hilbert space. But I have some hope to do it." It is not difficult to see how he meant to get this central decomposition of the various dimension functions (cf. Page 294). But whether he succeeded in getting a satisfactory decomposition for the lattices themselves is not known to me.

Von Neumann reviewed his previous work on continuous geometries in four colloquium lectures delivered before the American Mathematical Society, in September, 1937, at Pennsylvania State College, State College, Pa., U.S.A. Later, he began to write a systematic account of his research on continuous geometry which he planned to publish as a book in the American Mathematical Society Colloquium Series. But his work in the theory of games, other interests, and the war, intervened. As the years went by, he finally decided that the Princeton lecture notes, at least, should be reproduced. This is now accomplished with the publication of the present book.

Kingston, Ontario, Canada
December 14, 1959

ISRAEL HALPERIN

Table of Contents

PART III

PART I

PART I · CHAPTER I

Foundations and Elementary Properties

The basis of our discussion is a class L of elements $a, b, c, \cdots$, two or more in number, together with a binary relation $<$ between pairs of elements of L. Unless otherwise specified, Axioms I—VI listed below will be assumed.

AXIOM I: Order.

I_1: $a < a$ for no element a.

I_2: $a < b < c$ implies $a < c$.

DEFINITION 1.1: $a > b$ means $b < a$; $a \leqq b$ means $a < b$ or $a = b$; $a \geqq b$ means $b \leqq a$.

AXIOM II. Continuity.

II_1: For every set $S \subset L$, there is an element $\Sigma(S)$ in L, which is a *least upper bound* of S, i.e.

(a) $\Sigma(S) \geqq a$ for every a in S,

(b) $x \geqq a$ for every a in S implies $x \geqq \Sigma(S)$.

II_2: For every set $S \subset L$ there is an element $\Pi(S)$ in L, which is a *greatest lower bound* of S, i.e.

(a) $\Pi(S) \leqq a$ for every a in S,

(b) $x \leqq a$ for every a in S implies $x \leqq \Pi(S)$.

COROLLARY: $\Sigma(S)$ and $\Pi(S)$ are unique.

PROOF: If $\Sigma_1(S)$, $\Sigma_2(S)$ satisfy (a), (b) of II_1, then since $\Sigma_1(S)$ satisfies (a) and $\Sigma_2(S)$ satisfies (b), we have $\Sigma_1(S) \geqq \Sigma_2(S)$. Similarly, $\Sigma_1(S) \leqq \Sigma_2(S)$. If $\Sigma_1(S) \neq \Sigma_2(S)$, we have both $\Sigma_1(S) > \Sigma_2(S)$ and $\Sigma_2(S) > \Sigma_1(S)$, so I_2 implies $\Sigma_1(S) > \Sigma_1(S)$, which is contradictory to I_1. In a similar manner, $\Pi(S)$ is unique.

DEFINITION 1.2: The elements $\Pi(L)$, $\Sigma(L)$ will be denoted by 0, 1 respectively. It follows from II_1,(*b*) and II_2,(*b*) that if Θ is the empty subset of L, $0 = \Sigma(\Theta)$, $1 = \Pi(\Theta)$.

DEFINITION 1.3: Let (a, b) denote the class consisting of the elements a, b. Then we define

$$a + b \equiv \Sigma(a, b), \quad ab \equiv a \cdot b \equiv \Pi(a, b).$$

COROLLARY: $a + a = a$, $aa = a$.

DEFINITION 1.4: Let Ω be any infinite Cantor aleph. Let there be

given a system of elements $(a_\alpha;\ \alpha < \Omega)$. We consider only the two cases
(i) $\alpha < \beta$ implies $a_\alpha \leqq a_\beta$,
(ii) $\alpha < \beta$ implies $a_\alpha \geqq a_\beta$.
In case (i) we define

$$\lim_{\alpha\to\Omega}{}^{*}\, a_\alpha \equiv \Sigma(a_\alpha;\ \alpha < \Omega),$$

where the right member denotes the least upper bound of the set of all $a_\alpha;\ \alpha < \Omega$; in case (ii) we define

$$\lim_{\alpha\to\Omega}{}^{*}\, a_\alpha \equiv \Pi(a_\alpha;\ \alpha < \Omega).$$

Corollary: Let (a_α) be a system satisfying (i). Then for every b, the system $(b + a_\alpha)$ satisfies (i) and

$$\lim_{\alpha\to\Omega}{}^{*}\,(b + a_\alpha) = b + \lim_{\alpha\to\Omega}{}^{*}\, a_\alpha.$$

Similarly if (a_α) satisfies (ii), (ba_α) satisfies (ii) and

$$\lim_{\alpha\to\Omega}{}^{*}\,(ba_\alpha) = b \lim_{\alpha\to\Omega}{}^{*}\, a_\alpha.$$

Proof: Consider the first part in which (i) holds. Clearly $\Sigma(b + a_\alpha;\ \alpha < \Omega) \geqq b + a_\alpha \geqq b$ and a_α for all $\alpha < \Omega$ (use II_1, (a)). Hence it is $\geqq \Sigma(a_\alpha;\ \alpha < \Omega)$ and $\geqq b + \Sigma(a_\alpha;\alpha < \Omega)$ (use II_1, (b)). On the other hand $b + \Sigma(a_\alpha;\alpha < \Omega) \geqq b$ and $\geqq \Sigma(a_\alpha;\alpha < \Omega) \geqq a_\alpha$ for all $\alpha < \Omega$ (use II_1, (a)). Hence it is $\geqq b + a_\alpha$ and $\geqq \Sigma(b + a_\alpha;\alpha < \Omega)$ (use II_1, (b)). Thus I_1 gives $\Sigma(b + a_\alpha;\alpha < \Omega) = b + \Sigma(a_\alpha;\ \alpha < \Omega)$, that is $\lim_{\alpha\to\Omega}{}^{*}\,(b + a_\alpha) = b + \lim_{\alpha\to\Omega}{}^{*}\, a_\alpha$. The second part is similarly proved.

Axiom III: Continuity of addition and multiplication.

III_1: Let Ω be an infinite aleph and $(a_\alpha;\ \alpha < \Omega)$ a system satisfying (ii) of Definition 1.4. Then

$$\lim_{\alpha\to\Omega}{}^{*}\,(b + a_\alpha) = b + \lim_{\alpha\to\Omega}{}^{*}\, a_\alpha.$$

III_2: Let Ω be an infinite aleph and $(a_\alpha;\ \alpha < \Omega)$ a system satisfying condition (i) of Definition 1.4. Then

$$\lim_{\alpha\to\Omega}{}^{*}\,(ba_\alpha) = b \lim_{\alpha\to\Omega}{}^{*}\, a_\alpha.$$

Axiom IV: Modularity.

IV_1: $a \leqq c$ implies $(a + b)c = a + bc$ for every b.

Axiom V: Complementation.

V_1: Corresponding to each element a of L there exists an element x in L such that

$$a + x = 1, \quad ax = 0.$$

The element x is referred to as an *inverse* of a. It is not assumed to be unique.

AXIOM VI: Irreducibility.

VI_1: If a has a unique inverse, then a is either 0 or 1.

It is to be noted that if in the axioms I—VI the relations $<$ and $>$ be interchanged, the statements thus obtained are equivalent to the original axioms, provided, of course, that the accompanying substitution

$$\begin{pmatrix} < & \Sigma & 0 & + \\ > & \Pi & 1 & \cdot \end{pmatrix} \tag{1}$$

be made. This invariance of the postulates is referred to as the *duality* of our theory. Because of it, any theorem possesses a *dual* theorem obtained from it by means of the substitution (1) and any other interchanges arising by subsequent definitions. The system L is seen to form a *lattice* in the terminology of G. Birkhoff.*

Several further remarks concerning the axioms are in order at this point. First, it should be observed that Axioms II_1 and II_2 are not independent. In fact if I be assumed, then II_1 and II_2 are equivalent. By duality, it suffices to show that II_1 implies II_2. Let S be any subset of L, and let T be the set of all elements $a \leqq x$ for every x in S. Then $\Sigma(T) \leqq x$ for every x in S, and if $a \leqq x$ for every x in S, then $a \in T$, whence $a \leqq \Sigma(T)$. Hence $\Pi(S) \equiv \Sigma(T)$ is effective in II_2. Both II_1 and II_2 have been stated as axioms for purposes of symmetry. Secondly, it is worthy of note that if a set L' is given which satisfies I but not II, then L' may be completed by a process** similar to the method of completing the rational numbers by Dedekind cuts, and the completion of L' will satisfy II. Thirdly, some explanation concerning the use of the term "irreducibility" in connection with Axiom VI is necessary. This will be given at the end of the chapter. Finally, although this axiom is included at the outset of the theory, no use will be made of it until the latter part of Chapter VI.

THEOREM 1.1:

(i) $a + b = b + a$, $ab = ba$, $(a + b) + c = a + (b + c)$, $(ab)c = a(bc)$.

(ii) $0 \leqq a \leqq 1$.

(iii) $a + 0 = a \cdot 1 = a$, $a \cdot 0 = 0$, $a + 1 = 1$.

(iv) $0 \neq 1$.

* G. Birkhoff, *Combinatorial relations in projective geometries*, Annals of Mathematics, vol. 36 (1935), pages 743—748.

** H. M. MacNeille, *Extensions of partially ordered sets*, Proceedings of the National Academy of Sciences, vol. 22 (1936), No. 1, pages 45—50.

(v) $a \geqq b$, $b \geqq a$ if and only if $a = b$.
(vi) $a + ba = a(b + a) = a$.
(vii) $a + b = a$ if and only if $a \geqq b$.
(viii) $ab = b$ if and only if $a \geqq b$.
(ix) If $a \geqq b$ and $c \geqq d$, then $a + c \geqq b + d$ and $ac \geqq bd$.

PROOFS: Parts (i), (ii), (v) are immediate, and their proofs will be omitted.

(iv) Suppose $0 = 1$. Then by (ii) $0 \leqq a \leqq 0$ for every a. By (v) $a = 0$ and L consists of but one element, contrary to the supposition that there are at least two elements.

(vi) Clearly $a + ba \geqq a$. But $a \geqq ba$, whence $a \geqq a + ba$. Hence by (v) $a = a + ba$. Dually, $a(b + a) = a$.

(vii) Clearly $a + b \geqq b$, hence $a + b = a$ implies $a \geqq b$. Conversely: $a \geqq b$ implies by II_1, (b), owing to $a \geqq a$, that $a \geqq a + b$. But $a + b \geqq a$, so I_1 gives $a + b = a$.

(viii) This is the dual of (vii).

(iii) This is immediate from (ii), (vii), (viii).

(ix) $a + c \geqq a \geqq b$, whence $a + c \geqq b$. Similarly $a + c \geqq d$. Therefore $a + c \geqq b + d$. Dually, $ac \geqq bd$.

THEOREM 1.2: If $(a + c)(b + d) = 0$, then $(a + b)(c + d) = ac + bd$.

PROOF: Define $e = a + c$, $f = b + d$. Then $a \leqq e$, $c \leqq e$, $b \leqq f$, $d \leqq f$, $ef = 0$. Now

$$\begin{aligned}(a + f)(c + f) &= (a + f)c + f \quad \text{(by IV)}\\ &= (a + f)ec + f\\ &= (a + fe)c + f \quad \text{(by IV), and}\end{aligned}$$

$$(a + f)(c + f) = ac + f. \tag{2}$$

Similarly,

$$(b + e)(d + e) = bd + e. \tag{3}$$

Also, by two applications of IV,

$$(a + f)(b + e) = a + f(b + e) = a + b, \tag{4}$$

and similarly,

$$(c + f)(d + e) = c + d. \tag{5}$$

Therefore,

$$\begin{aligned}(a + b)(c + d) &= (a + f)(b + e)(c + f)(d + e) \quad \text{(by (4), (5))}\\ &= (ac + f)(bd + e) \quad \text{(by (2), (3))}\\ &= ac + bd,\end{aligned}$$

the final step being proved in a manner similar to the proof of (4).

DEFINITION 1.5: If $a \leqq b$, we define $L(a, b)$ to be the class of all elements x in L such that $a \leqq x \leqq b$.

COROLLARY: If L satisfies Axioms I—IV, then $L(a, b)$ satisfies these axioms, whenever $a \leqq b$. (Of course, $L(a,b)$ fails to have two or more elements if $a = b$. —Ed.)

PROOF: Axiom I is obviously satisfied by $L(a, b)$. To prove Axiom II, let $S \subset L(a, b)$. If S is non-empty, then $\Sigma(S)$ (as defined in L) is an effective least upper bound of S, and lies in $L(a, b)$. For, let $x \in S$. Then $x \leqq \Sigma(S)$, $a \leqq x$, whence $a \leqq \Sigma(S)$. Also $b \geqq y$ for every y in S, whence $b \geqq \Sigma(S)$. If $S = \Theta$, $\Sigma(S)$ is 0 and does not in general lie in $L(a, b)$. But in this case the element a is effective as least upper bound of S (in $L(a, b)$). This, together with dual considerations, proves II. Axioms III and IV are immediate. Hence in passing from L to $L(a, b)$, the operations of addition and multiplication as well as the order relation are preserved, while 0, 1 are replaced by a, b, respectively.

THEOREM 1.3: If $a \leqq c \leqq b$, then there exists an element x in $L(a, b)$ such that $c + x = b$, $cx = a$.

PROOF: By V there is an inverse y to c, whence $c + y = 1$, $cy = 0$. Define $x \equiv yb + a$. Hence

$$c + x = c + (yb + a) = c + yb = (c + y)b = b,$$
$$cx = c(yb + a) = a + cby = a + cy = a.$$

COROLLARY: If L satisfies Axioms I—V, and if $a \leqq b$, then $L(a, b)$ also satisfies these axioms.

DEFINITION 1.6: An element x will be called an *inverse of a in b* in case $a + x = b$ and $ax = 0$.

COROLLARY: If $a \leqq b$ then there exists an inverse of a in b, and conversely.

PROOF: This results from setting $a = 0$ in Theorem 1.3. The converse is obvious.

THEOREM 1.4: If x is an inverse of a in b, and y is an inverse of b in c, then $x + y$ is an inverse of a in c.

PROOF: By hypothesis, $x + a = b$, $y + b = c$, $ax = yb = 0$. Hence $(x + y) + a = c$. Also, with the help of Theorem 1.2 it is seen that

$$a(x + y) = (a + 0)(x + y) = ax + 0 \cdot y = 0.$$

Hence $x + y$ is an inverse of a in c.

COROLLARY:

(i) Let $ax = 0$. Then there exists $y \geqq x$ such that y is an inverse of a.

(ii) Let $a + x = 1$. Then there exists $y \leqq x$ such that y is an inverse of a.

PROOF: (i) Clearly x is an inverse of a in $a + x$. Let z be an inverse of $a + x$. Then $y = x + z$ is an inverse of a by Theorem 1.4. Also $y \geqq x$, whence y is effective.

(ii) This is dual to (i).

LEMMA 1.1: Let x and a be given. Then there exist elements u and z such that $u \leqq a$, $az = 0$, and $x = u + z$.

PROOF: Define $u \equiv ax$. Let z be an inverse of u in x (Corollary to Definition 1.6), $x \equiv u + z$. Then $uz = 0$, $z \leqq x$, whence $az = axz = uz = 0$.

LEMMA 1.2: Let there be given two elements a, b with the following properties:

(i) $ab = 0$,

(ii) Every element x is expressible in the form $x = u + v$, with $u \leqq a$, $v \leqq b$. Then the elements u, v are unique and are equal to ax, bx, respectively. Moreover, a and b have unique inverses equal to b and a respectively.

PROOF: Suppose $x = u + v$ for every x, with $u \leqq a$, $v \leqq b$. Then $ax = a(u + v) = u + av$ by IV. But $av \leqq ab = 0$. Hence $ax = u$; similarly $bx = v$. Then $x = ax + bx$, whence $1 = a + b$. Therefore a and b are mutual inverses.

Let us now suppose that b_1 is an inverse of a. Then $b_1 = ab_1 + bb_1$, $b_1a = 0$, $b_1 + a = 1$. Hence $b_1 = bb_1$, whence $b_1 \leqq b$. Thus by IV

$$b_1 = b_1 + ab = (a + b_1)b = b.$$

This proves that a has but one inverse, viz., b; similarly, a is the unique inverse of b.

Axiom VI is referred to as an axiom of irreducibility, although its statement does not resemble the notion generally called irreducibility in algebraic theories. However, we shall now formulate a property Δ that bears a close resemblance to the ordinary notion of irreducibility, and shall show its equivalence to VI.

THEOREM 1.5: Suppose that L satisfies Axioms I—V. Let L be said to satisfy Δ in case for every pair of elements a, b such that $ab = 0$ and such that every element x is expressible as a sum $u + v$, with $u \leqq a$, $v \leqq b$, it is so that $a = 0$ and $b = 1$, or $a = 1$ and $b = 0$. Then L satisfies VI if and only if it satisfies Δ.

PROOF: All the results thus far obtained are valid, since none depends on Axiom VI. Suppose VI is satisfied. Then if $ab = 0$ and every x is expressible as $u + v$, $u \leqq a$, $v \leqq b$, we have by Lemma 1.2 that b is the unique inverse of a. Hence by VI, $a = 0$ or $a = 1$, and $b = 1$ or $b = 0$ in the respective cases. This proves Δ.

Conversely, suppose that an element a has a unique inverse b, whence $a + b = 1$, $ab = 0$. By Lemma 1.1 to every x there correspond elements u, z, with $u \leqq a$, $u + z = x$, $az = 0$. Now there is an inverse b' of a with $b' \geqq z$, by the corollary to Theorem 1.4. Hence $b' = b$, and $b \geqq z$. Thus every x is expressible in the form $u + v$, with $u \leqq a$, $v \leqq b$. Therefore $a = 0$ or $a = 1$ by Δ.

Independence

In what follows arbitrary sets of indices will frequently enter the discussion. These sets will be denoted by I, J, K, $\cdots$, while the indices will be denoted by ρ, σ, τ, $\cdots$. Set theoretical summation and intersection will be denoted by the symbols $\mathfrak{S}$, $\mathfrak{P}$ respectively, in order to avoid confusion with the symbols Σ, Π already in use.

DEFINITION 2.1: If I is any set, and if $(a_\sigma;\ \sigma\in I)$ is a system of elements in L, the system (a_σ) is said to be independent, in notation $(a_\sigma;\ \sigma\in I)\perp$, in case

$$(1)\qquad \Sigma(a_\sigma;\ \sigma\in J)\Sigma(a_\sigma;\ \sigma\in K)=0,$$

for every pair of non-intersecting subsets J, K of I. In (1) the notation $\Sigma(a_\sigma;\ \sigma\in J)$ denotes the least upper bound (defined in Axiom II) of the class of all elements $a_\sigma;\ \sigma\in J$.

COROLLARY: If $(a_\sigma;\ \sigma\in I)\perp$, then $(a_\sigma;\ \sigma\in I_0)\perp$ for every $I_0\subset I$.

THEOREM 2.1: Let I_1, I_2 be two non-intersecting sets. If $(a_\sigma;\sigma\in I_1)\perp$, $(a_\sigma;\ \sigma\in I_2)\perp$ and $\Sigma(a_\sigma;\ \sigma\in I_1)\Sigma(a_\sigma;\ \sigma\in I_2)=0$, then $(a_\sigma;\ \sigma\in\mathfrak{S}(I_1,I_2))\perp$.

PROOF: Let J, K be two non-intersecting subsets of $\mathfrak{S}(I_1, I_2)$. Then define

$$J_1\equiv\mathfrak{P}(J,I_1),\ J_2\equiv\mathfrak{P}(J,I_2),$$
$$K_1\equiv\mathfrak{P}(K,I_1),\ K_2\equiv\mathfrak{P}(K,I_2).$$

Thus $J=\mathfrak{S}(J_1,J_2)$, $K=\mathfrak{S}(K_1,K_2)$ and $\mathfrak{P}(J_1,K_1)=\mathfrak{P}(J_2,K_2)=\Theta$. If we define

$$a\equiv\Sigma(a_\sigma;\ \sigma\in J_1),\ b\equiv\Sigma(a_\sigma;\ \sigma\in J_2),$$
$$c\equiv\Sigma(a_\sigma;\ \sigma\in K_1),\ d\equiv\Sigma(a_\sigma;\ \sigma\in K_2),$$

we have $ac=0$, since J_1, $K_1\subset I_1$ and $(a_\sigma;\ \sigma\in I_1)\perp$ by hypothesis. Similarly $bd=0$. Now from

$$a\leqq\Sigma(a_\sigma;\ \sigma\in I_1),\ c\leqq\Sigma(a_\sigma;\ \sigma\in I_1)$$

we conclude

$$a+c\leqq\Sigma(a_\sigma;\ \sigma\in I_1),$$

and similarly

$$b+d\leqq\Sigma(a_\sigma;\ \sigma\in I_2).$$

Hence $(a+c)(b+d)=0$ and by Theorem 1.2,

$$(a+b)(c+d)=ac+bd=0.$$

This states that

$$\Sigma(a_\sigma;\ \sigma\in J)\Sigma(a_\sigma;\ \sigma\in K)=0,$$

whence $(a_\sigma;\ \sigma\in\mathfrak{S}(I_1,\ I_2))\perp$.

COROLLARY: Let $(I_i;\ i=1,\cdots,n)$ be a finite disjoint system of sets. If $(a_\sigma;\ \sigma\in I_i)\perp$ $(i=1,\cdots,n)$, and if $(\Sigma(a_\sigma;\ \sigma\in I_i);\ i=1,\cdots,n)\perp$, then $(a_\sigma;\ \sigma\in\mathfrak{S}(I_i;\ i=1,\cdots,n))\perp$.

PROOF: The proof will be made by induction on n. The case $n=1$ is obvious, and the case $n=2$ is treated in Theorem 2.1. Suppose the corollary is true for $n=1,\cdots,m$. Consider a system of sets $(I_i;i=1,\cdots,m+1)$ and $m+1$ systems of elements $(a_\sigma;\ \sigma\in I_i)$ $(i=1,\cdots,m+1)$ of L satisfying the hypotheses. The systems $(a_\sigma;\ \sigma\in I_i)$ $(i=1,\cdots,m)$ also satisfy the hypotheses, whence we may conclude

$$(a_\sigma;\ \sigma\in\mathfrak{S}\ (I_i;i=1,\cdots,m))\perp.$$

Now

$$\Sigma(a_\sigma;\ \sigma\in\mathfrak{S}(I_i;\ i=1,\cdots,m))=\Sigma(\Sigma(a_\sigma;\ \sigma\in I_i);\ i=1,\cdots,m)$$

as may be readily verified. Hence by the independence of $(\Sigma(a_\sigma;\ \sigma\in I_i);$ $i=1,\cdots,m+1)$ it follows that

$$\Sigma(a_\sigma;\ \sigma\in I_{m+1})\Sigma(a_\sigma;\ \sigma\in\mathfrak{S}(I_i;\ i=1,\cdots,m))=0.$$

By the induction hypothesis, $(a_\sigma;\ \sigma\in I_{m+1})\perp$. Therefore, by Theorem 2.1, $(a_\sigma;\ \sigma\in\mathfrak{S}\ (I_i;\ i=1,\cdots,m+1))\perp$. This completes the induction.

THEOREM 2.2: Let $(a_i;\ i=1,\cdots,n)$ be a finite system of elements in L. Then $(a_i;\ i=1,2,\cdots,n)\perp$ if and only if *

$$(a_1+\cdots+a_i)a_{i+1}=0 \qquad (i=1,\cdots,n-1). \tag{2}$$

PROOF: Condition (2) is obviously necessary. Its sufficiency is proved by induction. Clearly $(a_1,\ a_2)\perp$, since $a_1a_2=0$. Suppose $2\leqq m<n$, and $(a_i;\ i=1,\cdots,m)\perp$. Define $I_1=(1,\cdots,m)$, $I_2=(m+1)$. Then $(a_{m+1})\perp$ and by hypothesis

$$(a_1+\cdots+a_m)a_{m+1}=0.$$

Thus Theorem 2.1 applies, and we conclude $(a_i;\ i=1,\cdots,m+1)\perp$. Hence by the induction, $(a_i;\ i=1,\cdots,n)\perp$.

It should be observed that while condition (2) seems to depend on the order of the elements in the system (a_i), its equivalence to the independence of the system, which is readily shown not to depend on order, shows

* The notations $a_1+\cdots+a_n$, $\sum_{i=1}^n a_i$ will frequently be used in place of their equivalent, $\Sigma\ (a_i;\ i=1,\cdots,n)$.

that (2) is equivalent to the condition resulting from (2) by effecting any permutation on the a_i. More precisely, if $i \to k_i$ is a permutation of the integers $1, \cdots, n$, then (2) is equivalent to

$$(a_{k_1} + \cdots + a_{k_i})a_{k_{i+1}} = 0 \qquad (i = 1, \cdots, n-1).$$

THEOREM 2.3: A system $(a_\sigma;\ \sigma \in I)$ is independent if and only if $(a_\sigma;\ \sigma \in I_0)$ is independent for every finite $I_0 \subset I$.

PROOF: Clearly $(a_\sigma;\ \sigma \in I)\perp$ implies $(a_\sigma;\ \sigma \in I_0)\perp$ for every finite $I_0 \subset I$ by the Corollary to Definition 2.1. To prove the converse, suppose $J, K \subset I$, with $\mathfrak{P}(J, K) = \Theta$. Then it must be shown that

$$\Sigma(a_\sigma;\ \sigma \in J)\Sigma(a_\sigma;\ \sigma \in K) = 0. \tag{3}$$

If (3) does not hold for every disjoint pair (J, K), then there is a pair (J, K) violating (3) such that

(A) the power of J is a minimum $= \aleph_1$,

(B) for power of $J = \aleph_1$, the power of K is a minimum $= \aleph_2$.

We note these facts:

(a) $\aleph_1 \leqq \aleph_2$, for since J, K violate (3), the same is true of K, J.

(b) $\aleph_2$ is infinite, for otherwise both $\aleph_1$ and $\aleph_2$ are finite, whence J, K are both finite sets and $I_0 = \mathfrak{S}(J, K)$ is finite. Hence $(a_\sigma;\ \sigma \in I_0)\perp$ by hypothesis, and

$$\Sigma(a_\sigma;\ \sigma \in J)\Sigma(a_\sigma;\ \sigma \in K) = 0,$$

contrary to our assumption.

(c) If J, K' violate (3), then if $\aleph_2'$ is the power of K', $\aleph_2' \geqq \aleph_2$.

Now let Ω be the smallest ordinal corresponding to $\aleph_2$. Clearly Ω is a limit ordinal. Evidently the system $(a_\sigma;\ \sigma \in K)$ can be replaced by the system $(a_\alpha;\ \alpha < \Omega)$. If we define $a \equiv \Sigma(a_\sigma;\ \sigma \in J)$, we have by the contrapositive of (c),

$$a\Sigma(a_\beta;\ \beta < \alpha) = 0 \quad (\alpha < \Omega).$$

Hence

$$\Sigma(a\Sigma(a_\beta;\ \beta < \alpha);\ \alpha < \Omega) = 0.$$

But $\alpha' < \alpha''$ implies $\Sigma(a_\beta;\ \beta < \alpha') \leqq \Sigma(a_\beta;\ \beta < \alpha'')$, whence Axiom III_2 applies, and we have

$$a\Sigma(\Sigma(a_\beta;\ \beta < \alpha);\ \alpha < \Omega) = 0.$$

It is easily seen that

$$\Sigma(\Sigma(a_\beta;\ \beta < \alpha);\ \alpha < \Omega) = \Sigma(a_\beta;\ \beta < \Omega),$$

since Ω is a limit ordinal. Hence

$$a\Sigma(a_\beta;\ \beta < \Omega) = 0,$$

or

$$\Sigma(a_\sigma;\ \sigma \in J)\Sigma(a_\sigma;\ \sigma \in K) = 0,$$

which contradicts the assumption that J, K violate (3). This proves that (3) holds, and therefore that $(a_\sigma;\ \sigma \in I)$ is independent.

THEOREM 2.4: Let J be any class and $(I_\rho;\ \rho \in J)$ any disjoint system of classes. If $(a_\sigma;\ \sigma \in I_\rho)\perp$ for every $\rho \in J$, and if $(\Sigma(a_\sigma;\ \sigma \in I_\rho);\ \rho \in J)\perp$, then $(a_\sigma;\ \sigma \in \mathfrak{S}(I_\rho;\ \rho \in J))\perp$.

PROOF: Let $K = (\sigma_1, \cdots, \sigma_n)$ be any finite subset of $I \equiv \mathfrak{S}(I_\rho;\ \rho \in J)$. It will be shown that $(a_\sigma;\ \sigma \in K)\perp$, whence the independence of $(a_\sigma;\ \sigma \in I)$ will follow by Theorem 2.3.

For every $i = 1, \cdots, n$ there is a unique class I_i such that $\sigma_i \in I_i$. Let the system (I_i) be replaced by the system $(K_j;\ j = 1, \cdots, m)$ of distinct classes. For every $j=1, \cdots, m$ there is a unique $\rho^{(j)}$ in J such that $K_j = I_{\rho^{(j)}}$, and each $I_{\rho^{(j)}}$ arises from but one j. Hence $(a_\sigma;\ \sigma \in K_j)\perp$ $(j = 1, \cdots, m)$; also, the system $(\Sigma(a_\sigma;\ \sigma \in K_j);\ j = 1, \cdots, m)$ may be represented as a subsystem of $(\Sigma(a_\sigma;\ \sigma \in I_\rho);\ \rho \in J)$ and is therefore independent. From this follows the independence of $(a_\sigma;\ \sigma \in \mathfrak{P}(K, K_j))$ $(j = 1, \cdots, m)$ and of $(\Sigma(a_\sigma;\ \sigma \in \mathfrak{P}(K, K_j));\ j = 1, \cdots, m)$. By the Corollary to Theorem 2.1, $(a_\sigma;\ \sigma \in K)\perp$, since $K = \mathfrak{S}(\mathfrak{P}(K, K_j);\ j = 1, \cdots, m)$. This completes the proof.

The theorem just proved is the extension of Theorem 2.1 and its corollary to arbitrary systems of sets.

THEOREM 2.5: A system $(a_\sigma;\ \sigma \in I)$ is independent if and only if for every $J_1, J_2 \subset I$,

$$\Sigma(a_\sigma;\ \sigma \in J_1)\Sigma(a_\sigma;\ \sigma \in J_2) = \Sigma(a_\sigma;\ \sigma \in \mathfrak{P}(J_1, J_2)). \tag{4}$$

PROOF: The sufficiency of (4) is evident, since (4) implies (1). To prove the necessity, define

$$K \equiv J_1 - \mathfrak{P}(J_1, J_2),\ a \equiv \Sigma(a_\sigma;\ \sigma \in \mathfrak{P}(J_1, J_2)),$$
$$b \equiv \Sigma(a_\sigma;\ \sigma \in K),\ c \equiv \Sigma(a_\sigma;\ \sigma \in J_2).$$

Since $\mathfrak{P}(J_2, K) = \Theta$, we have $bc = 0$. By IV,

$$\Sigma(a_\sigma;\ \sigma \in J_1)\Sigma(a_\sigma;\ \sigma \in J_2) = (a + b)c = a + bc =$$
$$= a = \Sigma(a_\sigma;\ \sigma \in \mathfrak{P}(J_1, J_2)).$$

COROLLARY: Let $(a_\sigma;\ \sigma \in I)\perp$. If $(I_i;\ i = 1, \cdots, n)$ is a finite system of subsets of I, then

$$\Pi(\Sigma(a_\sigma;\ \sigma \in I_i);\ i = 1, \cdots, n) = \Sigma(a_\sigma;\ \sigma \in \mathfrak{P}(I_i;\ i = 1, \cdots, n)). \tag{5}$$

PROOF: The case $n = 1$ is trivial, and the case $n = 2$ is treated in Theorem 2.5. Suppose that (5) holds for $n = m$, where $m \geqq 2$. Then

$$\begin{aligned}
&\Pi(\Sigma(a_\sigma;\ \sigma \in I_i);\ i = 1, \cdots, m+1) = \\
&\quad = \Pi(\Sigma(a_\sigma;\ \sigma \in I_i);\ i = 1, \cdots, m) \cdot \Sigma(a_\sigma;\ \sigma \in I_{m+1}) \\
&\quad = \Sigma(a_\sigma;\ \sigma \in \mathfrak{P}(I_i;\ i = 1, \cdots, m)) \cdot \Sigma(a_\sigma;\ \sigma \in I_{m+1}) \\
&\quad = \Sigma(a_\sigma;\ \sigma \in \mathfrak{P}(I_i;\ i = 1, \cdots, m+1)),
\end{aligned}$$

the last equality holding by Theorem 2.5. Hence (5) holds for all values of n.

THEOREM 2.6: Let $(a_\sigma; \sigma \in I)\perp$. If $(I_\rho; \rho \in J)$ is any system of subsets of I, then

$$\Pi(\Sigma(a_\sigma;\ \sigma \in I_\rho);\ \rho \in J) = \Sigma(a_\sigma;\ \sigma \in \mathfrak{P}(I_\rho;\ \rho \in J)). \tag{6}$$

PROOF: By the Corollary to Theorem 2.5, (6) holds if J is finite. Suppose (6) does not hold for all sets J. Then there is one, which we denote by J, of minimum power for which (6) fails. Let Ω be the smallest ordinal corresponding to this power. It is a limit ordinal, and J may be replaced by the set of all ordinals $\alpha < \Omega$. Our hypotheses yield that

$$\Pi(\Sigma(a_\sigma;\ \sigma \in I_\beta);\ \beta < \gamma) = \Sigma(a_\sigma;\ \sigma \in \mathfrak{P}(I_\beta;\ \beta < \gamma)) \tag{7}$$

holds for $\gamma < \Omega$ and fails for $\gamma = \Omega$. We shall reach a contradiction by proving that (7) holds for $\gamma = \Omega$. Now

$$b \equiv \Pi(\Sigma(a_\sigma;\ \sigma \in I_\alpha);\ \alpha < \Omega) = \Pi(\Pi(\Sigma(a_\sigma;\ \sigma \in I_\beta);\ \beta < \gamma);\ \gamma < \Omega),$$

as can be easily verified since Ω is a limit ordinal. By applying (7), we have

$$b = \Pi(\Sigma(a_\sigma;\ \sigma \in \mathfrak{P}(I_\beta;\ \beta < \alpha));\ \alpha < \Omega).$$

Similarly,

$$c \equiv \Sigma(a_\sigma;\ \sigma \in \mathfrak{P}(I_\alpha;\ \alpha < \Omega)) = \Sigma(a_\sigma;\ \sigma \in \mathfrak{P}(\mathfrak{P}(I_\beta;\ \beta < \alpha);\ \alpha < \Omega)).$$

For each $\alpha < \Omega$ define $J_\alpha \equiv \mathfrak{P}(I_\beta;\ \beta < \alpha)$. Then

$$\begin{aligned}
b &= \Pi(\Sigma(a_\sigma;\ \sigma \in J_\alpha);\ \alpha < \Omega), \\
c &= \Sigma(a_\sigma;\ \sigma \in \mathfrak{P}(J_\alpha;\ \alpha < \Omega)),
\end{aligned}$$

and $\alpha' < \alpha''$ implies $J_{\alpha'} \supset J_{\alpha''}$.

Thus in the problem of proving (7) for $\gamma = \Omega$, i.e. of proving $b = c$, we have replaced the sets I_α by a monotonic decreasing system J_α. If we define

$$J^\Omega \equiv \mathfrak{P}(J_\alpha;\ \alpha < \Omega),\ J^\alpha \equiv J_\alpha - J^\Omega \qquad (\alpha < \Omega),$$

the sets J^α are monotonic decreasing and have the empty set as intersection. Hence $\alpha' < \alpha''$ implies

$$\Sigma(a_\sigma;\ \sigma \in J^{\alpha'}) \geqq \Sigma(a_\sigma;\ \sigma \in J^{\alpha''}).$$

Since

$$\Sigma(a_\sigma;\ \sigma \in J_\alpha) = \Sigma(a_\sigma;\ \sigma \in J^\Omega) + \Sigma(a_\sigma;\ \sigma \in J^\alpha),$$

it follows that

$$b = \Pi(\Sigma(a_\sigma;\ \sigma \in J^\Omega) + \Sigma(a_\sigma;\ \sigma \in J^\alpha);\ \alpha < \Omega).$$

Hence by III_1, $b = c + a$, where

$$a = \Pi(\Sigma(a_\sigma;\ \sigma \in J^\alpha);\ \alpha < \Omega).$$

In order to prove that $b = c$, we show that $a = 0$, which is our original problem with the sets I_α replaced by a decreasing system J^α which have the empty set as intersection. Now if $\beta < \Omega$,

$$\begin{aligned}\Pi(\Sigma(a_\sigma;\ \sigma \in J^\alpha);\ \alpha < \Omega) \cdot \Sigma(a_\sigma;\ \sigma \in (I - J^\beta))&\\ \leqq \Sigma(a_\sigma;\ \sigma \in J^\beta) \cdot \Sigma(a_\sigma;\ \sigma \in (I - J^\beta))&\\ = 0,&\end{aligned}$$

since $(a_\sigma;\ \sigma \in I)\perp$ and $\mathfrak{P}(J^\beta, I - J^\beta) = \Theta$. Hence

$$\Sigma(a \cdot \Sigma(a_\sigma;\ \sigma \in (I - J^\beta));\ \beta < \Omega) = 0.$$

But $\alpha' < \alpha''$ implies $I - J^{\alpha'} \subset I - J^{\alpha''}$ whence

$$\Sigma(a_\sigma;\ \sigma \in (I - J^{\alpha'})) \leqq \Sigma(a_\sigma;\ \sigma \in (I - J^{\alpha''})),$$

and we have by III_2

$$a\Sigma(\Sigma(a_\sigma;\ \sigma \in (I - J^\beta));\ \beta < \Omega) = 0.$$

But

$$\begin{aligned}\Sigma(\Sigma(a_\sigma;\ \sigma \in (I - J^\beta));\ \beta < \Omega) &=\\ &= \Sigma(a_\sigma;\ \sigma \in \mathfrak{S}((I - J^\beta);\ \beta < \Omega))\\ &= \Sigma(a_\sigma;\ \sigma \in (I - \mathfrak{P}(J^\beta;\ \beta < \Omega)))\\ &= \Sigma(a_\sigma;\ \sigma \in I).\end{aligned}$$

Thus $a\Sigma(a_\sigma;\ \sigma \in I) = 0$; but obviously $a \leqq \Sigma(a_\sigma;\ \sigma \in I)$, and consequently $a = 0$. This completes the proof that $b = c$, from which it is seen that (7) holds for $\gamma = \Omega$.

The theorem just proved is the extension of Theorem 2.5 and its corollary to arbitrary systems of subsets of I.

THEOREM 2.7: A system $(a_\sigma;\ \sigma \in I)\perp$ if and only if the class S of all $\Sigma(a_\sigma;\ \sigma \in J)$, $J \subset I$, is isomorphic to the Boolean algebra $\mathscr{I}$ of the subsets of I, where isomorphism means a transformation of $\mathscr{I}$ into S carrying the operations $\mathfrak{S}$, $\mathfrak{P}$ into Σ, Π respectively.

PROOF: Let us first consider the forward implication. Clearly $\sum_J(\Sigma(a_\sigma;\ \sigma \in J))$, where J ranges over any subset $\mathscr{I}_1$ of $\mathscr{I}$ is equal to $\Sigma(a_\sigma;\ \sigma \in \mathfrak{S}(J;\ J \in \mathscr{I}_1))$, which shows that the operation $\mathfrak{S}$ in $\mathscr{I}$ is carried by the correspondence $J \to \Sigma(a_\sigma;\ \sigma \in J)$ into the operation Σ in S. It remains to prove the similar statement concerning $\mathfrak{P}$ and Π, i.e., to prove

$$\Pi(\Sigma(a_\sigma;\ \sigma \in J);\ J \in \mathscr{I}_1) = \Sigma(a_\sigma;\ \sigma \in \mathfrak{P}(J;\ J \in \mathscr{I}_1)). \tag{8}$$

This is a special case of Theorem 2.6, where the J appearing in the statement of that theorem is $\mathscr{I}_1$ and $I_J = J$ for every $J \in \mathscr{I}_1$.

To prove the converse, we see that (8) implies (1) by letting $\mathscr{I}_1$ range over the class of all pairs of disjoint subsets of $\mathscr{I}$.

LEMMA 2.1: Let $(a_\sigma;\ \sigma \in I)\perp$, and let $\rho \in I$, $\rho' \in I$ with $\rho \neq \rho'$, and

$$x_\rho \leqq a_\rho + a_{\rho'}\,,\ a_{\rho'}x_\rho = 0. \tag{9}$$

Then $(b_\sigma;\ \sigma \in I)\perp$, where $b_\sigma = a_\sigma (\sigma \neq \rho)$, $b_\rho = x_\rho$.

PROOF: Define $I_1 \equiv I - (\rho, \rho')$, $I_2 \equiv (\rho, \rho')$. Now $(b_\sigma;\ \sigma \in I_1)\perp$, since $\sigma \in I_1$ implies $b_\sigma = a_\sigma$. Also $(b_\sigma;\ \sigma \in I_2)\perp$ by (9). Finally,

$$\begin{aligned} &\Sigma(b_\sigma;\ \sigma \in I_1)\Sigma(b_\sigma;\ \sigma \in I_2) = \\ &\quad = \Sigma(a_\sigma;\ \sigma \in I_1)(x_\rho + a_{\rho'}) \\ &\quad \leqq \Sigma(a_\sigma;\ \sigma \in I_1)(a_\rho + a_{\rho'}) \\ &\quad = \Sigma(a_\sigma;\ \sigma \in I_1)\Sigma(a_\sigma;\ \sigma \in I_2) = 0. \end{aligned}$$

Hence by Theorem 2.1, $(b_\sigma;\ \sigma \in I)\perp$.

COROLLARY: Let $(a_\sigma;\ \sigma \in I)\perp$, and let $J = (\rho_1, \cdots, \rho_n)$, K be two finite disjoint subsets of I (J consisting of n elements). Suppose there is given a mapping $\rho \to \rho'$ of J on K, such that $K = (\rho_1', \cdots, \rho_n')$, where the ρ_i' are not necessarily distinct, and suppose

$$x_{\rho_i} \leqq a_{\rho_i} + a_{\rho_i'}, \quad a_{\rho_i'}x_{\rho_i} = 0 \qquad (i = 1, \cdots, n). \tag{10}$$

Then $(b_\sigma;\ \sigma \in I)\perp$, where $b_\sigma = a_\sigma (\sigma \notin J)$, $b_{\rho_i} = x_{\rho_i}$ $(i = 1, \cdots, n)$.

PROOF: Define for each $i = 1, \cdots, n$

$$J_i \equiv (\rho_1, \cdots, \rho_i),\ b_\sigma^{(i)} \equiv a_\sigma (\sigma \notin J_i),\ b_{\rho_j}^{(i)} \equiv x_{\rho_j} \qquad (j = 1, \cdots, i).$$

The system $(b_\sigma^{(1)};\ \sigma \in I)$ is independent by Lemma 2.1. Suppose $1 \leqq m < n$ and $(b_\sigma^{(m)};\ \sigma \in I)\perp$. Then by applying Lemma 2.1 to this system, with $\rho = \rho_{m+1}$, $\rho' = \rho_{m+1}'$, we find $(c_\sigma;\ \sigma \in I)\perp$, where $c_\sigma = b_\sigma^{(m)}$ $(\sigma \neq \rho_{m+1})$, $c_{\rho_{m+1}} = x_{\rho_{m+1}}$. Hence $c_\sigma = b_\sigma^{(m+1)}$ $(\sigma \in I)$, and $(b_\sigma^{(m+1)};\ \sigma \in I)\perp$, whence by induction $(b_\sigma;\ \sigma \in I)\perp$, since $b_\sigma^{(n)} = b_\sigma$ $(\sigma \in I)$.

THEOREM 2.8: Let $(a_\sigma;\ \sigma \in I)\perp$, and let J, K be two non-empty disjoint subsets of I, with a mapping $\rho \to \rho'$ of J on K, not necessarily one-to-one. Suppose that for each $\rho \in J$,

$$x_\rho \leqq a_\rho + a_{\rho'},\quad a_{\rho'} x_\rho = 0. \tag{11}$$

Then $(b_\sigma;\ \sigma \in I)\perp$, where $b_\sigma = a_\sigma(\sigma \notin J)$, $b_\rho = x_\rho(\rho \in J)$.

PROOF: Let I_0 be any finite subset of I. It suffices to prove $(b_\sigma;\ \sigma \in I_0)\perp$ by Theorem 2.3. Obviously $J_1 = \mathfrak{P}(J, I_0)$ is finite, and we may write $J_1 = (\rho_1, \cdots, \rho_n)$. Define $K_1 \equiv (\rho_1', \cdots, \rho_n')$. Clearly J_1, K_1 are disjoint and equation (10) is satisfied. Hence by the Corollary to Lemma 2.1, $(c_\sigma;\ \sigma \in I)\perp$, where $c_\sigma = a_\sigma(\sigma \notin J_1)$, $c_{\rho_i} = x_{\rho_i}$ $(i = 1, \cdots, n)$. We have also that $(c_\sigma;\ \sigma \in I_0)\perp$. But $\sigma \in I_0$ implies $c_\sigma = b_\sigma$ and therefore $(b_\sigma; \sigma \in I_0)\perp$.

NOTE: In the applications of Theorem 2.8 the conditions

$$a_\rho + x_\rho = a_{\rho'} + x_\rho = a_\rho + a_{\rho'},$$
$$a_\rho x_\rho = a_{\rho'} x_\rho = 0 \qquad (\rho \in J)$$

are satisfied, instead of (11). These clearly imply (11) however.

PART I · CHAPTER III

Perspectivity and Projectivity. Fundamental Properties

DEFINITION 3.1: An element a is said to be *perspective* to an element b $(a \sim b)$ in case a and b have a common inverse.

COROLLARY: The relation $\sim$ is reflexive and symmetric.

The relation $\sim$ is also transitive, although the proof of this fact will not be given until later in Chapter V, after a considerable amount of preliminary theory has been developed (see the Corollary to Theorem 5.16; for a proof that the relation $\sim$ is transitive without assuming Axiom VI, see these Notes, Part III, Theorem 2.3. - Ed.).

LEMMA 3.1: If $a \sim b$, i.e., if there is an element x such that $a+x=b+x=1$, $ax = bx = 0$, then for every $c_1 \geqq a + b$, $c_2 \leqq ab$ there is an element y such that $a + y = b + y = c_1$, $ay = by = c_2$.

PROOF: Define $y \equiv c_1x + c_2$. Then

$$\begin{aligned} a + y = a + (c_2 + c_1x) &= a + c_1x \\ &= (a + x)c_1 \quad \text{(by IV)} \\ &= c_1, \end{aligned}$$

and $b + y = c_1$ similarly. Also

$$\begin{aligned} ay = a(c_1x + c_2) &= c_2 + ac_1x \quad \text{(by IV)} \\ &= c_2 + 0 = c_2, \end{aligned}$$

and similarly, $by = c_2$.

LEMMA 3.2: Let a, b be given. If there exists an element x such that $a + x = b + x$, $ax = bx$, then there is an element w which is inverse to both a and b, i.e. $a \sim b$.

PROOF:

I. Suppose $a + x = b + x = 1$, $ax = bx$. By the Corollary to Definition 1.6 there is an inverse w of ax in x. Hence $w + ax = x$, $wax = 0$, whence also $aw = 0$ since $w \leqq x$. But

$$1 = a + x = a + w + ax = a + w,$$

and w is an inverse of a. Since $ax = bx$, w is also an inverse of b by symmetry.

II. Suppose $a + x = b + x$, $ax = bx = 0$. Then by dualizing I, we find that a, b have a common inverse.

III. Consider the general case $a + x = b + x$, $ax = bx$. By the Corollary to Theorem 1.3, any result deduced for L from Axioms I—V is true also for $L(c, d)$, where $c \leqq d$. Hence by applying the result obtained in I to the lattice $L(0, a + x)$, we see that there is an element y such that $a + y = b + y = a + x$, $ay = by = 0$, since 1 must be replaced by $a + x$ in passing from L to $L(0, a + x)$. Now an application of II yields the desired common inverse of a and b.

DEFINITION 3.2: We shall write $u \sim v$ (mod x) in a, b to mean $u \leqq a$, $v \leqq b$, $u + x = v + x$.

THEOREM 3.1: Let a, b be given. Then for each of the following sets of equations it is so that $a \sim b$ is equivalent to the existence of a solution of that set:

(a) $a + x = b + x = c_1$, $ax = bx = c_2$ $(c_1 \geqq a + b,\ c_2 \leqq ab)$,
(b) $a + x = b + x$, $ax = bx$,
(c) $a + x = b + x$, $ax = bx = 0$,
(d) $a + x = b + x = a + b$, $ax = bx = ab$,
(e) $a + x = b + x = 1$, $ax = bx = 0$.

PROOF: We shall write $(i) \to (j)$ if when (i) has a solution (j) also has a solution. Clearly $(a) \to (c) \to (b)$, $(a) \to (d) \to (b)$. By Lemma 3.2 $(b) \to (e)$, and by Lemma 3.1 $(e) \to (a)$. Hence

$$(a) \begin{matrix} \nearrow (c) \searrow \\ \searrow (d) \nearrow \end{matrix} (b) \to (e) \to (a).$$

The proof is completed by observing that $a \sim b$ is equivalent to the existence of a solution of (e) by Definition 3.1.

THEOREM 3.2: Let a, b, x be given. Then the equations $a + x = b + x$, $ax = bx = 0$ are equivalent to

(a) for every $u \leqq a$ there is a unique element v_u such that $u \sim v_u$ (mod x) in a, b, and
(b) for every $v \leqq b$ there is a unique element u_v such that $u_v \sim v$ (mod x) in a, b.

NOTE: Since the equations in the hypothesis are precisely equations (c) of Theorem 3.1, it follows from that theorem that $a \sim b$ is equivalent to the existence of an element x satisfying conditions (a), (b) of Theorem 3.2.

PROOF: Only (a) will be proved since (b) follows from it by symmetry.

Let $u \leqq a$, and define $v_u \equiv b(u + x)$. Then $v_u \leqq b$, and

$$\begin{aligned} v_u + x = b(u + x) + x &= (b + x)(u + x) \qquad \text{(by IV)} \\ &= (a + x)(u + x) \\ &= u + x, \end{aligned}$$

whence $u \sim v_u \pmod{x}$ in a, b. To prove the uniqueness of v_u, let $u \sim v \pmod{x}$ in a, b. Then since $v \leqq b$ and

$$v + x = u + x = v_u + x,$$

we have with the help of IV,

$$v = v + bx = (v + x)b = (v_u + x)b = v_u + bx = v_u.$$

Conversely, let (a), (b) hold. By (a) we have an element $v_a \leqq b$ such that $a + x = v_a + x$. But $v_a + x \leqq b + x$, whence $a + x \leqq b + x$. Similarly, by (b) we have $b + x \leqq a + x$, and therefore $a + x = b + x$. Now by (a) there is a unique element v_0 such that

(1) $$v_0 \leqq b, \ x = 0 + x = v_0 + x.$$

But $v_0 = 0$ and $v_0 = bx$ both satisfy (1), and hence by the uniqueness of v_0, $bx = 0$. Similarly, $ax = 0$.

Definition 3.3: Two lattices L_1, L_2 (i.e., systems satisfying I and a weaker form of II which requires the existence of greatest lower bounds and least upper bounds for finite subsets) will be said to be isomorphic if a one-to-one correspondence exists between L_1 and L_2 which preserve addition and multiplication and hence order.

Corollary: If L_1 and L_2 are isomorphic, and if L_1 has a zero element 0_1 (which satisfies, by definition, $0_1 \leqq a$ for every a in L_1), then L_2 has a zero element 0_2, and the two zero elements correspond. Dually, if L_1 has a unit element 1_1 (such that $1_1 \geqq a$ for every a in L_1), then L_2 has a unit 1_2 and the two units correspond.

Theorem 3.3: Let $a \sim b$. Then $L(0, a)$ and $L(0, b)$ are isomorphic. If x is any element such that $a + x = b + x$, $ax = bx = 0$, then the equations

(2) $$v = b(u + x), \ u = a(v + x)$$

define such an isomorphic correspondence between $L(0, a)$ and $L(0, b)$. If u, v correspond under (2), then $u \sim v \pmod{x}$ in a, b.

Proof: By Theorem 3.2 we see that there is a one-to-one correspondence between $L(0, a)$ and $L(0, b)$ which is defined by (2), and that $u \sim v \pmod{x}$ in a, b for every pair u, v of corresponding elements. It remains to prove that the correspondence (2) preserves addition and multiplication. If u_1, v_1 and u_2, v_2 are two pairs of corresponding elements under (2), then $u_1 \sim v_1$, $u_2 \sim v_2$ (mod x) in a, b. It follows then that $u_1 + u_2 \sim v_1 + v_2$ (mod x) in a, b, whence (2) preserves addition. We have also that

$$(u_1 + x)(u_2 + x) = (v_1 + x)(v_2 + x).$$

But

$$\begin{aligned}(u_1 + x)(u_2 + x) &= x + u_2(u_1 + x) \qquad \text{(by IV)} \\ &= x + u_2 u_1 + 0 \cdot x \\ &= x + u_1 u_2,\end{aligned}$$

the second equality holding by Theorem 1.2, which applies since $(u_2 + u_1)(0 + x) \leqq ax = 0$. Similarly $(v_1 + x)(v_2 + x) = x + v_1 v_2$, and $u_1 u_2 \sim v_1 v_2$ (mod x) in a, b. This proves that (2) preserves multiplication.

COROLLARY: The elements a and b correspond under (2). An element u is self-corresponding under (2) if and only if $u \leqq ab$.

PROOF: The first part is obvious since $a \sim b$ (mod x) in a, b. Suppose $u \sim u$ (mod x) in a, b. Clearly $u \leqq a$, $u \leqq b$, whence $u \leqq ab$. Conversely, if $u \leqq ab$, then $u \leqq a$, and $u \leqq b$, whence $u \sim u$ (mod x) in a, b, and u is self-corresponding.

DEFINITION 3.4: If $a + x = b + x$, $ac = bx = 0$, the isomorphism (2) is called the *perspective isomorphism of* $L(0, a)$ *and* $L(0, b)$ *with axis* x.

DEFINITION 3.5: An element a is said to be projective to an element b ($a \approx b$) in case there exists a finite sequence $(a_0, \cdots, a_n)$ such that

$$a_0 = a, \; a_n = b, \; a_{i-1} \sim a_i \quad (i = 1, \cdots, n).$$

COROLLARY 1: The relation $\approx$ is reflexive, symmetric and transitive.

COROLLARY 2: If $a \sim b$, then $a \approx b$.

Let $a \approx b$ by means of the sequence $(a_0, \cdots, a_n)$. Then by Theorem 3.3 there is a perspective isomorphism P_i between $L(0, a_{i-1})$ and $L(0, a_i)$ $(i = 1, \cdots, n)$. Hence the product $P_1 \cdots P_n$ defines an isomorphism between $L(0, a)$ and $L(0, b)$. This is called a *projective isomorphism* of $L(0, a)$ and $L(0, b)$.

THEOREM 3.4: If $(a, b, c)\perp$, then $a \sim b$, $b \sim c$ implies $a \sim c$.

PROOF: By Theorem 3.1 there exist x, z such that

$$a + z = b + z = a + b, \; az = bz = 0,$$

$$b + x = c + x = b + c, \; bx = cx = 0.$$

Define $y \equiv (a + c)(x + z)$. Now

$$\begin{aligned} a + y &= a + (a + c)(x + z) \\ &= (a + c)(a + x + z) \qquad \text{(by IV)} \\ &= (a + c)(a + b + c) = a + c. \end{aligned}$$

Similarly, $c + y = a + c$. Also,

$$\begin{aligned} ay &= a(a + c)(x + z) \\ &= a(x + z) \\ &\leqq (a + b)(x + z) = z + (a + b)x \quad \text{(by IV)}. \end{aligned}$$

But $(a, b, x)\perp$ by Theorem 2.8. Hence $(a + b)x = 0$, and $ay \leqq z$. Thus $ay \leqq az = 0$, and $ay = 0$. Similarly, $cy = 0$. This completes the proof of $a \sim c$.

COROLLARY 1: For the elements a, b, c, x, y, z of the theorem we have

$$\begin{aligned} a + z = b + z = a + b, &\qquad b + x = c + x = b + c, \\ az = bz = ab = 0, &\qquad bx = cx = bc = 0, \end{aligned}$$
$$a + y = c + y = a + c,$$
$$ay = cy = ac = 0.$$

PROOF: These results are contained in the proof of Theorem 3.4.

COROLLARY 2: For the elements a, b, c, x, y, z of the theorem, $u \sim v$ (mod z) in a, b, $v \sim w$ (mod x) in b, c implies $u \sim w$ (mod y) in a, c.

PROOF: By hypothesis $u + z = v + z$, $v + x = w + x$. Therefore

$$\begin{aligned} u + y &= u + (a + c)(x + z) \\ &= (a + c)(u + x + z) \quad \text{(by IV)} \\ &= (a + c)(v + x + z) \\ &= (a + c)(w + x + z) \\ &= w + (a + c)(x + z) \quad \text{(by IV)} \\ &= w + y. \end{aligned}$$

THEOREM 3.5: If $(a, b, c)\perp$ and $a \sim b$, then $a + c \sim b + c$.

PROOF: By hypothesis and Theorem 3.1 there exists an x such that

$$a + x = b + x = a + b,$$
$$ax = bx = 0.$$

Clearly $(a + c) + x = (b + c) + x = a + b + c$. Now since $(a, b, c)\perp$, we have $(a + b)c = 0$, whence $(a + x)c = 0$. Hence by Theorem 1.2,

$$(a + c)x = ax + c \cdot 0 = 0.$$

Similarly $(b + c)x = 0$, and $a + c \sim b + c$.

THEOREM 3.6: Let I be any class, and $(a_\sigma;\ \sigma \in I)$ any independent system. If J and K are two non-intersecting subsets of I with a one-to-one correspondence $\rho \leftrightarrow \rho'$ between them, such that $a_\rho \sim a_{\rho'}$ for each pair ρ, ρ' of corresponding elements, $\rho \in J$, $\rho' \in K$, then

$$\Sigma(a_\rho;\ \rho \in J) \sim \Sigma(a_{\rho'};\ \rho' \in K).$$

PROOF: By the hypothesis that $a_\rho \sim a_{\rho'}$ and Theorem 3.1, there exists for each ρ in J an element x_ρ such that

$$\begin{aligned} a_\rho + x_\rho = a_{\rho'} + x_\rho = a_\rho + a_{\rho'}, \\ a_\rho x_\rho = a_{\rho'} x_\rho = 0. \end{aligned} \tag{3}$$

Hence it is clear that

$$\Sigma(a_\rho;\ \rho \in J) + x = \Sigma(a_{\rho'};\ \rho' \in K) + x,$$

where $x = \Sigma(x_\rho;\ \rho \in J)$. By Theorem 2.8, $(b_\sigma;\ \sigma \in I)\perp$, where $b_\sigma = a_\sigma(\sigma \notin J)$, $b_\rho = x_\rho(\rho \in J)$. Thus

$$\Sigma(b_\rho;\ \rho \in J)\Sigma(b_\sigma;\ \sigma \in K) = 0,$$

or

$$x\Sigma(a_{\rho'};\ \rho' \in K) = 0.$$

Similarly,

$$x\Sigma(a_\rho;\ \rho \in J) = 0,$$

and the theorem follows by Theorem 3.1(c).

THEOREM 3.7: If $a \sim b$, $a \leqq b$, then $a = b$.

PROOF: Since $a \sim b$, there is a perspective isomorphism P of $L(0, a)$ and $L(0, b)$. By the Corollary to Theorem 3.3, since $a = ab$, we have that a corresponds to itself under P. But a also corresponds to b, whence $b = a$.

THEOREM 3.8: Let $(a_i; i = 1, 2, \cdots)$ be an infinite independent sequence. If for every $i = 1, 2, \cdots, a_i \sim a_{i+1}$, then each $a_i = 0$.

PROOF: By Theorem 3.4, $a_1 \sim a_n$ $(n = 2, 3, \cdots)$, whence there is a sequence $(x_n;\ n = 2, 3, \cdots)$ such that

$$a_1 + x_n = a_n + x_n = a_1 + a_n,\ a_1 x_n = a_n x_n = 0 \quad (n > 1). \tag{4}$$

Define

$$c \equiv \sum_{i=2}^{\infty} x_i,\quad b_n \equiv \sum_{i=n+1}^{\infty} a_i,\quad r_n \equiv c + b_n \qquad (n \geqq 1).$$

Since $x_{n+1} + a_{n+1} = x_{n+1} + a_1$, $r_n \geqq a_1$. Hence

(5) $$\prod_{n=1}^{\infty} r_n \geqq a_1.$$

Now $b_n \geqq b_{n+1}$, whence III_1 applies, and

$$\prod_{n=1}^{\infty} r_n = c + \prod_{n=1}^{\infty} b_n.$$

But since $(a_i)\perp$, it follows that $\prod_{n=1}^{\infty} b_n = 0$ by Theorem 2.6, whence by (5)

(6) $$\sum_{i=2}^{\infty} x_i \geqq a_1.$$

By (4) we may apply Theorem 2.8 to the system $(a_i;\ i = 1, 2, \cdots)$, letting I be the class of all positive integers, J the class of all integers $\geqq 2$, and K the class consisting of 1 alone. This yields $(b_i;\ i = 1, 2, \cdots)\perp$, where $b_1 = a_1$, $b_i = x_i$ $(i \geqq 2)$. Hence

$$a_1 \sum_{i=2}^{\infty} x_i = 0,$$

and therefore $a_1 = 0$ by (6). Replacing the sequence $(a_1, a_2, \cdots)$ by $(a_i, a_{i+1}, \cdots)$, which satisfies the hypotheses and to which the result just proved therefore applies, we have $a_i = 0$ $(i = 1, 2, \cdots)$.

THEOREM 3.9: If $a \sim c \sim b$, $bc = 0$, $a \leqq b$, then $a = b$.

PROOF: Define $b_1 \equiv a$, and let b_1' be an inverse of b_1 in b (cf. the Corollary to Definition 1.6). We have

(7) $$b_1 + b_1' = b,\ b_1 b_1' = 0.$$

Now since $b \sim c$, there is a perspective isomorphism $P(b, c)$ of $L(0, b)$ and $L(0, c)$. This carries b_1, b_1' into elements c_1, c_1' of $L(0, c)$. Similarly, there is a perspective isomorphism $P(c, a)$ of $L(0, c)$ and $L(0, a)$. This carries c_1, c_1' into elements a_1, a_1' of $L(0, a)$. We define

$$b_2 \equiv a_1,\quad b_2' \equiv a_1'.$$

In general, if b_i, b_i' have been defined, let c_i, c_i' be their images under $P(b, c)$, a_i, a_i' the images of c_i, c_i' under $P(c, a)$, and $b_{i+1} \equiv a_i$, $b_{i+1}' \equiv a_i'$. The sequences thus defined have the following properties:

(a) b_{i+1}' is an inverse of b_{i+1} in b_i,

(b) $b_i' \leqq b$, $c_i' \leqq c$,

(c) $b_i' \sim c_i' \sim b_{i+1}'$ $\qquad (i = 1, 2, \cdots)$.

By (a), $b'_{i+1} \leqq b_i$, $b_{i+1} \leqq b_i$. Hence $b'_{i+j} \leqq b_i$ for every i, j, and so $b'_k \leqq b_i$ for all $k > i$.

Thus $\sum_{k>i} b'_k \leqq b_i$, and $b'_i \sum_{k>i} b'_k \leqq b'_i b_i$. But $b'_i b_i = 0$ by (7) and (a). Hence for every i, l,

$$b'_i(b'_{i+1} + \cdots + b'_{i+l}) = 0. \tag{8}$$

Let n be any positive integer. For each $i < n$, let $l = n - i$. Hence by (8)

$$b'_i(b'_{i+1} + \cdots + b'_n) = 0 \qquad (i = 1, \cdots, n-1). \tag{9}$$

This proves $(b'_1, \cdots, b'_n)\perp$ $(n = 1, 2, \cdots)$ by Theorem 2.2. Hence by Theorem 2.3 $(b'_i;\ i = 1, 2, \cdots)\perp$. By hypothesis $bc = 0$, whence we see from (b) that $(b'_i + b'_{i+1})c'_i = 0$. Therefore $(b'_i, c'_i, b'_{i+1})\perp$ $(i = 1, 2, \cdots)$, and by Theorem 3.4, (c) implies $b'_i \sim b'_{i+1}$. Hence $b'_i = 0$ by Theorem 3.8, and in particular $b'_1 = 0$. But then from (7) we have $b_1 = b$, or $a = b$.

Theorem 3.10: If $a \sim b \sim c$, $ab = 0$, $bc = 0$, $b \leqq a + c$, then $a \sim c$.

Proof: Define $c' \equiv c(a + b)$. Now $a + b \leqq a + c$, since $b \leqq a + c$. Hence

$$\begin{aligned} a + b = (a + b)(a + c) &= a + c(a + b) \qquad \text{(by IV)} \\ &= a + c'. \end{aligned} \tag{10}$$

Now since $ac' \leqq c'$ there is an inverse c'' of ac' in c'. Thus $ac' + c'' = c'$, $ac'c'' = 0$. In fact $ac'' = ac'c'' = 0$ since $c'' \leqq c'$. Hence by (10),

$$a + b = a + c' = a + (ac' + c'') = a + c'', \tag{11}$$

and

$$ab = ac'' = 0, \tag{12}$$

whence $c'' \sim b$. We have therefore

$$c'' \sim b \sim c,\ c'' \leqq c,\ bc = 0,$$

and hence $c'' = c$ by Theorem 3.9. Thus we obtain from (11) and (12),

$$a + b = a + c,\ ab = ac = 0. \tag{13}$$

Interchanging a and c in (13), as we may do, since a and c enter symmetrically in the hypothesis, and combining the result with (13), we have

$$a + b = b + c = c + a,\ ab = bc = ca = 0,$$

from which we may infer $a \sim c$.

PART I · CHAPTER IV

Perspectivity by Decomposition

DEFINITION 4.1: A triple (a, b, c) is said to be *perspective by decomposition* ($(a, b, c)pd$), in case there exist three independent sequences $(a_i; i = 1, 2, \cdots)$, (b_i), (c_i), such that $a_i \sim b_i \sim c_i \sim a_i$ $(i = 1, 2, \cdots)$, and

$$a = \sum_{i=1}^{\infty} a_i, \qquad b = \sum_{i=1}^{\infty} b_i, \qquad c = \sum_{i=1}^{\infty} c_i.$$

COROLLARY 1: If $a \sim b \sim c \sim a$, then $(a, b, c)pd$.

PROOF: Define $a_1 \equiv a$, $a_i \equiv 0$ $(i = 2, 3, \cdots)$. Then $(a_i; i = 1, 2, \cdots)\perp$ and $a = \sum_{i=1}^{\infty} a_i$. Proceed similarly for b, c. Clearly $a_i \sim b_i \sim c_i \sim a_i$ $(i = 1, 2, \cdots)$, and $(a, b, c)pd$.

(The converse to Corollary 1 also holds, i.e. if $(a, b, c)pd$ then $a \sim b \sim c \sim a$ but the proof of this fact will not be given until later; it follows from Theorem 6.9 (iii)'' and Lemma 7.1 - Ed.)

COROLLARY 2: If $a \sim b$, then $(a, b, b)pd$.

PROOF: Obviously $a \sim b \sim b \sim a$, since $\sim$ is reflexive and symmetric. Hence by Corollary 1, $(a, b, b)pd$.

COROLLARY 3: The relation pd is symmetric, i.e. $(a_1, a_2, a_3)pd$ implies $(a_{k_1}, a_{k_2}, a_{k_3})pd$ for every permutation $i \to k_i$ of the integers 1, 2, 3.

LEMMA 4.1: Let a, b, c be given. Suppose

$$a = \sum_{i=1}^{\infty} a_i, \qquad b = \sum_{i=1}^{\infty} b_i, \qquad c = \sum_{i=1}^{\infty} c_i,$$

where each sequence $(a_i; i = 1, 2, \cdots)$, (b_i), (c_i) is independent. If for each $i = 1, 2, \cdots$, $(a_i, b_i, c_i)pd$, then $(a, b, c)pd$.

PROOF: By hypothesis,

$$a_i = \sum_{j=1}^{\infty} a_{ij}, \qquad b_i = \sum_{j=1}^{\infty} b_{ij}, \qquad c_i = \sum_{j=1}^{\infty} c_{ij},$$

where for every i each of the sequences $(a_{ij}; j = 1, 2, \cdots)$, (b_{ij}), (c_{ij}) is independent. Hence by Theorem 2.4, the double sequences $(a_{ij}; i, j = 1, 2, \cdots)$, (b_{ij}), (c_{ij}) are independent. Furthermore,

$$a_{ij} \sim b_{ij} \sim c_{ij} \sim a_{ij} \qquad (i, j = 1, 2, \cdots).$$

Therefore, if we renumber the double sequences (a_{ij}), (b_{ij}), (c_{ij}) into simple sequences (a'_i), (b'_i), (c'_i), respectively, we have

$$a'_i \sim b'_i \sim c'_i \sim a'_i \qquad (i = 1, 2, \cdots),$$

$(a'_i;\ i = 1, 2, \cdots)\perp$, $(b'_i)\perp$, $(c'_i)\perp$,

$$a = \sum_{i=1}^{\infty} a'_i, \quad b = \sum_{i=1}^{\infty} b'_i, \quad c = \sum_{i=1}^{\infty} c'_i,$$

and $(a, b, c)pd$.

COROLLARY: If $(a_1, b_1, c_1)pd$, $(a_2, b_2, c_2)pd$, and $a_1a_2 = b_1b_2 = c_1c_2 = 0$, and $a = a_1 + a_2$, $b = b_1 + b_2$, $c = c_1 + c_2$, then $(a, b, c)pd$.

PROOF: Define $a_i \equiv b_i \equiv c_i \equiv 0$ $(i = 3, 4, \cdots)$. Then Lemma 4.1 applies and $(a, b, c)pd$.

LEMMA 4.2: Let two sequences $(a_i;\ i = 1, 2, \cdots)$, $(a'_i;\ i = 1, 2, \cdots)$ be such that

(1) $$a_i a'_i = 0,\ a_{i+1} \leqq a'_i,\ a'_{i+1} \leqq a'_i \qquad (i = 1, 2, \cdots).$$

Then $(a_i;\ i = 1, 2, \cdots)\perp$. If $a_0 \equiv \prod_{i=1}^{\infty} a'_i$, then $(a_i;\ i = 0, 1, \cdots)\perp$.

PROOF: By (1) we have $a_m \leqq a'_e$ $(m > l)$, whence

$$\sum_{m=l+1}^{\infty} a_m \leqq a'_l \qquad (l = 1, 2, \cdots),$$

and

$$a_l \sum_{m=l+1}^{\infty} a_m \leqq a_l\, a'_l = 0 \qquad (l = 1, 2, \cdots).$$

Thus

$$a_l(a_{l+1} + \cdots + a_{l+k}) \leqq a_l \sum_{m=l+1}^{\infty} a_m = 0 \quad (l, k = 1, 2, \cdots),$$

and we may infer

$$a_i(a_{i+1} + \cdots + a_n) = 0 \quad (i < n,\ n = 2, 3, \cdots),$$

and hence for each $n = 1, 2, \cdots$, $(a_1, \cdots, a_n)\perp$ by Theorem 2.2. Thus by Theorem 2.3 $(a_i;\ i = 1, 2, \cdots)\perp$, since every finite subsequence of this sequence is contained in one of the form $(a_1, \cdots, a_n)$, and is hence independent. In order to prove the second part, it clearly suffices to prove

(2) $$a_0(a_1 + \cdots + a_n) = 0 \qquad (n = 1, 2, \cdots).$$

Now by (1), $a'_1 a_1 = 0$. Suppose $a'_i(a_1 + \cdots + a_i) = 0$. Then define

$$a \equiv 0,\ b \equiv a'_{i+1},\ c \equiv a_1 + \cdots + a_i,\ d \equiv a_{i+1}.$$

Obviously,

$$\begin{aligned}(a+c)(b+d) &= (a_{i+1}+a'_{i+1})(a_1+\cdots+a_i)\\ &\leqq a'_i(a_1+\cdots+a_i) \qquad \text{(by (1))}\\ &= 0,\end{aligned}$$

and Theorem 1.2 applies, yielding $(a+b)(c+d) = ac+bd$, or

$$a'_{i+1}(a_1+\cdots+a_{i+1}) = a'_{i+1}a_{i+1} = 0.$$

This induction proves (2), since $a_0(a_1+\cdots+a_n) \leqq a'_n(a_1+\cdots+a_n) = 0$.

LEMMA 4.3. If $a \sim b \sim c$ and $(a+c)b = 0$, then $(a, b, c)pd$.

PROOF: Define $a'_0 \equiv a$, $b'_0 \equiv b$, $c'_0 \equiv c$, $a'_1 \equiv ac$, and let a_1 be an inverse of a'_1 in a. Let $P(a, b)$ and $P(b, c)$ be perspective isomorphisms of $L(0, a)$, $L(0, b)$ and $L(0, b)$, $L(0, c)$, respectively. By $P(a, b)$ we obtain images b_1, b'_1 in $L(0, b)$ of a_1, a'_1; and by $P(b, c)$ we obtain images c_1, c'_1 in $L(0, c)$ of b_1, b'_1. If a_i, b_i, c_i, a'_i, b'_i, c'_i have been defined, put $a'_{i+1} \equiv a'_i c'_i$, let a_{i+1} be an inverse of a'_{i+1} in a'_i, and let b_{i+1}, b'_{i+1} be the images of a_{i+1}, a'_{i+1} under $P(a, b)$, and c_{i+1}, c'_{i+1} the images of b_{i+1}, b'_{i+1} under $P(b, c)$. This construction gives rise to six sequences $(a_i;\ i = 1, 2, \cdots)$, (b_i), (c_i), $(a'_i;\ i = 0, 1, \cdots)$, (b'_i), (c'_i), which have in particular the following properties;

(a) $a'_{i+1} = a'_i c'_i \quad (i = 0, 1, \cdots)$,

(b) a_{i+1} is inverse to a'_{i+1} in a'_i,
b_{i+1} is inverse to b'_{i+1} in b'_i,
c_{i+1} is inverse to c'_{i+1} in $c'_i \quad (i = 0, 1, \cdots)$,

(c) $a_i \sim b_i \sim c_i \quad (i = 1, 2, \cdots)$, $a'_i \sim b'_i \sim c'_i \quad (i = 0, 1, \cdots)$.

From (b) we have immediately

(3) $$a'_{i+1} \leqq a'_i \leqq a, \quad b'_{i+1} \leqq b'_i \leqq b, \quad c'_{i+1} \leqq c'_i \leqq c \quad (i = 0, 1, \cdots),$$

and

(4) $$a_{i+1} \leqq a'_i, \quad b_{i+1} \leqq b'_i, \quad c_{i+1} \leqq c'_i \quad (i = 0, 1, \cdots),$$

also, from (3), (4) and (a) there results

$$a_{i+1}c_{i+1} \leqq a_{i+1}c'_i = a'_i a_{i+1} c'_i = a_{i+1}a'_{i+1} = 0 \quad (i = 0, 1, \cdots),$$

whence $a_i c_i = 0 \quad (i = 1, 2, \cdots)$. But we also have by (4) and the hypothesis,

$$(a_i + c_i)b_i \leqq (a+c)b = 0,$$

and we may conclude $(a_i, b_i, c_i)\perp \ (i = 1, 2, \cdots)$ by Theorem 2.2. Therefore we infer from (c) that $a_i \sim c_i$ by Theorem 3.4, and we have

(5) $$a_i \sim b_i \sim c_i \sim a_i \qquad (i = 1, 2, \cdots).$$

Now if $a_0 \equiv \prod_{i=1}^{\infty} a'_i$, $b_0 \equiv \prod_{i=1}^{\infty} b'_i$, $c_0 \equiv \prod_{i=1}^{\infty} c'_i$, we have by an application of Lemma 4.2,

(6) $$(a_i;\ i = 0, 1, \cdots)\perp,\ (b_i;\ i = 0, 1, \cdots)\perp,\ (c_i;\ i = 0, 1, \cdots)\perp.$$

From (b) it follows readily that

$$a = a'_1 + a_1 = a'_2 + a_2 + a_1 = \cdots = a'_n + \sum_{i=1}^{n} a_i \qquad (n = 1, 2, \cdots).$$

Hence, since each $a_i \leqq a$,

$$a = a'_n + \sum_{i=1}^{\infty} a_i \qquad (n = 1, 2, \cdots),$$

and also

$$a = \prod_{n=1}^{\infty} (a'_n + \sum_{i=1}^{\infty} a_i).$$

By (3), III_1 applies, and $a = \prod_{n=1}^{\infty} a'_n + \sum_{i=1}^{\infty} a_i = \sum_{i=0}^{\infty} a_i$. This, together with similar considerations for b, c, yields

(7) $$a = \sum_{i=0}^{\infty} a_i, \quad b = \sum_{i=0}^{\infty} b_i, \quad c = \sum_{i=0}^{\infty} c_i.$$

Now by (a), $a'_{m+1} \leqq c'_m$, whence

$$\prod_{i=1}^{\infty} a'_i \leqq \prod_{i=1}^{\infty} c'_i$$

and $a_0 \leqq c_0$. Let x be the axis of the perspective isomorphism $P(a, b)$. Then $a'_i + x = b'_i + x$ $(i = 1, 2, \cdots)$, and by III_1, $a_0 + x = b_0 + x$. This together with $a_0 x \leqq ax = 0$, $b_0 x \leqq bx = 0$ implies $a_0 \sim b_0$. Similarly, $b_0 \sim c_0$. Finally,

$$b_0 c_0 \leqq bc \leqq b(a + c) = 0.$$

Thus Theorem 3.9 applies, and we have $a_0 = c_0$, whence

(8) $$a_0 \sim b_0 \sim c_0 \sim a_0.$$

But (5), (8), (6), (7) tell us that $(a, b, c)pd$ by Definition 4.1, and the lemma is proved.

LEMMA 4.4: If $a \sim b \sim c$ and $ab = cb = 0$, then $(a, b, c)pd$.

PROOF: Let $P(a, b)$ and $P(b, c)$ be perspective isomorphisms of $L(0, a)$, $L(0, b)$ and $L(0, b)$, $L(0, c)$, respectively. Define $a'_0 \equiv a$, $b'_0 \equiv b$, $c'_0 \equiv c$, $b'_1 \equiv (a + c)b$, and let b_1 be an inverse of b'_1 in b. Let c_1, c'_1 be the images in $L(0, c)$ of b_1, b'_1 under $P(b, c)$, and a_1, a'_1 the images in $L(0, a)$ of b_1,

b'_1 under $P(a, b)$. If $a_i, b_i, c_i, a'_i, b'_i, c'_i$ have been defined, put $b'_{i+1} \equiv (a'_i + c'_i)b'_i$ and let b_{i+1} be an inverse of b'_{i+1} in b'_i. Also, let c_{i+1}, c'_{i+1} be the images of b_{i+1}, b'_{i+1} under $P(b, c)$ and let a_{i+1}, a'_{i+1} be the images of b_{i+1}, b'_{i+1} under $P(a, b)$. This construction gives rise to six sequences $(a_i;\ i = 1, 2, \cdots)$, (b_i), (c_i), $(a'_i;\ i = 0, 1, \cdots)$, (b'_i), (c'_i), which have in particular the following properties:

(a) $b'_{i+1} = (a'_i + c'_i)b'_i \qquad (i = 0, 1, \cdots)$,

(b) a_{i+1} is inverse to a'_{i+1} in a'_i,
b_{i+1} is inverse to b'_{i+1} in b'_i,
c_{i+1} is inverse to c'_{i+1} in $c'_i \qquad (i = 0, 1, \cdots)$,

(c) $a_i \sim b_i \sim c_i \quad (i = 1, 2, \cdots), \qquad a'_i \sim b'_i \sim c'_i \quad (i = 0, 1, \cdots)$.

From (b) it follows readily that equations (3) and (4) hold, and we have by (a)

$$(a_{i+1} + c_{i+1})b_{i+1} \leqq (a'_i + c'_i)b'_i = b'_{i+1},$$

whence by (b)

$$(a_{i+1} + c_{i+1})b_{i+1} \leqq b_{i+1}b'_{i+1} = 0 \qquad (i = 0, 1, \cdots). \tag{9}$$

If $a_0 \equiv \prod_{i=1}^{\infty} a'_i$, $b_0 \equiv \prod_{i=1}^{\infty} b'_i$, $c_0 \equiv \prod_{i=1}^{\infty} c'_i$, we have, by Lemma 4.2,

$$(a_i;\ i = 0, 1, \cdots)\perp, \ (b_i;\ i = 0, 1, \cdots)\perp, \ (c_i;\ i = 0, 1, \cdots)\perp. \tag{10}$$

As in the proof of Lemma 4.3,

$$a = \sum_{i=0}^{\infty} a_i, \ b = \sum_{i=0}^{\infty} b_i, \ c = \sum_{i=0}^{\infty} c_i. \tag{11}$$

By (9), $(a_i + c_i)b_i = 0$ $(i = 1, 2, \cdots)$. This together with (c) yields

$$(a_i, b_i, c_i)pd \qquad (i = 1, 2, \cdots) \tag{12}$$

by Lemma 4.3. Now by (a), $b'_{i+1} \leqq a'_i + c'_i$ $(i = 1, 2, \cdots)$, whence $b_0 \leqq a'_i + c'_i$ $(i = 1, 2, \cdots)$, and

$$b_0 \leqq f \equiv \prod_{i=1}^{\infty} (a'_i + c'_i).$$

If we define

$$d \equiv \prod_{n=1}^{\infty} \prod_{m=1}^{\infty} (a'_n + c'_m),$$

then clearly $f \geqq d$, since each factor in f occurs in d. Consider any factor $a'_n + c'_m$ in d. If $m \geqq n$, $a'_n + c'_m \geqq a'_m + c'_m \geqq f$, and if $m < n$, $a'_n + c'_m \geqq a'_n + c'_n \geqq f$, whence $d \geqq f$, and therefore $d = f$. Thus $b_0 \leqq d$. But

$$\begin{aligned} a_0 + c_0 &= \prod_{n=1}^{\infty} a'_n + \prod_{m=1}^{\infty} c'_m \\ &= \prod_{n=1}^{\infty} \Big(\prod_{m=1}^{\infty} c'_m + a'_n \Big) \quad \text{(by III}_1\text{)} \\ &= \prod_{n=1}^{\infty} \prod_{m=1}^{\infty} (a'_n + c'_m) \quad \text{(by III}_1\text{)} \\ &= d. \end{aligned}$$

Hence $b_0 \leqq a_0 + c_0$. We have also $a_0 b_0 \leqq ab = 0$, $c_0 b_0 \leqq cb = 0$. Finally, if x is the axis of perspectivity of $P(a, b)$, then $a'_i + x = b'_i + x$ $(i = 1, 2, \cdots)$, whence by III_1, $a_0 + x = b_0 + x$. This, together with $a_0 x \leqq ax = 0$, $b_0 x \leqq bx = 0$, implies $a_0 \sim b_0$. Similarly $b_0 \sim c_0$, and Theorem 3.10 applies, yielding $a_0 \sim b_0 \sim c_0 \sim a_0$, whence $(a_0, b_0, c_0)pd$ by the Corollary to Definition 4.1. Combining this result with (10), (11), (12), we may infer by Lemma 4.1, $(a, b, c)pd$.

Lemma 4.5: If $a \sim b \sim c$ and $ab = 0$, then $(a, b, c)pd$.

Proof: Let b' be an inverse of bc in b. Thus

$$b = bc + b', \quad b'c = b'bc = 0. \tag{13}$$

Let $P(a, b)$ and $P(b, c)$ be perspective isomorphisms of $L(0, a)$, $L(0, b)$ and $L(0, b)$, $L(0, c)$ respectively, and let a_1, c_1 be the images of bc under $P(a, b)$, $P(b, c)$, respectively, and a_2, c_2 the images of b' under $P(a, b)$, $P(b, c)$ respectively. By the Corollary to Theorem 3.3, $c_1 = bc$. Also, a_1, a_2 are inverses in a, and c_1, c_2, i.e., bc, c_2, are inverses in c, whence

$$a_1 + a_2 = a, \quad a_1 a_2 = 0, \quad bc + c_2 = c \tag{14}$$

and $(a_1, bc, bc)pd$ by Corollary 2 to Definition 4.1, since $a_1 \sim bc$. Now $a_2 \sim b' \sim c_2$, and $a_2 b' \leqq ab = 0$, $b'c_2 \leqq b'c = 0$, so that $a_2 b' = b'c_2 = 0$. Hence by Lemma 4.4, $(a_2, b', c_2)pd$, and the Corollary to Lemma 4.1 applies by (13) and (14), yielding $(a, b, c)pd$.

Lemma 4.6: If $a \sim b \sim c$ and $bc = 0$, then $(a, b, c)pd$.

Proof: By hypothesis, $c \sim b \sim a$, $cb = 0$, whence Lemma 4.5 applies, and $(c, b, a)pd$. Thus $(a, b, c)pd$ by Corollary 3 to Definition 4.1.

Theorem 4.1: If $a \sim b \sim c$, then $(a, b, c)pd$.

Proof: This follows from Lemma 4.6 in precisely the same manner as Lemma 4.5 follows from Lemma 4.4.

Lemma 4.7: If $(a_i;\ i = 0, \cdots, n)$ is a finite sequence such that $a_{i-1} \sim a_i$ $(i = 1, \cdots, n)$, $a_0 a_n = 0$, then $a_0 \sim a_n$.

Proof: The lemma is obvious for $n = 0, 1$. Let $m \geqq 1$ and suppose that it is true for $n = 1, \cdots, m$. Then consider a sequence $(a_0, \cdots, a_{m+1})$ of

$m+2$ terms satisfying the hypotheses. Since $a_0 \sim a_1 \sim a_2$, we have by Theorem 4.1 that $(a_0, a_1, a_2)pd$. Hence a_0, a_1, a_2 are expressible in the form

$$a_i = \sum_{j=1}^{\infty} a_{ij} \qquad (i = 0, 1, 2),$$

where $(a_{ij};\ j = 1, 2, \cdots)\perp$ for $i = 0, 1, 2$, and

$$a_{0j} \sim a_{1j} \sim a_{2j} \sim a_{0j} \quad (j = 1, 2, \cdots). \tag{15}$$

Thus if $m = 1$, a_0 and a_{m+1} are expressed as the sums of independent sequences, corresponding terms of the sequences being perspective. Suppose $m > 1$ and let P_i be a perspective isomorphism of $L(0, a_i)$ and $L(0, a_{i+1})$ $(i = 2, \cdots, m)$. The independent elements $a_{2j} \leqq a_2$ are thus mapped by P_2 into independent elements $a_{3j} \leqq a_3$, which in turn are mapped by P_3 into independent elements $a_{3j} \leqq a_3$, etc.; finally, the $a_{mj} \leqq a_m$ are mapped by P_m into $a_{m+1,j} \leqq a_{m+1}$. Hence $(a_{ij};\ j = 1, 2, \cdots)\perp$ $(i = 3, \cdots, m+1)$, and we have

$$a_i = \sum_{j=1}^{\infty} a_{ij} \qquad (i = 3, \cdots, m+1).$$

Moreover,

$$a_{0j} \sim a_{1j} \sim a_{2j} \sim a_{3j} \sim \cdots \sim a_{m+1,j} \qquad (j = 1, 2, \cdots). \tag{16}$$

But by (15), $a_{0j} \sim a_{2j}$, and we may compress (16) into

$$a_{0j} \sim a_{2_j} \sim a_{3_j} \sim \cdots \sim a_{m+1,j} \qquad (j = 1, 2, \cdots); \tag{17}$$

obviously $a_{0j}a_{m+1,j} \leqq a_0a_{m+1} = 0$. Hence the induction hypothesis applies to the sequences of length m in (17), and we may conclude $a_{0j} \sim a_{m+1,j}$. Therefore, in the case $m > 1$ also, a_0 and a_{m+1} are expressed as sums of independent sequences with corresponding pairs of terms perspective. Since

$$\sum_{j=1}^{\infty} a_{0j} \sum_{j=1}^{\infty} a_{m+1,j} = a_0 a_{m+1} = 0,$$

the combined system $(a_{01}, a_{02}, \cdots, a_{m+1,1}, a_{m+1,2}, \cdots)$ is independent, and $a_0 \sim a_{m+1}$ by Theorem 3.6. This completes the induction.

THEOREM 4.2: If $ab = 0$, then $a \approx b$ if and only if $a \sim b$.

PROOF: It has already been noted (Corollary 2, Definition 3.5) that $a \sim b$ implies $a \approx b$. Suppose $a \approx b$. Then there is a sequence $(a_0, \cdots, a_n)$ with $a_0 = a$, $a_n = b$, $a_{i-1} \sim a_i$ $(i = 1, \cdots, n)$. Since $a_0a_n = 0$, we have by Lemma 4.7, $a_0 \sim a_n$, i.e., $a \sim b$.

THEOREM 4.3: Let $(a_i; i = 1, 2, \cdots)$ be an infinite independent sequence. If for every $i = 1, 2, \cdots$, $a_i \approx a_{i+1}$, then each $a_i = 0$.

PROOF: Clearly the pair (a_i, a_{i+1}) is independent, whence $a_i a_{i+1} = 0$ $(i = 1, 2, \cdots)$. Therefore by Theorem 4.2, $a_i \sim a_{i+1}$ for $i = 1, 2, \cdots$, and by Theorem 3.8 each $a_i = 0$.

THEOREM 4.4: If $a \approx b$, $a \leqq b$, then $a = b$.

PROOF: Define $b_1' \equiv a \leqq b$, and let b_1 be an inverse of b_1' in b. If P is a projective isomorphism of $L(0, a)$ and $L(0, b)$, we obtain images a_1, a_1' in $L(0, a)$ of b_1, b_1' under P. If a_i, a_i', b_i, b_i' have been defined, put

(a) $$b_{i+1} \equiv a_i, \quad b_{i+1}' \equiv a_i',$$

and let a_{i+1}, a_{i+1}' be the images of b_{i+1}, b_{i+1}' under P in $L(0, b)$. Thus there are defined four infinite sequences $(a_i; i = 1, 2, \cdots)$, (b_i), $(a_i'; i = 0, 1, \cdots)$, (b_i'), which have the following properties:

(b) a_i is inverse to a_i' in a_{i-1}' $(a_0' = a)$,
b_i is inverse to b_i' in b_{i-1}' $(b_0' = b)$ $\qquad (i = 1, 2, \cdots)$,

(c) $a_i \approx b_i$, $a_i' \approx b_i'$ $\qquad (i = 1, 2, \cdots)$.

From (b) we conclude immediately

$$b_i b_i' = 0, \quad b_{i+1} \leqq b_i', \quad b_{i+1}' \leqq b_i' \quad (i = 1, 2, \cdots).$$

But these conditions imply $(b_i; i = 1, 2, \cdots)\perp$ by Lemma 4.2. Now by (c) and (a) we have $b_i \approx a_i = b_{i+1}$ $(i = 1, 2, \cdots)$. Hence by Theorem 4.3, $b_i = 0$ $(i = 1, 2, \cdots)$. In particular, $b_1 = 0$. Since b_1 is an inverse of a in b, $a = b + b_1 = b$.

PART I · CHAPTER V

Distributivity. Equivalence of Perspectivity and Projectivity

DEFINITION 5.1:

a) $(a, b, c)D$ means

(1) $$(a + b)c = ac + bc;$$

b) $(a, b)D$ means $(a, b, c)D$ for every c in L, i.e., (1) holds for every choice of c;

c) $(a)D$ means $(a, b)D$ for every b, i.e., (1) holds for every choice of b, c.

COROLLARY: $(a, b, c)D$ is equivalent to $(b, a, c)D$; $(a, b)D$ is equivalent to $(b, a)D$.

THEOREM 5.1: The relations D defined in Definition 5.1 are self-dual and symmetric.

PROOF: It suffices to prove the theorem for the ternary relation D defined in part a), since it then automatically follows for the binary and unary relations of b) and c). Let D' be the dual relation to the ternary relation D, so that $(a, b, c)D'$ in case $ab + c = (a + c)(b + c)$. It will first be shown that $(a, b, c)D$ implies $(b, c, a)D'$. Suppose $(a, b, c)D$, i.e., that (1) holds. Then

$$\begin{aligned}(b + a)(c + a) &= a + (a + b)c \quad \text{(by IV)}\\ &= a + ac + bc\\ &= a + bc,\end{aligned}$$

whence $(b, c, a)D'$. Dually, $(a, b, c)D'$ implies $(b, c, a)D$. Hence from a statement of the form $(a, b, c)D$ or $(a, b, c)D'$ we may infer one obtained by simultaneously interchanging D and D' and performing the cyclic permutation of the letters which replaces each letter by its right neighbor. Hence, by three applications of this principle,

(2) $$(a, b, c)D \rightarrow (b, c, a)D' \rightarrow (c, a, b)D \rightarrow (a, b, c)D'.$$

From (2) we may conclude that D is self-dual, since $(a, b, c)D$ implies $(a, b, c)D'$, and hence by duality $(a, b, c)D'$ implies $(a, b, c)D$. We have also from (2) that $(a, b, c)D$ implies $(c, a, b)D$; this states that a statement of the form $(a, b, c)D$ implies the statement obtained from it by performing the cyclic permutation on the letters which replaces each letter by its left neighbor. By this result and the Corollary to Definition 5.1, we have

$$(a,b,c)D\to(c,a,b)D\to(b,c,a)D\to(c,b,a)D\to(a,c,b)D\to(b,a,c)D\to(a,b,c)D,$$

which proves the symmetry of D.

Tacit use of Theorem 5.1 will be made constantly in the sequel, the frequency of its application prohibiting references to it.

In the discussion of Axiom VI in Chapter I, it was seen (Theorem 1.5) that this axiom is equivalent to a property which resembles algebraic irreducibility. Since this notion and the accompanying notion of direct sum are vital in what follows, we shall give at this point precise definitions of them.

Definition 5.2: The lattice L is said to be the direct sum of $L(0, a)$ and $L(0, b)$ ($L = L(0, a) \oplus L(0, b)$) in case $ab = 0$ and each element x of L is expressible in the form $x = u + v$, with $u \leqq a$, $v \leqq b$. We shall say that L is reducible if there exist elements a, b, both distinct from 0, 1, such that $L = L(0, a) \oplus L(0, b)$, and irreducible if it is not reducible. Irreducibility is the property denoted in Chapter I by Δ.

Corollary: If $L = L(0, a) \oplus L(0, b)$, then

(a) $a + b = 1$, and

(b) each element x is expressible uniquely in the form $x = u + v$, $u \leqq a$, $v \leqq b$; in fact $u = ax$, $v = bx$.

Proof: (a) Since 1 is of the form $u + v$, we have $1 = u + v \leqq a + b$, whence $a + b = 1$. (b) See Lemma 1.2.

Theorem 5.2: $L = L(0, a) \oplus L(0, b)$ if and only if a is inverse to b and $(a, b)D$.

Proof: By Definition 5.2, $L = L(0, a) \oplus L(0, b)$ implies $ab = 0$, $a + b = 1$ and $x = u + v = ax + bx$ for every x. Hence $x = 1 \cdot x = (a + b)x = ax + bx$ and $(a, b)D$. Conversely, let $ab = 0$, $a + b = 1$, $(a + b)x = ax + bx$ for each x. Then $x = ax + bx = u + v$, with $u = ax \leqq a$, $v = bx \leqq b$.

Theorem 5.3: The following statements are equivalent to each other:

(a) a has an inverse b, for which $(a, b)D$.

(b) a has a unique inverse.

(c) $(a)D$.

PROOF: We prove that $(a) \leftrightarrows (b)$, and that $(a) \leftrightarrows (c)$.

$(a) \to (b)$: See Theorem 5.2 and Lemma 1.2.

$(b) \to (a)$: See Theorems 1.5 and 5.2.

$(a) \to (c)$: By Theorem 5.2, $L = L(0, a) \oplus L(0, b)$. Hence for every $x, y, x = u + v, y = u_1 + v_1$, with $u, u_1 \leqq a$ and $v, v_1 \leqq b$. Consequently $x + y = (u + u_1) + (v + v_1)$, where $u + u_1 \leqq a$ and $v + v_1 \leqq b$. Thus by Lemma 1.2 (or by the Corollary to Definition 5.2), $u = ax$, $u_1 = ay$, $u + u_1 = a(x + y)$, and therefore $a(x + y) = ax + ay$. This proves $(a)D$.

$(c) \to (a)$: Let b be an inverse of a. Since $(a)D$, therefore $(a, b)D$.

THEOREM 5.4: If $(a)D$, $(b)D$, and if a' is the unique inverse of a, then $(a + b)D$, $(ab)D$ and $(a')D$.

PROOF: $ab(x + y) = a(bx + by) = abx + aby$, whence $(ab)D$. By duality $(a + b)D$. The inverse a' of a is unique by Theorem 5.3, (b), (c), and both $(a)D$ and $(a')D$ are equivalent to $(a, a')D$ by Theorem 5.3, (a), (c).

By Theorem 5.4 we see that the set of all elements x such that $(x)D$ forms a Boolean algebra D. Axiom VI states that D consists of the elements 0, 1 alone. Although we shall devote our attention to systems in which VI is satisfied, one might study systems in which VI fails, the essential difficulties arising from the fact that the algebra D is considerably more complicated. (For such a discussion of systems in which VI fails, see Part III of these Notes - Ed.)

THEOREM 5.5: Suppose $ab = 0$. If $(a, b)D$ is false, (i.e. if not $(a, b)D$) then there exist elements a_1, b_1, both different from 0, such that $a_1 \leqq a$, $b_1 \leqq b$ and $a_1 \sim b_1$.

PROOF: The denial of $(a, b)D$ states that there exists an element x such that

$$x(a + b) \neq xa + xb. \tag{3}$$

Now $xa \leqq x(a + b)$, $xb \leqq x(a + b)$, whence $xa + xb \leqq x(a + b)$, and therefore $xa + xb < x(a + b)$ by (3). Hence if u is an inverse of $xa + xb$ in $x(a+b)$, then $u \neq 0$. Now $u \leqq x(a+b) \leqq a+b$, and $ua \leqq x \cdot a(a+b) = xa$. Thus $ua = ua \cdot xa = u \cdot xa \leqq u(xa + xb) = 0$; similarly $ub = 0$. Now define

$$a_1 \equiv a(u + b) \leqq a, \quad b_1 \equiv b(u + a) \leqq b.$$

Then $a_1 u \leqq au = 0$, $b_1 u \leqq bu = 0$, and

$$\begin{aligned} a_1 + u &= a(u + b) + u \\ &= (u + a)(u + b) \qquad \text{(by IV)}; \end{aligned}$$

similarly $b_1 + u = (u + a)(u + b)$. Therefore

$$b_1 + u = a_1 + u, \ a_1u = b_1u = 0,$$

and $a_1 \sim b_1$ by Theorem 3.1. Suppose now $a_1 = 0$. Then $(a+0)(b+u) = 0$, and Theorem 1.2 applies, yielding

$$u = (a + b)(0 + u) = a \cdot 0 + bu = 0 + 0 = 0,$$

contrary to $u \neq 0$. Hence $a_1 \neq 0$, and $b_1 \neq 0$ similarly. This completes the proof.

THEOREM 5.6: If $(a, b)D$, $a' \leqq a$, $b' \leqq b$, $a' \cdot ab = b' \cdot ab = 0$, then $(a', b')D$, $a'b' = 0$.

PROOF: By hypothesis, $a' \leqq a$; therefore $a'b = a'ab = 0$. Similarly, $b' \leqq b$, $b'a = b'ba = 0$. Now for every x

$$\begin{aligned}(a' + b')x &= (a + b)(a' + b')x = \\ &= a(a' + b')x + b(a' + b')x \qquad \text{(since } (a, b)D) \\ &= (a' + ab')x + (b' + ba')x \qquad \text{(by IV)} \\ &= a'x + b'x,\end{aligned}$$

and hence $(a', b')D$. Moreover, $a'b' \leqq a'b = 0$.

THEOREM 5.7: $(a, b)D$, $ab = 0$ if and only if

(4) $\qquad a_1 \leqq a, \ b_1 \leqq b, \ a_1 \sim b_1$ implies $a_1 = b_1 = 0$.

PROOF: First suppose that (4) holds. Then $a_1 = b_1 = ab$ is effective in the hypothesis of (4), and hence $ab = 0$. If $(a, b)D$ is false, then by Theorem 5.5 there exist a_1, b_1 such that $a_1, b_1 \neq 0$, $a_1 \leqq a$, $b_1 \leqq b$, $a_1 \sim b_1$. But this is the denial of (4), whence $(a, b)D$.

Conversely, let $ab = 0$, $(a, b)D$, and $a_1 \leqq a$, $b_1 \leqq b$, $a_1 \sim b_1$. Then by Theorem 3.1 there exists x such that

$$a_1 + x = b_1 + x = a_1 + b_1, \ a_1x = b_1x = 0.$$

Hence $x \leqq a_1 + b_1$, and $(a_1, b_1)D$ by Theorem 5.6. Thus

$$x = x(a_1 + b_1) = xa_1 + xb_1 = 0 + 0 = 0.$$

Then $a_1 = b_1 \leqq ab = 0$, so $a_1 = b_1 = 0$.

COROLLARY: $(a, b)D$, $ab = 0$ if and only if

(5) $\qquad a_1 \leqq a, \ b_1 \leqq b, \ a_1 \approx b_1$ implies $a_1 = b_1 = 0$.

PROOF: The forward implication is evident since if $ab = 0$, then also

$a_1 b_1 = 0$, and $a_1 \approx b_1$ implies $a_1 \sim b_1$ by Theorem 4.2, whence Theorem 5.7 applies, yielding $a_1 = b_1 = 0$. Conversely (5) implies (4) and hence implies $(a, b)D$, $ab = 0$, by Theorem 5.7.

We are now in a position to clarify to some extent the motivation of the present discussion. Since this theory is similar to the theory of sets, one would expect in it an analogue of the theory of comparability. It is natural in the light of Theorem 4.4, which states that $a \leqq b \approx a$ implies $a = b$, to define $a \prec b$ as the existence of b' such that $a \approx b' < b$. (The corresponding relation in set theory is "a has smaller power than b".) One expects the relations $a \prec b$, $a \approx b$, $b \prec a$ to be exhaustive. This matter is, however, closely connected with the question of irreducibility, as is seen by the following consideration. Suppose that L is reducible, i.e., is a direct sum $L(0, a) \oplus L(0, b)$, $a \neq 0$, $b \neq 0$. Then by Theorem 5.2, $(a, b)D$, and $a_1 \leqq a$, $b_1 \leqq b$, $a_1 \approx b_1$ implies $a_1 = b_1 = 0$ by the Corollary to Theorem 5.7. If now $a \prec b$, there exists $b_1 \leqq b$ with $a \approx b_1$, and $a = b_1 = 0$, which is impossible. Similarly neither $a \approx b$ nor $b \prec a$ holds. This shows that Axiom VI must hold if comparability holds. The converse to this, viz., that VI implies comparability, is one of the essential results to be obtained at the end of the present chapter.

Lemma 5.1: If $(a_1, b)D$, $(a_2, b)D$, $a_1 b = a_2 b = 0$, then $(a_1 + a_2, b)D$ and $(a_1 + a_2)b = 0$.

Proof: For every x

$$\begin{aligned} (a_1 + a_2 + x)b &= a_1 b + (a_2 + x)b && \text{(since } (a_1, b)D) \\ &= a_1 b + a_2 b + xb && \text{(since } (a_2, b)D) \\ &= (a_1 + a_2)b + xb && \text{(since } (a_2, b)D). \end{aligned}$$

Hence $(a_1 + a_2, b)D$. Clearly $(a_1 + a_2)b = a_1 b + a_2 b = 0$.

Lemma 5.2: Let Ω be any ordinal and $(a_\alpha;\ \alpha < \Omega)$ a system such that $\alpha < \beta$ implies $a_\alpha \leqq a_\beta$. If

$$(a_\alpha, b)D,\ a_\alpha b = 0 \qquad (\alpha < \Omega),$$

then

$$(\lim_{\alpha \to \Omega}{}^{*} a_\alpha, b)D,\ (\lim_{\alpha \to \Omega}{}^{*} a_\alpha)b = 0.$$

Proof:

$$\begin{aligned} (\lim_{\alpha \to \Omega}{}^{*} a_\alpha)b &= \lim_{\alpha \to \Omega}{}^{*} b a_\alpha && \text{(by III}_2) \\ &= \lim_{\alpha \to \Omega}{}^{*} 0 = 0. \end{aligned}$$

Also, for every x

$$\begin{aligned}(b + \lim^*_{\alpha\to\Omega} a_\alpha)x &= (\lim^*_{\alpha\to\Omega} (b + a_\alpha))x \\ &= \lim^*_{\alpha\to\Omega} (b + a_\alpha)x \qquad \text{(by III}_2\text{)} \\ &= \lim^*_{\alpha\to\Omega} (bx + a_\alpha x) \\ &= bx + \lim^*_{\alpha\to\Omega} a_\alpha x = bx + (\lim^*_{\alpha\to\Omega} a_\alpha)x \qquad \text{(by III}_2\text{)}\end{aligned}$$

and the proof is complete.

THEOREM 5.8: If $(a, b)D$, $ab = 0$, there exists $a^* \geqq a$ such that $(a^*, b)D$, $a^*b = 0$ and a^* is maximal with this property, i.e., if $a' \geqq a^*$, $(a', b)D$, $a'b = 0$, then $a' = a^*$.

PROOF: For the purposes of this proof we shall write $(x, y)P$ in case $(x, y)D$ and $xy = 0$. Let Ω be the first ordinal corresponding to the first power greater than the power of L. Define a_β for every ordinal $\beta < \Omega$ by induction: define $a_0 \equiv a$ and if a_α has been defined with $(a_\alpha, b)P$ for all $\alpha < \beta$, proceed as follows:

I. Let β have a predecessor α_0 so that $\beta = \alpha_0 + 1$. If a_{α_0} is maximal such that $(a_{\alpha_0}, b)P$, define $a_\beta \equiv a_{\alpha_0}$. In the contrary case, there exists $a_\beta > a_{\alpha_0}$ such that $(a_\beta, b)P$.

II. Let β be a limit ordinal. Then define $a_\beta \equiv \Sigma(a_\alpha;\ \alpha < \beta)$, whence $(a_\beta, b)P$ by Lemma 5.2.

If $\alpha < \beta < \Omega$ and a_α is not maximal such that $(a_\alpha, b)P$, then it is evident that $a_\alpha < a_\beta$. If for every $\alpha < \Omega$, a_α fails to be maximal such that $(a_\alpha, b)P$, then the a_α are all distinct, and their class has power greater than the power of L, which is impossible. Hence there exists an ordinal α_1 for which a_{α_1} is maximal such that $(a_{\alpha_1}, b)P$. We put $a^* = a_{\alpha_1}$. It is evident that $a^* \geqq a$ and a^* is effective in the theorem.

COROLLARY 1: If $(a', b)D$, $a'b = 0$, then $a' \leqq a^*$.

PROOF: By Lemma 5.1, $(a^* + a', b)D$, $(a^* + a')b = 0$; therefore $a^* + a' > a^*$ is impossible. But $a^* + a' \geqq a^*$; hence $a^* + a' = a^*$, that is $a' \leqq a^*$.

COROLLARY 2: The element a^* such that $(a^*, b)D$, $a^*b = 0$, and such that a^* is maximal with this property, exists and is unique.

PROOF: The existence follows from the theorem. If a_1^* is another such element, then by Corollary 1, $a_1^* \leqq a^*$; similarly $a^* \leqq a_1^*$. Therefore $a^* = a_1^*$.

LEMMA 5.3: If $(a, b)D$, $ab = 0$, and if c is such that there exists a' with $c \sim a' \leqq a$, then $(c, b)D$, $cb = 0$.

Proof: Assume $c_1 \leqq c$, $b_1 \leqq b$, $c_1 \sim b_1$. Since $c \sim a'$, there exists a perspective isomorphism of $L(0, c)$ and $L(0, a')$, which carries c_1 into an element $a_1 \leqq a'$. Thus

$$b_1 \sim c_1 \sim a_1 \leqq a' \leqq a,$$

that is $a_1 \leqq a$, $b_1 \leqq b$, $a_1 \approx b_1$. Therefore the Corollary to Theorem 5.7 gives $a_1 = b_1 = 0$. Now $c_1 \sim a_1 = 0$, $c_1 = 0$. Thus we have $c_1 = b_1 = 0$ under these assumptions. Therefore Theorem 5.7 yields $(c, b)D$, $cb = 0$.

Theorem 5.9: If $(a, b)D$, $ab = 0$, and if $a^* \geqq a$ is the maximal element such that $(a^*, b)D$, $a^*b = 0$ (cf. Theorem 5.8 and its Corollary 2), then $(a^*)D$.

Proof: Let b^* be inverse to a^* and consider $a_1 \leqq a^*$, $b_1 \leqq b^*$, $a_1 \sim b_1$. We have $(a^*, b)D$, $a^*b = 0$ and $b_1 \sim a_1 \leqq a^*$; therefore Lemma 5.3 gives $(b_1, b)D$, $b_1 b = 0$. Now Corollary 1 to Theorem 5.8 implies $b_1 \leqq a^*$. But at the same time $b_1 \leqq b^*$, so that $b_1 \leqq a^*b^* = 0$, $a_1 \sim b_1 = 0$; that is $a_1 = b_1 = 0$. Therefore Theorem 5.7 yields $(a^*, b^*)D$, and by Theorem 5.3, (a), (c), we have $(a^*)D$.

At this point we assume Axiom VI, the irreducibility of L. Unless otherwise specified, all six axioms will be assumed henceforth.

Theorem 5.10: If $(a, b)D$, $ab = 0$, then $a = 0$ or $b = 0$.

Proof: Let $a^* \geqq a$ be maximal with the property $(a^*, b)D$, $a^*b = 0$. Hence $(a^*)D$ by Theorem 5.9, i.e., a^* has a unique inverse by Theorem 5.3, (a), (b). Axiom VI then yields that $a^* = 0$ or $a^* = 1$. If $a^* = 0$, then $a = 0$; if $a^* = 1$, then

$$b = 1 \cdot b = a^*b = 0.$$

Theorem 5.11: $(a, b)D$ if and only if $a \leqq b$ or $b \leqq a$.

Proof: Suppose $a \leqq b$ or $b \leqq a$. If $a \leqq b$, $a + b = b$ and $ax \leqq bx$, $ax + bx = bx$. Therefore

$$(a + b)x = bx = ax + bx.$$

Thus $(a, b)D$. The case $b \leqq a$ is similarly treated. Conversely, suppose $(a, b)D$. Let a_1 be an inverse of ab in a, and b_1 an inverse of ab in b; thus $a_1 \leqq a$, $b_1 \leqq b$, and $a_1 \cdot ab = b_1 \cdot ab = 0$. But $(a_1, b_1)D$, $a_1 b_1 = 0$, by Theorem 5.6, whence $a_1 = 0$ or $b_1 = 0$ by Theorem 5.10. Hence either

$$a = ab + a_1 = ab \leqq b,$$

or

$$b = ab + b_1 = ab \leqq a,$$

and the proof is complete.

THEOREM 5.12: If L satisfies Axioms I—VI, and if $a \leqq b$, then $L(a, b)$ also satisfies these axioms.

PROOF: It has already been shown (Corollary to Theorem 1.3) that Axioms I—V hold for $L(a, b)$ if they hold for L. It remains to prove that if L is irreducible then $L(a, b)$ is also irreducible.

I. $a = 0$. It will be shown that if $L(0, b) = L(0, c) \oplus L(0, d)$, then $c = 0$ or $d = 0$. Now $cd = 0$ by the definition of direct sum; by the Corollary to Definition 5.2, $c + d = b$ and $x \leqq b$ implies $x = cx + dx$. Let y be any element in L. Then $y(c + d) \leqq c + d$, and

$$y(c + d) = y(c + d)(c + d) = y(c + d)c + y(c + d)d$$
$$= yc + yd,$$

whence $(c, d)D$. Hence $c = 0$ or $d = 0$ by Theorem 5.10.

II. $b = 1$. The lattice $L(a, 1)$ is irreducible by duality.

III. Consider now the general case. Since $L(0, b)$ is irreducible, it satisfies all the axioms, and hence property of L deducible from these axioms is possessed also by $L(0, b)$. In particular, the property proved in II above is possessed by $L(0, b)$. But this is precisely the statement that $L(a, b)$ is irreducible, since 1 must be replaced by b when passing from L to $L(0, b)$.

THEOREM 5.13: If $ab = 0$, $a \neq 0$, $b \neq 0$, then there exist a', b' such that $0 \neq a' \leqq a$, $0 \neq b' \leqq b$, and $a' \sim b'$.

PROOF: If $(a, b)D$, then $a = 0$ or $b = 0$ by Theorem 5.10, contrary to the hypothesis. Hence $(a, b)D$ is false. Theorem 5.5 then applies, yielding the desired result.

THEOREM 5.14: Let $ab = 0$. Then either there exists b' such that $a \sim b' \leqq b$, or there exists a' such that $b \sim a' \leqq a$.

PROOF: Let Ω be the first ordinal corresponding to the first power greater than that of L. We shall proceed to construct two independent transfinite systems $(a_\alpha;\ \alpha < \alpha_0)$, $(b_\alpha;\ \alpha < \alpha_0)$, such that $a_\alpha \leqq a$, $b_\alpha \leqq b$, $a_\alpha \sim b_\alpha$ $(\alpha < \alpha_0 < \Omega)$, and such that either $\Sigma(a_\alpha;\ \alpha < \alpha_0) = a$ or $\Sigma(b_\alpha;\ \alpha < \alpha_0) = b$. Suppose $(a_\beta;\ \beta < \alpha)\perp$, $(b_\beta;\ \beta < \alpha)\perp$ have been defined, and that $a_\beta \leqq a$, $b_\beta \leqq b$, $a_\beta \sim b_\beta$ $(\beta < \alpha)$. It is then desired to define a_α, if possible. If either $\Sigma(a_\beta;\ \beta < \alpha) = a$ or $\Sigma(b_\beta;\ \beta < \alpha) = b$, then neither a_α nor b_α is to be defined. In the contrary case when neither of these conditions holds, there exist inverses c_α and d_α of $\Sigma(a_\beta;\ \beta < \alpha)$ and $\Sigma(b_\beta;\ \beta < \alpha)$ in a and b respectively, such that $c_\alpha \neq 0$, $d_\alpha \neq 0$. Then $c_\alpha d_\alpha \leqq ab = 0$, and there exist a_α, b_α such that $0 \neq a_\alpha \leqq c_\alpha$, $0 \neq b_\alpha \leqq d_\alpha$ and $a_\alpha \sim b_\alpha$ by Theorem 5.13. If α' is such that all the a_α (and hence all the b_α) are defined for $\alpha < \alpha'$, then clearly $a_\alpha \Sigma(a_\beta;\ \beta < \alpha) = 0$ for every $\alpha < \alpha'$. Let J_0 be a finite set of $\alpha < \alpha'$ and denote the elements of J_0 (ordered increasingly) by $\alpha_1 < \cdots < \alpha_n$. Then $a_{\alpha_i}(a_{\alpha_1} + \cdots + a_{\alpha_{i-1}}) \leqq a_{\alpha_i}\Sigma(a_\beta;\ \beta < \alpha_i) = 0$ for $i = 2, \cdots, n$, so that $(a_{\alpha_1}, \cdots, a_{\alpha_n})\perp$, that is $(a_\alpha;\ \alpha \in J_0)\perp$. Thus Theorem 2.3 gives $(a_\alpha;\ \alpha < \alpha')\perp$. Similarly $(b_\alpha; \alpha < \alpha')\perp$. Clearly $0 \neq a_\alpha \leqq a$, $0 \neq b_\alpha \leqq b$, $a_\alpha \sim b_\alpha$ $(\alpha < \alpha')$; hence all the a_α are distinct, and similarly all the b_α are distinct. Now a_α cannot be defined for every $\alpha < \Omega$ for then the set of all the a_α would have power greater than that of L, which is impossible. Hence there exists a smallest α_0 for which a_{α_0} is undefined. This means, however, that $a' \equiv \Sigma(a_\alpha;\ \alpha < \alpha_0) = a$ or $b' \equiv \Sigma(b_\alpha;\ \alpha < \alpha_0) = b$, since otherwise a_{α_0} and b_{α_0} would be defined. We have already seen that $(a_\alpha;\ \alpha < \alpha_0)\perp$, $a_\alpha \neq 0$ $(\alpha < \alpha_0)$, that similar statements hold for the b_α and that $a_\alpha \sim b_\alpha$ $(\alpha < \alpha_0)$. Now by Theorem 2.1 the combined system $(a_\alpha, b_\alpha;\ \alpha < \alpha_0)$ is independent, because $ab = 0$. Therefore $a' \sim b'$ by Theorem 3.6. If $a' = a$, then $a \sim b' \leqq b$; if $b' = b$, then $b \sim a' \leqq a$. Since either $a' = a$ or $b' = b$, the desired conclusion follows.

THEOREM 5.15: For every pair of elements a, b either

(a) there exists b' such that $a \sim b' \leqq b$, or
(b) there exists a' such that $b \sim a' \leqq a$.

PROOF: Let a_1 and b_1 be inverses of ab in a and b respectively. Then $ba_1 = baa_1 = 0$, $ab_1 = abb_1 = 0$, and $a_1b_1 \leqq a_1b = 0$. Now by Theorem 2.2, $(ab, a_1, b_1)\perp$, since $(ab + a_1)b = ab = 0$ and $ab \cdot a_1 = 0$. Since $a_1b_1 = 0$, we have by Theorem 5.14 that

(a) there exists b_1' such that $a_1 \sim b_1' \leqq b_1$, or
(b) there exists a_1' such that $b_1 \sim a_1' \leqq a_1$.

In the first case $(ab, a_1, b_1')\perp$ and $ab + a_1 \sim ab + b_1'$ by Theorem 3.5. Define $b' \equiv ab + b_1'$. Then

$$a = ab + a_1 \sim b' \leqq ab + b_1 = b.$$

Similar consideration in the second case leads to an element a' such that $b \sim a' \leqq a$, and the theorem is proved.

THEOREM 5.16: $a \approx b$ if and only if $a \sim b$.

PROOF: It has already been proved (Corollary 2, Definition 3.5) that $a \sim b$ implies $a \approx b$. Suppose $a \approx b$. Now by Theorem 5.15 either

(a) there exists b' such that $a \sim b' \leqq b$, or
(b) there exists a' such that $b \sim a' \leqq a$.

In case (a), $b \approx a \sim b' \leqq b$, whence $b \approx b' \leqq b$, and by Theorem 4.4

$b = b'$. Hence $a \sim b$. Case (b) is similarly treated and leads to the same result.

COROLLARY: The relation $\sim$ is transitive.

PROOF: By Theorem 5.16, the relations $\approx$, $\sim$ are equivalent. Hence $\sim$ is transitive, since it has been observed already (Corollary 1, Definition 3.5) that $\approx$ is transitive.

The discussion just concluded shows that the relation $\sim$ is an equivalence, being reflexive, symmetric and transitive, and hence that it divides L into mutually exclusive equivalence classes. These classes will be studied in detail in the following chapter.

PART I · CHAPTER VI

Properties of the Equivalence Classes

DEFINITION 6.1:

(a) $a \prec b$ means that there exists b^* such that

$$a \sim b^* < b;$$

(b) $a \succ b$ means that there exists b^* such that

$$a \sim b^* > b.$$

COROLLARY: The relations $\prec$, $\succ$ are dual to each other.

PROOF: This is evident, since the relation $\sim$ is self-dual and the relations $>$, $<$ are dual to each other.

THEOREM 6.1:

(a) $a \succ b$ if and only if $b \prec a$.
(b) If $a \sim a'$, $b \sim b'$, then $a \prec b$ is equivalent to $a' \prec b'$.
(c) $a \prec b$, $b \prec c$ implies $a \prec c$.
(d) For every pair a, b, one and only one of the three relations $a \succ b$, $a \sim b$, $a \prec b$ holds.

PROOF:

(a) Let $a \succ b$. Then there exists b^* such that $a \sim b^* > b$. Let a^* be a perspective image of b in $L(0, a)$. Then $b \sim a^* < a$, and $b \prec a$. The converse is dual.

(b) It suffices to show that $a \prec b$ implies $a' \prec b'$. If $a \prec b$, there exists b^* such that $a \sim b^* < b$. Let b_1^* be a perspective image of b^* in $L(0, b')$. Then $a' \sim a \sim b^* \sim b_1^* < b'$, whence $a' \prec b'$, since the relation $\sim$ is transitive.

(c) By hypothesis there exist b^*, c^* with

$$a \sim b^* < b \sim c^* < c.$$

Since $b \sim c^*$, $b^* < b$, there is a perspective image of b^*, $c^{**} < c$, and $a \prec c$.

(d) By Theorem 5.15 and (a) of the present theorem one of the three relations $a \succ b$, $a \sim b$, $a \prec b$ must hold. If two of them hold simultaneously we have $b \prec a$, $a \sim b$, or $a \prec b$, $a \sim b$, or $a \prec b$, $b \prec a$. Using (a), (b), (c), we obtain $a \prec a$ in each case; that is, the existence of a^* such that $a \sim a^* < a$, which contradicts Theorem 3.7.

Let us consider a lattice $L(a, b)$. Since it satisfies the same axioms as L satisfies, the relations of perspectivity and projectivity may be defined in $L(a, b)$ in the same way as in L. Now let $c, d \in L(a, b)$. We propose to investigate the question whether or not $L(a, b)$-perspectivity of c and d is equivalent to L-perspectivity of c and d. The latter means by Theorem 3.1 the existence of $x \in L$ such that

$$c + x = d + x = b, \quad cx = dx = a.$$

But this means the existence of a common inverse in $L(a, b)$ of c, d, i.e., $L(a, b)$-perspectivity. Hence perspectivity in L and perspectivity in $L(a, b)$ coincide. For projectivity this is not so obvious. However, since perspectivity and projectivity are equivalent (in $L(a, b)$ as well as in L), our result extends also to projectivity.

Definition 6.2: Let A_a denote the class of all elements x such that $x \sim a$, and let $\mathscr{L}$ denote the class of all A_a, $a \in L$.

Corollary: The system $(A_a; a \in L)$ is a mutually exclusive and exhaustive partition of L into subclasses, i.e.

(a) $A_a = A_b$ if and only if $a \sim b$;
(b) $A_a \neq A_b$ implies $\mathfrak{P}(A_a, A_b) = \Theta$;
(c) $a \in A_a$ for every $a \in L$.

Proof: This follows from the fact that the relation $\sim$ is reflexive, symmetric and transitive.

The elements of $\mathscr{L}$ will be denoted by $A, B, C, \cdots$.

Definition 6.3:

(a) $A < B$ means that there exist $a \in A$, $b \in B$ with $a \prec b$.
(b) $A > B$ means that there exist $a \in A$, $b \in B$ with $a \succ b$.

Corollary 1: $A < B$ is equivalent to $B > A$; the relations $>$, $<$ are dual to each other.

Proof: This follows from Theorem 6.1(a) and the Corollary to Definition 6.1.

Corollary 2: $A < B$ if and only if $a \prec b$ for every a, b such that $a \in A$, $b \in B$.

Proof: This is obvious by Theorem 6.1 (b).

Corollary 3: $A < B$ if and only if for each $a \in A$ there exists $b \in B$ such that $a < b$.

Proof: Suppose $A < B$, and let b_1 be any element of B. Then $a \prec b_1$, by Corollary 2, and there exists b with $b_1 \sim b > a$, whence $b \in B$.

Conversely, let a be any element of A and b the corresponding element of B with $a < b$. Then $a \prec b$ and $A < B$ by definition.

COROLLARY 4: $A < B$ if and only if there exist $a \in A$, $b \in B$ with $a < b$.

PROOF: This is obvious from Corollary 3.

COROLLARY 5: For every A, B one and only one of the relations $A > B$, $A = B$, $A < B$ holds.

PROOF: This is evident by Theorem 6.1 (d), considering Definition 6.2, Corollary (a), and Corollary 2 above.

COROLLARY 6: $\mathscr{L}$ is linearly ordered by the relations $<$, $>$.

PROOF: The relation $<$ is transitive by Theorem 6.1 (c) and Corollary 2 above; $>$ is transitive by Corollary 1. The linear ordering follows by Corollaries 1, 5.

LEMMA 6.1: If $a \sim b$ and a', b' are inverses of a, b, respectively, then $a' \sim b'$.

PROOF: Since $a \sim b$, there exists x with

$$a + x = b + x = 1, \; ax = bx = 0.$$

But $a' + a = 1$, $a'a = 0$, whence $a' \sim x$. Similarly, $b' \sim x$, and $a' \sim b'$ by the transitivity of $\sim$.

LEMMA 6.2: If $a \leqq c$, $b \leqq c$, $a \sim b$, and if a', b' are inverses in c of a, b, respectively, then $a' \sim b'$.

PROOF: This follows from Lemma 6.1 by replacing L by $L(0, c)$.

LEMMA 6.3: If $a \leqq c$, $b \leqq d$, $a \sim b$, $c \sim d$, and if a', b' are inverses of a, b in c, d, respectively, then $a' \sim b'$.

PROOF: Let b_1, b_1' be perspective images in $L(0, c)$ of b, b', so that $b \sim b_1 \leqq c$, $b' \sim b_1' \leqq c$ and b_1' is inverse to b_1 in c. Hence $a \sim b_1$ and Lemma 6.2 applies, yielding $a' \sim b_1'$. But $b_1' \sim b'$, whence $a' \sim b'$.

LEMMA 6.4: If $ab = 0$ and c is an inverse of $a + b$, then a is an inverse of $b + c$, and b is an inverse of c in $b + c$.

PROOF: Since $ab = 0$, $(a + b)c = 0$, it follows that $(a, b, c)\perp$. Hence $a(b + c) = 0$, and by hypothesis $a + b + c = 1$; therefore a is an inverse of $b + c$. Also since $bc = 0$, b is inverse to c in $b + c$.

LEMMA 6.5: If $ab = 0$, $ef = 0$, $a \sim e$, $b \sim f$, then $a + b \sim e + f$.

PROOF: Let c, g be inverses of $a + b$, $e + f$ respectively. Then by Lemma 6.4, $b + c$ is inverse to a, and $f + g$ is inverse to e. Thus $b + c \sim f + g$ by Lemma 6.1. Also by Lemma 6.4, b is inverse to c in $b + c$, and f is inverse to g in $f + g$. Therefore $c \sim g$ by Lemma 6.3, and $a + b \sim e + f$ by Lemma 6.1.

DEFINITION 6.4: We shall say that $A + B$ exists in case there exist $a \in A$, $b \in B$ such that $ab = 0$. When $A + B$ exists it is defined as the unique class C which is equal to A_{a+b} for every a, b such that $a \in A$, $b \in B$, $ab = 0$ (the existence and uniqueness of C following from Lemma 6.5). Thus $A + B$ depends on A and B only.

We shall say that $A - B$ exists in case there exist $a \in A$, $b \in B$ such that $a \geqq b$, i.e., in case $A \geqq B$. When existent, $A - B$ is defined as the unique class C which is equal to $A_{b'}$, for every b' such that there exist $a \in A$, $b \in B$, $b \leqq a$, for which b' is inverse to b in a (the existence and uniqueness of C following from Lemma 6.3). Thus $A - B$ depends only on A and B.

We agree that the assertion of any property of $A + B$ means first the assertion of the existence of $A + B$, and secondly that the class $A + B$ as defined in Definition 6.4 has the stated property. In many cases where the first part is trivial its proof will be omitted.

The same convention will be adopted for $A - B$.

Definition 6.5: We define $\theta \equiv A_0$, $\dagger \equiv A_1$.

Corollary 1: $\theta \leqq A \leqq \dagger$ for every A in $\mathscr{L}$.

Corollary 2: $\dagger - A$ exists for every A.

Theorem 6.2:

(a) $A - B$ exists if and only if $A \geqq B$,
(b) $A + B$ exists if and only if $\dagger - A \geqq B$.

Proof: Part (a) is obvious. To prove (b), suppose first that $A + B$ exists. Then there exist a, b with $a \in A$, $b \in B$, $ab = 0$. By the Corollary to Theorem 1.4, there is an inverse b_1 of a such that $b_1 \geqq b$. Hence $b_1 \in (\dagger - A)$, and $\dagger - A \geqq B$. Conversely, let $\dagger - A \geqq B$. If a is any element of A and if b_1 is an inverse of a, then there exists $b \in B$ such that $b_1 \geqq b$. Hence $ab \leqq ab_1 = 0$.

Theorem 6.3:

(i) If $A + B$ exists, then $A + B = B + A$.
(ii) If $A + B$, $(A + B) + C$ exist, then $(A + B) + C = A + (B + C)$.
(iii) $A + \theta = A$.
(iv) If $A - B$ exists, then $(A - B) + B = A$.
(v) If $A + B$ exists, then $(A + B) - B = A$.
(vi) If $A + B$ exists, then $A \leqq A + B$.
(vii) The equation $A + X = B$ has a solution if and only if $A \leqq B$, in which case the solution is unique.
(viii) $A < B$ is equivalent to the existence of $X \neq \theta$ such that $B = A + X$.
(ix) If $A + C$, $B + C$ exist, then $A \gtreqless B$ are equivalent to $A + C \gtreqless B + C$ respectively.

(x) If $A \leqq C$, $B \leqq D$, and if $C + D$ exists, then $A + B$ exists and $A + B \leqq C + D$.

(xi) $A \gtreqless B$ are equivalent to $\dagger - A \lesseqgtr \dagger - B$ respectively.

PROOF:

(i) This is trivial.

(ii) Since $A + B$ exists, there are elements $a \in A$, $b \in B$ with $ab = 0$, and since $(A + B) + C$ exists, there are elements $c \in C$, $d \in (A + B)$ with $cd = 0$, $d \sim a + b$. Let a_1, b_1 be perspective images of a, b in $L(0, d)$. Then $a_1 \in A$, $b_1 \in B$, $a_1 b_1 = 0$; also $d = a_1 + b_1$. Now $b_1 c \leqq cd = 0$, and $B + C$ exists; also $a_1(b_1 + c) = 0$ by Theorem 1.2, whence $A + (B + C)$ exists. Finally, $A + (B + C) = A_{a_1+(b_1+c)} = A_{(a_1+b_1)+c} = (A + B) + C$.

(iii) Clearly $A + \theta$ exists, since $a \cdot 0 = 0$ for every $a \in A$. Hence $A + \theta = A_{a+0} = A_a = A$.

(iv) Since $A - B$ exists, there are elements $a \in A$, $b \in B$ with $a \geqq b$. Let b_1 be an inverse of b in a. Then $bb_1 = 0$, $b + b_1 = a$, and

$$A = A_a = A_{b+b_1} = A_b + A_{b_1} = B + (A - B).$$

(v) Let $ab = 0$, $a \in A$, $b \in B$. Then

$$A + B = A_{a+b}, \ (A + B) - B = A_{a+b} - B = A_a = A,$$

since b is inverse to a in $a + b$.

(vi) This is obvious, since always $a \leqq a + b$.

(vii) Suppose $A+X = B$ has a solution. Then by (vi), $A \leqq A+X = B$. On the other hand, if $A \leqq B$, then by Theorem 6.2(a), $B - A$ exists, and $X \equiv B - A$ is effective in $A + X = B$ by (iv). Suppose now that X is a solution. Then $B - A = (A + X) - A = X$ by (v), whence the solution is unique.

(viii) $A \leqq B$ is equivalent to the existence of X with $B = A + X$ by (vii). Also, by (vii) (the uniqueness of X), $A = B$ if and only if $X = \theta$. Hence $A < B$ if and only if $X \neq \theta$.

(ix) By (viii), $A < B$ is equivalent to the existence of $X \neq \theta$ with $B = A + X$. Hence $B + C = (A + X) + C = (A + C) + X$ (by (i), (ii)), i.e., $A + C < B + C$ by (viii). Interchanging A, B, we see that $A > B$ implies $A + C > B + C$. Finally, $A = B$ obviously implies $A + C = B + C$. Hence $A \gtreqless B$ imply $A + C \gtreqless B + C$ respectively. Since we have exhaustive disjunctions in both hypotheses and conclusions, the converse implications hold also.

(x) By (vii), C and D are of the form $C = A + X$, $D = B + Y$ with $X, Y \neq \theta$. We assumed the existence of $C + D = (A + X) + (B + Y)$. Hence by (i), (ii),

$$(X + A) + (B + Y) = X + (A + (B + Y)) = X + ((A + B) + Y) =$$
$$= X + (Y + (A + B)) = (A + B) + (X + Y)$$

whence $A + B$ exists, and $A + B \leqq C + D$ by (vii).

(xi) By (ix), $A < B$ and $\dagger - A \leqq \dagger - B$ imply

$$\dagger = A + (\dagger - A) < B + (\dagger - A) \leqq B + (\dagger - B) = \dagger$$

and $\dagger < \dagger$, which is impossible. Hence by Corollary 5 to Definition 6.3, $A < B$ implies $\dagger - A > \dagger - B$. Interchanging A, B yields that $A > B$ implies $\dagger - A < \dagger - B$. Clearly $A = B$ implies $\dagger - A = \dagger - B$. Hence $A \gtreqless B$ imply $\dagger - A \lesseqgtr \dagger - B$ respectively. Since we have exhaustive disjunctions in both hypotheses and conclusions, the converse implications hold also.

Definition 6.6: Put $0 \cdot A \equiv \theta$. If $(n - 1)A$ has been defined, put $nA \equiv (n - 1)A + A$ if $(n - 1)A + A$ exists. Otherwise nA is undefined.

(Corollary: $n(A + B) = nA + nB$ and $(n + m)A = nA + mA$ (in each case, both sides exist if either side exists) - Ed.)

Theorem 6.4: nA is defined for every $n \geqq 0$ if and only if $A = \theta$.

Proof: The reverse implication is trivial. Conversely, let nA be defined for every n. Then suppose that $a_1, \cdots, a_n$ are elements of A such that $(a_1, \cdots, a_n)\perp$. Then $(a_1 + \cdots + a_n) \in nA$. Since $(n + 1)A$ exists, $A \leqq \dagger - nA$ by Theorem 6.2(b). Now let x_n be an inverse of $a_1 + \cdots + a_n$, whence $x_n \in (\dagger - nA)$. Consequently there exists $a_{n+1} \in A$ with $a_{n+1} \leqq x_n$ by the dual of Corollary 3, Definition 6.3. Since $(a_1 + \cdots + a_n)a_{n+1} \leqq (a_1 + \cdots + a_n)x_n = 0$, and $(a_1, \cdots, a_n)\perp$, it follows that $(a_1, \cdots, a_{n+1})\perp$. Moreover, $a_1, \cdots, a_{n+1} \in A$. Thus if a_1 is any element of A, we have an infinite sequence $(a_i;\ i = 1, 2, \cdots)$ such that $a_i \in A$, $(a_1, \cdots, a_i)\perp$ $(i = 1, 2, \cdots)$. Thus by Theorem 2.3 $(a_i;\ i = 1, 2, \cdots)\perp$. We have also $a_1 \sim a_2 \sim \cdots$, whence $a_1 = 0$ by Theorem 3.8, and $A = A_{a_1} = \theta$.

Theorem 6.5: Let $A \neq \theta$, B be given. Then there exist uniquely an integer $n \geqq 0$ and a class $B_1 < A$ such that $B = nA + B_1$.

Proof: We shall establish the existence indirectly. Assume the theorem false. It will be proved by induction that $nA \leqq B$ for every n. This is obvious for $n = 0$. Suppose it true for $n = m \geqq 0$. Then there exists B_1 such that $mA + B_1 = B$ by Theorem 6.3 (vii). Now $B_1 \geqq A$, since otherwise the theorem is true. Hence by Theorem 6.3 (vii) there exists B_2 with $B_1 = A + B_2$. Thus $mA + (A + B_2) = B$, $mA + A \leqq B$, whence $(m + 1)A$ is defined, and $\leqq B$. Thus all nA, $n \geqq 0$, are defined, and so $A = \theta$ by Theorem 6.4, contrary to the hypothesis. This proves the existence.

Let $B = n_1A + B_1 = n_2A + B_2$, $B_1 < A$, $B_2 < A$, $n_1 \neq n_2$. By symmetry it suffices to consider $n_1 < n_2$. Hence there exists $m \geqq 0$ with $n_2 = n_1 + 1 + m$, and

$$n_2A + B_2 = (n_1 + 1 + m)A + B_2 = n_1A + (A + mA + B_2),$$

whence $B_1 = A + mA + B_2$ by Theorem 6.3 (vii). Thus $B_1 \geqq A$, contrary to $B_1 < A$. Hence $n_1 = n_2$. Then $B = n_1A + B_1 = n_2A + B_2 = n_1A + B_2$ and $B_1 = B_2$ by Theorem 6.3 (vii).

DEFINITION 6.7: Let $A \neq \theta$, B be given. The unique integer $n \geqq 0$ such that B is of the form $B = nA + B_1$, $B_1 < A$ will be denoted by $[B : A]$. Thus $B = [B : A]A + B_1$.

DEFINITION 6.8: An element A of $\mathscr{L}$ is said to be *minimal* (A min) in case $A > \theta$ and there exists no element B such that $A > B > \theta$.

THEOREM 6.6: If $A \neq \theta$ is not minimal, then there exists $B \neq \theta$ such that $2B \leqq A$.

PROOF: If A is not minimal, there exists B_1 with $\theta < B_1 < A$. By Theorem 6.3 (viii), A may be expressed in the form $B_1 + B_2$, $B_2 \neq \theta$. Since $B_1 \neq \theta$, $B_2 < A$. Hence if B is defined as the smaller of B_1, B_2 (use Corollary 5 to Definition 6.3), we have, by Theorem 6.3 (vii), (x),

$$A = B_1 + B_2 \geqq B + B = 2B.$$

DEFINITION 6.9: A minimal sequence (A_i) of elements $\neq \theta$ in $\mathscr{L}$ is one containing but one element A_1 which is minimal, or containing a denumerable infinitude of elements such that $2A_{i+1} \leqq A_i$ $(i = 1, 2, \cdots)$.

COROLLARY: There exists a minimal sequence.

At this point we fix attention on a minimal sequence (A_i) which will remain fixed throughout the ensuing discussion.

LEMMA 6.6: For every A, B, C for which the symbols are defined,

(a) $[A : C] + [B : C] \leqq [A + B : C] \leqq [A : C] + [B : C] + 1$.

(b) $[A : B] \cdot [B : C] \leqq [A : C] < ([A : B] + 1)([B : C] + 1)$.

PROOF:

(a) Put $p \equiv [A : C]$, $q \equiv [B : C]$. Hence

$$A = pC + A_1, \quad B = qC + B_1, \quad A_1 < C, \; B_1 < C.$$

Therefore $A + B = (p + q)C + (A_1 + B_1)$. Now

$$A_1 + B_1 = rC + A_2,$$

where $r \geqq 0$, $A_2 < C$. If $r \geqq 2$, then $A_1 + B_1 \geqq 2C$, contrary to $A_1 < C$, $B_1 < C$. Hence $r = 0$ or 1, and

$$A + B = (p + q)C + A_2, \text{ or}$$
$$A + B = (p + q + 1)C + A_2.$$

Thus $[A+B:C]=p+q$ or $p+q+1$; but

$$p+q \leqq \begin{Bmatrix} p+q \\ p+q+1 \end{Bmatrix} \leqq p+q+1,$$

whence (a) follows.

(b) Put $p \equiv [A:B]$, $q \equiv [B:C]$. Then

$$A = pB + B_1, \quad B = qC + C_1, \quad B_1 < B, \; C_1 < C,$$

and $A = pqC + (pC_1 + B_1)$. But $pC_1 + B_1$ may be expressed in the form $rC + C_2$, $C_2 < C$, whence $A = (pq+r)C + C_2$, and $[A:C] = pq + r$. Now $r < p + q + 1$. For suppose $r \geqq p + q + 1$; then

$$\begin{aligned} r &= [pC_1 + B_1 : C] \\ &= [\underbrace{C_1 + \cdots + C_1 +}_{p} B_1 : C] \\ &\leqq p[C_1 : C] + [B_1 : C] + p \quad \text{(by (a))} \\ &\leqq 0 + q + p = q + p, \end{aligned}$$

which is a contradiction. Consequently

$$pq \leqq pq + r < pq + p + q + 1 = (p+1)(q+1),$$

and (b) follows.

COROLLARY:

(a) If $A \leqq B$ and $C \neq \theta$, then $[A:C] \leqq [B:C]$,

(b) If $\theta < B \leqq A$, then

$$[C:B] \geqq [C:A].$$

PROOF:

(a) There exists B_1 with $B = A + B_1$. Hence $[A:C] + [B_1:C] \leqq [B:C]$, and $[A:C] \leqq [B:C]$.

(b) Put $n = [C:A]$. Then $C = nA + A_1$, $A_1 < A$, whence $C \geqq nB + A_1$. Hence $[C:A] = n \leqq [C:B]$.

LEMMA 6.7: If $A \neq \theta$, then $\lim_{i\to\infty} [A:A_i] = \infty$ in the case when the minimal sequence (A_i) is infinite.

PROOF: We shall prove first that $[A:A_i]$ cannot be zero for every i. For suppose $[A:A_i] = 0$ $(i = 1, 2, \cdots)$. Then $A = 0 \cdot A_i + B_i$, $B_i < A_i$, and $A = B_i < A_i$, whence $A < A_i$ $(i = 1, 2, \cdots)$. By Lemma 6.6 (b), if H is any element of $\mathscr{L}$,

$$[H:A_{i+1}] \geqq [H:A_i][A_i:A_{i+1}] \geqq 2[H:A_i],$$

and by induction

$$(1) \qquad [H:A_{i+j}] \geqq 2^j[H:A_i] \qquad (i, j = 1, 2, \cdots).$$

But $[\dagger : A_1] \geqq [\dagger : \dagger] = 1$ by (b) of the Corollary to Lemma 6.6. Put $i = 1$, $j = k - 1$, $H = \dagger$. Then $[\dagger : A_k] \geqq 2^{k-1}$ $(k = 1, 2, \cdots)$. Since $A < A_i$, we have again by (b) of the Corollary to Lemma 6.6,

$$[\dagger : A] \geqq [\dagger : A_i] \geqq 2^{i-1} \qquad (i = 1, 2, \cdots),$$

which is impossible since $[\dagger : A]$ is a non-negative integer. Hence there exists i_0 such that $[A : A_{i_0}] \geqq 1$, and (1) yields

$$[A : A_i] \geqq 2^{i-i_0} \qquad (i \geqq i_0).$$

Consequently $\lim_{i\to\infty} [A : A_i] = \infty$.

THEOREM 6.7: Let $A \neq \theta$, $B \neq \theta$ be given. Then

$$\lim_{i\to\infty} \frac{[B : A_i]}{[A : A_i]}$$

exists and is > 0, $< +\infty$. (If (A_i) consists of one element A_1 min, we mean by $\lim_{i\to\infty}$ the value at $i = 1$).

PROOF:

I. Let $A_i = A_1$ be minimal. Then $A_1 \neq \theta$ and B is of the form $[B : A_1]A_1 + B_1$, $B_1 < A_1$. Hence $B_1 = \theta$, and $B = [B : A_1]A_1$. Now $[B : A_1] \neq 0$ since $B \neq \theta$. Similarly $A = [A : A_1]A_1$, whence $[A : A_1] \neq 0$. Hence $\lim_{i\to\infty} \dfrac{[B : A_i]}{[A : A_i]} = \dfrac{[B : A_1]}{[A : A_1]}$ exists and has the desired properties.

II. Let (A_i) be infinite and minimal. By Lemma 6.7, $[B : A_i]$ and $[A : A_i]$ have the limit $+\infty$, whence each is zero for at most a finite number of values of i. By Lemma 6.6,

$$[B : A_{i+j}] \leqq ([B : A_i] + 1)([A_i : A_{i+j}] + 1),$$
$$[A : A_{i+j}] \geqq [A : A_i][A_i : A_{i+j}] \qquad (i, j = 1, 2, \cdots),$$

whence

$$\frac{[B : A_{i+j}]}{[A : A_{i+j}]} \leqq \frac{[B : A_i] + 1}{[A : A_i]} \cdot \frac{[A_i : A_{i+j}] + 1}{[A_i : A_{i+j}]}$$
$$\leqq \frac{[B : A_i] + 1}{[A : A_i]} \cdot \left(1 + \frac{1}{[A_i : A_{i+j}]}\right)$$
$$\leqq \frac{[B : A_i] + 1}{[A : A_i]} \left(1 + \frac{1}{2^j}\right) \qquad \text{(by (1))}.$$

Therefore

$$\overline{\lim_{j\to\infty}} \frac{[B : A_{i+j}]}{[A : A_{i+j}]} \leqq \frac{[B : A_i] + 1}{[A : A_i]},$$

that is

$$\overline{\lim_{k\to\infty}} \frac{[B : A_k]}{[A : A_k]} \leqq \frac{[B : A_i] + 1}{[A : A_i]},$$

so that

$$\overline{\lim_{k\to\infty}} \frac{[B : A_k]}{[A : A_k]} \leq \underline{\lim_{i\to\infty}} \frac{[B : A_i] + 1}{[A : A_i]}$$
$$= \underline{\lim_{i\to\infty}} \frac{[B : A_i]}{[A : A_i]}$$

since $1/[A : A_i]$ tends to zero by Lemma 6.7. But this implies the existence of the desired limit and shows that the limit is $\geqq 0$. It is finite also, since it is less than or equal to $([B : A_i] + 1)/[A : A_i]$, and since this is finite if i is sufficiently great. Since the same reasoning applies to the reciprocal $[A : A_i]/[B : A_i]$, this fraction also has a finite limit, whence the limit of $[B : A_i]/[A : A_i]$ is actually positive. This completes the proof.

Definition 6.10: If $A \neq \theta$, $B \neq \theta$, we define

$$(B : A) \equiv \lim_{i\to\infty} \frac{[B : A_i]}{[A : A_i]}.$$

Theorem 6.8: Let A, B, C be different from θ. Then

(a) $(A : A) = 1$,
(b) $(A : B) = (B : A)^{-1}$,
(c) $(A : C) = (A : B)(B : C)$,
(d) $(A + B : C) = (A : C) + (B : C)$,
(e) $A > B$ implies $(A : C) > (B : C)$.

Proof:

(a) This is obvious, since $[A : A_i]/[A : A_i] = 1$ for i sufficiently large.

(b) $$(A : B) = \lim_{i\to\infty} \frac{[A : A_i]}{[B : A_i]} = \lim_{i\to\infty} \left(\frac{[B : A_i]}{[A : A_i]}\right)^{-1} = (B : A)^{-1}.$$

(c) $$(A : C) = \lim_{i\to\infty} \frac{[A : A_i]}{[C : A_i]} = \lim_{i\to\infty} \frac{[A : A_i]}{[B : A_i]} \lim_{i\to\infty} \frac{[B : A_i]}{[C : A_i]} = (A : B)(B : C).$$

(d) I. Let A_1 be minimal, and put $p \equiv [A : A_1]$, $q \equiv [B : A_1]$, $s \equiv [C : A_1]$. Then $A = pA_1$, $B = qA_1$, $C = sA_1$. Thus $A + B = (p + q)A_1$, and

$$(A + B : C) = \frac{[A + B : A_1]}{[C : A_1]} = \frac{p + q}{s} = \frac{p}{s} + \frac{q}{s} = (A : C) + (B : C).$$

II. Let (A_i) be infinite $(i = 1, 2, \cdots)$. By Lemma 6.6,

$$[A : A_i] + [B : A_i] \leqq [A + B : A_i] \leqq [A : A_i] + [B : A_i] + 1.$$

If we now divide through by $[C : A_i]$ and let i tend to infinity, each term has a limit, and we have

$$(A : C) + (B : C) \leqq (A + B : C) \leqq (A : C) + (B : C) + 0,$$

from which (d) follows.

(e) There exists B_1 with $A = B + B_1$, $B_1 \neq \theta$. Then $(A : C) = = (B : C) + (B_1 : C) > (B : C)$.

DEFINITION 6.11: We define

$$(2) \qquad D(a) \equiv \begin{cases} (A_a : \dagger) & \text{if } a \neq 0, \\ 0 & \text{if } a = 0. \end{cases}$$

THEOREM 6.9:

(i) $D(a) \geqq 0$, $D(a) \leqq 1$, $D(0) = 0$, $D(1) = 1$.

(ii) $D(a + b) + D(ab) = D(a) + D(b)$.

(iii) $a \sim b$ implies $D(a) = D(b)$.

(iii)′ $c > a$ implies $D(c) > D(a)$.

(iii)″ $a \gtrsim b$, $a \sim b$, $a \lesssim b$ are equivalent to $D(a) \gtreqless D(b)$ respectively.

PROOF:

(i) These properties are evident.

(ii) We shall first prove that $ab = 0$ implies $D(a + b) = D(a) + D(b)$. If either $a = 0$ or $b = 0$ the statement is obvious. In the contrary case we argue as follows. Since $ab = 0$, $A_{a+b} = A_a + A_b$. Hence $(A_{a+b} : \dagger) = = (A_a : \dagger) + (A_b : \dagger)$ by Theorem 6.8 (d), and $D(a + b) = D(a) + D(b)$.

Now let us consider the general case. If x is an inverse of ab in b, then $ab + x = b$, $ax = abx = 0$. Hence

$$D(ab) + D(x) = D(b);$$

but $a + b = a + ab + x = a + x$, whence

$$D(a) + D(x) = D(a + x) = D(a + b),$$

and we obtain (ii) by elimination of $D(x)$.

(iii) Now $a = 0$ implies $b \sim a = 0$, whence $b = 0$ and $D(a) = D(b) = 0$. The same is true if $b = 0$. If $a, b \neq 0$, then $D(a) = (A_a : \dagger) = (A_b : \dagger) = = D(b)$.

(iii)′ If $c > a$, there exists an inverse b of a in c, so that $c = a + b$,

$ab = 0$, $b \neq 0$. Hence

$$D(c) = D(a) + D(b) > D(a),$$

since $D(b) = (A_b : \dagger) > 0$.

(iii)'' Suppose $a \prec b$. Then there exists a' with $a \sim a' < b$, and $D(a) = D(a') < D(b)$ by (iii), (iii)'. Because of Theorem 6.1(a), interchanging a, b shows that $a \succ b$ implies $D(a) > D(b)$. Clearly $a \sim b$ implies $D(a) = D(b)$ by (iii). So $a \begin{smallmatrix}\succ\\ \sim\\ \prec\end{smallmatrix} b$ imply $D(a) \gtreqless D(b)$ respectively. Since we have exhaustive disjunctions in both hypotheses and conclusions, the converse implications hold also.

PART I · CHAPTER VII

Dimensionality

We first prove a strengthened form of the additivity of $D(a)$ (Theorem 6.9 (ii)):

LEMMA 7.1: $(a_1, \cdots, a_n)\perp$ implies

$$D(a_1 + \cdots + a_n) = D(a_1) + \cdots + D(a_n).$$

PROOF: For $n = 1$ this is trivial. Suppose the lemma holds for $n = m$. If $(a_1, \cdots, a_{m+1})\perp$, then $(a_1, \cdots, a_m)\perp$, and $(a_1 + \cdots + a_m)a_{m+1} = 0$, whence by Theorem 6.9 (ii)

$$\begin{aligned} D(a_1 + \cdots + a_{m+1}) &= D(a_1 + \cdots + a_m) + D(a_{m+1}) \\ &= D(a_1) + \cdots + D(a_m) + D(a_{m+1}), \end{aligned}$$

and the lemma holds for $n = m + 1$.

LEMMA 7.2: Let a be any element of L and $(a_i; i = 1, 2, \cdots)$ a sequence, finite or denumerably infinite such that

$$\sum_i D(a_i) \leqq D(a);$$

then there exists a sequence $(a_i'; i = 1, 2, \cdots)\perp$ such that $a_i \sim a_i' \leqq a$ $(i = 1, 2, \cdots)$.

PROOF: The desired sequence will be constructed by induction. Assume that $a_1', \cdots, a_{n-1}'$ have been defined, and that $(a_1', \cdots, a_{n-1}')\perp$, $a_i \sim a_i' \leqq a$ $(i = 1, \cdots, n - 1)$. For $n = 1$ this is vacuously true. We proceed now to the construction of a_n'. Since $a_i' \leqq a$ $(i = 1, \cdots, n-1)$, $a_1' + \cdots + a_{n-1}' \leqq a$. Hence there exists an inverse x_n of $a_1' + \cdots + a_{n-1}'$ in a. Then $(a_1' + \cdots + a_{n-1}')x_n = 0$, and therefore $(a_1', \cdots, a_{n-1}', x_n)\perp$; also $a_1' + \cdots + a_{n-1}' + x_n = a$. Hence by Lemma 7.1,

$$\begin{aligned} D(a) &= D(a_1') + \cdots + D(a_{n-1}') + D(x_n) \\ &= D(a_1) + \cdots + D(a_{n-1}) + D(x_n). \end{aligned}$$

But by hypothesis

$$D(a) \geqq \sum_i D(a_i) \geqq D(a_1) + \cdots + D(a_n),$$

whence $D(a_n) \leqq D(x_n)$, and $a_n \lesssim x_n$ by Theorem 6.9 (iii)''. Thus there exists a_n' such that $a_n \sim a_n' \leqq x_n$, and

$$(a_1' + \cdots + a_{n-1}')a_n' \leqq (a_1' + \cdots + a_{n-1}')x_n = 0,$$

whence $(a_i'; i = 1, \cdots, n)\perp$. In this manner we obtain the sequence $(a_i'; i = 1, 2, \cdots)$, where $a_i \sim a_i' \leqq a$ $(i = 1, 2, \cdots)$. Since $(a_1', \cdots, a_n')\perp$ $(n = 1, 2, \cdots)$, we have $(a_i'; i = 1, 2, \cdots)\perp$ by Theorem 2.3.

THEOREM 7.1: If $(a_i; i = 1, 2, \cdots)$ is an independent sequence, finite or denumerably infinite, then

$$\text{(1)} \qquad D(\sum_i a_i) = \sum_i D(a_i).$$

PROOF: If (1) holds for each $a_i \neq 0$, it holds when this condition is not satisfied, since introduction of zero terms in either member of (1) leaves the equation unaltered. Hence we may assume $a_i \neq 0$, and thus $D(a_i) > 0$. If the sequence (a_i) is finite, (1) follows from Lemma 7.1. Let (a_i) be infinite. Since $(a_i; i = 1, 2, \cdots)\perp$, we have

$$(a_1 + \cdots + a_n) \sum_{i=n+1}^{\infty} a_i = 0 \qquad (n = 1, 2, \cdots);$$

but $(a_1, \cdots, a_n)\perp$, whence $(a_1, \cdots, a_n, \sum_{i=n+1}^{\infty} a_i)\perp$ by Theorem 2.1. Hence Lemma 7.1 yields

$$\text{(2)} \qquad D(\sum_{i=1}^{\infty} a_i) = [D(a_1) + \cdots + D(a_n)] + D(\sum_{i=n+1}^{\infty} a_i).$$

If $n \to \infty$, the left member of (2) is constant, while the first term in the right member converges (increasing properly) to $\sum_{i=1}^{\infty} D(a_i)$. Hence the second term (decreasing properly) converges to a number $\varepsilon \geqq 0$:

$$\lim_{n\to\infty} D(\sum_{n+1}^{\infty} a_i) = \varepsilon,$$

and

$$\text{(3)} \qquad D(\sum_{i=1}^{\infty} a_i) = \sum_{i=1}^{\infty} D(a_i) + \varepsilon.$$

Suppose $\varepsilon \neq 0$, i.e., $\varepsilon > 0$. Since $\sum_{i=1}^{\infty} D(a_i)$ converges, $\lim_{n\to\infty} \sum_{i=n+1}^{\infty} D(a_i) = 0$, and also $\lim_{m\to\infty} D(a_m) = 0$. Hence there exists m such that $0 < D(a_m) < \varepsilon$; then there exists $n \geqq m$ such that $\sum_{i=n+1}^{\infty} D(a_i) \leqq D(a_m)$. Thus by applying Lemma 7.2 to the sequence $(a_{n+i}; i = 1, 2, \cdots)$ and $a = a_m$, we see that there exists a sequence $(a_i'; i = 1, 2, \cdots)\perp$ such that $a_{n+i} \sim a_i' \leqq a_m$ $(i = 1, 2, \cdots)$. But since $(a_{n+i}; i = 1, 2, \cdots)\perp$, $(a_i'; i = 1, 2, \cdots)\perp$, we have

$$\sum_{i=1}^{\infty} a_{n+i} \cdot \sum_{i=1}^{\infty} a_i' \leqq \left(\sum_{i=1}^{\infty} a_{n+i}\right) a_m = 0,$$

whence $(a_{n+1}, a_1', a_{n+2}, a_2', \cdots) \perp$ by Theorem 2.1. Thus $a_{n+i} \sim a_i'$ $(i = 1, 2, \cdots)$ yields $\sum_{i=1}^{\infty} a_{n+i} \sim \sum_{i=1}^{\infty} a_i'$ by Theorem 3.6. Therefore, since $\sum_{i=1}^{\infty} a_i' \leqq a_m$,

$$D\left(\sum_{i=1}^{\infty} a_{n+i}\right) = D\left(\sum_{i=1}^{\infty} a_i'\right) \leqq D(a_m) \leqq \varepsilon.$$

But this contradicts

$$D\left(\sum_{i=1}^{\infty} a_{n+i}\right) > \lim_{p\to\infty} D\left(\sum_{i=1}^{\infty} a_{p+i}\right) = \varepsilon$$

and we must have $\varepsilon = 0$.

Theorem 7.2: Let $(a_i;\ i = 1, 2, \cdots)$ be a denumerably infinite sequence such that $i < j$ implies $a_i \leqq a_j$. Then

$$D(\lim_{n\to\infty}{}^* a_n) = \lim_{n\to\infty} D(a_n)$$

Proof: Define $a_0 \equiv 0$, and let x_n be an inverse of a_{n-1} in a_n $(n = 1, 2, \cdots)$. For $1 \leqq m < n$ we have $x_m \leqq a_m \leqq a_{n-1}$, whence $x_1 + \cdots + x_{n-1} \leqq a_{n-1}$, and $(x_1 + \cdots + x_{n-1})x_n \leqq a_{n-1}x_n = 0$. This yields, for $n = 1, \cdots, p$, $(x_1, \cdots, x_p) \perp$ by Theorem 2.2. Since this holds for $p = 1, 2, \cdots$, $(x_i;\ i = 1, 2, \cdots) \perp$ by Theorem 2.3. Now

$$\begin{aligned} a_n = a_{n-1} + x_n = a_{n-2} + x_{n-1} + x_n = \cdots &= a_0 + x_1 + \cdots + x_n \\ = x_1 + \cdots + x_n & \qquad (n = 1, 2, \cdots); \end{aligned}$$

consequently

$$\lim_{n\to\infty}{}^* a_n = \sum_{i=1}^{\infty} a_i = \sum_{i=1}^{\infty} x_i.$$

By Theorem 7.1 we conclude

$$\begin{aligned} D(\lim_{n\to\infty}{}^* a_n) &= D\left(\sum_{i=1}^{\infty} x_i\right) = \sum_{i=1}^{\infty} D(x_i) = \lim_{n\to\infty} \left(D(x_1) + \cdots + D(x_n)\right) \\ &= \lim_{n\to\infty} D(x_1 + \cdots + x_n) = \lim_{n\to\infty} D(a_n). \end{aligned}$$

Definition 7.1: Let Δ denote the range of $D(a)$, where a varies over L.

Lemma 7.3:

(i) $0, 1 \in \Delta$.

(ii) $\alpha, \beta \in \Delta$, $\alpha \leqq \beta$ implies $(\beta - \alpha) \in \Delta$.

(iii) $\alpha_i \in \Delta$ $(i = 1, 2, \cdots)$, $\alpha_i \leqq \alpha_j$ $(i < j)$ implies $(\lim_{n\to\infty} \alpha_n) \in \Delta$.

PROOF:

(i) $D(0) = 0$, $D(1) = 1$, whence $0, 1 \in \Delta$.

(ii) Let $D(a) = \alpha$, $D(b) = \beta$; then $D(a) \leqq D(b)$, and $a \precsim b$. Thus there exists a' with $a \sim a' \leqq b$. Let x be inverse to a' in b. Since $a'x = 0$, $a' + x = b$, we have

$$D(b) = D(a' + x) = D(a') + D(x) = D(a) + D(x),$$

and $D(x) = D(b) - D(a) = \beta - \alpha$. Hence $(\beta - \alpha) \in \Delta$.

(iii) Let a_1 be any element such that $D(a_1) = \alpha_1$. If a_n with $D(a_n) = \alpha_n$ has been defined, let a'_{n+1} be any element such that $D(a'_{n+1}) = \alpha_{n+1}$. Since $D(a_n) \leqq D(a'_{n+1})$, $a_n \precsim a'_{n+1}$, and there exists a_{n+1} such that $a'_{n+1} \sim a_{n+1} \geqq a_n$, whence $D(a_{n+1}) = \alpha_{n+1}$. Thus we have a sequence $a_1 \leqq a_2 \leqq \cdots$ such that $D(a_n) = \alpha_n$ $(n = 1, 2, \cdots)$. Then by Theorem 7.2, $D(\lim^*_{n\to\infty} a_n) = \lim_{n\to\infty} \alpha_n$, whence $(\lim_{n\to\infty} \alpha_n) \in \Delta$.

LEMMA 7.4: The only sets Δ' of real numbers x, $0 \leqq x \leqq 1$ which possess the properties (i), (ii), (iii) of Lemma 7.3 are the following:

Cases $N = 1, 2, \cdots$: The set Δ_N consists of all real numbers n/N, $n = 0, 1, \cdots, N$.

Case ∞: The set Δ_∞ consists of all real numbers x, $0 \leqq x \leqq 1$.

PROOF: By property (i), Δ' contains elements $\neq 0$; these elements have a greatest lower bound δ. We consider the two possibilities $\delta > 0$, $\delta = 0$.

a) $\delta > 0$: Elements $\alpha \in \Delta'$, $\alpha \neq 0$ which are $< 2\delta$ must exist. If two different ones existed, say $\alpha, \beta, \alpha < \beta$, then we would have $\delta \leqq \alpha < \beta < 2\delta$, $0 < \beta - \alpha < \delta$, and by property (ii) $\beta - \alpha \in \Delta'$. This contradicts the definition of δ. Therefore exactly one $\alpha \in \Delta'$, $\alpha \neq 0$ with $\alpha < 2\delta$ exists. This necessitates $\alpha = \delta$, and so $\delta \in \Delta'$.

Consider now any $\alpha \in \Delta'$. Then $\alpha \geqq 0$, and there exists an integer $n \geqq 0$ such that $n\delta \leqq \alpha < (n + 1)\delta$. By applying inductively property (ii), we see that α, $\alpha - \delta$, $\alpha - 2\delta, \cdots, \alpha - n\delta$ all belong to Δ'. But $\alpha - n\delta < \delta$ whence $\alpha - n\delta \not> 0$ by the definition of δ, and consequently $\alpha - n\delta = 0$. Hence $\alpha = n\delta$. By applying this to $\alpha = 1$, we see that there exists an integer $N \geqq 0$ such that $1 = N\delta$. Hence $N \neq 0$ and $\delta = 1/N$. Every $\alpha \in \Delta'$ has then the form n/N $(n = 0, 1, 2, \cdots)$, and since $\alpha \leqq 1$, $n \leqq N$, whence n is one of the numbers $0, 1, 2, \cdots, N$. Conversely, since $1/N$, $1 \in \Delta'$, inductive application of property (ii) shows that 1, $N - 1/N$, $N - 2/N, \cdots, 1/N$, 0 all belong to Δ'. Thus Δ' is the set of all n/N, $n = 0, 1, 2, \cdots, N$, and we have the situation described by Case N.

b) $\delta = 0$: Consider any two numbers α, β such that $0 \leqq \alpha < \beta \leqq 1$. Since $\beta - \alpha > 0$ and $\delta = 0$, there exists $\gamma \in \Delta'$, $\gamma \neq 0$, such that $\gamma < \beta - \alpha$.

Hence $0 < \gamma < \beta - \alpha$, and there is an integer $n \geqq 0$ such that $n\gamma \leqq \beta < (n+1)\gamma$. Now $n\gamma \leqq \beta \leqq 1$, and $\gamma, 1, \in \Delta'$. Consequently by repeated application of property (ii) we have that

$$1, 1-\gamma, 1-2\gamma, \cdots, 1-n\gamma$$

all belong to Δ'; one further application shows that

$$n\gamma = 1 - (1 - n\gamma) \in \Delta'$$

also. Furthermore, $n\gamma \leqq \beta$, and $n\gamma = (n+1)\gamma - \gamma > \beta - (\beta - \alpha) = \alpha$. Define $\alpha' \equiv n\gamma$. Then $\alpha' \in \Delta'$, $\alpha < \alpha' \leqq \beta$. Consider now any x with $0 \leqq x \leqq 1$. If $x = 0$, $x \in \Delta'$ by property (i). Suppose therefore that $x \neq 0$, i.e., $x > 0$. Then there exists a sequence $0 \leqq \alpha_1 < \alpha_2 < \cdots < x$ such that $\lim_{n\to\infty} \alpha_n = x$. Since $0 \leqq \alpha_n < \alpha_{n+1} \leqq 1$ $(n = 1, 2, \cdots)$, we have the existence of $(\alpha_n'; n = 1, 2, \cdots)$ for which

$$\alpha_n < \alpha_n' \leqq \alpha_{n+1}, \quad \alpha_n' \in \Delta \quad (n = 1, 2, \cdots).$$

Thus $\alpha_1' < \alpha_2' < \cdots$, and $\lim_{n\to\infty} \alpha_n' = x$. Hence by property (iii), $x \in \Delta'$. Consequently Δ' contains every x with $0 \leqq x \leqq 1$, and thus coincides with the interval. This establishes Case ∞.

THEOREM 7.3: The range Δ must be one of the sets $\Delta_1, \Delta_2, \cdots, \Delta_\infty$ defined in Lemma 7.4. We say accordingly that Case $1, 2, \cdots, \infty$ respectively holds for L.

PROOF: This results by combining Lemmas 7.3 and 7.4.

It is easy to see now, by using the axiomatic treatment of projective geometry of G. Birkhoff or that of K. Menger, that Case N corresponds to N-dimensional, projective geometry $(N = 1, 2, \cdots)$. This will be discussed somewhat later. Case ∞, however, corresponds to an entirely new system, to the study of which the subsequent chapters will be devoted.

We have obtained the dimension function $D(a)$ by a constructive process, in the course of which the principle of duality was not always observed, and which contained many details of which no trace appears in the properties of $D(a)$ ultimately obtained. It is essential therefore to show that $D(a)$ can be characterized by these properties alone, independently of its construction.

DEFINITION 7.2: A real-valued function $D'(a)$ whose domain is L is called an (unnormalized) dimension function in case

(a) the range of $D'(a)$ has either an upper bound or a lower bound,
(b) $D'(a+b) + D'(ab) = D'(a) + D'(b)$.

THEOREM 7.4: Every function

$$D'(a) = \gamma_1 D(a) + \gamma_2 (\gamma_1, \gamma_2 \text{ real numbers})$$

is a dimension function, and every dimension function is of this form.

PROOF: We observe first that if $D''(a)$ is a dimension function, then every $D'(a) = \gamma_1 D''(a) + \gamma_2 (\gamma_1, \gamma_2$ real numbers) is also a dimension function. Now $D(a)$ is a dimension function, since it satisfies (a) because $D(a) \geqq 0$ and (b) by Theorem 6.9(ii). Hence every $D'(a)$ of the theorem is also a dimension function. We need then prove only the converse, i.e., that every dimension function is of the form $\gamma_1 D(a) + \gamma_2$.

If a, x are inverses, then (b) yields

$$D'(a) + D'(x) = D'(a + x) + D'(ax) = D'(1) + D'(0),$$

and $D'(a) = D'(1) + D'(0) - D'(x)$. So if $a \sim b$, then $D'(a) = D'(b)$. Thus by Theorem 6.9 (iii)'', $D(a) = D(b)$ implies $D'(a) = D'(b)$, and so $D'(a)$ is a function of $D(a)$:

$$D'(a) = f(D(a)),$$

where $f(x)$ has the domain Δ. Assume $x, y \in \Delta$, $x + y \leqq 1$. Then by Theorem 6.8 (d), (e) there exist a, b with $D(a) = x$, $D(b) = y$, $ab = 0$. Then $D(a + b) = D(a) + D(b) = x + y$, so that

$$\begin{aligned} f(x + y) + f(0) &= f(D(a + b)) + f(D(ab)) \\ &= D'(a + b) + D'(ab) = D'(a) + D'(b) \\ &= f(x) + f(y). \end{aligned}$$

Thus if $g(x) \equiv f(x) - f(0)$, $g(x)$ satisfies

(4) $$g(x + y) = g(x) + g(y) \qquad (x, y \in \Delta,\ x + y \leqq 1).$$

We consider now Cases 1, 2, $\cdots$ and Case ∞ separately. Define $\gamma_2 \equiv f(0)$, $\gamma_1 \equiv g(1)$.

a) Case N ($N = 1, 2, \cdots$): Equation (4) yields

$$g\left(\frac{n}{N}\right) = g\Big(\underbrace{\frac{1}{N} + \cdots + \frac{1}{N}}_{n}\Big) = \underbrace{g\left(\frac{1}{N}\right) + \cdots + g\left(\frac{1}{N}\right)}_{n} = ng\left(\frac{1}{N}\right)$$

for $n = 0, 1, \cdots, N$. Putting $n = N$ we have $Ng(1/N) = g(1) = \gamma_1$, $g(1/N) = \gamma_1\, 1/N$, whence $g(n/N) = ng(1/N) = \gamma_1\, n/N$. Hence $g(x) = \gamma_1 x$ for every $x \in \Delta$.

b) Case ∞: For every $p = 1, 2, \cdots$ and $n = 0, 1, \cdots, p$ we have

$$g\left(\frac{n}{p}\right) = g\Big(\underbrace{\frac{1}{p} + \cdots + \frac{1}{p}}_{n}\Big) = \underbrace{g\left(\frac{1}{p}\right) + \cdots + g\left(\frac{1}{p}\right)}_{n} = ng\left(\frac{1}{p}\right).$$

Putting $n = p$ we have $pg(1/p) = g(1) = \gamma_1$, $g(1/p) = \gamma_1\, 1/p$, whence

$g(n/p) = ng\,(1/p) = \gamma_1 n/1p$. Thus for every rational $x \in \Delta$, we have $g(x) = \gamma_1 x$.

Now suppose that there exists a real x such that $g(x) \neq \gamma_1 x$, i.e., $g(x) \gtrless \gamma_1 x$. Then $g(1-x) = g(1) - g(x) \lesseqgtr \gamma_1(1-x)$, and we may assume that $g(x) < \gamma_1 x$, since in the contrary case x may be replaced by $1-x$ where this condition is satisfied. Now there exists an integer $k \geqq 1$ with $g(x) \leqq \gamma_1 x - 1/k$. To every integer $m \geqq 1$ there corresponds a rational number x' such that $x \leqq x' \leqq x + 1/mk$. Now $0 \leqq x' - x \leqq 1/mk$ so that $x' - x,\ 2(x'-x), \cdots, mk(x'-x)$ are all $\geqq 0$, $\leqq 1$ and hence in Δ. Moreover,

$$g(x'-x) = g(x') - g(x) = \gamma_1 x' - g(x) \geqq \gamma_1 x - \left(\gamma_1 x - \frac{1}{k}\right) = \frac{1}{k}.$$

Therefore $g(mk(x'-x)) = mkg(x'-x) \geqq m$ for every m, and the range of $g(x)$ has no upper bound. Since $g(1-x) = 1 - g(x)$, this range has also no lower bound. The same is true for $f(x) = g(x) + f(0)$, and thus also for $D'(x)$. But this contradicts (a). Hence $g(x) = \gamma_1 x$ for every $x \in \Delta$. Thus

$$f(x) = g(x) + f(0) = \gamma_1 x + \gamma_2 \qquad (x \in \Delta)$$

so that

$$D'(a) = f(D(a)) = \gamma_1 D(a) + \gamma_2 \qquad (a \in L).$$

Corollary 1: Among all (unnormalized) dimension functions the normalization $D'(0) = 0$, $D'(1) = 1$ characterizes $D'(a) = D(a)$ uniquely.

The notion of an (unnormalized) dimension function is clearly self-dual. But the above normalization which characterizes $D(a)$ is not self-dual, since dualization interchanges the elements 0, 1, of L, while it does not affect the numbers 0, 1. Thus dualization carries $D(a)$ into $1 - D(a)$. We can thus infer immediately the dual of Theorem 7.2:

Theorem 7.5: $a_1 \geqq a_2 \geqq \cdots$ implies $D(\lim^*_{n\to\infty} a_n) = \lim_{n\to\infty} D(a_n)$.

Corollary: Both Theorems 7.2 and 7.5 hold for every dimension function $D'(a)$.

Proof: This is immediate from Theorems 7.2 and 7.5 in the light of Theorem 7.4.

Another inference is as follows:

Theorem 7.6: The case which holds for L (in the sense of Lemma 7.4) is unchanged by dualization.

Proof: Dualization replaces $D(a)$ by $1 - D(a)$, and thus leaves its range Δ (cf. Lemma 7.4) unchanged.

PART II

PART II · CHAPTER I

Theory of Ideals and Coordinates in Projective Geometry

We shall study in this and succeeding chapters the problem of introducing coordinates in a continuous geometry L_∞. Since the procedure in the continuous case is suggested by a solution of the problem for finite-dimensional geometries L_n, we shall first outline the method for these latter cases.

Let us consider first a projective geometry L_n of $n-1$ dimensions, $n \geqq 4$. It is well known that there exists an associative division algebra $\mathfrak{D}$ (cf. Definition 1.2) which bears a close relation to L_n. The algebra $\mathfrak{D}$ is obtained by defining operations of addition and multiplication in the class of all points of a given fixed line of L_n; the details of this construction are to be found in Veblen and Young, *Projective Geometry, I,* Chapter VI (pp. 141, ff.), and they will not be reproduced here. The definition of $\mathfrak{D}$ involves in addition to an arbitrary line, a choice of three distinct points on the line; however alteration of any of these arbitrary elements leads to an algebra isomorphic to $\mathfrak{D}$. We now describe briefly the relation between L_n and $\mathfrak{D}$.

Let us consider the class of all n-dimensional vectors $(x_i; i = 1, \cdots, n)$ with elements in $\mathfrak{D}$. If (x_i) and (y_i) are given, the existence of $z \in \mathfrak{D}$ with $x_i = y_i z$, $z \neq 0$ defines a relation between (x_i) and (y_i); this relation is easily seen to be an equivalence. A class of vectors equivalent under this relation is called a *right ratio* *, and the right ratio determined by a vector (x_i) will be denoted by $(x_1 : x_2 : \cdots : x_n)_r$. It may be shown that the class of all right ratios excluding $(0 : 0 : \cdots : 0)_r$ is in one-to-one correspondence with the class of all points $p \in L_n$. This correspondence has the property that if $a \in L$ and if P is the set of all points $p \leqq a$, then there exists a finite set of linear equations

$$(1) \qquad \sum_{j=1}^{n} \alpha_{ij} x_j = 0 \qquad (i = 1, \cdots, m)$$

such that the points corresponding to all solutions of (1) comprise exactly

* Left ratios may be similarly defined (by placing the proportionality factor z on the left).

the set P; moreover, every set of right linear equations (1) represents an element $a \in L_n$ in this manner. An equivalent description of this property is the following. Consider $a \in L_n$ and let P be the set of all points $p \leqq a$. Define Γ as the class of all n-dimensional vectors representing the points of P plus the zero vector. The Γ has the properties

$$(2) \qquad (x_i),\ (y_i) \in \Gamma \text{ implies } (x_i + y_i) \in \Gamma,$$

$$(3) \qquad (x_i) \in \Gamma \text{ implies } (x_i z) \in \Gamma \text{ for every } z \in \mathfrak{D};$$

moreover every right linear set Γ satisfying (2) and (3) represents an element $a \in L_n$ in this manner.

It should be remarked that left ratios could be used in a similar way; equation (1) must be replaced by $\sum_{j=1}^{n} x_j \alpha_{ji} = 0$, and in the conclusion of (3) $(x_i z)$ must be replaced by (zx_i). Of course the same division algebra $\mathfrak{D}$ will lead to two different lattices L_n and L'_n if we use respectively right and left ratios, and these need not be isomorphic (cf. the Corollaries 1, 2 to Lemma 2.2). On the other hand the same lattice L_n if described by means of right ratios in $\mathfrak{D}$ will be described by left ratios of another division algebra $\mathfrak{D}'$, $\mathfrak{D}'$ being obtained from $\mathfrak{D}$ by leaving its elements and the addition operation $x + y$ unchanged, but replacing the multiplication operation xy by its transpose yx. (In the abstract algebraic terminology $\mathfrak{D}'$ is called the inverse of $\mathfrak{D}$; cf. van der Waerden, *Moderne Algebra*, II, pages 169—170. The lattice L'_n is the dual of L_n; cf. the Corollaries 1,2 to Lemma 2.2.)

On account of the essential use of points in this formulation of the parameterization of the lattice L_n, some alterations will be necessary before the generalization to continuous geometry can be made. We shall therefore give an equivalent formulation in which points are not used.

DEFINITION 1.1: A *ring* * is a set $\mathfrak{R}$ of elements $x, y, z, \cdots$ together with two binary operations $+$, $\cdot$ (the symbol for multiplication generally being suppressed) with the following properties:

(a) $x + y = y + x$,
(b) $(x + y) + z = x + (y + z)$,
(c) $(xy)z = x(yz)$,
(d) $x(y + z) = xy + xz$,
(e) $(y + z)x = yx + zx$,
(f) For every x, y there exists z such that $x + z = y$,
(g) There exists a unit 1 such that $x \cdot 1 = 1 \cdot x = x$ for every x.

It is well known that the element z in (f) is uniquely determined by

* What we call a ring is generally referred to as a "ring with unit."

x, y; that therefore a unique operation $y - x$ is defined; that $0 = x - x$ is independent of x; that $0 + x = x$, $x \cdot 0 = 0 \cdot x = 0$; that if $-x = 0 - x$, then $y - x = y + (-x)$; and that the unit 1 is unique. Proofs of these properties and their immediate consequences will be omitted.

DEFINITION 1.2: A ring $\mathfrak{R}$ is a *division algebra*, or *field*, if for every $x \neq 0$ there exists y such that $xy = yx = 1$.

It is well known that the element y is uniquely determined by x; it will be denoted by x^{-1}.

COROLLARY: If $\mathfrak{R}$ is any ring, and if $\mathfrak{R}_n$ is the class of all square matrices of order n with elements in $\mathfrak{R}$, then $\mathfrak{R}_n$ forms a ring with respect to the usual definitions of $+$, $\cdot$:

$$(x_{ij}) + (y_{ij}) = (x_{ij} + y_{ij}), \quad (x_{ij})(y_{ij}) = \left(\sum_{k=1}^{n} x_{ik}y_{kj}\right).$$

The unit is the identity matrix 1_n of order n.

It is well known that even for a division algebra $\mathscr{D}$, $\mathscr{D}_n$ is only a ring (i.e., is not a division algebra) if $n \neq 1$.

DEFINITION 1.3: If $\mathfrak{R}$ is a ring, and if $\mathfrak{a} \subset \mathfrak{R}$, then $\mathfrak{a}$ is a *right ideal* in case $x + y \in \mathfrak{a}$, $xz \in \mathfrak{a}$ when $x, y \in \mathfrak{a}$, $z \in \mathfrak{R}$; $\mathfrak{a}$ is a *left ideal* in case $x + y \in \mathfrak{a}$, $zx \in \mathfrak{a}$ when $x, y \in \mathfrak{a}$, $z \in \mathfrak{R}$. Finally, $\mathfrak{a}$ is an *ideal* in case $\mathfrak{a}$ is both a right ideal and a left ideal. The set of all right ideals is denoted by $R_{\mathfrak{R}}$, and the set of all left ideals by $L_{\mathfrak{R}}$.

COROLLARY 1: If $R \subset R_{\mathfrak{R}}$ is any class of right ideals, the intersection * $\mathfrak{P}(R)$ of all ideals in R is again a right ideal. If $\mathfrak{a} \subset \mathfrak{R}$, then $\mathfrak{a}$ has a unique least extension $\mathfrak{a}_r$ which is a right ideal.

PROOF: The first part is obvious. To prove the second part, we consider the class of all right ideals containing $\mathfrak{a}$ and define $\mathfrak{a}_r$ as the intersection of this class. This class is non-empty, since the entire ring $\mathfrak{R}$ belongs to it; thus $\mathfrak{a}_r \supset \mathfrak{a}$, and $\mathfrak{a}_r$ is the smallest right ideal containing $\mathfrak{a}$.

COROLLARY 2: If $R \subset R_{\mathfrak{R}}$ is any class of right ideals, there exists both a smallest right ideal (join or least upper bound of R) containing every element of R and a greatest right ideal (intersection or greatest lower bound of R) contained in every element of R. Thus $R_{\mathfrak{R}}$ is a continuous lattice to the relation $\subset$ and the operations thus defined. The zero (smallest element) of $R_{\mathfrak{R}}$ is $(0)_r = (0)$, and the unit (greatest element) is $(1)_r = \mathfrak{R}$.

PROOF: The right ideal $(\mathfrak{S}(R))_r$ is effective as join of R. The remainder of the corollary is obvious if we recall that a continuous lattice** is a

* As heretofore, $\mathfrak{P}$ represents set-theoretical intersection and $\mathfrak{S}$ set-theoretical addition.

** A *lattice* L satisfies Axiom I for continuous geometry and Axiom II for finite subsets S of L. It is said to be partially ordered with respect to the relation $<$ (or $\leqq$, here written $\subset$) used in its definition.

partially ordered set satisfying axioms I, II for Continuous Geometry (Part I, page 1).

NOTE 1: We shall use the symbols $\mathfrak{a} \cup \mathfrak{b}$ and $\mathfrak{a} \cap \mathfrak{b}$ respectively for $(\mathfrak{S}(\mathfrak{a}, \mathfrak{b}))_r$ and $\mathfrak{P}(\mathfrak{a}, \mathfrak{b})$ when $\mathfrak{a}$, $\mathfrak{b}$ are right ideals.

NOTE 2: Since there is an obvious symmetry between right ideals and left ideals, we shall frequently state only one of two symmetric theorems or definitions.

DEFINITION 1.4: A *principal right ideal* is one of the form $(x)_r$. The class of all principal right ideals will be denoted by $\bar{R}_{\mathfrak{R}}$, and the class of all principal left ideals by $\bar{L}_{\mathfrak{R}}$.

We return now to the consideration of a geometry L_n and its associated division algebra $\mathscr{D}$ and propose the problem of finding all right ideals and all principal right ideals in $\mathscr{D}_n$. Let a class $\mathfrak{a}$ of matrices $X \in \mathscr{D}_n$, where

$$X = \begin{pmatrix} x_{11} & \cdots & x_{1n} \\ \cdot & \cdots & \cdot \\ x_{n1} & \cdots & x_{nn} \end{pmatrix} = (x_{ij}),$$

be a right ideal. Let Γ be the vector set composed of the first columns of all matrices $X \in \mathfrak{a}$. Certain inferences may now be made.

(A) The set Γ is the set of all columns of all matrices $X \in \mathfrak{a}$.

PROOF: We shall prove a more general property. Consider any matrix $X \in \mathfrak{a}$ and let p, q be two fixed integers between 1 and n. Then if Y is the matrix (y_{ij}), where

$$y_{pq} = 1, \; y_{ij} = 0 \qquad ((i, j) \neq (p, q)),$$

we see that XY is the matrix with q^{th} column equal to the pth column of X, and with zeros elsewhere; moreover $XY \in \mathfrak{a}$. Thus in particular any column of X is the first column of a suitable matrix of $\mathfrak{a}$, and the statement is proved.

(B) The set $\mathfrak{a}$ consists of all matrices of $\mathscr{D}_n$ whose columns are in Γ.

PROOF: Clearly any matrix in $\mathfrak{a}$ has its columns in Γ by (A). Let X be any matrix with columns in Γ. Then

$$X = X_1 + \cdots + X_n,$$

where X_i has i^{th} column equal to that of X and zeros elsewhere. The ith column of X belongs to Γ by (A); hence a matrix $X_i' \in \mathfrak{a}$ exists, the first column of which coincides with the ith column of X, i.e. of X_i. All other columns of X_i are 0. Hence, as in the proof of (A), $X_i = X_i' \cdot (y_{kl})$, where $y_{li} = 1$, $y_{kl} = 0$ $((k, l) \neq (1, i))$, and therefore $X_i \in \mathfrak{a}$. Thus $X \in \mathfrak{a}$.

(C) Properties (2), (3) hold for Γ, i.e., Γ is a linear vector space.

PROOF: If (x_i), $(y_i) \in \Gamma$, there exist matrices X, Y having (x_i), (y_i), respectively, as first columns. Then $Z = X + Y \in \mathfrak{a}$, and the first column of Z, viz., $(x_i + y_i)$ belongs to Γ. Now if $(x_i) \in \Gamma$ and $z \in \mathscr{D}$, let $X \in \mathfrak{a}$ have (x_i) as first column. Then $Z \in \mathfrak{a}$, where $Z = X \cdot (z1)$, 1 being the nth order identity matrix. Therefore $(x_i z) \in \Gamma$, since $(x_i z)$ is the first column of Z.

We are now in a position to determine all right ideals.

(D) Any linear space Γ of n-dimensional vectors (x_i) gives rise to a right ideal $\mathfrak{a}$ related to Γ in the manner described in (B).

PROOF: Let Γ be given, and define $\mathfrak{a}$ as the class of all square matrices with columns in Γ. Then if X, $Y \in \mathfrak{a}$, clearly $X + Y \in \mathfrak{a}$, since each column of $X + Y$ is the sum of columns of X and Y. Moreover, if $X \in \mathfrak{a}$, $Y \in \mathscr{D}_n$, then $XY \in \mathfrak{a}$, since each column of XY is a right linear combination of all the columns of X and thus belongs to Γ by (2) and (3).

(E) The class of all linear spaces Γ of n-dimensional vectors is in one-to-one correspondence with the class of all right ideals $\mathfrak{a}$, corresponding pairs being related as described in (B). Moreover, the correspondence carries the operations $\cup$, $\cap$ for ideals into linear join and intersection for linear spaces respectively.

PROOF: The first part is evident from the foregoing considerations. Let $\Gamma \rightleftarrows \mathfrak{a}$, $\mathfrak{J} \rightleftarrows \mathfrak{b}$. Then $\mathfrak{a} \cap \mathfrak{b}$ is the class of all matrices with columns in Γ and $\mathfrak{J}$, i.e. the intersection of Γ and $\mathfrak{J}$, which we denote by $\Gamma \cap \mathfrak{J}$, whence $\mathfrak{a} \cap \mathfrak{b} \rightleftarrows \Gamma \cap \mathfrak{J}$. Clearly $\mathfrak{a} \cup \mathfrak{b}$ is the class of all $X + Y$, $X \in \mathfrak{a}$, $Y \in \mathfrak{b}$, i.e. the class of all matrices with columns in the linear join of Γ and $\mathfrak{J}$ which we shall denote by $\Gamma \cup \mathfrak{J}$, and $\mathfrak{a} \cup \mathfrak{b} \rightleftarrows \Gamma \cup \mathfrak{J}$.

(F) Every right ideal is also a principal right ideal.

PROOF: Let $\mathfrak{a}$ be a right ideal, and Γ its corresponding linear space. Since Γ contains the zero vector, it is the join of n of its vectors. If X is a matrix having these vectors as its columns, then $(X)_r$ is the class of all XY, $Y \in \mathscr{D}_n$, whence the columns of $(X)_r$ comprise all vectors of Γ, i.e. $(X)_r \rightleftarrows \Gamma$; thus $(X)_r = \mathfrak{a}$.

These considerations show that if L_n is the lattice of an $(n - 1)$-dimensional projective geometry $(n \geqq 4)$, and if $\mathscr{D}$ is the corresponding division algebra and $\mathscr{D}_n$ the matrix ring generated by $\mathscr{D}$, then L_n is isomorphic to the lattice of all (principal) right ideals, since the lattice L_n is clearly isomorphic to the lattice of all linear vector sets Γ.

If we wish to generalize this to the case of a continuous geometry L_∞, we are confronted with the problem of finding a ring $\mathfrak{R}$ associated with L_∞, such that its right ideals are in some plausible way isomorphic to L_∞. Since $\mathfrak{R}$ must be more general than a matrix ring $\mathscr{D}_n$ $(n = 1, 2, \cdots)$

over a division algebra $\mathcal{D}$, it will be necessary to face the possibility that not all ideals are principal; therefore the question arises: which right ideals of $\mathfrak{R}$ form a set isomorphic to L_∞? The set $R_\mathfrak{R}$ of all right ideals is, by Corollary 2 after Definition 1.3, a (continuous) lattice; hence L_∞ could be isomorphic to some sublattice $R'_\mathfrak{R}$ of $R_\mathfrak{R}$. Inasmuch as L_∞ is a *complemented* lattice (cf. Axiom V for Continuous Geometry Part I, p. 2), it will be necessary also for $R'_\mathfrak{R}$ to be one. Hence for every right ideal $\mathfrak{a} \in R'_\mathfrak{R}$ a right ideal $\mathfrak{b} \in R'_\mathfrak{R}$ with $\mathfrak{a} \cup \mathfrak{b} = \mathfrak{R}$, $\mathfrak{a} \cap \mathfrak{b} = (0)$ should exist.

We shall see in Theorem 2.1 that if two right ideals $\mathfrak{a}$, $\mathfrak{b}$ are inverses, i.e., if $\mathfrak{a} \cup \mathfrak{b} = \mathfrak{R}$, $\mathfrak{a} \cap \mathfrak{b} = (0)$, then both are principal. Hence it is natural to conclude that $R'_\mathfrak{R}$ should consist of all *principal right ideals*.

Now the set $\bar{R}_\mathfrak{R}$ of all principal right ideals is in general not a lattice.* Thus the rings $\mathfrak{R}$ which we shall consider will be strongly restricted, in that their sets $\bar{R}_\mathfrak{R}$ will be complemented lattices. We shall see in the next chapter, however, that the last condition for $\mathfrak{R}$, and even somewhat less, implies everything that we need. Indeed, we shall prove the following: If for every principal right ideal $\mathfrak{a} \subset \mathfrak{R}$ ($\mathfrak{R}$ being perfectly arbitrary, subject only to Definition 1.1), there exists a right ideal $\mathfrak{b}$ in $\mathfrak{R}$, which is an inverse of $\mathfrak{a}$, i.e., $\mathfrak{a} \cup \mathfrak{b} = \mathfrak{R}$, $\mathfrak{a} \cap \mathfrak{b} = (0)$, then $\bar{R}_\mathfrak{R}$ is a complemented lattice. We shall show, furthermore, that the corresponding condition for principal left ideals and left ideals is equivalent to the above condition for principal right ideals and right ideals. Thus the set $\bar{L}_\mathfrak{R}$ of all principal left ideals is also a complemented lattice. Finally we shall show that these conditions are equivalent to a simple algebraic property of $\mathfrak{R}$, and that in the case where $\mathfrak{R}$ satisfies a chain condition for right (or left) ideals, they are equivalent to the classical notion of semi-simplicity (cf. Appendix 1, Chapter II).

* This is best seen from the classical example, to which the (Kummerian) ideal-theory owes its origin. If $\mathfrak{R}$ is the (commutative) ring of all $p+q\sqrt{-5}$ ($p, q = 0, \pm 1, \pm 2, \cdots$), then $\mathfrak{a} = (3)_r$ and $\mathfrak{b} = (1+2\sqrt{-5})_r$ are principal, but $\mathfrak{a} \cup \mathfrak{b}$ (their join, or least upper bound, or greatest common divisor) is not principal.

PART II · CHAPTER II

Theory of Regular Rings

Throughout the chapter we shall suppose $\mathfrak{R}$ to be a given fixed ring (cf. Definition 1.1).

DEFINITION 2.1: An element $e \in \mathfrak{R}$ is said to be *idempotent* in case $e^2 = e \cdot e = e$.

COROLLARY:

(i) e is idempotent if and only if $1 - e$ is idempotent.

(ii) $(e)_r$ is the set of all x such that $x = ex$.

(iii) The principal right ideals $(e)_r$ and $(1 - e)_r$, where e is idempotent, are mutual inverses.

(iv) If $(e)_r = (f)_r$, $(1 - e)_r = (1 - f)_r$, where e, f are idempotents, then $e = f$.

PROOF: (i) Let e be idempotent. Then

$$(1 - e)^2 = (1 - e)(1 - e) = 1 - 2e + e^2 = 1 - 2e + e = 1 - e,$$

and $(1 - e)$ is idempotent. The converse follows since $e = 1 - (1 - e)$.

(ii) If $x = ex$, then x is of the form eu and hence belongs to $(e)_r$. Conversely, if $x \in (e)_r$, x is of the form $x = eu$, and $ex = e(eu) = eu = x$, whence x is of the form ex.

(iii) First, $(e)_r \cup (1 - e)_r$ contains the element $e + (1 - e) = 1$, whence $(e)_r \cup (1-e)_r = (1)_r$. Now if $x \in (e)_r \cap (1-e)_r$, $x = ex = (1-e)x$. Therefore $x - ex = x$, $ex = 0$, and $x = 0$. Thus $(e)_r \cap (1 - e)_r = (0)$, and $(e)_r$ is inverse to $(1 - e)_r$.

(iv) Since $(e)_r = (f)_r$, $ef = f$ by (ii), whence $(1 - e)f = 0$. But since $(1 - e)_r = (1 - f)_r$, we may replace e, f herein by $(1 - e)$, $(1 - f)$. Thus $e(1 - f) = 0$, i.e., $ef = e$, and $e = f$.

THEOREM 2.1: Two right ideals $\mathfrak{a}$, $\mathfrak{b}$ are inverses if and only if there exists an idempotent e such that $\mathfrak{a} = (e)_r$, $\mathfrak{b} = (1 - e)_r$. This property characterizes uniquely the idempotent e.

PROOF: The reverse implication and the uniqueness of e follow from parts (iii) and (iv) of the Corollary to Definition 2.1. Let $\mathfrak{a}$, $\mathfrak{b}$ be inverse right ideals. Then there exist elements x, y with $x + y = 1$, $x \in \mathfrak{a}$, $y \in \mathfrak{b}$ (note: $\mathfrak{a} \cup \mathfrak{b} = (x + y;\ x \in \mathfrak{a},\ y \in \mathfrak{b})$ if $\mathfrak{a}$, $\mathfrak{b}$ are right ideals - Ed.) If $z \in \mathfrak{a}$, then $xz + yz = z$, whence $yz = z - xz$. Since $z, xz \in \mathfrak{a}$, $yz \in \mathfrak{a}$. But $yz \in \mathfrak{b}$ since $y \in \mathfrak{b}$; hence $yz = 0$. Thus $z = xz \in (x)_r$ for every $z \in \mathfrak{a}$, and $\mathfrak{a} \subset (x)_r$. But $x \in \mathfrak{a}$, whence $\mathfrak{a} = (x)_r$. Similarly $\mathfrak{b} = (y)_r = (1 - x)_r$, since $x + y = 1$. Finally, since $z = xz$ for every $z \in \mathfrak{a}$, this holds for $z = x$, and x is idempotent. Therefore $e = x$ is effective in the theorem.

THEOREM 2.2: The following statements are equivalent:

(α) Every principal right ideal $(a)_r$ has an inverse right ideal.
(β) For every a there exists an idempotent e such that $(a)_r = (e)_r$.
(γ) For every a there exists an element x such that $axa = a$.
(β') For every a there exists an idempotent f such that $(a)_l = (f)_l$.
(α') Every principal left ideal $(a)_l$ has an inverse left ideal.

PROOF: By Theorem 2.1, (α) is equivalent to (β). Now (β) is equivalent to $ea = a$ and the existence of x with $e = ax$. Hence (β) implies the existence of x with $axa = a$, i.e., (β) implies (γ). If (γ) holds, then define $e \equiv ax$. Then $e^2 = ax \cdot ax = axa \cdot x = ax = e$, and e is idempotent. Moreover $e \in (a)_r$, whence $(e)_r \subset (a)_r$, and $a \in (e)_r$ (since $ea = a$), whence $(e)_r \supset (a)_r$. Thus $(a)_r = (e)_r$. The equivalence of (α'), (β'), (γ) is right-left symmetric to that of (α), (β), (γ). Hence the proof is complete.

DEFINITION 2.2: The ring $\mathfrak{R}$ is said to be *regular* in case $\mathfrak{R}$ possesses any one of the equivalent properties of Theorem 2.2.

NOTE: See the Appendices 1, 2 to Chapter II for the connection between regularity and the classical notion of "semi-simplicity".

DEFINITION 2.3: For every right ideal $\mathfrak{a} \subset \mathfrak{R}$ we define

$$\mathfrak{a}^l \equiv (y;\ z \in \mathfrak{a} \rightarrow yz = 0);$$

for every left ideal $\mathfrak{b} \subset \mathfrak{R}$ we define

$$\mathfrak{b}^r \equiv (z;\ y \in \mathfrak{b} \rightarrow yz = 0).$$

COROLLARY: $\mathfrak{a}^l$ is a left ideal, and $\mathfrak{b}^r$ is a right ideal. The transformation $\mathfrak{a} \rightarrow \mathfrak{a}^l$ maps $R_{\mathfrak{R}}$ on a part of $L_{\mathfrak{R}}$, and the transformation $\mathfrak{b} \rightarrow \mathfrak{b}^r$ maps $L_{\mathfrak{R}}$ on a part of $R_{\mathfrak{R}}$.

LEMMA 2.1: Let $\mathfrak{a}$, $\mathfrak{b}$ be right ideals. Then

(i) $\mathfrak{a} \subset \mathfrak{b}$ implies $\mathfrak{a}^l \supset \mathfrak{b}^l$;
(ii) $\mathfrak{a} \subset \mathfrak{a}^{lr}$ $(\equiv (\mathfrak{a}^l)^r)$;
(iii) $\mathfrak{a}^l = \mathfrak{a}^{lrl}$.

(The right-left symmetric results will be denoted by (i)′, (ii)′, (iii)′; this convention will be used generally whenever convenient.)

PROOF: (i) If $y \in \mathfrak{b}^l$, $z \in \mathfrak{b}$ implies $yz = 0$. Hence $z \in \mathfrak{a}$ implies $yz = 0$, and $y \in \mathfrak{a}^l$, whence $\mathfrak{b}^l \subset \mathfrak{a}^l$.

(ii) Let $u \in \mathfrak{a}$, and consider $y \in \mathfrak{a}^l$. Now $z \in \mathfrak{a}$ implies $yz = 0$. Hence in particular $yu = 0$, and $u \in \mathfrak{a}^{lr}$.

(iii) Since $\mathfrak{a} \subset \mathfrak{a}^{lr}$, $\mathfrak{a}^l \supset \mathfrak{a}^{lrl}$ by (i). But by (ii)′ with $\mathfrak{a}$ replaced by $\mathfrak{a}^l$, $\mathfrak{a}^l \subset \mathfrak{a}^{lrl}$. Hence $\mathfrak{a}^l = \mathfrak{a}^{lrl}$.

Henceforth we shall assume that $\mathfrak{R}$ is regular.

LEMMA 2.2:

(i) If $\mathfrak{a} = (e)_r$ (e idempotent) is a principal right ideal, there exists a principal left ideal $\mathfrak{c} = (1 - e)_l$ such that $\mathfrak{a} = \mathfrak{c}^r$.

(ii) If $\mathfrak{a}$ is a principal right ideal, then $\mathfrak{a} = \mathfrak{a}^{lr}$.

(iii) If $\mathfrak{a}$ is a principal right ideal, $\mathfrak{a}^l$ is a principal left ideal.

PROOF: (i) Let $\mathfrak{a} = (e)_r$, e idempotent. Now

$$\begin{aligned}(e)_r &= (x;\ (1 - e)x = 0)\\ &= (x;\ u \in \mathfrak{R} \rightarrow u(1 - e)x = 0)\\ &= (x;\ y \in (1 - e)_l \rightarrow yx = 0)\\ &= \mathfrak{c}^r \qquad\qquad (\mathfrak{c} = (1 - e)_l).\end{aligned}$$

(ii) By (i), $\mathfrak{a} = \mathfrak{c}^r$, whence $\mathfrak{a}^{lr} = \mathfrak{c}^{rlr} = \mathfrak{c}^r = \mathfrak{a}$.

(iii) $\mathfrak{a} = \mathfrak{c}^r$, with $\mathfrak{c}$ a principal left ideal, whence $\mathfrak{a}^l = \mathfrak{c}^{rl} = \mathfrak{c}$ (by (ii′)). Hence $\mathfrak{a}^l$ is a principal left ideal.

COROLLARY 1: The mappings $\mathfrak{a} \rightarrow \mathfrak{a}^l$, $\mathfrak{b} \rightarrow \mathfrak{b}^r$ define a one-to-one correspondence between $\bar{R}_{\mathfrak{R}}$ and $\bar{L}_{\mathfrak{R}}$; they are inverse transformations between these sets.

PROOF: By part (iii) of Lemma 2.2, $\mathfrak{a} \rightarrow \mathfrak{a}^l$ carries all elements of $\bar{R}_{\mathfrak{R}}$ into $\bar{L}_{\mathfrak{R}}$, and the symmetric property holds for $\mathfrak{b} \rightarrow \mathfrak{b}^r$ by (iii)′. The two mappings are inverses by (ii) and (ii)′. Hence they define a one-to-one correspondence between $\bar{R}_{\mathfrak{R}}$ and $\bar{L}_{\mathfrak{R}}$.

COROLLARY 2: The sets $\bar{R}_{\mathfrak{R}}$ and $\bar{L}_{\mathfrak{R}}$ are anti-isomorphic under the correspondence $\mathfrak{a} \rightarrow \mathfrak{a}^l$.

PROOF: This is obvious by Corollary 1 and Lemma 2.1 (i).

THEOREM 2.3: If $\mathfrak{a}$, $\mathfrak{b}$ are two principal right ideals then their join $\mathfrak{a} \cup \mathfrak{b}$ is also a principal right ideal.

PROOF: We first note that $\mathfrak{a}$ is of the form $(e)_r$ and $\mathfrak{b}$ is of the form $(b)_r$, e being idempotent. Then $(b)_r = (by;\ y \in \mathfrak{R})$ and

$$\begin{aligned}\mathfrak{a} \cup \mathfrak{b} &= (ex + by;\ x,\ y \in \mathfrak{R})\\ &= (ex + eby + (1 - e)by;\ x,\ y \in \mathfrak{R})\\ &= (e(x + by) + (1 - e)by;\ x,\ y \in \mathfrak{R})\\ &= (ex' + (1 - e)by;\ x',\ y \in \mathfrak{R})\\ &= \mathfrak{a} \cup \mathfrak{b}_1,\end{aligned}$$

where $\mathfrak{b}_1 = ((1-e)b)_r$. Define $c \equiv (1-e)b$. Then $\mathfrak{a} \cup \mathfrak{b} = (e)_r \cup (c)_r$, where

$$ec = e(1-e)b = (e-e^2)b = 0.$$

Hence $\mathfrak{a} \cup \mathfrak{b} = (e)_r \cup (f')_r$, where f' is an idempotent of the form $f' = cx$, whence $ef' = 0$. Now define $f \equiv f'(1-e)$. Then

$$ff' = f'(1-e) \cdot f' = f'(f' - ef') = f'f' = f'.$$

Consequently,

$$f^2 = f \cdot f'(1-e) = f'(1-e) = f,$$

and f is idempotent. Moreover, $(f)_r = (f')_r$, since $(f)_r \subset (f')_r$ (because $f = f'(1-e)$), and $(f')_r \subset (f)_r$ because $f' = ff'$. Hence $\mathfrak{a} \cup \mathfrak{b} = (e)_r \cup (f)_r$, and $ef = ef'(1-e) = 0$, $fe = f'(1-e)e = 0$. Since $(e+f) \in (e)_r \cup (f)_r$, $(e+f)_r \subset (e)_r \cup (f)_r$. But $(e+f)e = e^2 + fe = e$, and $(e+f)_r \supset (e)_r$; also $(e+f)f = ef + f^2 = f$, and $(e+f)_r \supset (f)_r$. Thus $(e+f)_r \supset (e)_r \cup (f)_r$, whence $\mathfrak{a} \cup \mathfrak{b} = (e+f)_r$. This completes the proof.

Corollary: The set $\bar{R}_{\mathfrak{R}}$ is closed respect to the operation $\cup$ defined in $R_{\mathfrak{R}}$.

Lemma 2.3: Let $\mathfrak{c}$, $\mathfrak{d}$ be right ideals. Then

$$(\mathfrak{c} \cup \mathfrak{d})^l = \mathfrak{c}^l \cap \mathfrak{d}^l.$$

Proof: By Definition 2.3,

$$\begin{aligned}\mathfrak{c}^l \cap \mathfrak{d}^l &= (y;\ z \in \mathfrak{c} \to yz = 0 \text{ and } z \in \mathfrak{d} \to yz = 0)\\ &= (y;\ z \in \mathfrak{S}(\mathfrak{c}, \mathfrak{d}) \to yz = 0)\\ &= (y;\ z \in (\mathfrak{c} \cup \mathfrak{d}) \to yz = 0),\end{aligned}$$

the last equality holding since $yz = 0$ for every $z \in \mathfrak{c} \cup \mathfrak{d}$ is clearly equivalent to $yz = 0$ for every $z \in \mathfrak{S}(\mathfrak{c}, \mathfrak{d})$. Hence $\mathfrak{c}^l \cap \mathfrak{d}^l = (\mathfrak{c} \cup \mathfrak{d})^l$.

Lemma 2.4: Let $\mathfrak{a}$, $\mathfrak{b}$ be principal right ideals. Then $\mathfrak{a} \cap \mathfrak{b}$ is a principal right ideal.

Proof: Now

$$\begin{aligned}\mathfrak{a} \cap \mathfrak{b} &= \mathfrak{a}^{lr} \cap \mathfrak{b}^{lr} && \text{(by Lemma 2.2 (ii))}\\ &= (\mathfrak{a}^l \cup \mathfrak{b}^l)^r && \text{(by Lemma 2.3}').\end{aligned}$$

But $\mathfrak{a}^l$, $\mathfrak{b}^l$ are principal left ideals by Lemma 2.2 (iii), whence $\mathfrak{a}^l \cup \mathfrak{b}^l$ is also a principal left ideal by Theorem 2.3′. Thus $\mathfrak{a} \cap \mathfrak{b}$ is a principal right ideal by Lemma 2.2 (iii)′.

Corollary: The set $\bar{R}_{\mathfrak{R}}$ is closed with respect to the operation $\cap$ defined in $R_{\mathfrak{R}}$.

Theorem 2.4: The set $\bar{R}_{\mathfrak{R}}$ is a complemented modular lattice, partially ordered by the relation $\subset$, the meet being $\cap$ and the join $\cup$; its zero is (0), and its unit is $(1)_r$.

NOTE: The definition of a lattice has been recalled in the proof of the Corollary 2 to Definition 1.3. A *modular* lattice is one which satisfies Axiom IV for Continuous Geometry (Part I, page 2), and a *complemented* lattice is one which satisfies Axiom V for Continuous Geometry (Part I, page 2).

PROOF: The fact that $\bar{R}_{\mathfrak{R}}$ is a lattice follows from the Corollary to Theorem 2.3 and the Corollary to Lemma 2.4, which state that $\bar{R}_{\mathfrak{R}}$ is a sublattice of $R_{\mathfrak{R}}$. The regularity of $\mathfrak{R}$ and Theorem 2.1 yield that $\bar{R}_{\mathfrak{R}}$ is complemented. The modularity is established as follows. Let $\mathfrak{a}$, $\mathfrak{b}$, $\mathfrak{c}$ be right ideals with $\mathfrak{a} \subset \mathfrak{c}$. Since $(\mathfrak{a} \cup \mathfrak{b}) \cap \mathfrak{c} \supset \mathfrak{a} \cap \mathfrak{c} = \mathfrak{a}$, $(\mathfrak{a} \cup \mathfrak{b}) \cap \mathfrak{c} \supset \mathfrak{b} \cap \mathfrak{c}$, it follows that $(\mathfrak{a} \cup \mathfrak{b}) \cap \mathfrak{c} \supset \mathfrak{a} \cup (\mathfrak{b} \cap \mathfrak{c})$. Now assume $x \in (\mathfrak{a} \cup \mathfrak{b}) \cap \mathfrak{c}$. Then $x \in \mathfrak{a} \cup \mathfrak{b}$; hence x is of the form $y+z$, $y \in \mathfrak{a}$, $z \in \mathfrak{b}$, and $x \in \mathfrak{c}$. Now $y \in \mathfrak{a} \subset \mathfrak{c}$, $z = x-y$, and x, $y \in \mathfrak{c}$. Thus $z \in \mathfrak{c}$, whence $z \in \mathfrak{b} \cap \mathfrak{c}$. Consequently $x \in \mathfrak{a} \cup (\mathfrak{b} \cap \mathfrak{c})$, and it results that $(\mathfrak{a} \cup \mathfrak{b}) \cap \mathfrak{c} \subset \mathfrak{a} \cup (\mathfrak{b} \cap \mathfrak{c})$. Therefore $(\mathfrak{a} \cup \mathfrak{b}) \cap \mathfrak{c} = \mathfrak{a} \cup (\mathfrak{b} \cap \mathfrak{c})$; (this argument establishes modularity of an arbitrary sublattice of $R_{\mathfrak{R}}$ - Ed.).

DEFINITION 2.4: The *center* $\mathscr{Z}$ of $\mathfrak{R}$ is the set of all elements $a \in \mathfrak{R}$ such that $ax = xa$ for every $x \in \mathfrak{R}$.

COROLLARY: The center $\mathscr{Z}$ of $\mathfrak{R}$ is a commutative subring of $\mathfrak{R}$ with the unit 1.

PROOF: Clearly $\mathscr{Z}$ is closed with respect to addition. Moreover, since $abx = axb = xab$ for a, $b \in \mathscr{Z}$, $x \in \mathfrak{R}$, $\mathscr{Z}$ is closed with respect to multiplication. Obviously $1 \in \mathscr{Z}$ since $a \cdot 1 = 1 \cdot a$ for every $a \in \mathfrak{R}$.

THEOREM 2.5: The center $\mathscr{Z}$ is regular.

PROOF: We shall prove that for every $a \in \mathscr{Z}$ there exists $y \in \mathscr{Z}$ such that $aya = a$. Since $\mathfrak{R}$ is regular, there exists x such that $axa = a$. Define $y = ax^2$. Then

$$aya = a \cdot ax^2 \cdot a = a \cdot (xa \cdot x) \cdot a \quad (\text{since } a \in \mathscr{Z})$$
$$= axa \cdot xa = a.$$

And for $u \in \mathfrak{R}$, $yu = ax^2 \cdot u = x^2ua = x^2u \cdot axa = x^2a^2 \cdot u \cdot x = x \cdot u \cdot axa \cdot x = xa^2 \cdot u \cdot x^2 = a \cdot u \cdot x^2 = u \cdot ax^2 = uy$; thus $y \in \mathscr{Z}$. This proves that $\mathscr{Z}$ is regular by Definition 2.2.

(Note: If $a \in \mathscr{Z}$ and $(a)_r = (e)_r$ with e idempotent, then necessarily, $e \in \mathscr{Z}$. - Ed.).

LEMMA 2.5: For every $a \in \mathscr{Z}$ the principal right ideal $(a)_r$ and the principal left ideal $(a)_l$ are the same. They will be denoted by $(a)_*$. Moreover, if $a \in \mathscr{Z}$, $(a)_*^l = (a)_*^r$, and the common value will be denoted by $(a)_*^*$.

Proof: We have, for $a \in \mathscr{Z}$,

$$(a)_r = (ax;\ x \in \mathfrak{R}) = (xa;\ x \in \mathfrak{R}) = (a)_l.$$

Also,

$$(a)^l_* = (x;\ xa = 0) = (x;\ ax = 0) = (a)^r_*.$$

Lemma 2.6:

(i) If $\mathfrak{a}$ is both a right and left ideal of the form $(e)_r$ (or $(e)_l$) with e an idempotent in $\mathfrak{R}$, then e is unique and in $\mathscr{Z}$, and $\mathfrak{a} = (e)_*$.
(ii) A principal right ideal $\mathfrak{a}$ is a left ideal if and only if there exists an idempotent $e \in \mathscr{Z}$ such that $\mathfrak{a} = (e)_*$.

Proof: (i) Let $\mathfrak{a} = (e)_r$ be a left ideal, e being idempotent. For every $y \in \mathfrak{R}$, $ye \in (e)_r$ whence $eye = ye$, $(1 - e)ye = 0$. Now for every $x \in \mathfrak{R}$, there is some $y \in \mathfrak{R}$ with $ex(1-e) \cdot y \cdot ex(1 - e) = ex\,(1-e)$. Since $(1 - e)ye = 0$, it follows that $ex(1 - e) = 0$, $ex = exe$ and hence $ex = xe$; this shows that $e \in \mathscr{Z}$ and that we may write $\mathfrak{a} = (e)_*$. To prove the uniqueness of e, let $\mathfrak{a} = (g)_r$, with g an idempotent in $\mathfrak{R}$. Then

$$g \in (g)_r = \mathfrak{a} = (e)_l, \quad e \in (e)_r = \mathfrak{a} = (g)_r,$$

whence $g = ge = e$.

(ii) The reverse implication is trivial. Suppose $\mathfrak{a}$ is a principal right ideal and also a left ideal. Then there exists an idempotent $e \in \mathfrak{R}$ such that $\mathfrak{a} = (e)_r$, since $\mathfrak{R}$ is regular. Then by part (i), $e \in \mathscr{Z}$ and $\mathfrak{a} = (e)_*$.

Definition 2.5: $\mathfrak{R}$ is said to be the *direct sum* of two subrings $\mathfrak{R}_1$, $\mathfrak{R}_2$ in case $\mathfrak{P}(\mathfrak{R}_1, \mathfrak{R}_2) = (0)$, every element $x \in \mathfrak{R}$ is expressible in the form $y + z$, $y \in \mathfrak{R}_1$, $z \in \mathfrak{R}_2$, and $yz = zy = 0$ for every $y \in \mathfrak{R}_1$, $z \in \mathfrak{R}_2$. $\mathfrak{R}$ is said to be irreducible in case it is the direct sum only of (0) and $\mathfrak{R}$.

Corollary: If $\mathfrak{R}$ is the direct sum of $\mathfrak{R}_1$ and $\mathfrak{R}_2$, then $\mathfrak{R}_1$ and $\mathfrak{R}_2$ are inverse ideals (both left and right ideals). Conversely, any two inverse ideals yield a direct sum decomposition of $\mathfrak{R}$.

Proof: If $y \in \mathfrak{R}_1$, $x \in \mathfrak{R}$, then $x = y' + z'$, $y' \in \mathfrak{R}_1$, $z' \in \mathfrak{R}_2$, and

$$yx = y(y' + z') = yy' + yz' = yy',$$

whence $\mathfrak{R}_1$ is a right ideal. Likewise,

$$xy = (y' + z')y = y'y + z'y = y'y,$$

whence $\mathfrak{R}_1$ is a left ideal. Thus $\mathfrak{R}_1$ is an ideal; similarly $\mathfrak{R}_2$ is an ideal. Moreover, we evidently have $\mathfrak{R}_1 \cup \mathfrak{R}_2 = \mathfrak{R}$, $\mathfrak{R}_1 \cap \mathfrak{R}_2 = (0)$. Conversely, two inverse ideals $\mathfrak{R}_1$, $\mathfrak{R}_2$ are obviously subrings. Furthermore, the condition $\mathfrak{R}_1 \cup \mathfrak{R}_2 = \mathfrak{R}$ means that every $x \in \mathfrak{R}$ is of the form $y + z$, $y \in \mathfrak{R}_1$, $z \in \mathfrak{R}_2$. Finally, since $yz \in \mathfrak{R}_1$, $\mathfrak{R}_2$ if $y \in \mathfrak{R}_1$, $z \in \mathfrak{R}_2$, $yz = 0$. Similarly $zy = 0$.

NOTE: The decomposition $x = y + z$, $y \in \mathfrak{R}_1$, $z \in \mathfrak{R}_2$ is unique, since if also $x = y' + z'$, then $y - y' = z' - z \in \mathfrak{R}_1$, $\mathfrak{R}_2$, and consequently $y - y' = z' - z = 0$. Hence $y = y'$, $z = z'$.

THEOREM 2.6: The only direct sum decompositions of $\mathfrak{R}$ are those of the form

$$\mathfrak{R} = (e)_* \cup (1 - e)_*, \tag{1}$$

where e is idempotent and in $\mathscr{Z}$.

PROOF: Clearly any decomposition of the form (1) is a direct sum decomposition. Let $\mathfrak{R} = \mathfrak{a} \cup \mathfrak{b}$, $(0) = \mathfrak{a} \cap \mathfrak{b}$, with $\mathfrak{a}$, $\mathfrak{b}$ ideals. Then by Theorem 2.1, there exists an idempotent $e \in \mathfrak{R}$ such that $\mathfrak{a} = (e)_r$, $\mathfrak{b} = (1 - e)_r$. But since $\mathfrak{a}$, $\mathfrak{b}$ are also left ideals, Lemma 2.6 (i) yields that $e \in \mathscr{Z}$, $\mathfrak{a} = (e)_*$, $\mathfrak{b} = (1 - e)_*$.

THEOREM 2.7: $\mathfrak{R}$ is irreducible if and only if $\mathscr{Z}$ is a field.

PROOF: Clearly $\mathfrak{R}$ is irreducible if and only if 0, 1 are the only idempotents in $\mathscr{Z}$ by Theorem 2.6. Suppose now that $\mathscr{Z}$ is a field. Then for every idempotent $e \in \mathscr{Z}$ we have $e(1 - e) = e - e^2 = 0$ whence $e = 0$ or $e = 1$. Conversely, if 0,1 are the only idempotents in $\mathscr{Z}$, then $\mathscr{Z}$, (0) are the only principal right ideals of $\mathscr{Z}$ by Theorem 2.1 applied to $\mathscr{Z}$ instead of $\mathfrak{R}$. If $a \in \mathscr{Z}$, $a \neq 0$, then the principal right ideal generated in $\mathscr{Z}$ by a cannot be (0), and therefore is $\mathscr{Z}$. Hence there exists $x \in \mathscr{Z}$ with $1 = ax = xa$. Then $a^{-1} \equiv x$ defines for each $a \neq 0$ of $\mathscr{Z}$ an inverse $a^{-1} \in \mathscr{Z}$, whence $\mathscr{Z}$ is a field.

DEFINITION 2.6: We define $Z_{\mathfrak{R}}$ as the intersection of $R_{\mathfrak{R}}$ and $L_{\mathfrak{R}}$; we define $\bar{Z}_{\mathfrak{R}}$ as the intersection of $\bar{R}_{\mathfrak{R}}$ and $\bar{L}_{\mathfrak{R}}$.

COROLLARY: The set $\bar{Z}_{\mathfrak{R}}$ is a sublattice of $\bar{R}_{\mathfrak{R}}$ (and also of $\bar{L}_{\mathfrak{R}}$). Its elements are precisely the ideals $(e)_*$ with $e \in \mathscr{Z}$ idempotent, and the correspondence between ideals in $\bar{Z}_{\mathfrak{R}}$ and idempotents $e \in \mathscr{Z}$ thus defined is one-to-one.

LEMMA 2.7: Let e be a given idempotent in $\mathfrak{R}$. Then the set of all idempotents $f \in \mathfrak{R}$ such that $(e)_r = (f)_r$ is exactly the set

$$(e + ey(1 - e); y \in \mathfrak{R}).$$

PROOF: First, $(e)_r = (f)_r$ if and only if $e = fe$, $f = ef$. Now these two equations themselves imply that f is idempotent:

$$f^2 = ff = f \cdot ef = fe \cdot f = ef = f.$$

Hence they alone characterize our elements f. Let us define x by the relation $f = e + x$. Then the relation $e = fe$, $f = ef$ means $e = (e + x)e$, $e + x = e(e + x)$, i.e., $xe = 0$, $ex = x$. The latter two equations clearly hold if x is of the form $x = ey(1 - e)$ and conversely imply $x = ey(1 - e)$ with $y = x$. Hence our elements f are given by $f = e + ey(1 - e)$.

LEMMA 2.8: An idempotent $e \in \mathfrak{R}$ is in $\mathfrak{Z}$ if and only if $ey(1-e) = 0$ for every $y \in \mathfrak{R}$.

PROOF: The forward implication is clear since $e \in \mathfrak{Z}$ implies $ey(1-e) = ye(1-e) = 0$. Suppose $ey(1-e) = 0$ for every $y \in \mathfrak{R}$. Then, as in the proof of Lemma 2.6, for every $x \in \mathfrak{R}$, there is some $y \in \mathfrak{R}$ such that $(1-e)xe = (1-e)xe \cdot y \cdot (1-e)xe = 0$, and hence $xe = exe = ex$. This proves that $e \in \mathfrak{Z}$.

LEMMA 2.9: A principal right ideal $\mathfrak{a}$ is uniquely represented in the form $\mathfrak{a} = (e)_r$, where $e \in \mathfrak{R}$ is idempotent, if and only if $e \in \mathfrak{Z}$.

PROOF: The idempotent $e \in \mathfrak{R}$ has the property that $(e)_r = (f)_r$ with f idempotent implies $f = e$. This means by Lemma 2.7, $ey(1-e) = 0$ for every $y \in \mathfrak{R}$, and hence $e \in \mathfrak{Z}$ by Lemma 2.8.

LEMMA 2.10: A principal right ideal $\mathfrak{a}$ has a unique principal right ideal inverse if and only if there is a unique idempotent $e \in \mathfrak{Z}$ with $\mathfrak{a} = (e)_*$.

PROOF: By Theorem 2.1 there is a one-to-one correspondence between the inverse right ideals $\mathfrak{b}$ of $\mathfrak{a}$ and the idempotents $e \in \mathfrak{R}$ with $\mathfrak{a} = (e)_r$ described by $\mathfrak{b} = (1-e)_r$. Hence $\mathfrak{a}$ has a unique inverse $\mathfrak{b}$ if and only if the idempotent $e \in \mathfrak{R}$ with $\mathfrak{a} = (e)_r$ is unique. But this means $e \in \mathfrak{Z}$ and $\mathfrak{a} = (e)_*$ by Lemma 2.9.

THEOREM 2.8: The set $\bar{Z}_{\mathfrak{R}}$ is the set of all principal right ideals which have unique inverses.

PROOF: This is clear by Lemma 2.10 and the Corollary to Definition 2.6.

THEOREM 2.9: The ring $\mathfrak{R}$ is irreducible if and only if the lattice $\bar{R}_{\mathfrak{R}}$ is irreducible.*

PROOF: By definition $\bar{R}_{\mathfrak{R}}$ is irreducible in case (0), $\mathfrak{R}$ are its only elements with unique inverses. But this means that (0), $\mathfrak{R}$ are the only elements in $\bar{Z}_{\mathfrak{R}}$, i.e., that 0, 1 are the only idempotents in $\mathfrak{Z}$ by the Corollary to Definition 2.6. But this is equivalent to the irreducibility of $\mathfrak{R}$ by Theorem 2.6.

THEOREM 2.10: The lattice $\bar{Z}_{\mathfrak{R}}$ is a Boolean algebra. The operation of complementation is the operation $\mathfrak{a}^*$ defined in Lemma 2.5. If e, f are idempotents in $\mathfrak{Z}$, then ef, $e + f - ef$ are idempotents in $\mathfrak{Z}$, and

$$(e)_* \cap (f)_* = (ef)_*,$$
$$(e)_* \cup (f)_* = (e + f - ef)_*.$$

PROOF: We have seen that $\bar{Z}_{\mathfrak{R}}$ is a complemented sublattice of $\bar{R}_{\mathfrak{R}}$ (and, of course, of $\bar{L}_{\mathfrak{R}}$). If $\mathfrak{a} = (e)_* \in \bar{Z}_{\mathfrak{R}}$, $e \in \mathfrak{Z}$ idempotent, then the

* A complemented lattice is irreducible in case its zero and unit are the only elements with unique inverses. Cf. Axiom VI for continuous geometry, Part I, page 3.

unique inverse $\mathfrak{b}$ of $\mathfrak{a}$ is $(1-e)_* \in \bar{Z}_{\mathfrak{R}}$. Now

$$\begin{aligned}(e)_*^* &= (x;\ y \in (e)_* \to xy = 0)\\ &= (x;\ u \in \mathfrak{R} \to xeu = 0)\\ &= (x;\ xe = 0)\\ &= (x;\ x(1-e) = x)\\ &= (1-e)_l = (1-e)_*,\end{aligned}$$

whence the operation $\mathfrak{a}^*$ is the operation of complementation. If e, f are idempotents in $\mathscr{Z}$, then $(ef)^2 = ef \cdot ef = eeff = ef$ whence ef is likewise an idempotent in $\mathscr{Z}$. And $1-(1-e)(1-f) = e+f-ef$ is one also. Since $ef \in (e)_*, (f)_*, (ef)_* \subset (e)_* \cap (f)_*$. But if $x \in (e)_* \cap (f)_*$, $ex = x = fx$, and $efx = x$, whence $x \in (ef)_*$. Thus $(ef)_* = (e) \cap (f)_*$. Now since the operation $\mathfrak{a}^*$ generates an anti-automorphism of $\bar{Z}_{\mathfrak{R}}$, we have

$$\begin{aligned}(e)_* \cup (f)_* &= ((e)_*^* \cap (f)_*^*)^*\\ &= ((1-e)_* \cap (1-f)_*)^*\\ &= (1-(1-e)(1-f))_*\\ &= (e+f-ef)_*.\end{aligned}$$

The fact that $\bar{Z}_{\mathfrak{R}}$ is a Boolean algebra follows from the property that each element has a unique inverse by the theory of lattices (cf. Part I, Theorem 5.3). However, the distributive law may be readily verified:

$$\begin{aligned}((e)_* \cup (f)_*) \cap (g)_* &= ((e+f-ef)g)_*\\ &= (eg+fg-efg)_*\\ &= (eg+fg-egfg)_*\\ &= ((e)_* \cap (g)_*) \cup ((f)_* \cap (g)_*),\end{aligned}$$

for e, f, g any idempotents in $\mathscr{Z}$.

DEFINITION 2.7: If e is an idempotent in $\mathfrak{R}$, we define $\mathfrak{R}(e)$ as the set of all $x \in \mathfrak{R}$ with $ex = xe = x$.

COROLLARY: $\mathfrak{R}(e)$ is a ring with e as its unit.

THEOREM 2.11: If $e \in \mathfrak{R}$ is idempotent, $\mathfrak{R}(e)$ is regular.

PROOF: Since $\mathfrak{R}$ is regular, for every $a \in \mathfrak{R}(e)$ there exists $y \in \mathfrak{R}$ such that $aya = a$. Since $a \in \mathfrak{R}(e)$, $ea = ae = a$. Define $x = eye$. Then $ex = xe = x$ whence $x \in \mathfrak{R}(e)$; moreover,

$$axa = a \cdot eye \cdot a = ae \cdot y \cdot ea = aya = a.$$

Thus $\mathfrak{R}(e)$ is regular.

THEOREM 2.12: Suppose given m linear equations

(2) $$\sum_{j=1}^{n} a_{ij}x_j = 0 \qquad (i = 1, \cdots, m).$$

in n variables $x_j \in \mathfrak{R}$ $(j = 1, \cdots, n)$, with coefficients $a_{ij} \in \mathfrak{R}$. Their general solution may be expressed with the help of n arbitrary parameters u_j $(j = 1, \cdots, n)$ thus:

(3) $$x_j = \sum_{k=1}^{j-1} c_{jk}u_k + e_j u_j \qquad (j = 1, \cdots, n).$$

Here c_{jk}, e_j are fixed elements of $\mathfrak{R}$, each e_j is idempotent, and $e_j c_{jk} = 0$, $c_{jk}e_k = c_{jk}$ (in matrix notation (3) becomes:

$$\begin{Vmatrix} x_1 \\ x_2 \\ \cdot \\ \cdot \\ \cdot \\ x_n \end{Vmatrix} = \begin{Vmatrix} e_1 & 0 & \cdots & 0 \\ c_{21} & e_2 & \cdots & 0 \\ \cdot & \cdot & \ddots & \cdot \\ \cdot & \cdot & & \cdot \\ \cdot & \cdot & & \cdot \\ c_{n1} & c_{n2} & \cdots & e_n \end{Vmatrix} \begin{Vmatrix} u_1 \\ u_2 \\ \cdot \\ \cdot \\ \cdot \\ u_n \end{Vmatrix}$$ - Ed.)

PROOF: We first replace (2) by an equivalent set of equations.

The join $(a_{1n})_l \cup \cdots \cup (a_{mn})_l$ of principal left ideals is a principal left ideal by Theorem 2.3′. Thus there exists an idempotent f such that

$$(f)_l = (a_{1n})_l \cup \cdots \cup (a_{mn})_l,$$

since $\mathfrak{R}$ is regular. Hence f is of the form $\sum_{i=1}^{m} \xi_i a_{in}$, where $\xi_i \in \mathfrak{R}$ $(i = 1, \cdots, m)$. Since $a_{in} \in (f)_l$, $a_{in}f = a_{in}$. Now (2) may be written in the form

(4) $$a_{in}x_n = -\sum_{j=1}^{n-1} a_{ij}x_j \qquad (i = 1, \cdots, m).$$

Multiplication of (4) on the left by ξ_i and summation with respect to i yields

$$fx_n = -\sum_{j=1}^{n-1} \left(\sum_{i=1}^{m} \xi_i a_{ij}\right) x_j,$$

and multiplication by f on the left gives

(5) $$fx_n = -\sum_{j=1}^{n-1} f\left(\sum_{i=1}^{m} \xi_i a_{ij}\right) x_j.$$

The left side of (4) is thus equal to

$$a_{in}x_n = a_{in}fx_n = -\sum_{j=1}^{n-1} a_{in}f\left(\sum_{k=1}^{m} \xi_k a_{kj}\right)x_j$$
$$= -\sum_{j=1}^{n-1} a_{in}\left(\sum_{p=1}^{m} \xi_p a_{pj}\right)x_j.$$

Thus (4) becomes

$$(6)\qquad \sum_{j=1}^{n-1}\left(a_{ij}-a_{in}\left(\sum_{p=1}^{m}\xi_p a_{pj}\right)\right)x_j=0 \qquad (i=1,\cdots,m).$$

This discussion shows that (2) is equivalent to (4), i.e., to (4) and (5), and hence to (5) and (6). But (5) is equivalent to

$$(7)\qquad x_n=-\sum_{j=1}^{n-1}f\left(\sum_{i=1}^{m}\xi_i a_{ij}\right)x_j+(1-f)u_n,$$

where u_n is arbitrary, because (5) implies (7) with $u_n=x_n$, and (7) with u_n arbitrary implies (5) by multiplication on the left by f. Define

$$e_n\equiv 1-f,\quad d_j\equiv -f\sum_{i=1}^{m}\xi_i a_{ij} \qquad (j=1,\cdots,n-1).$$

Then (7) becomes

$$(8)\qquad x_n=\sum_{j=1}^{n-1}d_j x_j+e_n u_n,$$

where e_n is idempotent, and $e_n\cdot d_j=(1-f)fd_j=0$. Now we have that (2) is equivalent to (6) and (8). If $n=1$, then (6) is vacuously satisfied, and (8) becomes $x_1=e_1u_1$, with e_1 an idempotent. This coincides with (3) and our theorem is proved for $n=1$. Suppose now that $n=2,3,\cdots$ and that the theorem holds for $n-1$. Then (6), representing m linear equations in the $n-1$ variables x_j $(j=1,\cdots,n-1)$, is completely solved by

$$(9)\qquad x_j=\sum_{k=1}^{j-1}c_{jk}u_k+e_ju_j \qquad (j=1,\cdots,n-1),$$

where e_j is idempotent, and $e_jc_{jk}=0$, $c_{jk}e_k=c_{jk}$. Now (8) becomes

$$(10)\qquad x_n=\sum_{k=1}^{n-1}c_{nk}u_k+e_nu_n,$$

where $c_{nk}=\sum_{j=k+1}^{n-1}d_jc_{jk}+d_ke_k$. Since $e_nd_j=0$ and $c_{jk}e_k=c_{jk}$, $e_k^2=e_k$, we have $e_nc_{nk}=0$, $c_{nk}e_k=c_{nk}$; moreover, e_n is idempotent. Hence (9) and (10) together give precisely (3), and our theorem is proved for n. Hence by induction the theorem is true for all values of n.

LEMMA 2.11: Let n variables x_j, $j=1,\cdots,n$, be expressed as linear functions of p arbitrary parameters y_i, $i=1,\cdots,p$:

$$(11)\qquad x_j=\sum_{i=1}^{p}a_{ji}y_i \qquad (j=1,\cdots,n),$$

(in matrix notation,

$$\begin{Vmatrix} x_1 \\ x_2 \\ \vdots \\ x_n \end{Vmatrix} = \begin{Vmatrix} a_{11} & \cdots & a_{1p} \\ \cdot & & \cdot \\ \cdot & \cdots & \cdot \\ \cdot & & \cdot \\ a_{n1} & \cdots & a_{np} \end{Vmatrix} \begin{Vmatrix} y_1 \\ \cdot \\ \cdot \\ \cdot \\ y_p \end{Vmatrix}$$ - Ed.).

Then an equivalent parameterization of the x_j with n arbitrary parameters u_j is the following:

(12) $$x_j = \sum_{k=1}^{j-1} c_{jk} u_k + e_j u_j \qquad (j = 1, \cdots, n),$$

the c_{jk}, e_j being fixed elements of $\mathfrak{R}$, each e_j idempotent, and $e_j c_{jk} = 0$, $c_{jk} e_k = c_{jk}$.

PROOF: We may write (11) in the form $x_j - \sum_{i=1}^{p} a_{ji} y_i = 0$ $(j = 1, \cdots, n)$; these are n linear equations for the $n + p$ variables x_j $(j = 1, \cdots, n)$ and y_i $(i = 1, \cdots, p)$. Hence we may apply Theorem 2.12 with n, m replaced by $n + p$, n, and the variables in the order specified above, then

$$x_j = \sum_{k=1}^{j-1} c_{jk} u_k + e_j u_j \qquad (j = 1, \cdots, n),$$

$$y_i = \sum_{k=1}^{n+i-1} c_{jk} u_k + e_{n+i} u_{n+i} \qquad (i = 1, \cdots, p)$$

represent the solution, where all e_h are idempotent, $e_h c_{hk} = 0$, $c_{hk} e_k = c_{hk}$, and the u_h $(h = 1, \cdots, n + p)$ are arbitrary parameters. Since the equations for the y_i have no bearing on the x_j, the equations for the x_j, which involve only the u_j $(j = 1, \cdots, n)$, are equivalent to (11). But these equations coincide with (12).

LEMMA 2.12: Let n be any integer $\geqq 1$. The every principal right ideal $\mathfrak{a}$ in $\mathfrak{R}_n$ (cf. the Corollary to Definition 1.2) can be described with the help of a set $\mathfrak{A}$ of n-dimensional vectors $(x_1, \cdots, x_n)$ $(x_i \in \mathfrak{R})$ in the following way:

(13) $(x_{ij}) \in \mathfrak{a}$ if and only if $(x_{1j}, \cdots, x_{nj}) \in \mathfrak{A}$ for $j = 1, \cdots, n$.

Conversely, $\mathfrak{a}$ determines $\mathfrak{A}$ by

(14) $(x_1, \cdots, x_n) \in \mathfrak{A}$ if and only if there exists $(x_{ij}) \in \mathfrak{a}$ with $x_{i1} = x_i$ $(i = 1, \cdots, n)$.

If $\mathfrak{a} = ((a_{ij}))_r$, then $\mathfrak{A}$ is the set of all $(x_1, \cdots, x_n)$ with

(15) $$x_i = \sum_{j=1}^{n} a_{ij} y_j \qquad (y_j \in \mathfrak{R}).$$

PROOF: Let $\mathfrak{a} = ((a_{ij}))_r$ be an arbitrary principal right ideal in $\mathfrak{R}_n$. Then the general element of $\mathfrak{a}$ is $(x_{ij}) = (a_{ij})(z_{ij}) = \left(\sum_{k=1}^{n} a_{ik} z_{kj}\right)$. This proves (14) for the sets $\mathfrak{A}$ of (15). (13) follows immediately from (14).

We repeat a definition from Chapter I:

DEFINITION 2.8: A set $\mathfrak{A}$ of n-dimensional vectors $(x_1, \cdots, x_n)$ $(x_i \in \mathfrak{R})$ is *right-linear* in case

$$(x_i),\ (y_i) \in \mathfrak{A} \text{ implies } (x_i + y_i) \in \mathfrak{A}, \tag{16}$$

$$(x_i) \in \mathfrak{A},\ z \in \mathfrak{R} \text{ implies } (x_i z) \in \mathfrak{A}. \tag{17}$$

If $\mathfrak{A}$, $\mathfrak{B}$ are two right linear sets of n-dimensional vectors, then $\mathfrak{A} \cap \mathfrak{B}$ is their intersection, and $\mathfrak{A} \cup \mathfrak{B}$ their linear sum, i.e., the set of all $(x_i + y_i)$ with $(x_i) \in \mathfrak{A}$, $(y_i) \in \mathfrak{B}$.

LEMMA 2.13: The set $\mathfrak{A}$ defined in Lemma 2.12 (15) is right-linear. If the principal right ideals $\mathfrak{a}$, $\mathfrak{b}$ in $\mathfrak{R}_n$ correspond to the sets $\mathfrak{A}$, $\mathfrak{B}$ of n-dimensional vectors in the sense of Lemma 2.12 (13), then $\mathfrak{a} \cap \mathfrak{b}$ corresponds to $\mathfrak{A} \cap \mathfrak{B}$ and $\mathfrak{a} \cup \mathfrak{b}$ corresponds to $\mathfrak{A} \cup \mathfrak{B}$.

PROOF: All parts of this lemma are readily verified.

The Lemmas 2.12, 2.13 will play an essential role in Appendix 3. We shall need them now only as preparations for Theorem 2.13.

THEOREM 2.13: For every $n = 1, 2, \cdots$, $\mathfrak{R}_n$ is regular.

PROOF: We shall find for every principal right ideal $\mathfrak{a} \in \mathfrak{R}_n$ an idempotent matrix $B = (b_{ij}) \in \mathfrak{R}_n$ such that $(B)_r = \mathfrak{a}$. We apply Lemma 2.12 to $\mathfrak{a}$ and then Lemma 2.11 to the parameterization (15) of $\mathfrak{A}$ in Lemma 2.12. Thus $(x_1, \cdots, x_n) \in \mathfrak{A}$ is equivalent to

$$x_i = \sum_{k=1}^{i-1} c_{ik} u_k + e_i u_i \qquad (i = 1, \cdots, n), \tag{18}$$

where the c_{ik}, e_i are fixed elements of $\mathfrak{R}$ as described in Lemma 2.11 and the u_i are arbitrary parameters. Now we define b_{ij} by the relations

$$b_{ij} \begin{cases} \equiv e_i & (i = j) \\ \equiv c_{ij} & (i > j) \\ \equiv 0 & (i < j). \end{cases}$$

Then (18) implies that $((b_{ij}))_r = \mathfrak{a}$ and it is easily verified that (b_{ij}) is idempotent.

For the purpose of the final theorem, which is a converse of Theorem 2.13, we drop the assumption that $\mathfrak{R}$ is regular.

THEOREM 2.14: Suppose given an integer $n = 1, 2, \cdots$. Then $\mathfrak{R}_n$ is regular if and only if $\mathfrak{R}$ is regular.

PROOF: If $\mathfrak{R}$ is regular, $\mathfrak{R}_n$ is regular by Theorem 2.13. Suppose $\mathfrak{R}_n$ is regular. Define $\bar{e} \equiv (\bar{e}_{ij})$, where

$$\bar{e}_{ij} = \begin{cases} 1 \text{ if } i = j = 1, \\ 0 \text{ otherwise.} \end{cases}$$

Then $(x_{ij}) \in \mathfrak{R}_n(\bar{e})$ clearly means $x_{ij} = 0$ of $i \neq 1$ or $j \neq 1$. Define $x \equiv x_{11}$; then this means $x_{ij} = x\bar{e}_{ij}$. We then have

$$(x\bar{e}_{ij}) = (0) = 0 \text{ is equivalent to } x = 0,$$
$$(x\bar{e}_{ij}) + (y\bar{e}_{ij}) = ((x + y)\bar{e}_{ij}),$$
$$(x\bar{e}_{ij}) \cdot (y\bar{e}_{ij}) = (xy\bar{e}_{ij}).$$

Thus $x \rightleftarrows (x\bar{e}_{ij})$ is an isomorphism of $\mathfrak{R}$ and $\mathfrak{R}_n(\bar{e})$. The regularity of $\mathfrak{R}_n$ implies the regularity of $\mathfrak{R}_n(\bar{e})$ by Theorem 2.11. Hence $\mathfrak{R}$ is regular also, since $\mathfrak{R}$ is isomorphic to $\mathfrak{R}_n(\bar{e})$.

APPENDIX 1

The reader who is not interested in abstract algebra may omit this appendix.

Given a ring $\mathfrak{R}$ (with unit 1), a condition of great importance in abstract algebra is the so-called "chain condition" or "minimum-maximum condition" for right ideals. (Cf. van der Waerden, *Moderne Algebra II*, page 151. In what follows this book will be quoted as "V. d. W., II"). Instead of formulating the condition for the lattice $R_{\mathfrak{R}}$ of all right ideals in $\mathfrak{R}$ we shall formulate it for any lattice L of elements $\mathfrak{a}$, $\mathfrak{b}$, $\mathfrak{c}$, $\cdots$, partially ordered with respect to $<$:

(1) Every non-empty set $S \subset L$ contains a minimal element $\mathfrak{a}$ and a maximal element $\mathfrak{b}$ such that $\mathfrak{c} \in S$ implies $\mathfrak{c} \nless \mathfrak{a}$, $\mathfrak{c} \ngtr \mathfrak{b}$.

It should be observed that $\mathfrak{a}(\mathfrak{b})$ is *weakly* minimal (maximal): $\mathfrak{c} \in S$ implies $\mathfrak{c} \nless \mathfrak{a}$ ($\ngtr \mathfrak{b}$), and not *strongly* minimal (maximal), which would mean: $\mathfrak{c} \in S$ implies $\mathfrak{c} \geqq \mathfrak{a}$ ($\leqq \mathfrak{b}$). This difference arises, of course, from the fact that S is partially, and not linearly, ordered by $<$.

The condition (1) is equivalent to

(2) Infinite descending sequences $\mathfrak{a}_1 > \mathfrak{a}_2 > \cdots$, and infinite ascending sequences $\mathfrak{a}_1 < \mathfrak{a}_2 < \cdots$ do not exist.

PROOF: It suffices to consider minimality and descending sequences $\mathfrak{a}_1 > \mathfrak{a}_2 > \cdots$; the other case then follows by duality. Assume (1); then if there exists an infinite descending sequence $\mathfrak{a}_1 > \mathfrak{a}_2 > \cdots$, the class S of all elements in such a sequence clearly has no minimal element contrary to (1). Hence (1) implies (2). Conversely, assume that a non-empty set S has no minimal element. Let $\mathfrak{a}_1$ be any element of S; since $\mathfrak{a}_1$

is not minimal, there exists $\mathfrak{a}_2 \in S$ with $\mathfrak{a}_2 < \mathfrak{a}_1$. We may choose $\mathfrak{a}_3$ any element $< \mathfrak{a}_2$, etc., and obtain an infinite descending sequence, contrary to (2). Thus (2) implies (1).

For systems L satisfying our axioms I—VI we have a unique dimension function $D(a)$, the range of which is either a set $\Delta_n = (0, 1/n, 2/n, \cdots, 1)$, with $n = 1, 2, \cdots$, or the set Δ_∞ of all $x \geqq 0, \leqq 1$. These possibilities correspond respectively to Cases 1, 2, $\cdots$, ∞ (cf. Part I, Theorem 7.3).

Since $\mathfrak{a}_1 > \mathfrak{a}_2 > \cdots$ implies $D(\mathfrak{a}_1) > D(\mathfrak{a}_2) > \cdots$, and $\mathfrak{a}_1 < \mathfrak{a}_2 < \cdots$ implies $D(\mathfrak{a}_1) < D(\mathfrak{a}_2) < \cdots$, condition (2) holds in Case 1, 2, $\cdots$. In Case ∞ we may define a decreasing sequence $\mathfrak{a}_i$ by induction: $\mathfrak{a}_1 \equiv 1$, where $D(\mathfrak{a}_1) = 1$; if $\mathfrak{a}_i$ is defined, with $D(\mathfrak{a}_i) = 1/i$, choose $\mathfrak{a}_{i+1}$ with $D(\mathfrak{a}_{i+1}) = 1/i + 1$, $\mathfrak{a}_{i+1} < \mathfrak{a}_i$. That this is possible follows from $1/i + 1 < 1/i = D(\mathfrak{a}_i)$. Hence there exists an infinite sequence $\mathfrak{a}_1 > \mathfrak{a}_2 > \cdots$. Likewise an increasing sequence may be constructed. This proves

(3) For systems L satisfying Axioms I—VI for Continuous Geometry, the chain condition holds if and only if Cases 1, 2, $\cdots$ hold.

An important algebraic notion is the "semi-simplicity" of a ring $\mathfrak{R}$. It is defined as follows:

(4) If $\mathfrak{a}$, $\mathfrak{b}$ are right ideals in $\mathfrak{R}$ then $\mathfrak{a} \otimes \mathfrak{b}$ is the smallest right ideal containing all xy with $x \in \mathfrak{a}$, $y \in \mathfrak{b}$ i.e., the set of all $\sum_{k=1}^{p} x_k y_k$, $x_k \in \mathfrak{a}$, $y_k \in \mathfrak{b}$ $(k = 1, \cdots, p)$, $p = 1, 2, \cdots$. For $n = 1, \cdots$, $\mathfrak{a}^n$ is defined by induction as $\mathfrak{a} \otimes \mathfrak{a} \otimes \cdots \otimes \mathfrak{a}$ with exactly n factors $\mathfrak{a}$. $\mathfrak{a}$ is *nilpotent* in case there exists $n (= 1, 2, \cdots)$ with $\mathfrak{a}^n = (0)$; $\mathfrak{R}$ is *semi-simple* in case no right ideal $\mathfrak{a} \neq (0)$ is nilpotent (i.e. $\mathfrak{a}^n = (0)$ implies $\mathfrak{a} = (0)$; equivalently, $\mathfrak{a}^2 = (0)$ implies $\mathfrak{a} = (0)$ - Ed.).

REMARK 1: If a right ideal $\mathfrak{a} \neq (0)$ is nilpotent, then there exists a nilpotent principal right ideal $\mathfrak{a}' \neq (0)$: Let a be any element of $\mathfrak{a}$ with $a \neq 0$, and define $\mathfrak{a}' \equiv (a)_r$. Then $a \in \mathfrak{a}'$ yields $\mathfrak{a}' \neq (0)$, and $\mathfrak{a}' \subset \mathfrak{a}$ yields the nilpotency of $\mathfrak{a}'$. Hence we may, in defining semi-simplicity, restrict ourselves to principal right ideals.

REMARK 2: Our definition (4) of semi-simplicity coincides with the definition of "rings without a radical" (V. d. W. II, page 155). These "rings without radical" are called loc. cit. "semi-simple" only if they satisfy the chain condition also. We prefer, however, to use "semi-simplicity" in the sense of (4) without reference to the chain condition.

We may now prove

(5) If $\mathfrak{R}$ is regular, $\mathfrak{R}$ is semi-simple.

Proof: Suppose $\mathfrak{R}$ is regular, and consider a right ideal $\mathfrak{a}$ in $\mathfrak{R}$. For every $a \in \mathfrak{a}$ there exists x with $axa = a$, whence $a = ax \cdot a \in \mathfrak{a} \otimes \mathfrak{a}$. Thus $\mathfrak{a} \subset \mathfrak{a} \otimes \mathfrak{a}$. Since $\mathfrak{a}$ is a right ideal, $\mathfrak{a} \otimes \mathfrak{a} \subset \mathfrak{a}$, and we have $\mathfrak{a} = \mathfrak{a} \otimes \mathfrak{a}$. Now by induction $\mathfrak{a} = \mathfrak{a} \otimes \mathfrak{a} \otimes \cdots \otimes \mathfrak{a} = \mathfrak{a}^n$ for every $n = 1, 2, \cdots$. Thus if $\mathfrak{a}$ is nilpotent, then $\mathfrak{a}^n = (0)$ for an existent $n = 1, 2, \cdots$, whence $\mathfrak{a} = (0)$, and $\mathfrak{a}$ is semi-simple.

(6) If $\mathfrak{R}$ is semi-simple, and if the chain condition holds for the lattice $R_{\mathfrak{R}}$ of all right ideals in $\mathfrak{R}$, then $\mathfrak{R}$ is regular.

Proof: Every minimal right ideal $\mathfrak{a}$ in $\mathfrak{R}$ can be represented in the form $(e)_r$, e idempotent (V. d. W. II, page 157, "Hilfssatz 3"), and this can be extended to all right ideals. (Cf. V. d. W. II, pages 158—160, "Satz 1". In this discussion the $\mathfrak{o}$ which is our $\mathfrak{R}$ must be replaced by our $\mathfrak{a}$; this does not seriously affect the proof.) Thus $\mathfrak{R}$ is regular.

Remark 1: It would suffice to assume the form (2) of the chain condition for the system $\bar{R}_{\mathfrak{R}}$ (*a priori* merely a partially ordered set) of all principal right ideals in $\mathfrak{R}$, as a closer inspection of the proofs shows. We shall, however, not pursue this question further.

Remark 2: It should be noted that the proof of (6) showed that every right ideal in $\mathfrak{R}$ is principal. This is a consequence of the chain condition and is not generally true without it.

There is no point in comparing our notion of regularity with semi-simplicity if the chain condition does not hold, since then semi-simplicity possesses none of its essential properties (cf. Appendix 2). Regularity is probably its reasonable extension to this case. Even in the special case when the chain condition holds, our condition $axa = a$ is quite useful. It characterizes semi-simplicity in a new way which has some advantages. Thus it makes the right-left symmetry of this notion immediately evident. Moreover, it is analogous to the requirement which characterizes division algebras, viz., the existence for every $a \neq 0$ of an inverse x with $ax = xa = 1$. Our $axa = a$ is a correctly weakened form of this (cf. E. H. Moore, *General Analysis, Part I*, 1935, pages 197—209, where this condition is considered in matrix rings).

APPENDIX 2

The discussions of this appendix pursue a double purpose. First, we wish to examine the behavior of operators in Hilbert Space from the point of view of Chapter II. Secondly, these results will serve to illustrate how the standard notions of "semi-simplicity" and "simplicity" (V. d. W. II, page 155) fail when the chain condition is omitted. The reader who is not familiar with the elements of the theory of operators may omit this appendix.

In what follows, $\mathfrak{H}$ will denote a complex Hilbert space and $\mathfrak{M}$ a ring of bounded operators in $\mathfrak{H}$ which contains 1. (Cf. J. v. Neumann, Math. Annalen, Vol. 102, 1929, page 388, or F. J. Murray and J. v. Neumann, Annals of Math., Vol. 37, 1936, pages 116, 127. The latter paper will be quoted in this appendix as "M. and N." A *ring* is a set of bounded operators in $\mathfrak{H}$ which contains, αA, A^*, $A + B$, AB — α any complex number — along with A, B, and is closed in a suitable topology for operators, cf. loc. cit. The requirement concerning the adjoint A^* of A is more than the usual requirements in algebra. It is directly connected with the "semi-simplicity" of $\mathfrak{M}$, as will be seen in (I) below.)

(I) Every ring $\mathfrak{M}$ of bounded operators in $\mathfrak{H}$ is semi-simple.

PROOF: It is well known that $A^*A = 0$ implies $A = 0$: If $A^*A = 0$, then $||Af||^2 = (Af, Af) = (A^*Af, f) = 0$, $Af = 0$ for every $f \in \mathfrak{H}$, whence $A = 0$. Therefore $A \neq 0$ implies $A^*A \neq 0$; moreover $A \neq 0$ implies $A^* \neq 0$, $AA^* \neq 0$. If A is Hermitian, this implication becomes $A \neq 0$ implies $A^2 \neq 0$, hence $A^{2m} = (\cdots (A^2)^2 \cdots)^2 \neq 0$, and $A^n \neq 0$ for $n \leqq 2^m$. Thus if A is Hermitian, $A \neq 0$ implies $A^n \neq 0$ for every $n \geqq 1$. Since AA^* is Hermitian for every A, $A \neq 0$ implies $AA^* \neq 0$, $(AA^*)^n \neq 0$. Now let $\mathfrak{a}$ be a right ideal $\neq (0)$ in $\mathfrak{M}$. Choose $A \in \mathfrak{a}$ with $A \neq 0$. Then $AA^* \in \mathfrak{a}$, $(AA^*)^n \in \mathfrak{a}^n$, $(AA^*) \neq 0$, whence $\mathfrak{a}^n \neq (0)$, and $\mathfrak{a}$ is not nilpotent. Hence $\mathfrak{M}$ is semi-simple.

(II) A necessary and sufficient condition that a ring $\mathfrak{M}$ of bounded operators in $\mathfrak{H}$ which contains 1 be regular is that $\mathfrak{M}$ possess a finite basis (with complex numbers as coefficients), i.e., that $\mathfrak{M}$ be a finite hypercomplex system over the division algebra of complex numbers (cf. V. d. W. II, page 149).

PROOF: The sufficiency is obvious, since if $\mathfrak{M}$ is a finite hypercomplex system, it satisfies the chain condition, is semi-simple by (I), and is therefore regular by Appendix 1 (6). The necessity will be proved in eight steps (i)—(viii). Suppose now that $\mathfrak{M}$ is regular.

(i) If $A \in \mathfrak{M}$, and $\mathfrak{A}$ is the range of A (in $\mathfrak{H}$), then $\mathfrak{A}$ is a closed set.

PROOF: There exists a bounded operator $X \in \mathfrak{M}$ with $AXA = A$, since $\mathfrak{M}$ is regular. Hence $AXAf = Af$ for every $f \in \mathfrak{H}$ i.e., $AXg = g$ for every $g \in \mathfrak{A}$. Conversely, $AXg = g$ implies $g \in \mathfrak{A}$ ($g = Af$ with $f = Xg$). Therefore $g \in \mathfrak{A}$ is equivalent to $AXg = g$, and hence $\mathfrak{A}$ is a closed set, since A, X are bounded.

A *projection* is a Hermitian idempotent, i.e., a bounded operator E with $E = E^* = E^2$.

(ii) An infinite sequence of projections $E_1, E, \cdots \in \mathfrak{M}$, with $E_i E_j = 0$ for $i \neq j$ and $E_i \neq 0$ for every i does not exist.

PROOF: Suppose such a sequence $E_1, E_2, \cdots$ does exist. For every $f \in \mathfrak{H}$ the elements $1/i\, E_i f$ $(i = 1, 2, \cdots)$ are mutually orthogonal, and

$$\sum_{i=1}^{\infty} \left\| \frac{1}{i} E_i f \right\|^2 = \sum_{i=1}^{\infty} \frac{1}{i^2} \| E_i f \|^2 \leqq \sum_{i=1}^{\infty} \| E_i f \|^2 \leqq \| f \|^2 < +\infty,$$

whence $\sum_{i=1}^{\infty} (1/i)\, E_i f$ converges, and $\|\sum_{i=1}^{\infty} (1/i)\, E_i f\|^2 = \sum_{i=1}^{\infty} \|(1/i)\, E_i f\|^2 \leqq \|f\|^2$. Thus

$$Af \equiv \sum_{i=1}^{\infty} \frac{1}{i} E_i f$$

defines a bounded operator A which is clearly in $\mathfrak{M}$. Now the range $\mathfrak{A}$ of A is obviously linear, and by (i) it is closed. Since $E_j f = A(jE_j f) \in \mathfrak{A}$, $\mathfrak{A}$ contains every $E_j f$, and hence every convergent sum $\sum_{j=1}^{\infty} E_j f_j$. Now since $E_j \neq 0$, there exists $\varphi_j = E_j f_j \neq 0$ such that $\|\varphi_j\| = 1/j$. The elements $\varphi_1, \varphi_2, \cdots$ are mutually orthogonal, and $\sum_{j=1}^{\infty} \|\varphi_j\|^2 = \sum_{j=1}^{\infty} 1/j^2 < +\infty$ therefore $\sum_{j=1}^{\infty} \varphi_j = \sum_{j=1}^{\infty} E_j f_j$ converges and is in $\mathfrak{A}$. Hence $\sum_{j=1}^{\infty} \varphi_j = Af = \sum_{j=1}^{\infty} 1/j\, E_j f$, and application of E_i yields $\varphi_i = 1/i\, E_i f$, whence $\| E_i f \| = i \|\varphi_i\| = 1$, $\|f\|^2 \geqq \sum_{i=1}^{\infty} \|E_i f\|^2 = \sum_{i=1}^{\infty} 1 = +\infty$, which is impossible.

(iii) An infinite sequence of projections $F_1, F_2, \cdots \in \mathfrak{M}$ with $F_1 < F_2 < \cdots$, or with $F_1 > F_2 > \cdots$, does not exist.

PROOF: If such a sequence $F_1, F_2, \cdots$ exists, the existence of the sequence $E_1, E_2, \cdots$ with $E_i = F_{i+1} - F_i$ or $E_i = F_i - F_{i+1}$ would contradict (ii).

(iv) $E \in \mathfrak{M}$ is *minimal* if it is a projection, if $E \neq 0$, and if $F \in \mathfrak{M}$, $F \leqq E$, $F \neq 0$ implies $F = E$. For every projection $E \in \mathfrak{M}$, $E \neq 0$, there exists a minimal projection $F \leqq E$.

PROOF: If no minimal $F \leqq E$ exists, then there is an infinite sequence $E = F_1 > F_2 > \cdots$ in $\mathfrak{M}$ contrary to (iii).

(v) There exists a finite number of minimal elements $E_1, \cdots, E_n$ with $E_i E_j = 0$ for $i \neq j$, $E_1 + \cdots + E_n = 1$.

PROOF: Suppose the statement to be false. Let $E_1, \cdots, E_m$ be minimal and $E_i E_j = 0$ for $i \neq j$. Then $E_1 + \cdots + E_m \neq 1$ by our hypothesis and by (iv) a minimal element $E_{m+1} \leqq 1 - E_1 - \cdots - E_m$ exists. Hence $E_i E_{m+1} = 0$ for $i \neq m+1$. Applying this result to $m = 0, 1, 2, \cdots$ successively, we construct an infinite sequence $E_1, E_2, \cdots$, with $E_i \in \mathfrak{M}$, $E_i \neq 0$ $(i = 1, 2, \cdots)$, and $E_i E_j = 0$ for $i \neq j$. This contradicts (ii).

The proofs of the remaining steps are taken from M. and N., pages **147—148.**

(vi) Let E be minimal. Then $A \in \mathfrak{M}$, $EA = AE = A$ implies the existence of a complex number α such that $A = \alpha E$.

PROOF: The elements A with $EA = AE = A$ form a ring $\mathfrak{M}_1$. A projection $F \in \mathfrak{M}_1$ would be in $\mathfrak{M}$ and $\leqq E$, and hence would be 0 or E. Since $\mathfrak{M}_1$ is generated by its projections (cf. J. v. Neumann, Math. Annalen, Vol. 102, 1929, page 392), $\mathfrak{M}_1$ consists of the elements αE only.

(vii) Let E, F be minimal. Then every $A \in \mathfrak{M}$ such that $EA = AF = A$ has the form $A = \alpha A_0$, where α is a complex number, and A_0 is a fixed element with $EA_0 = A_0 F = A_0$.

PROOF: If all the elements A (with the property $EA = AF = A$) are 0, define $A_0 \equiv 0$. If not, then choose any A_0 in the set which is not 0. If A, B are in the set, then $AB^* \in \mathfrak{M}: EAB^* = AB^*$, $AB^*E = A(EB)^* = AB^*$, and by (vi) $AB^* = \beta E$, where β is a fixed complex number. Similarly $A^*B = \gamma F$. So in particular $A_0^*A_0 = \gamma_0 F$, and $A_0 \neq 0$, $A_0^*A_0 \neq 0$ yields $\gamma_0 \neq 0$. Also $AA_0^* = \beta_0 E$, whence

$$A = AF = A \cdot \frac{1}{\gamma_0} A_0^* A_0 = \frac{1}{\gamma_0} \beta_0 E A_0 = \frac{\beta_0}{\gamma_0} A_0.$$

(viii) If $E_1, \cdots, E_m$ are minimal elements of $\mathfrak{M}$ with $E_1 + \cdots + E_m = 1$, $E_i E_j = 0$ for $i \neq j$ (existent by (v)), then $\mathfrak{M}$ contains at most m^2 linearly independent elements (complex numbers being the coefficients).

PROOF: If $A \in \mathfrak{M}$,

$$A = \left(\sum_{i=1}^{m} E_i\right) A \left(\sum_{j=1}^{m} E_j\right) = \sum_{i,j=1}^{m} E_i A E_j = \sum_{i,j=1}^{m} A_{ij},$$

where $A_{ij} = E_i A E_j$. Clearly $E_i A_{ij} = A_{ij} E_j = A_{ij}$, there exists by (vii) a fixed A_{ij}^0 for every $i, j = 1, \cdots, m$ such that $A_{ij} = \alpha_{ij} A_{ij}^0$ (α_{ij} being a complex number). Hence $A = \sum_{i,j=1}^{m} \alpha_{ij} A_{ij}^0$.

(III) Every right ideal $\mathfrak{a} \neq (0)$ in $\mathfrak{M}$ contains a projection $E \neq 0$.

PROOF: Since $\mathfrak{a} \neq (0)$, there exists $A \in \mathfrak{M}$ such that $A \neq 0$. Now there exist operators B, $U \in \mathfrak{M}$ with $A = BU$, $B = AU^*$, B Hermitian and definite, and U bounded; hence $B \in \mathfrak{a}$, $B \neq 0$. (For the spectrum-theoretical material which follows, the reader is referred to M. H. Stone, *Linear Transformations in Hilbert Space*, Amer. Math. Soc. Coll. Series, 1932.) If $E(\lambda)$ is the decomposition of unity of B, then $E(\lambda) = 0$ for $\lambda < 0$ since B is definite, and we cannot have $E(\lambda) = 1$ for every $\lambda > 0$, since $B \neq 0$. Select $\lambda_0 > 0$ such that $E(\lambda_0) \neq 1$ and define

$$f(\lambda) \equiv \begin{cases} 1 & (\lambda \geqq \lambda_0) \\ 0 & (\lambda < \lambda_0), \end{cases} \qquad g(\lambda) \equiv \begin{cases} 1/\lambda & (\lambda \geqq \lambda_0) \\ 0 & (\lambda < \lambda_0). \end{cases}$$

Then $Bg(B) = f(B) = 1 - E(\lambda_0)$, $g(B) \in \mathfrak{M}$, whence $1 - E(\lambda_0) \in \mathfrak{a}$. Thus $E = 1 - E(\lambda_0)$ meets our requirements.

This discussion shows that a ring $\mathfrak{M}$ of bounded operators is regular if and only if it possesses a finite basis — a case completely treated by the classical theory of finite hypercomplex number systems (the preceding discussion actually assumed that $\mathfrak{M}$ contains the operator 1, but it is not difficult to alter the discussion to avoid this assumption - Ed.). Nevertheless, it is always semi-simple by (I). By (III) it is even true that every right ideal $\mathfrak{a} \neq (0)$ in $\mathfrak{M}$ contains an idempotent $\neq 0$; however, if $\mathfrak{M}$ is not regular, not every $\mathfrak{a} \neq (0)$ can be generated by such an idempotent.

For semi-simple systems with chain condition *irreducibility* and *simplicity* are equivalent. For rings of operators the situation is as follows:

(IV) $\mathfrak{M}$ is *irreducible* — i.e., the only pair of inverse right and left ideals in $\mathfrak{M}$ is (0), $\mathfrak{M}$ — if and only if its center consists of the operators $\alpha 1$ only (α being any complex number), that is, if and only if $\mathfrak{M}$ is a *factor* in the terminology of M. and N. (page 138).

PROOF: Suppose $\mathfrak{M}$ is irreducible, and that $\mathfrak{M}$ is not a factor. Its center is a ring, which is thus not generated by the idempotents 0, 1. Hence it contains an idempotent $E \neq 0, 1$ (cf. the proof of (II), (vi)). The elements $A \in \mathfrak{M}$ with $AE = A$ form a right ideal, and since $AE = A$ means $EA = A$, they form a left ideal as well. Similarly the elements A with $EA = 0$ (i.e., $AE = 0$) form an ideal, and these two ideals so obtained are inverses. For $EA = A$, $EA = 0$ implies $A = 0$, and every $A \in \mathfrak{M}$ is of the form $A = A_1 + A_2$, $EA_1 = A_1$, $EA_2 = 0$, with $A_1 = EA$, $A_2 = (1 - E)A$. Since $E \neq 0, 1$, both ideals are not (0), inasmuch as they contain respectively the elements E and $1 - E$, both $\neq 0$. Hence $\mathfrak{M}$ is not irreducible, contrary to the hypotheses.

Suppose, conversely, that $\mathfrak{M}$ is a factor, and let $\mathfrak{a}$, $\mathfrak{b}$ be two inverse (right and left) ideals in $\mathfrak{M}$, both $\neq (0)$. By (III) there exist projections $E, F \in \mathfrak{M}$, both $\neq 0$, with $E \in \mathfrak{a}$, $F \in \mathfrak{b}$; moreover, there exist projections $E_1, F_1 \in \mathfrak{M}$, both $\neq 0$, with $E_1 \leqq E$, $F_1 \leqq F$, and such that $E_1 = UU^*$, $F_1 = U^*U$ with $U \in \mathfrak{M}$ (cf. M. and N., p. 153). Thus $E_1 = EE_1 \in \mathfrak{a}$, $F_1 = FF_1 \in \mathfrak{b}$, but since $U^*E_1U = U^*UU^*U = F_1F_1 = F_1$, $E_1 \in \mathfrak{a}$ implies $F_1 \in \mathfrak{a}$. Hence $F_1 = 0$, contrary to the property $F_1 \neq 0$. Therefore $\mathfrak{a}$ or $\mathfrak{b}$ is (0), and $\mathfrak{b}$ or $\mathfrak{a}$ is $\mathfrak{M}$, whence $\mathfrak{M}$ is irreducible.

(V) A factor $\mathfrak{M}$ is simple if and only if it belongs to one of the cases (I_n) $(n = 1, 2, \cdots)$, (II_1), (III_∞) (cf. M. and N., page 172).

PROOF: Let the factor $\mathfrak{M}$ be simple, and suppose the conclusion fails; then $\mathfrak{M}$ belongs to one of the cases (I_∞), (II_∞). Let $\mathfrak{a}$ be the set of all

$A \in \mathfrak{M}$ such that the closure of the range of A, [Range A], has a finite relative dimension in $\mathfrak{M}$ (M. and N., page 168). Since Range $A \supset$ Range AX, [Range A] $\supset$ [Range AX], whence $A \in \mathfrak{a}$ implies $AX \in \mathfrak{a}$. Now inasmuch as [Range A] and [Range A^*] have the same relative dimension (M. and N., page 162), we see that $A \in \mathfrak{a}$ implies $A^* \in \mathfrak{a}$. Hence $XA = (A^*X^*) \in \mathfrak{a}$, if $A \in \mathfrak{a}$. Thus $\mathfrak{a}$ is a right and left ideal. Now [Range 1] $= \mathfrak{H}$ has infinite relative dimension, whence $1 \notin \mathfrak{M}$, and there exists a projection $E \in \mathfrak{M}$, $E \neq 0$, for which [Range E] has a finite dimension, so that $E \in \mathfrak{a}$. (These facts follow from the definition of cases (I_∞) and (II_∞), for which see loc. cit.) Consequently, $\mathfrak{a} \neq 0$, $\mathfrak{a} \neq \mathfrak{M}$, and $\mathfrak{M}$ is not simple.

To prove the converse, suppose that $\mathfrak{M}$ belongs to one of the cases (I_n), (II_1), (III_∞) and that $\mathfrak{M}$ is not simple. Select a right and left ideal $\mathfrak{a} \neq (0)$, $\mathfrak{M}$ in $\mathfrak{M}$. By (III) a projection $E \in \mathfrak{a}$, $E \neq 0$ exists. Let D be the relative dimension function in $\mathfrak{M}$, and define $\varepsilon \equiv D(E) > 0$. If $F \in \mathfrak{M}$, $D(F) \leqq D(E)$, then there exists E_1 with $F \sim E_1 \leqq E$, i.e., $F = U^*U$, $E_1 = UU^*$, $U \in \mathfrak{M}$ (M. and N., page 168). Now $E \in \mathfrak{a}$ yields $E_1 = EE_1 \in \mathfrak{a}$, and since $U^*E_1U = U^*UU^*U = FF = F$, also $F \in \mathfrak{a}$.

In case (III_∞) necessarily $D(E) = +\infty$; hence $D(E) = D(1)$, and $1 \in \mathfrak{a}$, $\mathfrak{a} = \mathfrak{M}$, contrary to our assumptions.

In cases (I_n) $(n = 1, 2, \cdots)$ $D(E) \geqq 1/n$, and there exists $F \in \mathfrak{M}$ with $D(F) = 1/n$. In case (II_1) we choose n so that $D(E) \geqq 1/n$, and $F \in \mathfrak{M}$, so that $D(F) = 1/n$. Now there exist n projections $F_1, \cdots, F_n$ with $F_1 + \cdots + F_n = 1$, $D(F_1) = \cdots = D(F_n) = 1/n$ in both cases. Since $D(F_i) = 1/n \leqq D(E)$, we have $F_i \in \mathfrak{a}$, whence $1 = F_1 + \cdots + F_n \in \mathfrak{a}$, $\mathfrak{a} = \mathfrak{M}$, again in contradiction to our assumptions. This completes the proof.

Remark: If we are merely interested in establishing that the ring B of all bounded operators in $\mathfrak{H}$ (class I_∞) is irreducible but not simple, then the above considerations become elementary, since our "relative dimension with respect to $\mathfrak{M}$" becomes common dimension. Other right and left ideals $\neq (0)$, B in B are these: all totally continuous operators, all operators of the E. Schmidt class. In view of (II), the only regular factors are those in cases (I_n), $n = 1, 2, \cdots$.

An operatorial example of a regular ring is the following:

(VI) Let $\mathfrak{M}$ be a factor of class (II_1), and $\mathcal{U}(\mathfrak{M})$ the set of all (not necessarily bounded) operators which are linear, closed, have a domain dense in $\mathfrak{H}$ and which "belong" to $\mathfrak{M}$ (Cf. M. and N., page 229. In considering $\mathcal{U}(\mathfrak{M})$ as a ring, the sum and the product of $A, B \in \mathcal{U}(\mathfrak{M})$ are to be defined not as the usual $A + B$, AB but as their closures $[A + B]$, $[AB]$. Cf. loc. cit.) Then $\mathcal{U}(\mathfrak{M})$ is a regular ring.

PROOF: If $A \in \mathcal{U}(\mathfrak{M})$, then $A = BU$, $B = AU^*$, where B, $U \in \mathcal{U}(\mathfrak{M})$, B is Hermitian and definite (M. and N., page 229). Define

$$f(\lambda) \equiv \begin{cases} 1 & (\lambda \neq 0) \\ 0 & (\lambda = 0), \end{cases} \qquad g(\lambda) \equiv \begin{cases} 1/\lambda & (\lambda \neq 0) \\ 0 & (\lambda = 0). \end{cases}$$

Then $f(B)$, $g(B) \in \mathcal{U}(\mathfrak{M})$. Since $\lambda f(\lambda) = \lambda$ for every λ, $[Bf(B)] = B$, and since $g(\lambda)\lambda = f(\lambda)$ for every λ, $[g(B)B] = f(B)$. Define $X \equiv [U^*g(B)] \in \mathcal{U}(\mathfrak{M})$ Then we have

$$[AXA] = [A \cdot U^*g(B) \cdot BU] = [AU^* \cdot g(B)B \cdot U] = [B \cdot f(B) \cdot U] = [BU] = A.$$

(Note that we omit all intermediate closure-brackets $[\cdot \cdot \cdot]$ when forming products of three or more factors, cf. loc. cit. above. Since $BU = A$ is closed, $[BU] = A$.) Hence $\mathcal{U}(\mathfrak{M})$ is regular.

This discussion shows that in order to make a non-trivial ring (i.e., one not in class (I_n), $n = 1, 2, \cdots$) of operators regular, it is necessary to add unbounded operators to it. Since this is possible in factors of class (II_1), but in no others (cf. M. and N., pages 122—123), only the former ones lead to an essentially new type of regular rings which do not satisfy the chain condition.

The continuous geometries are, as we shall see in the chapters which follow, intrinsically connected with regular rings without chain condition. They are a very wide generalization of the $\mathcal{U}(\mathfrak{M})$ of factors $\mathfrak{M}$ of class (II_1), the role of complex numbers being played by arbitrary division algebras.

APPENDIX 3

We shall discuss here the structure of right ideals in a matrix ring $\mathfrak{R}_n$ ($\mathfrak{R}$ being a regular ring) somewhat more fully than was necessary in order to prove Theorem 2.13.

(I) The right ideals $\mathfrak{a}$ in $\mathfrak{R}_n$ and the right linear sets $\mathfrak{A}$ of n-dimensional vectors $(x_1, \cdots, x_n)$ with $x_i \in \mathfrak{R}$ (cf. Definition 2.8) are in a one-to-one correspondence, corresponding elements being linked by the following relations:

(13) $(x_{ij}) \in \mathfrak{a}$ if and only if $(x_{1j}, \cdots, x_{nj}) \in \mathfrak{A}$ for $j = 1, \cdots, n$;

(14) $(x_1, \cdots, x_n) \in \mathfrak{A}$ if and only if there exists $(x_{ij}) \in \mathfrak{a}$ with $x_{i1} = x_i$ $(i = 1, \cdots, n)$.

PROOF: We may make the proof by employing steps (A)—(E) of Chapter I, in which the division algebra $\mathcal{D}$ is to be replaced by the ring $\mathfrak{R}$. (Property (F) does not hold here, since in its proof use was made of the fact that $\mathcal{D}$ is a division algebra.)

NOTE: This is a generalization of Lemma 2.12, in which the same property was stated for principal right ideals.

(II) If the right ideals $\mathfrak{a}$, $\mathfrak{b}$ in $\mathfrak{R}_n$ correspond respectively to the sets $\mathfrak{A}$, $\mathfrak{B}$ of n-dimensional vectors, then $\mathfrak{a} \cap \mathfrak{b}$ corresponds to $\mathfrak{A} \cap \mathfrak{B}$ and $\mathfrak{a} \cup \mathfrak{b}$ corresponds to $\mathfrak{A} \cup \mathfrak{B}$.

PROOF: Both parts are immediately verified.

NOTE: This is a generalization of Lemma 2.13.

(III) The set $\mathfrak{A}$ corresponds to a principal right ideal $\mathfrak{a}$ if and only if there exists (a_{ij}) $(i, j = 1, \cdots, n)$ such that $\mathfrak{A}$ is the set of all vectors $(x_1, \cdots, x_n)$ for which

$$x_i = \sum_{j=1}^{n} a_{ij} u_j, \qquad (u_1, \cdots, u_n \in \mathfrak{R}).$$

PROOF: The proof is the same as that of Lemma 2.12, (15), with $\mathfrak{a} = ((a_{ij}))_r$.

(IV) The set $\mathfrak{A}$ corresponds to a principal right ideal $\mathfrak{a}$ if and only if there exists an integer m and a system a_{ij} $(i = 1, \cdots, n;\ j = 1, \cdots, m)$ such that $\mathfrak{A}$ is the set of all vectors $(x_1, \cdots, x_n)$ for which

$$x_i = \sum_{j=1}^{m} a_{ij} u_j, \qquad (u_1, \cdots, u_m \in \mathfrak{R}).$$

PROOF: The necessity of the condition is clear by (III) with $m = n$. To prove the sufficiency, suppose first that $m = 1$. Then $x_i = a_{i1} u_1$, with u_1 arbitrary; this is a special case of (III) with $a_{i2}, \cdots, a_{in}$ all zero. Hence for the case $m = 1$, $\mathfrak{A}$ corresponds to a principal right ideal. Now let m be any positive integer. Then define $\mathfrak{A}_k$ as the linear set represented by $x_i = a_{ik} u_k$, with u_k the parameter. The corresponding right ideal $\mathfrak{a}_k$ is principal by our consideration of the case $m=1$. Now $\mathfrak{A} = \mathfrak{A}_1 \cup \cdots \cup \mathfrak{A}_m$, whence $\mathfrak{a} = \mathfrak{a}_1 \cup \cdots \cup \mathfrak{a}_m$ by (II), and $\mathfrak{a}$ is principal by Theorem 2.3.

Clearly properties (I)—(IV) hold also for left ideals, provided that the roles of "right" and "left" in $\mathfrak{R}_n$ be interchanged. This interchange implies the interchange of roles of "right" and "left" in $\mathfrak{R}$, and of the indices i and j in the matrices (x_{ij}) in $\mathfrak{R}_n$, i.e., of the rows and columns of these matrices. (The latter interchange is necessitated by the definition of multiplication of matrices.) Thus (I) will characterize the left ideals $\mathfrak{b}$ of $\mathfrak{R}_n$ by means of left linear sets of n-dimensional vectors (defined by Definition 2.8 with the roles of "left" and "right" interchanged). The details of this discussion will be omitted.

(V) If $\mathfrak{b}$ is a left ideal in $\mathfrak{R}_n$, corresponding to the set $\mathfrak{B}$ of n-dimensional vectors, then the right ideal $\mathfrak{a} = \mathfrak{b}^r$ corresponds to the set $\mathfrak{A}$ of n-dimensional vectors consisting of all vectors $(x_1, \cdots, x_n)$ such that

$$\sum_{i=1}^{n} y_i x_i = 0$$

for every $(y_1, \cdots, y_n) \in \mathfrak{B}$.

Proof: Clearly $(x_{ij}) \in \mathfrak{a} = \mathfrak{b}^r$ means $(y_{ij})(x_{ij}) = 0$, i.e. $\sum_{k=1}^{n} y_{ik} x_{kj} = 0$ $(i, j = 1, \cdots, n)$ for every $(y_{ij}) \in \mathfrak{b}$. The latter condition means that $(y_{i1}, \cdots, y_{in}) \in \mathfrak{B}$ $(i = 1, \cdots, n)$. Hence $(x_{ij}) \in \mathfrak{a}$ if and only if $\sum_{k=1}^{n} y_k x_{kj} = 0$ for every $(y_1, \cdots, y_n) \in \mathfrak{B}$. In view of the definition of $\mathfrak{A}$, this means that $(x_1, \cdots, x_n) \in \mathfrak{A}$ is equivalent to $\sum_{k=1}^{n} y_k x_k = 0$ for every $(y_1, \cdots, y_n) \in \mathfrak{B}$.

We may now prove a "dual" to (IV):

(VI) $\mathfrak{A}$ corresponds to a principal right ideal $\mathfrak{a}$ if and only if there exist $m = 1, 2, \cdots$ and a_{ij} $(i = 1, \cdots, n; j = 1, \cdots, m)$ such that $(x_1, \cdots, x_n) \in \mathfrak{A}$ is equivalent to

$$\sum_{i=1}^{n} a_{ij} x_i = 0 \qquad (j = 1, \cdots, m).$$

Proof: $\mathfrak{a}$ is a principal right ideal if and only if there exists a principal left ideal $\mathfrak{b}$ such that $\mathfrak{a} = \mathfrak{b}^r$. Let $\mathfrak{B}$ correspond to $\mathfrak{b}$. By (V), $(x_1, \cdots, x_n) \in \mathfrak{A}$ is equivalent to $\sum_{i=1}^{n} y_i x_i = 0$ for every $(y_1, \cdots, y_n) \in \mathfrak{B}$. By the right-left symmetric result to (IV), $(y_1, \cdots, y_n) \in \mathfrak{B}$ means

$$y_i = \sum_{j=1}^{m} u_j a_{ij} \qquad (u_1, \cdots, u_m \in \mathfrak{R})$$

for a suitable integer $m \geqq 1$ and suitable a_{ij} $(i = 1, \cdots, n;$ $j = 1, \cdots, m)$. Hence $(x_1, \cdots, x_n) \in \mathfrak{A}$ means $\sum_{i=1}^{n} (\sum_{j=1}^{m} u_j a_{ij}) x_i = 0$, i.e., $\sum_{j=1}^{m} u_j (\sum_{i=1}^{n} a_{ij} x_i) = 0$ for all values of $u_1, \cdots, u_m$, i.e., $\sum_{i=1}^{n} a_{ij} x_i = 0$ $(j = 1, \cdots, m)$.

Note: By Lemma 2.11 we may specialize m, a_{ij} in (IV) without any loss of generality thus: $m = n$, $a_{ii} = e_i$ idempotent, $a_{ij} = 0$ for $i < j$, $e_i a_{ij} = 0$, $a_{ij} e_j = a_{ij}$ for $i > j$. Hence the same specialization of m, a_{ij} is permissible in (VI) without loss of generality.

Properties (IV), (VI) and the note following (VI) show how the principal right ideals in $\mathfrak{R}_n$ are characterized by the possibility of representing them parametrically or by a finite number of linear equations in a space of n-dimensional vectors $(x_1, \cdots, x_n)$, $(x_i \in \mathfrak{R})$. This representation will be important in succeeding chapters.

PART II · CHAPTER III

Order of a Lattice and of a Regular Ring

In this and succeeding chapters we shall frequently be concerned with a given complemented, modular lattice, which we shall denote by L. The elements of L will be denoted by $\mathfrak{a}$, $\mathfrak{b}$, $\mathfrak{c}$, etc., the meet by $\cap$, and the join by $\cup$. Since many of the results secured in Part I were obtained without requiring Axiom III, Axiom VI, or Axiom II for infinite sets, we shall use such results freely without repeating definitions, theorems and discussions given there. Hence we shall presuppose the algebraic material contained in Chapters I, II, III, V of Part I.

DEFINITION 3.1: Let L be a complemented, modular lattice with zero 0 and unit 1. By a *basis* for L is meant a system $(\mathfrak{a}_i;\ i = 1, \cdots, n)$ of n elements of L such that

$$(1) \qquad (\mathfrak{a}_i;\ i = 1, \cdots, n) \perp, \quad \mathfrak{a}_1 \cup \cdots \cup \mathfrak{a}_n = 1.$$

A basis is *homogeneous* if its elements are pairwise perspective:

$$(2) \qquad \mathfrak{a}_i \sim \mathfrak{a}_j \quad (i, j = 1, \cdots, n).$$

The number n of elements in a basis is called the *order* of the basis.

DEFINITION 3.2: A complemented, modular lattice L is said to have *order m* in case it has a homogeneous basis of order m.

NOTE: There is of course no implication that the order m of a lattice is unique. For example, three-dimensional projective geometry has orders 1, 2, 4, the bases being in the respective cases the entire space 1; any two skew lines, and any four non-coplanar points; any continuous geometry L_∞ has order m for every $m = 1, 2, \cdots$.

We shall now determine all bases and all homogeneous bases of L, when L is the lattice $\bar{R}_{\mathfrak{R}}$ associated with a regular ring $\mathfrak{R}$.

LEMMA 3.1: If idempotents $e_1, \cdots, e_n$ are *independent*, i.e., if they satisfy

$$(3) \qquad e_i e_j = \begin{cases} e_i & (i = j) \\ 0 & (i \neq j), \end{cases}$$

then

$$(e_{i_1})_r \cup \cdots \cup (e_{i_\nu})_r = (e_{i_1} + \cdots + e_{i_\nu})_r$$

for every set $i_1, \cdots, i_\nu$ of distinct integers on the range $1, \cdots, n$.

PROOF: Clearly if $\rho = 1, \cdots, \nu$, and if $e = e_{i_1} + \cdots + e_{i_\nu}$, then

$$ee_{i_\rho} = e_{i_\rho}e = e_{i_\rho},$$

whence $e^2 = e\sum_{\rho=1}^{\nu} e_{i\rho} = \sum_{\rho=1}^{\nu} e_{i\rho} = e$, and e is idempotent. Since $e \in (e_{i_1})_r \cup \cdots \cup (e_{i_\nu})_r$, $(e)_r \subset (e_{i_1})_r \cup \cdots \cup (e_{i_\nu})_r$. But $e_{i_\rho} = ee_{i_\rho}$, and $e_{i\rho} \in (e)_r$. Thus $(e_{i\rho})_r \subset (e)_r$, whence $(e_{i_1})_r \cup \cdots \cup (e_{i_\nu})_r \subset (e)_r$. Consequently, $(e_{i_1})_r \cup \cdots \cup (e_{i_\nu})_r = (e)_r$.

LEMMA 3.2: Let $\mathfrak{R}$ be a regular ring and $L = \bar{R}_{\mathfrak{R}}$ the lattice of principal right ideals in $\mathfrak{R}$. The system $(\mathfrak{a}_i;\ i = 1, \cdots, n)$ is independent if and only if there exist n idempotents $e_1, \cdots, e_n$ with

$$(3) \qquad e_i e_j = \begin{cases} e_i & (i = j) \\ 0 & (i \neq j), \end{cases}$$

and such that $\mathfrak{a}_i = (e_i)_r$.

PROOF: The forward implication is proved by induction on n. Clearly for $n = 1$ the result is trivial. Suppose it to hold for $n (n \geqq 1)$, and consider a system $(\mathfrak{a}_1, \cdots, \mathfrak{a}_{n+1})\perp$. Now

$$(\mathfrak{a}_1 \cup \cdots \cup \mathfrak{a}_n) \cap \mathfrak{a}_{n+1} = (0),$$

whence there is an inverse $\mathfrak{b}$ of $\mathfrak{a}_{n+1}$ such that $\mathfrak{b} \geqq \mathfrak{a}_1 \cup \cdots \cup \mathfrak{a}_n$ (Part I, Corollary to Theorem 1.4(i)). Since $(\mathfrak{a}_1, \cdots, \mathfrak{a}_n)\perp$, there exist by the induction hypothesis n idempotents e'_i satisfying (3) and such that $\mathfrak{a}_i = (e'_i)_r$ $(i = 1, \cdots, n)$. Moreover, by Theorem 2.1 there exists an idempotent e with $\mathfrak{b} = (e)_r$, $\mathfrak{a}_{n+1} = (1 - e)_r$. Define

$$e_i \equiv e'_i e \quad (i = 1, \cdots, n), \quad e_{n+1} \equiv 1 - e;$$

we shall show that the e_i $(i = 1, \cdots, n+1)$ satisfy the conditions of the lemma. Since $e'_i \in \mathfrak{a}_i$, we have $e'_i \in \mathfrak{b} = (e)_r$, whence $ee'_i = e'_i$ $(i = 1, \cdots, n)$. Thus

$$e_i^2 = (e'_i e)(e'_i e) = e'_i(ee'_i)e = e'_i e'_i e = e'_i e = e_i,$$

and e_i is idempotent for $i = 1, \cdots, n$; obviously e_{n+1} is idempotent. Now $e_i = e'_i e \in (e'_i)_r$; also $e_i e'_i = (e'_i e)e'_i = e'^2_i = e'_i$, whence $e'_i \in (e_i)_r$ $(i = 1, \cdots, n)$. Consequently $\mathfrak{a}_i = (e'_i)_r = (e_i)_r$ for $i = 1, \cdots, n$: we had $\mathfrak{a}_{n+1} = e(_{n+1})_r$. Finally, for $i, j = 1, \cdots, n$, $i \neq j$,

$$e_i e_j = (e'_i e)(e'_j e) = e'_i(ee'_j)e = e'_i e'_j e = 0;$$

$e_i e_{n+1} = e'_i e(1 - e) = 0$, $e_{n+1}e_i = (1 - e)e'_i e = e'_i e - ee'_i e = e'_i e - e'_i e = 0$ $(i = 1, \cdots, n)$. This establishes (3), and hence also the desired result for $n + 1$. By induction the lemma holds for every n.

To prove the converse, let $\mathfrak{a}_i = (e_i)_r$, where the elements e_i satisfy (3).

Let k be any integer on the range $1, \cdots, n-1$. Evidently $e_1 + \cdots + e_{k+1}$ is an idempotent e, whence by Lemma 3.1,

$$(e_1)_r \cup \cdots \cup (e_k)_r = (e_1 + \cdots + e_k)_r = (e - e_{k+1})_r,$$

and $e - e_{k+1}$ is, of course, idempotent. Now let $x \in (e - e_{k+1})_r \cap (e_{k+1})_r$; then $x = ex - e_{k+1}x = e_{k+1}x$, whence

$$\begin{aligned} x = (e - e_{k+1})e_{k+1}x &= (ee_{k+1} - e_{k+1})x \\ &= (e_{k+1} - e_{k+1})x \qquad \text{(by (3))} \\ &= 0. \end{aligned}$$

Thus we have proved

$$(\mathfrak{a}_1 \cup \cdots \cup \mathfrak{a}_k) \cap \mathfrak{a}_{k+1} = (0) \qquad (k = 1, \cdots, n-1),$$

from which it follows that $(\mathfrak{a}_i;\ i = 1, \cdots, n)\perp$ by Part I, Theorem 2.2.

Corollary: The condition $\mathfrak{a}_1 \cup \cdots \cup \mathfrak{a}_n = 1$ is equivalent to $e_1 + \cdots + e_n = 1$.

Proof: By Lemma 2.1 and $\mathfrak{R} = (1)_r$, $\mathfrak{a}_1 \cup \cdots \cup \mathfrak{a}_n = \mathfrak{R}$ is equivalent to $(e_1 + \cdots + e_n)_r = (1)_r$, i.e., to $1 \in (e_1 + \cdots + e_n)_r$, whence to $1 = (e_1 + \cdots + e_n) \cdot 1 = e_1 + \cdots + e_n$.

Theorem 3.1: If $\mathfrak{R}$ is a regular ring, then $(\mathfrak{a}_1, \cdots, \mathfrak{a}_n)$ is a basis for the lattice $L = \bar{R}_{\mathfrak{R}}$ of principal right ideals of $\mathfrak{R}$ if and only if there exist idempotents $e_1, \cdots, e_n$ with $\mathfrak{a}_i = (e_i)_r$, and such that

$$(4) \qquad e_ie_j = \begin{cases} e_i & (i = j) \\ 0 & (i \neq j), \end{cases} \qquad e_1 + \cdots + e_n = 1.$$

Proof: This is evident from Lemma 3.2 and its Corollary.

Lemma 3.3: If in a modular lattice L, $\mathfrak{b}$, $\mathfrak{c}$ are both inverses of a given element $\mathfrak{a}$ such that $\mathfrak{b} \lesseqgtr \mathfrak{c}$, then $\mathfrak{b} = \mathfrak{c}$.

Proof: By symmetry, we may assume $\mathfrak{b} \leq \mathfrak{c}$. Then

$$\begin{aligned} \mathfrak{c} = (\mathfrak{b} \cup \mathfrak{a}) \cap \mathfrak{c} &= \mathfrak{b} \cup (\mathfrak{a} \cap \mathfrak{c}) \qquad \text{(by IV)} \\ &= \mathfrak{b} \cup 0 = \mathfrak{b}. \end{aligned}$$

Definition 3.3: If $\mathfrak{a}_i$, $\mathfrak{a}_j$ are two elements in a complemented modular lattice L such that $\mathfrak{a}_i \cap \mathfrak{a}_j = 0$, we define L_{ij} as the class of all inverses $\mathfrak{b}_{ij}$ of $\mathfrak{a}_j$ in $\mathfrak{a}_i \cup \mathfrak{a}_j$.

Lemma 3.4: Let $L = \bar{R}_{\mathfrak{R}}$ be the lattice of principal right ideals of a regular ring $\mathfrak{R}$. Let $\mathfrak{a}_i, \mathfrak{a}_j \in L$ with $\mathfrak{a}_i \cap \mathfrak{a}_j = (0)$, and let e_i, e_j be idempotents such that $\mathfrak{a}_i = (e_i)_r$, $\mathfrak{a}_j = (e_j)_r$, $e_je_i = 0$ (cf. Lemma 3.2). Then the set L_{ij} is precisely the set of all $(e_i - v)_r$, with $v \in \mathfrak{R}$, and

$$(5) \qquad e_jv = ve_i = v.$$

Moreover, each inverse $\mathfrak{b}_{ij}$ determines uniquely the element $v \,\epsilon\, \mathfrak{R}$ associated with it.

PROOF: Let $\mathfrak{b}_{ij} \,\epsilon\, L_{ij}$. Since $e_i \,\epsilon\, \mathfrak{a}_i \subset \mathfrak{a}_j \cup \mathfrak{b}_{ij}$, there exist $v \,\epsilon\, \mathfrak{a}_j$, $s \,\epsilon\, \mathfrak{b}_{ji}$ with $e_i = v + s$. Now $(s)_r \subset \mathfrak{b}_{ij}$. But

$$\mathfrak{a}_j \cap (s)_r \subset \mathfrak{a}_j \cup \mathfrak{b}_{ij} = \mathfrak{a}_i \cup \mathfrak{a}_j;$$

also $\mathfrak{a}_j \cup (s)_r \supset \mathfrak{a}_j$. Since $v + s = e_i \,\epsilon\, \mathfrak{a}_j \cup (s)_r$, whence $\mathfrak{a}_i = (e_i)_r \subset \mathfrak{a}_j \cup (s)_r$, we have $\mathfrak{a}_j \cup (s)_r \supset \mathfrak{a}_i \cup \mathfrak{a}_j$. Hence $\mathfrak{a}_j \cup (s)_r = \mathfrak{a}_i \cup \mathfrak{a}_j$. Since $\mathfrak{a}_j \cap (s)_r = (0)$, we see that $(s)_r \subset \mathfrak{b}_{ij}$ is an inverse of $\mathfrak{a}_j$ in $\mathfrak{a}_i \cup \mathfrak{a}_j$; by Lemma 3.3 $\mathfrak{b}_{ij} = (s)_r = (e_i - v)_r$. Inasmuch as $v \,\epsilon\, \mathfrak{a}_j$, $e_j v = v$. Since $e_i - v \,\epsilon\, \mathfrak{b}_{ij}$, $(e_i - v)(1 - e_i) \,\epsilon\, \mathfrak{b}_{ij}$; but

$$(e_i - v)(1 - e_i) = -v(1 - e_i) \,\epsilon\, \mathfrak{a}_j,$$

whence $-v(1 - e_i) = 0$, and $v = v e_i$. Thus (5) is established.

Conversely, let us consider $\mathfrak{b}_{ij} = (e_i - v)_r$, where $v \,\epsilon\, \mathfrak{R}$ satisfies (5); we must show that $\mathfrak{b}_{ij} \,\epsilon\, L_{ij}$. Suppose $x \,\epsilon\, \mathfrak{a}_j \cap \mathfrak{b}_{ij}$. Then $x = e_j x$, $x = (e_i - v)z$, with $z \,\epsilon\, \mathfrak{R}$. Hence $x = e_j(e_i - v)z = -e_j v z = -vz$, whence $x = (e_i - v)z = -vz$, $e_i z = 0$, and therefore $x = -vz = -v e_i \cdot z = -v \cdot e_i z = 0$. Thus $\mathfrak{a}_j \cap \mathfrak{b}_{ij} = (0)$. Now since e_j, $v \,\epsilon\, \mathfrak{a}_j$, $e_i \,\epsilon\, \mathfrak{a}_i$,

$$\mathfrak{a}_j \cup \mathfrak{b}_{ij} = (e_j)_r \cup (e_i - v)_r \subset \mathfrak{a}_i \cup \mathfrak{a}_j.$$

But $(e_j)_r \cup (e_i - v)_r \supset \mathfrak{a}_j$, and since $z = e_i z = vz + (e_i - v)z \,\epsilon\, (e_j)_r \cup (e_i - v)$, for every $z \,\epsilon\, \mathfrak{a}_i$, we have $(e_j)_r \cup (e_i - v)_r \supset \mathfrak{a}_i$. Thus $\mathfrak{a}_j \cup \mathfrak{b}_{ij} = (e_j)_r \cup (e_i - v)_r = \mathfrak{a}_i \cup \mathfrak{a}_j$, and $\mathfrak{b}_{ij} \,\epsilon\, L_{ij}$.

Finally, let v, v' satisfy (5) and suppose $(e_i - v)_r = (e_i - v')_r$. Then since $e_i - v$, $e_i - v' \,\epsilon\, \mathfrak{b}_{ij}$, it follows that

$$v - v' = (e_i - v') - (e_i - v) \,\epsilon\, \mathfrak{b}_{ij}.$$

But $v - v' \,\epsilon\, \mathfrak{a}_j$, since v, $v' \,\epsilon\, \mathfrak{a}_j$ by (5). Thus $v - v' = 0$, i.e., $v = v'$, and v is unique.

The lemma just proved establishes a one-to-one correspondence $\mathfrak{b}_{ij} = (e_i - v)_r$ between all elements $v \,\epsilon\, \mathfrak{R}$ subject to the conditions (5) and all inverses $\mathfrak{b}_{ij}$ of $\mathfrak{a}_j$ in $\mathfrak{a}_i \cup \mathfrak{a}_j$. The notations $\mathfrak{a}_i$, $\mathfrak{a}_j$ were used since in applying the lemma we shall be dealing with independent systems $(\mathfrak{a}_1, \cdots, \mathfrak{a}_n)$ of which $\mathfrak{a}_i$, $\mathfrak{a}_j$ are members.

DEFINITION 3.4: Three elements $\mathfrak{a}$, $\mathfrak{b}$, $\mathfrak{c}$ of a complemented modular lattice L are said to be in the relation $\mathscr{C}$ $((\mathfrak{a}, \mathfrak{b}, \mathfrak{c})\mathscr{C})$ in case

$$\mathfrak{a} \cup \mathfrak{b} = \mathfrak{b} \cup \mathfrak{c} = \mathfrak{c} \cup \mathfrak{a}, \quad \mathfrak{a} \cap \mathfrak{b} = \mathfrak{b} \cap \mathfrak{c} = \mathfrak{c} \cap \mathfrak{a} = 0,$$

i.e., in case each element is an axis of perspectivity of the other two in their join.

LEMMA 3.5: Let $L = \bar{R}_{\mathfrak{R}}$ be the lattice of principal right ideals of a regular ring $\mathfrak{R}$. Let $\mathfrak{a}_i, \mathfrak{a}_j \in L$ with $\mathfrak{a}_i \cap \mathfrak{a}_j = (0)$, $\mathfrak{a}_i \sim \mathfrak{a}_j$, and let e_i, e_j be idempotents such that $\mathfrak{a}_i = (e_i)_r$, $\mathfrak{a}_j = (e_j)_r$, $e_i e_j = e_j e_i = 0$ (cf. Lemma 3.2). Then each element $\mathfrak{c}_{ij} = \mathfrak{c}_{ji}$ such that $(\mathfrak{a}_i, \mathfrak{a}_j, \mathfrak{c}_{ij})\mathscr{C}$ can be represented in the form

$$\mathfrak{c}_{ij} = (e_i - v)_r = (e_j - u)_r \qquad (u, v \in \mathfrak{R}), \tag{6}$$

where

$$e_j v = v e_i = v, \quad e_i u = u e_j = u, \quad uv = e_i, \quad vu = e_j. \tag{7}$$

Conversely, if u, v satisfy (7), then there exists $\mathfrak{c}_{ij}$ satisfying (6), and $(\mathfrak{a}_i, \mathfrak{a}_j, \mathfrak{c}_{ij})\mathscr{C}$.

PROOF: If $(\mathfrak{a}_i, \mathfrak{a}_j, \mathfrak{c}_{ij})\mathscr{C}$, then by Lemma 3.4 there exist uniquely u, v such that $e_j v = v e_i = v$, $e_i u = u e_j = u$, $\mathfrak{c}_{ij} = (e_i - v)_r = (e_j - u)_r$. By the last condition, $e_i - v = (e_j - u)z$, $e_j - u = (e_i - v)w$, where $z, w \in \mathfrak{R}$. Then $e_j(e_i - v) = e_j(e_j - u)z$, whence $-v = e_j z$. Hence

$$e_i - v = (e_j - u)z = (e_j - u)e_j z = -(e_j - u)v = -v + uv,$$

and $e_i = uv$. Similarly $vu = e_j$.

Conversely, if $e_j v = v e_i = v$, $e_i u = u e_j = u$, then $(e_i - v)_r \in L_{ij}$, $(e_j - u)_r \in L_{ji}$. The equations $uv = e_i$, $vu = e_j$ then imply $e_i - v = -(e_j - u)v$, $e_j - u = -(e_i - v)u$, i.e., $(e_i - v)_r = (e_j - u)_r$. Hence we may write $\mathfrak{c}_{ij} = (e_i - v)_r = (e_j - u)_r$, and since $\mathfrak{c}_{ij} \in L_{ij}, L_{ji}$, $(\mathfrak{a}_i, \mathfrak{a}_j, \mathfrak{c}_{ij})\mathscr{C}$.

DEFINITION 3.5: A system of n^2 elements s_{ij} $(i, j = 1, \cdots, n)$ of a regular ring $\mathfrak{R}$ is called a system of *matrix units* in case

$$\text{(8)} \quad \begin{aligned} &\text{(a)} \quad s_{ij}s_{kh} = \begin{cases} s_{ih} & \text{if } j = k \\ 0 & \text{if } j \neq k, \end{cases} \\ &\text{(b)} \quad s_{11} + \cdots + s_{nn} = 1. \end{aligned}$$

COROLLARY: If elements s_{ij} $(i, j = 1, \cdots, n)$ are matrix units, then the s_{ii} $(i = 1, \cdots, n)$ are n independent idempotents.

PROOF: Clearly, by (8)(a), $s_{ii}s_{ii} = s_{ii}$ and $i \neq j$ implies $s_{ii}s_{jj} = 0$, whence the s_{ii} are independent idempotents.

LEMMA 3.6: Let $\bar{R}_{\mathfrak{R}}$ be the lattice of principal right ideals in a regular ring $\mathfrak{R}$. Then $(\mathfrak{a}_i; i = 1, \cdots, n)$ is a homogeneous basis for $\bar{R}_{\mathfrak{R}}$ if and only if there exist n^2 matrix units $s_{ij} \in \mathfrak{R}$ $(i, j = 1, \cdots, n)$ with $\mathfrak{a}_i = (s_{ii})_r$.

PROOF: Suppose that $(\mathfrak{a}_i; i = 1, \cdots, n)$ is a homogeneous basis, that is $(\mathfrak{a}_i; i = 1, \cdots, n)\perp$, $\mathfrak{a}_i \cup \cdots \cup \mathfrak{a}_n = \mathfrak{R}$, $\mathfrak{a}_i \sim \mathfrak{a}_j$ $(i, j = 1, \cdots, n)$. Let us observe at this point that the condition $\mathfrak{a}_i \sim \mathfrak{a}_j$ $(i, j = 1, \cdots, n)$ is equivalent to $\mathfrak{a}_1 \sim \mathfrak{a}_h$ $(h = 2, \cdots, n)$ since the relation $\sim$ is transitive in independent systems (Part I, Theorem 3.4). Now by Theorem 3.1

there exist n independent idempotents e_i with $\mathfrak{a}_i = (e_i)_r$, $e_1 + \cdots + e_n = 1$. By Lemma 3.5, as applied to $\mathfrak{a}_h \sim \mathfrak{a}_1$, there exist elements u_h, v_h $(h = 2, \cdots, n)$ such that, if we define $u_1 \equiv v_1 \equiv e_1$,

$$(9) \quad e_h u_h = u_h e_1 = u_h, \quad e_1 v_h = v_h e_h = v_h, \quad u_h v_h = e_h, \quad v_h u_h = e_1 \quad (h = 1, \cdots, n).$$

We define $s_{ij} \equiv u_i v_j$. Now if $j \neq k$,

$$s_{ij} s_{kh} = u_i v_j u_k v_h = u_i v_j e_j e_k u_k v_h = 0;$$

also

$$s_{ij} s_{jh} = u_i v_j u_j v_h = u_i e_1 v_h = u_i v_h = s_{ih},$$

whence (8)(a) is established. Clearly $s_{ii} = u_i v_i = e_i$ by (9), and (8)(b) is proved.

Conversely, let s_{ij} $(i, j = 1, \cdots, n)$ be n^2 elements satisfying (8) such that $\mathfrak{a}_i = (s_{ii})_r$. Define $e_i \equiv s_{ii}$, $u_i \equiv s_{i1}$, $v_i \equiv s_{1i}$ $(i = 1, \cdots, n)$. Then $i \neq j$ implies $e_i e_j = 0$, and $e_i e_i = e_i$ by virtue of (8)(a); hence the e_i are independent idempotents. Moreover, by (8)(b), $e_1 + \cdots + e_n = 1$. Thus by Theorem 3.1 we have the result that $(\mathfrak{a}_i;\ i = 1, \cdots, n)$ is a basis for $\bar{R}_{\mathfrak{R}}$. Now the relations (9) are readily verified; for example,

$$e_h u_h = s_{hh} s_{h1} = s_{h1} = u_h,$$

$$u_h v_h = s_{h1} s_{1h} = s_{hh} = e_h.$$

Thus by Lemma 3.5, $\mathfrak{a}_h \sim \mathfrak{a}_1$ $(h = 2, \cdots, n)$, whence, as we have observed, $\mathfrak{a}_i \sim \mathfrak{a}_j$ $(i, j = 1, \cdots, n)$. (Or, by directly applying Lemma 3.5 with $u = s_{ij}$, $v = s_{ji}$, $\mathfrak{c}_{ij}$ given by (6), i.e., $\mathfrak{c}_{ij} = (e_i - s_{ji})_r = (e_j - s_{ij})_r$, we have $\mathfrak{a}_i \sim \mathfrak{a}_j$ $(i, j = 1, \cdots, n)$.) Hence the basis $(\mathfrak{a}_i;\ i = 1, \cdots, n)$ is homogeneous.

NOTE: We make the following observation for later use. If we suppose a homogeneous basis to be given, then our matrix units s_{ij} are obtained with the help of the u_h, v_h $(h = 2, \cdots, n)$, which in turn originate from the application of Lemma 3.5 to $\mathfrak{a}_h, \mathfrak{a}_1$. For this purpose we select an axis of perspectivity $\mathfrak{c}_{h1}$ of $\mathfrak{a}_h, \mathfrak{a}_1$, and then u_h, v_h are determined by $\mathfrak{c}_{h1} = (e_h - v_h)_r = (e_1 - u_h)_r$. Since

$$s_{h1} = u_h v_1 = u_h e_1 = u_h, \quad s_{1h} = u_1 v_h = e_1 v_h = v_h,$$

we have $\mathfrak{c}_{h1} = (e_h - s_{1h})_r = (e_1 - s_{h1})_r$. It is to be observed that while the $\mathfrak{c}_{h1}$ could be prescribed arbitrarily, the axes of perspectivity $\mathfrak{c}_{ij} = (e_i - s_{ji})_r = (e_j - s_{ij})_r$ of $\mathfrak{a}_i, \mathfrak{a}_j$ $(i, j = 1, \cdots, n,\ i \neq j)$ mentioned at the end of the proof of Lemma 3.6 are determined by the s_{ij}, i.e., by the $\mathfrak{c}_{h1}$. We shall see later how the $\mathfrak{c}_{h1}$ determine (geometrically) the entire system $\mathfrak{c}_{ij}$.

THEOREM 3.2: The lattice $\bar{R}_{\mathfrak{R}}$, $\mathfrak{R}$ being a regular ring, is of order n if and only if there exist n^2 matrix units $s_{ij} \in \mathfrak{R}$ $(i, j = 1, \cdots, n)$.

PROOF: This is obvious by Lemma 3.6.

THEOREM 3.3: The lattice $\bar{R}_{\mathfrak{R}}$, $\mathfrak{R}$ being a regular ring, is of order n if and only if there exists a regular ring $\mathfrak{S}$ such that $\mathfrak{R}$ is isomorphic to the matrix ring $\mathfrak{S}_n$.

PROOF: We observe first that the converse implication is trivial since if $\mathfrak{R} = \mathfrak{S}_n$ the matrices $s_{kh} = (s^{ij}_{kh})$, where

$$s^{ij}_{kh} = \begin{cases} I & ((i, j) = (k, h)) \\ 0 & ((i, j) \neq (k, h)), \end{cases}$$

I being the unit of $\mathfrak{S}$, obviously satisfy (8). Suppose now that $\bar{R}_{\mathfrak{R}}$ is of order n, and let the s_{kh} be matrix units in $\mathfrak{R}$, existent by virtue of Theorem 3.2. Then $(\mathfrak{a}_i;\ i = 1, \cdots, n)$ with $\mathfrak{a}_i = (e_i)_r$, $e_i = s_{ii}$, is a homogeneous basis of $\bar{R}_{\mathfrak{R}}$ by Lemma 3.6. Now define $\mathfrak{S} \equiv \mathfrak{R}(e_1)$. We shall prove that the correspondence

$$x \rightleftarrows (x^{ij}) \qquad (x \in \mathfrak{R},\ (x^{ij}) \in \mathfrak{S}_n)$$

defined by the equations

$$(10) \qquad x^{ij} = s_{1i} x s_{j1} \qquad (i, j = 1, \cdots, n)$$

$$(11) \qquad x = \sum_{i, j=1}^{n} s_{i1} x^{ij} s_{1j}$$

is one-to-one and an isomorphism. Clearly (10) carries elements $x \in \mathfrak{R}$ into elements x^{ij} of $\mathfrak{R}(e_1)$, i.e., into matrices $(x^{ij}) \in \mathfrak{S}_n$, since

$$e_1 x^{ij} = e_1 s_{1i} x s_{j1} = s_{11} s_{1i} x s_{j1} = s_{1i} x s_{j1} = x^{ij},$$
$$x^{ij} e_1 = s_{1i} x s_{j1} e_1 = s_{1i} x s_{j1} s_{11} = s_{1i} x s_{j1} = x^{ij};$$

moreover (11) carries matrices of $\mathfrak{S}_n$ into elements of $\mathfrak{R}$. To prove the one-to-one character of the correspondence, we shall show that (10) and (11) are mutually inverse transformations. We have, in fact, for $x \in \mathfrak{R}$,

$$\begin{aligned} \sum_{i, j} s_{i1} s_{1i} x s_{j1} s_{1j} &= \sum_{i,j} s_{ii} x s_{jj} \\ &= \Big(\sum_i x_{ii}\Big) x \Big(\sum_j s_{jj}\Big) \\ &= x; \end{aligned}$$

for $(x^{ij}) \in \mathfrak{S}_n$ we have

$$s_{1i} \Big(\sum_{i', j'} s_{i'1} x^{i'j'} s_{1j'}\Big) s_{j1} = e_1 x^{ij} e_1 = x^{ij}.$$

This establishes the one-to-one feature of the correspondence. Equation (10) shows that addition in $\mathfrak{R}$ corresponds to addition of matrices in $\mathfrak{S}_n$. Finally, let $x \rightleftarrows (x^{ij})$, $y \rightleftarrows (y^{ij})$, i.e.,

$$x = \sum_{i,j} s_{i1} x^{ij} s_{1j}, \quad y = \sum_{i',j'} s_{i'1} y^{i'j'} s_{1j'}.$$

Then

$$\begin{aligned} xy &= \sum_{i,j,i',j'} s_{i1} x^{ij} s_{1j} s_{i'1} y^{i'j'} s_{1j'} \\ &= \sum_{i,j,j'} s_{i1} x^{ij} e_1 y^{jj'} s_{1j'} \\ &= \sum_{i,j'} s_{i1} \big(\sum_j x^{ij} y^{jj'}\big) s_{1j'} \\ &= \sum_{i,j} s_{i1} z^{ij} s_{1j}, \end{aligned}$$

where $z^{ij} = \sum_k x^{ik} y^{kj}$, i.e. $(z^{ij}) = (x^{ij})(y^{ij})$. Hence multiplication in $\mathfrak{R}$ corresponds to multiplication of matrices in $\mathfrak{S}_n$, and the isomorphism of $\mathfrak{R}$ and $\mathfrak{S}_n$ is established. By Theorem 2.11, $\mathfrak{S}$ is regular.

DEFINITION 3.6: A regular ring $\mathfrak{R}$ is said to be of order n in case the lattice $\bar{R}_{\mathfrak{R}}$ is of order n.

We have seen in Theorems 3.2, 3.3 that $\mathfrak{R}$ is of order n if and only if it possesses n^2 matrix units, i.e., if and only if it is isomorphic to a matrix ring of order n over another regular ring $\mathfrak{S}$. When $\mathfrak{R}$ is of order n we shall frequently identify it with $\mathfrak{S}_n$ and shall represent its elements as matrices with elements in $\mathfrak{S}$. Thus it is readily seen that the matrix units in $\mathfrak{R}$ (assuming $\mathfrak{R}$ to be of order n) are the matrices $s_{kh} = (s^{ij}_{kh})$ with

$$s^{ij}_{kh} = \begin{cases} I & ((i, j) = (k, h)) \\ 0 & ((i, j) \neq (k, h)), \end{cases}$$

where I is the unit of $\mathfrak{S}$. The homogeneous basis $(\mathfrak{a}_k;\ k = 1, \cdots, n)$ is then defined by $\mathfrak{a}_k = (e_k)_r$, $e_k = s_{kk}$. The elements $v \in \mathfrak{R}$ satisfying $e_h v = v e_k = v$, i.e., those which define an inverse $\mathfrak{b}_{kh} = (e_k - v)_r$ of $\mathfrak{a}_h$ in $\mathfrak{a}_k \cup \mathfrak{a}_h$ (in other words, $\mathfrak{b}_{kh} \in L_{kh}$ [cf. Lemma 3.4]), are now seen to be the matrices

$$v = v_{kh} = (v^{ij}_{kh}),$$

with $v^{ij}_{kh} = 0$ for $(i, j) \neq (h, k)$. These matrices correspond with the elements $\beta \in \mathfrak{S}$ by the equations

$$v^{ij}_{kh} = \begin{cases} \beta & (i, j) = (h, k) \\ 0 & (i, j) \neq (h, k). \end{cases} \tag{12}$$

Finally we determine the sets $\mathfrak{A}$ of n-dimensional vectors which correspond to certain principal right ideals $\mathfrak{a}$ in $\mathfrak{R} = \mathfrak{S}_n$ in the sense of Lemma 2.12. (This connection was analyzed closely in Appendix 3 to Chapter II.)

By Lemma 2.12 we have for $\mathfrak{a} = ((a_{ij}))_r$ a set $\mathfrak{A}$ which consists of all $x_1, \cdots, x_n$ with $x_i = \sum_{j=1}^n a_{ij} u_j$, with arbitrary $u_1, \cdots, u_n \in \mathfrak{S}$. If $\mathfrak{R}$ is of order n and s_{kh} $(k, h = 1, \cdots, n)$ are matrix units, we have for $\mathfrak{a} = \bar{\mathfrak{a}}_k = (s_{kk})_r$,

$$a_{ij} = s_{kk}^{ij} = \begin{cases} I & ((i, j) = (k, k)) \\ 0 & ((i, j) \neq (k, k)) \end{cases}, \quad \text{whence } x_i = \begin{cases} u_k & (i = k) \\ 0 & (i \neq k). \end{cases}$$

We see that $\bar{\mathfrak{c}}_{kh} = (s_{kk} - s_{hk})_r = (s_{hh} - s_{kh})_r$ is an axis of perspectivity of $\bar{\mathfrak{a}}_k$ and $\bar{\mathfrak{a}}_h$ (by combining Lemma 3.5 (6), with k, h instead of i, j, and with $e_k = s_{kk}$, $e_h = s_{hh}$, $u = s_{kh}$, $v = s_{hk}$, and the last remark in the proof of Lemma 3.6), and we may therefore use

$$a_{ij} = s_{kk}^{ij} - s_{hk}^{ij} = \begin{cases} I & ((i, j) = (k, k)) \\ -I & ((i, j) = (h, k)) \\ 0 & \text{(otherwise)}. \end{cases}$$

This yields

$$x_i = \begin{cases} u_k & (i = k) \\ -u_k & (i = h) \\ 0 & \text{(otherwise)}. \end{cases}$$

The general element of L_{kh} (cf. above) is $\mathfrak{b}_{kh} = (s_{kk} - v)_r$, with v as in (12), whence

$$a_{ij} = \begin{cases} I & (i, j) = (k, k) \\ -\beta & (i, j) = (h, k) \\ 0 & \text{(otherwise)}, \end{cases}$$

and

$$x_i = \begin{cases} u_k & (i = k) \\ -\beta u_k & (i = h) \\ 0 & \text{(otherwise)}. \end{cases}$$

We may summarize these results in the following manner:

α) The set $\mathfrak{A} = \mathfrak{A}_k$ of $\bar{\mathfrak{a}}_k$ is the set of all vectors

$$(0, \cdots, 0, \xi, 0, \cdots, 0);$$

β) The set $\mathfrak{A} = \mathfrak{A}_{kh}$ of $\bar{\mathfrak{c}}_{kh}$ is the set of all vectors

$$(0, \cdots, 0, \xi, 0, \cdots, 0, -\xi, 0, \cdots, 0).$$

γ) The set $\mathfrak{A} = \mathfrak{B}_{kh}$ of the general element $\mathfrak{b}_{kh} \in L_{kh}$ is the set of all vectors

$$(0, \cdots, 0, \xi, 0, \cdots, 0, -\beta\xi, 0, \cdots, 0).$$

(In these vectors the first non-zero element appears in place k, and the second in place h. Moreover, ξ is an arbitrary element of $\mathfrak{S}$, β the fixed element of $\mathfrak{S}$ defined by (12).)

NOTE: In many proofs, discussions and constructions in the following chapters it will be necessary to analyze rather complicated expressions ($\cup$, $\cap$-polynomials) in elements of a complemented modular lattice in which a homogeneous basis ($\bar{\mathfrak{a}}_i$; $i = 1, \cdots, n$), with axes of perspectivity $\bar{\mathfrak{c}}_{ij}$ ($i, j = 1, \cdots, n$, $i \neq j$), is given. In all these cases the reader will do well to compute such expressions and to follow up the various transformations to which they are subjected for the special case where $L = \bar{R}_{\mathfrak{R}}$, $\mathfrak{R} = \mathfrak{S}_n$ and the $\bar{\mathfrak{a}}_i$, $\bar{\mathfrak{c}}_{ij}$ are as given above. Our expressions for the vectorial representation of the $\bar{\mathfrak{a}}_i$, $\bar{\mathfrak{c}}_{ij}$ and the $\mathfrak{b}_{ij} \in L_{ij}$ are to be used. (A typical procedure of this type is the proof of Theorem 4.2, but there, of course, the situation is not special.)

PART II · CHAPTER IV

Isomorphism Theorems

We shall prove in this chapter and the succeeding ones (V to XIV inclusive) three results which we state here only roughly: (a) Any automorphism of a regular ring $\mathfrak{R}$ leaving each principal right ideal invariant is the identity; (b) A lattice-isomorphism between two lattices $\bar{R}_{\mathfrak{R}}$ and $\bar{R}_{\mathfrak{R}'}$ ($\mathfrak{R}$ and $\mathfrak{R}'$ being regular rings) is generated by a ring-isomorphism between $\mathfrak{R}$ and $\mathfrak{R}'$; (c) Corresponding to a complemented modular lattice L there exists a regular ring $\mathfrak{R}$ such that $\bar{R}_{\mathfrak{R}}$ is lattice-isomorphic to L.

We observe first that (a) and (b) are uniqueness theorems for (b) and (c), respectively. For, let in (b) two ring-isomorphisms $\mathscr{I}$, $\mathscr{J}$ between $\mathfrak{R}$ and $\mathfrak{R}'$ generate the lattice-isomorphisms between $\bar{R}_{\mathfrak{R}}$ and $\bar{R}_{\mathfrak{R}'}$. Then the quotient $\mathscr{I}\mathscr{J}^{-1}$ is an automorphism of $\mathfrak{R}$ leaving each element of $\bar{R}_{\mathfrak{R}}$ invariant and is therefore the identity by (a). Hence $\mathscr{I} = \mathscr{J}$. Suppose in (c) that two rings $\mathfrak{R}$, $\mathfrak{R}'$ are such that L is isomorphic to $\bar{R}_{\mathfrak{R}}$ and $\bar{R}_{\mathfrak{R}'}$. Then clearly $\bar{R}_{\mathfrak{R}}$ and $\bar{R}_{\mathfrak{R}'}$ are isomorphic, whence by (b) $\mathfrak{R}$ and $\mathfrak{R}'$ are isomorphic.

Now it should be observed that neither statement (a) nor (b) nor (c) is true without further assumptions. In the case of (a), if we take $\mathfrak{R}$ to be a division algebra, we see that every automorphism leaves the only two right ideals (0), $\mathfrak{R}$ invariant; we shall see that it is sufficient, however, to assume that $\mathfrak{R}$ has order $\geqq 2$ in order to prove (a). An exception to (b) is seen to be the following. Let L be a one-dimensional projective geometry L_2 with denumerably many points. It is easily shown that L is isomorphic to $\bar{R}_{\mathfrak{R}}$ where $\mathfrak{R}$ is the ring of matrices of order 2 over any division algebra in one-to-one correspondence with the set of all points but one of L_2. It is well known that it is possible to construct two non-isomorphic rings $\mathfrak{R}$, $\mathfrak{R}'$ with this property. It will be seen, however, that if $\mathfrak{R}$ has order $\geqq 3$, then (b) holds. Finally, it is obvious from the existence of non-Desarguesian plane geometries that if L is a plane geometry L_3, there will not necessarily exist a ring $\mathfrak{R}$ such that $\bar{R}_{\mathfrak{R}}$ is isomorphic to L. Here we shall need the assumption that the order of L is $\geqq 4$, in addition to a further property which will be explicitly stated later. (The 'further' property was the transitivity of perspectivity; but while lecturing, Professor von Neumann developed the proof given in Chapters V to XIV, which does *not* assume this further property — Ed.)

THEOREM 4.1: If $\mathfrak{R}$ is a regular ring of order $n \geqq 2$, then the only automorphism $a \rightleftarrows a'$ of $\mathfrak{R}$, such that $(a)_r = (a')_r$ for every $a \in \mathfrak{R}$, is the identity.

PROOF: We first show that every idempotent $e \in \mathfrak{R}$, is left invariant by the given automorphism (obviously 0, 1 are invariant). Let $e \rightleftarrows e'$. Then $e^2 \rightleftarrows e'^2$, whence $e = e^2$ implies $e' = e'^2$, and e' is idempotent. Now clearly $1 - e \rightleftarrows 1 - e'$, whence $(e)_r = (e')_r$, $(1 - e)_r = (1 - e')_r$, and it follows that $e = e'$ by the Corollary to Definition 2.1 (4). Since $\mathfrak{R}$ has order n, there exist matrix units s_{ij} $(i, j = 1, \cdots, n)$ such that

$$s_{ij}\, s_{kh} = \begin{cases} s_{ih} & (j = k) \\ 0 & (j \neq k), \end{cases}$$

$\sum_{i=1}^n s_{ii} = 1$, $s_{ii} = e_i$ being independent idempotents, and such that $\bar{\mathfrak{a}}_i \equiv (s_{ii})_r$ $(i = 1, \cdots, n)$ defines a homogeneous basis for $\bar{R}_{\mathfrak{R}}$. Now if $\mathfrak{b}_{ij} \in L_{ij}$, $\mathfrak{b}_{ij} = (e_i - v)_r$, where

$$(1) \qquad e_j v = v e_i = v \qquad (i \neq j)$$

by Lemma 3.4. Now by the hypotheses on the automorphism of $\mathfrak{R}$, $\mathfrak{b}_{ij} = (e_i - v)_r = (e_i - v')_r$, where $v \rightleftarrows v'$, since e_i is invariant, as we have just shown. But the conditions (1) clearly hold for v' also; $e_j v' = v' e_i = v'$, whence by Lemma 3.4, $v' = v$. Hence every v satisfying (1) is invariant; in particular, s_{ji} is invariant if $i \neq j$. The same is true if $i = j$, since then $s_{ii} = e_i$ is idempotent. Consider now $x' \in \mathfrak{R}(e_1)$. Since $e_1 \cdot x's_{12} = e_1 x' \cdot s_{12} = x's_{12}$, $x's_{12} \cdot e_2 = x' \cdot s_{12}e_2 = x's_{12}$, whence $x's_{12}$ satisfies (1) for $i = 2$, $j = 1$, and $x's_{12}$ is therefore invariant under the automorphism. Now s_{21} is also invariant, and hence $x's_{12} \cdot s_{21} = x' \cdot s_{12}s_{21} = x'e_1 = x'$ is invariant. Since every element $x \in \mathfrak{R}$ is of the form $x = \sum_{i,j=1}^n s_{i1} x^{ij} s_{1j}$, with $x^{ij} \in \mathfrak{R}(e_1)$ by Theorem 3.2 and its proof, and since all the x^{ij} as well as s_{i1}, s_{1j} have been shown to be invariant, we see that every element x is invariant, i.e., $x = x'$ if $x \rightleftarrows x'$, $x \in \mathfrak{R}$. Thus the given automorphism is the identity.

LEMMA 4.1: In a modular lattice L, if $(\mathfrak{a}, \mathfrak{b}, \mathfrak{c})\mathscr{C}$, i.e., if

$$\mathfrak{a} \cup \mathfrak{b} = \mathfrak{b} \cup \mathfrak{c} = \mathfrak{c} \cup \mathfrak{a},$$

$$\mathfrak{a} \cap \mathfrak{b} = \mathfrak{b} \cap \mathfrak{c} = \mathfrak{c} \cap \mathfrak{a} = 0,$$

then the three perspective isomorphisms thus generated

(a) between $L(0, \mathfrak{a})$ and $L(0, \mathfrak{b})$ by the axis $\mathfrak{c}$,
(b) between $L(0, \mathfrak{b})$ and $L(0, \mathfrak{c})$ by the axis $\mathfrak{a}$,
(c) between $L(0, \mathfrak{c})$ and $L(0, \mathfrak{a})$ by the axis $\mathfrak{b}$,

have the property that the product of any two gives the third. If $\mathfrak{u} \leqq \mathfrak{a}$, $\mathfrak{v} \leqq \mathfrak{b}$, $\mathfrak{w} \leqq \mathfrak{c}$ correspond under the isomorphisms, then $(\mathfrak{u}, \mathfrak{v}, \mathfrak{w})\mathscr{C}$.

PROOF: Consider $\mathfrak{u} \leqq \mathfrak{a}$. The images of $\mathfrak{u}$ in $L(0, \mathfrak{b})$, $L(0, \mathfrak{c})$ are $\mathfrak{v}$, $\mathfrak{w}$, respectively, where

$$\mathfrak{v} = \mathfrak{b} \cap (\mathfrak{u} \cup \mathfrak{c}), \quad \mathfrak{w} = \mathfrak{c} \cap (\mathfrak{u} \cup \mathfrak{b}).$$

Now

$$\mathfrak{u} \cup \mathfrak{v} = \mathfrak{u} \cup (\mathfrak{b} \cap (\mathfrak{u} \cup \mathfrak{c})) = (\mathfrak{u} \cup \mathfrak{b}) \cap (\mathfrak{u} \cup \mathfrak{c}) \quad \text{(by IV)},$$

and similarly $\mathfrak{w} \cup \mathfrak{u} = (\mathfrak{u} \cup \mathfrak{b}) \cap (\mathfrak{u} \cup \mathfrak{c})$. Thus $\mathfrak{u} \cup \mathfrak{v} = \mathfrak{u} \cup \mathfrak{w}$. Since $\mathfrak{u} \leqq \mathfrak{a}$, $\mathfrak{a} \cup (\mathfrak{u} \cup \mathfrak{v}) = (\mathfrak{a} \cup \mathfrak{u}) \cup \mathfrak{v} = \mathfrak{a} \cup \mathfrak{v}$, and similarly $\mathfrak{a} \cup (\mathfrak{u} \cup \mathfrak{w}) = \mathfrak{a} \cup \mathfrak{w}$. Hence also $\mathfrak{a} \cup \mathfrak{v} = \mathfrak{a} \cup \mathfrak{w}$. We had the relations $\mathfrak{v} \leqq \mathfrak{b}$, $\mathfrak{w} \leqq \mathfrak{c}$; thus $\mathfrak{v} \sim \mathfrak{w}$ (mod $\mathfrak{a}$) in $\mathfrak{b}$, $\mathfrak{c}$. This proves that $\mathfrak{v}$, $\mathfrak{w}$ correspond and hence that the product of any two isomorphisms is the third. The symmetry between $\mathfrak{u}, \mathfrak{v}, \mathfrak{w}$ shows that in addition to $\mathfrak{u} \cup \mathfrak{v} = \mathfrak{u} \cup \mathfrak{w}$, which we have established, and the obvious relation $\mathfrak{u} \cap \mathfrak{v} \leqq \mathfrak{a} \cap \mathfrak{b} = 0$, we have also $\mathfrak{u} \cup \mathfrak{v} = \mathfrak{v} \cup \mathfrak{w}$, and $\mathfrak{u} \cap \mathfrak{w} = \mathfrak{v} \cap \mathfrak{w} = 0$. Hence $\mathfrak{u} \cup \mathfrak{v} = \mathfrak{u} \cup \mathfrak{w} = \mathfrak{v} \cup \mathfrak{w}$, $\mathfrak{u} \cap \mathfrak{v} = \mathfrak{u} \cap \mathfrak{w} = \mathfrak{v} \cap \mathfrak{w} = 0$, whence $(\mathfrak{u}, \mathfrak{v}, \mathfrak{w})\mathscr{C}$.

LEMMA 4.2: Let L be a complemented modular lattice of order $n \geqq 2$, and let $\bar{\mathfrak{a}}_i$ $(i = 1, \cdots, n)$ be a homogeneous basis for L. Then every element of L is expressible as a polynomial in elements from the L_{ij}, with the help of the operations $\cup$, $\cap$ of the lattice L.

PROOF: We divide the proof into steps. (Cf. the note at the end of Chapter III.)

(a) Let $\mathfrak{u} \leqq \bar{\mathfrak{a}}_i$; then we prove that $\mathfrak{u}$ is expressible in the form described in the lemma. Let $\bar{\mathfrak{a}}_j$ be any element of the basis distinct from $\bar{\mathfrak{a}}_i$, and let $\mathfrak{c}_{ij}$ be an inverse of $\bar{\mathfrak{a}}_i$, $\bar{\mathfrak{a}}_j$ in $\bar{\mathfrak{a}}_i \cup \bar{\mathfrak{a}}_j$. Then $(\bar{\mathfrak{a}}_i, \bar{\mathfrak{a}}_j, \bar{\mathfrak{c}}_{ij})\mathscr{C}$, and by Lemma 4.1 $(\mathfrak{u}, \mathfrak{v}, \mathfrak{w})\mathscr{C}$, $\mathfrak{v}$, $\mathfrak{w}$ being the images of $\mathfrak{u}$ in $L(0, \bar{\mathfrak{a}}_j)$, $L(0, \bar{\mathfrak{c}}_{ij})$. Let $\mathfrak{u}'$ be inverse to $\mathfrak{u}$ in $\bar{\mathfrak{a}}_i$. Then the images $\mathfrak{v}'$, $\mathfrak{w}'$ of $\mathfrak{u}'$ are clearly inverse to $\mathfrak{v}$, $\mathfrak{w}$ in $\bar{\mathfrak{a}}_j$, $\bar{\mathfrak{c}}_{ij}$, respectively. Now we shall show that $\mathfrak{w} \cup \mathfrak{v}'$ is inverse to $\bar{\mathfrak{a}}_i$ in $\bar{\mathfrak{a}}_i \cup \bar{\mathfrak{a}}_j$. Evidently

$$\begin{aligned}
(2) \qquad \bar{\mathfrak{a}}_i \cup \mathfrak{w} \cup \mathfrak{v}' &= \mathfrak{u} \cup \mathfrak{u}' \cup \mathfrak{w} \cup \mathfrak{v}' \\
&= \mathfrak{u} \cup \mathfrak{w} \cup \mathfrak{u}' \cup \mathfrak{v}' \\
&= \mathfrak{u} \cup \mathfrak{v} \cup \mathfrak{u}' \cup \mathfrak{v}' \quad \text{(since } (\mathfrak{u}, \mathfrak{v}, \mathfrak{w})\mathscr{C}) \\
&= \mathfrak{u} \cup \mathfrak{u}' \cup \mathfrak{v} \cup \mathfrak{v}' \\
&= \bar{\mathfrak{a}}_i \cup \bar{\mathfrak{a}}_j.
\end{aligned}$$

Now $(\bar{\mathfrak{a}}_i \cup \mathfrak{v}') \cap \mathfrak{w} \leqq \bar{\mathfrak{c}}_{ij}$, whence

$$(\bar{\mathfrak{a}}_i \cup \mathfrak{v}') \cap \mathfrak{w} = (\bar{\mathfrak{a}}_i \cup \mathfrak{v}') \cap \bar{\mathfrak{c}}_{ij} \cap \mathfrak{w} = \mathfrak{w}' \cap \mathfrak{w} = 0;$$

also $\bar{\mathfrak{a}}_i \cap \mathfrak{v}' \leqq \bar{\mathfrak{a}}_i \cap \bar{\mathfrak{a}}_j = 0$, whence $(\bar{\mathfrak{a}}_i, \mathfrak{v}', \mathfrak{w}) \perp$ (by Part I, Theorem 2.2). Hence $\bar{\mathfrak{a}}_i \cap (\mathfrak{w} \cup \mathfrak{v}') = 0$, and $\mathfrak{b}_{ji} \equiv \mathfrak{w} \cup \mathfrak{v}' \in L_{ji}$. Thus

$$\mathfrak{u} = (\mathfrak{w} \cup \bar{\mathfrak{a}}_j) \cap \bar{\mathfrak{a}}_i = ((\mathfrak{w} \cup \mathfrak{v}') \cup \bar{\mathfrak{a}}_j) \cap \bar{\mathfrak{a}}_i = (\mathfrak{b}_{ji} \cup \bar{\mathfrak{a}}_j) \cap \bar{\mathfrak{a}}_i.$$

Of course $\bar{\mathfrak{a}}_j \in L_{ji}$, $\bar{\mathfrak{a}}_i \in L_{ij}$, and the proof is complete.

(b) If $m = 2, \cdots, n$, and $\mathfrak{c}$ is an inverse of $\bar{\mathfrak{a}}_1 \cup \cdots \cup \bar{\mathfrak{a}}_{m-1}$ in $\bar{\mathfrak{a}}_1 \cup \cdots \cup \bar{\mathfrak{a}}_m$, then $\mathfrak{c}$ may be represented in the form

$$\mathfrak{c} = \prod_{i=1}^{m-1}{}_{\cap} (\mathfrak{c}_{mi} \cup \sum_{\substack{j=1 \\ j \neq i}}^{m-1}{}_{\cup} \bar{\mathfrak{a}}_i), \tag{3}$$

where $\mathfrak{c}_{mi} \in L_{mi}$ $(i = 1, \cdots, m-1)$. (In (3) the symbol $\Pi_{\cap}$ represents the extension of the binary operation $\cap$ to multiplication of any number of factors, and $\Sigma_{\cup}$ represents the similar extension of $\cup$. When, as in (3), the summation extends from $j = 1$ to $j = m-1$, $j \neq i$, we shall abbreviate the symbol by Σ'.) To establish (b), we define

$$\mathfrak{c}_{mi} \equiv (\mathfrak{c} \cup \Sigma' \bar{\mathfrak{a}}_j) \cap (\bar{\mathfrak{a}}_i \cup \bar{\mathfrak{a}}_m) \qquad (i = 1, \cdots, m-1), \tag{4}$$

and prove first that $\mathfrak{c}_{mi}$ is in L_{mi}. Now

$$\begin{aligned} \bar{\mathfrak{a}}_i \cup \mathfrak{c}_{mi} &= \bar{\mathfrak{a}}_i \cup [(\mathfrak{c} \cup \Sigma' \bar{\mathfrak{a}}_j) \cap (\bar{\mathfrak{a}}_i \cup \bar{\mathfrak{a}}_m)] \\ &= (\bar{\mathfrak{a}}_i \cup \mathfrak{c} \cup \Sigma' \bar{\mathfrak{a}}_j) \cap (\bar{\mathfrak{a}}_i \cup \bar{\mathfrak{a}}_m) \qquad \text{(by IV)} \\ &= (\mathfrak{c} \cup \sum_{j=1}^{m-1}{}_{\cup} \bar{\mathfrak{a}}_j) \cap (\bar{\mathfrak{a}}_i \cup \bar{\mathfrak{a}}_m) \\ &= (\bar{\mathfrak{a}}_1 \cup \cdots \cup \bar{\mathfrak{a}}_m) \cap (\bar{\mathfrak{a}}_i \cup \bar{\mathfrak{a}}_m) = \bar{\mathfrak{a}}_i \cup \bar{\mathfrak{a}}_m; \end{aligned}$$

moreover,

$$\Sigma' \bar{\mathfrak{a}}_j \cap \bar{\mathfrak{a}}_i = 0, \ (\Sigma' \bar{\mathfrak{a}}_j \cup \bar{\mathfrak{a}}_i) \cap \mathfrak{c} = (\bar{\mathfrak{a}}_1 \cup \cdots \cup \bar{\mathfrak{a}}_{m-1}) \cap \mathfrak{c} = 0,$$

and $(\mathfrak{c}, \Sigma' \bar{\mathfrak{a}}_j, \bar{\mathfrak{a}}_i) \perp$, whence

$$\mathfrak{c}_{mi} \cap \bar{\mathfrak{a}}_i = (\mathfrak{c} \cup \Sigma' \bar{\mathfrak{a}}_j) \cap \bar{\mathfrak{a}}_i = 0.$$

Hence $\mathfrak{c}_{mi}$ is an inverse of $\bar{\mathfrak{a}}_i$ in $\bar{\mathfrak{a}}_i \cup \bar{\mathfrak{a}}_m$, i.e., $\mathfrak{c}_{mi} \in L_{mi}$ for $i = 1, \cdots, m-1$. It remains to prove (3). We have

$$\begin{aligned} \prod_{i=1}^{m-1}{}_{\cap} (\mathfrak{c}_{mi} \cup \Sigma' \bar{\mathfrak{a}}_j) &= \prod_{i=1}^{m-1}{}_{\cap} [\{(\mathfrak{c} \cup \Sigma' \bar{\mathfrak{a}}_j) \cap (\bar{\mathfrak{a}}_i \cup \bar{\mathfrak{a}}_m)\} \cup \Sigma' \bar{\mathfrak{a}}_j] \\ &= \prod_{i=1}^{m-1}{}_{\cap} [(\mathfrak{c} \cup \Sigma' \bar{\mathfrak{a}}_j) \cap (\bar{\mathfrak{a}}_i \cup \bar{\mathfrak{a}}_m \cup \Sigma' \bar{\mathfrak{a}}_j)] \qquad \text{(by IV)} \end{aligned}$$

$$= \prod_{i=1}^{m-1}{}_{\cap}[(\mathfrak{c} \cup \Sigma'\bar{\mathfrak{a}}_j) \cap 1]$$

$$= \prod_{i=1}^{m-1}{}_{\cap}(\mathfrak{c} \cup \Sigma'\bar{\mathfrak{a}}_j) = \mathfrak{c}$$

the last equality holding by Part I, Theorem 2.6 since $(\bar{\mathfrak{a}}_1, \cdots, \bar{\mathfrak{a}}_{m-1}) \perp$ and $(\bar{\mathfrak{a}}_1 \cup \cdots \cup \bar{\mathfrak{a}}_{m-1}) \cap \mathfrak{c} = 0$ yield $(\bar{\mathfrak{a}}_1, \cdots, \bar{\mathfrak{a}}_{m-1}, \mathfrak{c}) \perp$ by Part I, Theorem 2.4, and since $\mathfrak{c}$ is the only term common to all factors in the product $\prod_{i=1}^{m-1}(\mathfrak{c} \cup \Sigma'\bar{\mathfrak{a}}_j)$.

(c) Let $m = 2, \cdots, n$, and let $\mathfrak{b} \leqq \bar{\mathfrak{a}}_1 \cup \cdots \cup \bar{\mathfrak{a}}_m$, $\mathfrak{b} \cap (\bar{\mathfrak{a}}_1 \cup \cdots \cup \bar{\mathfrak{a}}_{m-1}) = 0$. Then $\mathfrak{b}$ is of the form

(5) $$\mathfrak{b} = \mathfrak{c} \cap (\bar{\mathfrak{a}}_1 \cup \cdots \cup \bar{\mathfrak{a}}_{m-1} \cup \mathfrak{b}_m),$$

where $\mathfrak{c}$ is inverse to $\bar{\mathfrak{a}}_1 \cup \cdots \cup \bar{\mathfrak{a}}_{m-1}$ in $\bar{\mathfrak{a}}_1 \cup \cdots \cup \bar{\mathfrak{a}}_m$, and $\mathfrak{b}_m \leqq \bar{\mathfrak{a}}_m$. In order to prove (c), we define

(6) $$\mathfrak{b}_m \equiv [\mathfrak{b} \cup (\bar{\mathfrak{a}}_1 \cup \cdots \cup \bar{\mathfrak{a}}_{m-1})] \cap \bar{\mathfrak{a}}_m;$$

hence $\mathfrak{b}_m \leqq \bar{\mathfrak{a}}_m$. Let $\mathfrak{b}'$ be inverse to $\mathfrak{b}_m$ in $\bar{\mathfrak{a}}_m$. Now define $\bar{\mathfrak{a}} \equiv \bar{\mathfrak{a}}_1 \cup \cdots \cup \bar{\mathfrak{a}}_{m-1}$; then

(7) $$\begin{aligned} \bar{\mathfrak{a}} \cup \mathfrak{b}_m &= \bar{\mathfrak{a}} \cup [(\mathfrak{b} \cup \bar{\mathfrak{a}}) \cap \bar{\mathfrak{a}}_m] \\ &= (\mathfrak{b} \cup \bar{\mathfrak{a}}) \cap (\bar{\mathfrak{a}} \cup \bar{\mathfrak{a}}_m) \qquad \text{(by IV)} \\ &= (\bar{\mathfrak{a}} \cup \mathfrak{b}) \cap \sum_{i=1}^{m}{}_{\cup}\, \bar{\mathfrak{a}}_i = \bar{\mathfrak{a}} \cup \mathfrak{b}, \end{aligned}$$

whence $\bar{\mathfrak{a}} \cup (\mathfrak{b} \cup \mathfrak{b}') = \bar{\mathfrak{a}} \cup (\mathfrak{b}_m \cup \mathfrak{b}') = \bar{\mathfrak{a}}_1 \cup \cdots \cup \bar{\mathfrak{a}}_m$. Now $\mathfrak{b} \cap \bar{\mathfrak{a}} = 0$ by hypothesis; moreover,

$$(\mathfrak{b} \cup \bar{\mathfrak{a}}) \cap \mathfrak{b}' = (\mathfrak{b} \cup \bar{\mathfrak{a}}) \cap \bar{\mathfrak{a}}_m \cap \mathfrak{b}' = \mathfrak{b}_m \cap \mathfrak{b}' = 0,$$

whence $(\bar{\mathfrak{a}}, \mathfrak{b}, \mathfrak{b}') \perp$ (Part I, Theorem 2.2), and $\bar{\mathfrak{a}} \cap (\mathfrak{b} \cup \mathfrak{b}') = 0$. These considerations show that $\mathfrak{b} \cup \mathfrak{b}'$ is inverse to $\bar{\mathfrak{a}}$ in $\bar{\mathfrak{a}}_1 \cup \cdots \cup \bar{\mathfrak{a}}_m$. Define $\mathfrak{c} \equiv \mathfrak{b} \cup \mathfrak{b}'$; then

$$\begin{aligned} \mathfrak{c} \cap (\bar{\mathfrak{a}}_1 \cup \cdots \cup \bar{\mathfrak{a}}_{m-1} \cup \mathfrak{b}_m) &= (\mathfrak{b} \cup \mathfrak{b}') \cap (\bar{\mathfrak{a}} \cup \mathfrak{b}_m) \\ &= (\mathfrak{b} \cup \mathfrak{b}') \cap (\bar{\mathfrak{a}} \cup \mathfrak{b}) \quad \text{(by (7))} \\ &= \mathfrak{b} \cup [\mathfrak{b}' \cap (\bar{\mathfrak{a}} \cup \mathfrak{b})] \\ &= \mathfrak{b}, \end{aligned}$$

the last equality holding because $(\bar{\mathfrak{a}}, \mathfrak{b}, \mathfrak{b}') \perp$.

(d) For every $m = 1, \cdots, n$ and $\mathfrak{b} \leqq \bar{\mathfrak{a}}_1 \cup \cdots \cup \bar{\mathfrak{a}}_m$, $\mathfrak{b}$ is expressible in the form described in the lemma. For $m = 1$, the statement is true by (a). Let $m > 1$, and suppose the statement true for $m - 1$. Then let

$\mathfrak{b} \leqq \bar{\mathfrak{a}}_1 \cup \cdots \cup \bar{\mathfrak{a}}_m$. Define $\mathfrak{b}_1' \equiv \mathfrak{b} \cap (\bar{\mathfrak{a}}_1 \cup \cdots \cup \bar{\mathfrak{a}}_{m-1})$, and let $\mathfrak{b}_2'$ be inverse to $\mathfrak{b}_1'$ in $\mathfrak{b}$. Then $\mathfrak{b} = \mathfrak{b}_1' \cup \mathfrak{b}_2'$, $\mathfrak{b}_1' \cap \mathfrak{b}_2' = 0$, $\mathfrak{b}_1' \leqq \bar{\mathfrak{a}}_1 \cup \cdots \cup \bar{\mathfrak{a}}_{m-1}$. Now since $\mathfrak{b}_2' \leqq \mathfrak{b}$, we have $\mathfrak{b}_2' \cap (\bar{\mathfrak{a}}_1 \cup \cdots \cup \bar{\mathfrak{a}}_{m-1}) = \mathfrak{b}_2' \cap \mathfrak{b} \cap (\bar{\mathfrak{a}}_1 \cup \cdots \cup \bar{\mathfrak{a}}_{m-1}) = \mathfrak{b}_2' \cap \mathfrak{b}_1' = 0$. By induction hypothesis, $\mathfrak{b}_1'$ is expressible in the desired form. Moreover, by (c), $\mathfrak{b}_2'$ is of the form

$$\mathfrak{b}_2' = \mathfrak{c} \cap (\bar{\mathfrak{a}}_1 \cup \cdots \cup \bar{\mathfrak{a}}_{m-1} \cup \mathfrak{b}_m),$$

$\mathfrak{b}_m \leqq \bar{\mathfrak{a}}_m$, $\mathfrak{c}$ inverse to $\bar{\mathfrak{a}}_1 \cup \cdots \cup \bar{\mathfrak{a}}_{m-1}$ in $\bar{\mathfrak{a}}_1 \cup \cdots \cup \bar{\mathfrak{a}}_m$. By (b), $\mathfrak{c}$ is a polynomial in the $\mathfrak{c}_{mi}$ and the $\bar{\mathfrak{a}}_i$, hence in elements of the L_{ij}; by (a), $\mathfrak{b}_m$ is also a polynomial in elements of the L_{ij}. Hence $\mathfrak{b}_2'$ is such a polynomial. Since $\mathfrak{b}_1'$ is of the same form, and $\mathfrak{b} = \mathfrak{b}_1' \cup \mathfrak{b}_2'$, we see that $\mathfrak{b}$ is a polynomial in elements of L_{ij} as was asserted.

The lemma follows from (d) by setting $m = n$, since every $\mathfrak{b} \in L$ satisfies the condition $\mathfrak{b} \leqq \bar{\mathfrak{a}}_1 \cup \cdots \cup \bar{\mathfrak{a}}_n = 1$.

Theorem 4.2: Let $\mathfrak{R}$, $\mathfrak{R}'$ be regular rings such that $\mathfrak{R}$ has order $n \geqq 3$ and such that $\bar{R}_{\mathfrak{R}}$ and $\bar{R}_{\mathfrak{R}'}$ are lattice-isomorphic. Then there exists a ring-isomorphism of $\mathfrak{R}$ and $\mathfrak{R}'$ which generates the given lattice-isomorphism.

Proof: (We shall need to make frequent use of the representation of principal right ideals by vector sets described and illustrated in Chapter III; hence we shall introduce some conventions which will serve to abbreviate this representation. Since all the vectors that we use are n-dimensional vectors, with zeros in each place except possibly in the ith, j^{th} and k^{th} places, such vectors will be written in the form (a, b, c), where a, b, c are the ith, j^{th} and k^{th} components, respectively. Moreover, we shall agree to use the notation $\mathfrak{a}$: (ξ, η, ζ) to indicate that the right ideal $\mathfrak{a}$ is represented by the set of all vectors (ξ, η, ζ) where $\xi, \eta, \zeta \in \mathfrak{S}$, $\mathfrak{S}$ being the fixed ring appearing in our discussions; thus Greek letters ξ, η, ζ, etc. will be variables generating the vector set, and α, β, γ, etc. will be fixed for each set of vectors.) Since $\mathfrak{R}$ has order n, that is $\bar{R}_{\mathfrak{R}}$ has order n, there exists a homogeneous basis $(\bar{\mathfrak{a}}_i;\ i = 1, \cdots, n)$ for $\bar{R}_{\mathfrak{R}}$. Let $\bar{\mathfrak{c}}_{h1}$ be an axis of perspectivity of $\bar{\mathfrak{a}}_h$, $\bar{\mathfrak{a}}_1$ $(h = 2, \cdots, n)$. Now by applying to the $\bar{\mathfrak{a}}_i$ and $\bar{\mathfrak{c}}_{h1}$ Lemma 3.6 and observing the Note following it, we see that there exists a system of n^2 matrix units s_{ij} $(i, j = 1, \cdots, n)$ in $\mathfrak{R}$ such that $\bar{\mathfrak{a}}_i = (e_i)_r$, $e_i = s_{ii}$, and $\bar{\mathfrak{c}}_{h1} = (e_h - s_{1h})_r = (e_1 - s_{h1})_r$. Let us denote the given lattice-isomorphism between $\bar{R}_{\mathfrak{R}}$ and $\bar{R}_{\mathfrak{R}'}$ by $\mathcal{J}$ and let $\bar{\mathfrak{a}}_i'$, $\bar{\mathfrak{c}}_{h1}'$ be the images in $\bar{R}_{\mathfrak{R}'}$ of $\bar{\mathfrak{a}}_i$, $\bar{\mathfrak{c}}_{h1}$ respectively under $\mathcal{J}$. Since $\mathcal{J}$ is a lattice-isomorphism, it follows that $(\bar{\mathfrak{a}}_i';\ i = 1, \cdots, n)$ is a homogeneous basis for $\bar{R}_{\mathfrak{R}'}$, and $\bar{\mathfrak{c}}_{h1}'$ is an axis of perspectivity of $\bar{\mathfrak{a}}_h'$, $\bar{\mathfrak{a}}_1'$ $(h = 2, \cdots, n)$. By applying again Lemma 3.6 and the Note following it, this time to the $\bar{\mathfrak{a}}_i'$ and $\bar{\mathfrak{c}}_{h1}'$, we see that there exists a system of n^2 matrix units $s'_{ij} \in \mathfrak{R}'$ $(i, j = 1, \cdots, n)$, such that $\bar{\mathfrak{a}}'_i = (e'_i)$, $e'_i = s'_{ii}$, and $\bar{\mathfrak{c}}'_{h1} = (e'_h - s'_{1h})_r = (e'_1 - s'_{h1})_r$.

We now apply the construction of Theorem **3.3** which represents $\mathfrak{R}$ as a matrix ring $\mathfrak{S}_n$, using in $\mathfrak{R}$ the matrix units s_{ij} $(i, j = 1, \cdots, n)$ already obtained. The proof of Theorem **3.3** shows that $\mathfrak{S} = \mathfrak{R}(e_1)$ and that the correspondence between $\mathfrak{R}$ and $\mathfrak{S}_n$ is given by equations (10), (11) of Chapter III. We may construct similarly $\mathfrak{S}'_n$ for $\mathfrak{R}'$ and the s'_{ij} $(i, j = 1, \cdots, n)$, whence $\mathfrak{S}' = \mathfrak{R}'(e'_1)$, and the correspondence between $\mathfrak{R}'$ and $\mathfrak{S}'_n$ is defined just as that between $\mathfrak{R}$ and $\mathfrak{S}_n$. Now principal right ideals of $\mathfrak{R}$ $(\mathfrak{R}')$ are principal right ideals in $\mathfrak{S}_n(\mathfrak{S}'_n)$, and hence they may be represented by n-dimensional vector sets from $\mathfrak{S}(\mathfrak{S}')$, the conventions described at the outset of this proof being applicable.

Let us consider two distinct members $\bar{\mathfrak{a}}_i$, $\bar{\mathfrak{a}}_j$ of our homogeneous basis (existent by virtue of the hypothesis $n \geqq 3$), and the set L_{ij} of all inverses $\mathfrak{b}_{ij}$ of $\bar{\mathfrak{a}}_j$ in $\bar{\mathfrak{a}}_i \cup \bar{\mathfrak{a}}_j$. Since $\bar{\mathfrak{a}}'_i$, $\bar{\mathfrak{a}}'_j$ correspond to $\bar{\mathfrak{a}}_i$, $\bar{\mathfrak{a}}_j$ under $\mathcal{J}$, the set L'_{ij} of all inverses $\mathfrak{b}'_{ij}$ of $\bar{\mathfrak{a}}'_j$ in $\bar{\mathfrak{a}}'_i \cup \bar{\mathfrak{a}}'_j$ corresponds under $\mathcal{J}$ to L_{ij}. Now by γ) at the end of Chapter III the vector set representing the general $\mathfrak{b}_{ij} \in L_{ij}$ has the form $(\xi, -\beta\xi, 0)$ $(\xi, \beta \in \mathfrak{S}, \xi$ arbitrary, β fixed); similarly the vector set representing the general $\mathfrak{b}'_{ij} \in L'_{ij}$ has the form $(\xi', -\beta'\xi', 0)$ $(\xi', \beta' \in \mathfrak{S}'$, ξ' arbitrary, β' fixed). The fact that the $\mathfrak{b}_{ij}$ and $\mathfrak{b}'_{ij}$ correspond in one-to-one manner under $\mathcal{J}$ yields the existence of a one-to-one correspondence between all elements $\beta \in \mathfrak{S}$ and all elements $\beta' \in \mathfrak{S}'$. We denote this correspondence by

$$\beta \underset{(i,j)}{\rightleftarrows} \beta'. \tag{8}$$

It should be observed that the correspondence (8) depends on i, j and is defined only for $i \neq j$, $i, j = 1, \cdots, n$.

By β) at the end of Chapter III, $\beta = I$ yields $\mathfrak{b}_{ij} = \bar{\mathfrak{c}}_{ij}$, and similarly $\beta' = I'$ yields $\mathfrak{b}'_{ij} = \bar{\mathfrak{c}}'_{ij}$. Since $\bar{\mathfrak{c}}_{h1}$, $\bar{\mathfrak{c}}'_{h1}$ $(h = 2, \cdots, n)$ correspond under $\mathcal{J}$, we have

$$I \underset{(h,1)}{\rightleftarrows} I' \text{ and } I \underset{(1,h)}{\rightleftarrows} I' \qquad (h = 2, \cdots, n). \tag{9}$$

Let us consider now three distinct members $\bar{\mathfrak{a}}_i$, $\bar{\mathfrak{a}}_j$, $\bar{\mathfrak{a}}_k$ of our basis, existent since $n \geqq 3$. Let $\mathfrak{b}_{ij} \in L_{ij}$, $\mathfrak{c}_{jk} \in L_{jk}$; then $\mathfrak{d}_{ik} \in L_{ik}$, where $\mathfrak{d}_{ik} = (\mathfrak{b}_{ij} \cup \mathfrak{c}_{jk}) \cap (\bar{\mathfrak{a}}_i \cup \bar{\mathfrak{a}}_k)$, as was shown in Part I, Theorem **3.4**. In fact, if

$$\mathfrak{b}_{ij} : (\xi, \; -\beta\xi, \; 0),$$
$$\mathfrak{c}_{jk} : (0, \; \eta, \; -\gamma\eta),$$

then

$$\mathfrak{b}_{ij} \cup \mathfrak{c}_{jk} : (\xi, \; \eta - \beta\xi, \; -\gamma\eta),$$
$$\bar{\mathfrak{a}}_i \cup \bar{\mathfrak{a}}_k : (\theta, \; 0, \; \omega),$$

whence

$$\mathfrak{d}_{ik} : (\xi,\ 0,\ -\gamma\beta\xi).$$

This calculation, together with the corresponding one for $\mathfrak{b}'_{ij}$, $\mathfrak{c}'_{jk}$, $\mathfrak{d}'_{ik}$ shows that

$$\text{(10)} \qquad \text{if } \beta \underset{(i,j)}{\rightleftarrows} \beta',\ \gamma \underset{(j,k)}{\rightleftarrows} \gamma' \text{ then } \gamma\beta \underset{(i,k)}{\rightleftarrows} \gamma'\beta'.$$

The consequences of (9) and (10) will now be examined. If $i \neq 1$, $j \neq 1$, $i \neq j$, define α' by $I \underset{(i,j)}{\rightleftarrows} \alpha'$. By (9) $I \underset{(j,1)}{\rightleftarrows} I'$. Hence by (10) $I \underset{(i,1)}{\rightleftarrows} \alpha'$; but $I \underset{(i,1)}{\rightleftarrows} I'$ by (9), whence $\alpha' = I'$. Thus we have for this case

$$\text{(11)} \qquad I \underset{(i,j)}{\rightleftarrows} I'.$$

If $i = 1$ or $j = 1$ but $i \neq j$, (11) is implied by (9). Hence (11) is established for every $i, j = 1, \cdots, n$ with $i \neq j$. Now by virtue of (11), (10) with $\beta = I$, $\beta' = I'$ yields

$$\text{(12)} \qquad \gamma \underset{(j,k)}{\rightleftarrows} \gamma' \text{ implies } \gamma \underset{(i,k)}{\rightleftarrows} \gamma';$$

moreover (10) with $\gamma = I$, $\gamma' = I'$ yields

$$\text{(13)} \qquad \beta \underset{(i,j)}{\rightleftarrows} \beta' \text{ implies } \beta \underset{(i,k)}{\rightleftarrows} \beta'.$$

Let us consider now two pairs (i, j), (k, h) with $i \neq j$, $k \neq h$, and let us suppose the $\beta \underset{(i,j)}{\rightleftarrows} \beta'$. If $i \neq h$, then $\beta \underset{(i,h)}{\rightleftarrows} \beta'$ results from an application of (13) (with i, j, h in place of i, j, k, provided that $j \neq h$ — but for $j = h$ this inference is obvious), and $\beta \underset{(k,h)}{\rightleftarrows} \beta'$ results from an application of (12) (with i, k, h in place of i, j, k, provided that $i \neq k$ — but for $i = k$ this inference is obvious). Thus we have proved that if $i \neq h$, then

$$\text{(14)} \qquad \beta \underset{(i,j)}{\rightleftarrows} \beta' \text{ implies } \beta \underset{(k,h)}{\rightleftarrows} \beta'.$$

By symmetry, the same is true if $j \neq k$. If neither $i \neq h$ nor $j \neq k$, then $i = h$, $j = k$. There exists $\ell \neq i, j$ (since $n \geqq 3$), and we see that $\beta \underset{(i,j)}{\rightleftarrows} \beta'$ implies $\beta \underset{(i,\ell)}{\rightleftarrows} \beta'$ by (14), which in turn implies $\beta \underset{(j,i)}{\rightleftarrows} \beta'$ by (14); but this is precisely $\beta \underset{(k,h)}{\rightleftarrows} \beta'$, since $j = k$, $i = h$. Hence (14) holds in all cases. By interchanging in (14) the pairs (i, j) and (k, h) we establish

$$\text{(15)} \qquad \beta \underset{(i,j)}{\rightleftarrows} \beta' \text{ is equivalent to } \beta \underset{(k,h)}{\rightleftarrows} \beta'$$

if $i \neq j$, $k \neq h$. Thus all the correspondences $\beta \underset{(i,j)}{\rightleftarrows} \beta'$ are equal and may be denoted simply by $\beta \rightleftarrows \beta'$. Then (10) shows that this correspondence carries multiplication in $\mathfrak{S}$ to multiplication in $\mathfrak{S}'$.

Let $\beta \rightleftarrows \beta'$, $\gamma \rightleftarrows \gamma'$,

$$\mathfrak{b}_{ij} : (\xi, \ -\beta\xi, \ 0), \ \mathfrak{c}_{ij} : (\eta, \ -\gamma\eta, \ 0),$$
$$\mathfrak{b}'_{ij} : (\xi', \ -\beta'\xi', \ 0), \ \mathfrak{c}'_{ij} : (\eta', \ -\gamma'\eta', \ 0).$$

Thus $\mathfrak{b}_{ij}$, $\mathfrak{b}'_{ij}$ as well as $\mathfrak{c}_{ij}$, $\mathfrak{c}'_{ij}$ correspond under the lattice-isomorphism $\mathcal{J}$ of $\bar{R}_{\mathfrak{R}}$ and $\bar{R}_{\mathfrak{R}'}$. Define $\mathfrak{d}_{ij}$ as the right ideal represented by $\beta + \gamma$:

$$\mathfrak{d}_{ij} : (\xi, \ -(\beta + \gamma)\xi, \ 0), \tag{16}$$

and similarly $\mathfrak{d}'_{ij}$:

$$\mathfrak{d}'_{ij} : (\xi', \ -(\beta' + \gamma')\xi', \ 0).$$

We shall prove that

$$\mathfrak{d}_{ij} = (\mathfrak{g} \cup \mathfrak{f}) \cap (\bar{\mathfrak{a}}_i \cup \bar{\mathfrak{a}}_j), \tag{17}$$

where

$$\mathfrak{g} = (\mathfrak{b}_{ij} \cup \bar{\mathfrak{c}}_{ik}) \cap (\bar{\mathfrak{a}}_j \cup \bar{\mathfrak{a}}_k),$$
$$\mathfrak{f} = (\mathfrak{c}_{ij} \cup \bar{\mathfrak{a}}_k) \cap (\bar{\mathfrak{a}}_j \cup \bar{\mathfrak{c}}_{ik}).$$

The proof consists in finding the vector sets representing $\mathfrak{g}$, $\mathfrak{f}$ and the right member of (17) and showing that this last set reduces to that in (16). We have

$$\mathfrak{b}_{ij} : (\xi, \ -\beta\xi, \ 0), \quad \bar{\mathfrak{c}}_{ik} : (\eta, \ 0, \ -\eta),$$

whence

$$\mathfrak{b}_{ij} \cup \bar{\mathfrak{c}}_{ik} : (\xi + \eta, \ -\beta\xi, \ -\eta).$$

Also

$$\bar{\mathfrak{a}}_j \cup \bar{\mathfrak{a}}_k : (0, \ \theta, \ \omega),$$

and it follows that

$$\mathfrak{g} : (0, \ -\beta\xi, \ \xi).$$

Similarly,

$$\mathfrak{c}_{ij} : (\xi, \ -\gamma\xi, \ 0), \quad \bar{\mathfrak{a}}_k : (0, \ 0, \ \eta),$$

whence

$$\mathfrak{c}_{ij} \cup \bar{\mathfrak{a}}_k : (\xi, \ -\gamma\xi, \ \eta).$$

Also

$$\bar{\mathfrak{a}}_j \cup \bar{\mathfrak{c}}_{ik} : (\phi, \ \psi, \ -\phi),$$

and we have

$$\mathfrak{f} : (\zeta, \ -\gamma\zeta, \ -\zeta).$$

Thus

$$\mathfrak{g} \cup \mathfrak{f} : (\zeta,\ -\beta\xi - \gamma\zeta,\ \xi - \zeta);$$

but

$$\bar{\mathfrak{a}}_i \cup \bar{\mathfrak{a}}_j : (\phi,\ \psi,\ 0).$$

Therefore

$$(\mathfrak{g} \cup \mathfrak{f}) \cap (\bar{\mathfrak{a}}_i \cup \bar{\mathfrak{a}}_j) : (\xi,\ -(\beta + \gamma)\xi,\ 0),$$

and consequently (17) is established. The corresponding calculation for $\mathfrak{d}'_{ij}$ shows that $\mathfrak{d}'_{ij}$ is expressible by the formula obtained from (17) by placing accents on $\mathfrak{b}_{ij}$, $\bar{\mathfrak{c}}_{ik}$, etc., throughout. Hence $\mathfrak{d}_{ij}$, $\mathfrak{d}'_{ij}$ correspond under the given lattice-isomorphism, along with $\mathfrak{b}_{ij}$, $\mathfrak{b}'_{ij}$ and $\mathfrak{c}_{ij}$, $\mathfrak{c}'_{ij}$, and therefore $\beta + \gamma \rightleftarrows \beta' + \gamma'$.

We have shown that the correspondence $\beta \rightleftarrows \beta'$ is an isomorphism of the rings $\mathfrak{S}$, $\mathfrak{S}'$; this isomorphism clearly generates an isomorphism $\mathscr{I}$ between $\mathfrak{R}$ and $\mathfrak{R}'$, and $\mathscr{I}$ induces a lattice-isomorphism $\bar{\mathscr{I}}$ between $\bar{R}_{\mathfrak{R}}$ and $\bar{R}_{\mathfrak{R}'}$. It remains to prove that $\bar{\mathscr{I}} = \mathscr{J}$, where $\mathscr{J}$ represents the given lattice-isomorphism. Now $\mathscr{J}^{-1}\bar{\mathscr{I}}$ is an automorphism of $\bar{R}_{\mathfrak{R}}$; moreover, by the construction of $\bar{\mathscr{I}}$ we see that the $\bar{\mathfrak{a}}_i$ and the $\mathfrak{b}_{ij} \in L_{ij}$ are all left invariant by $\mathscr{J}^{-1}\bar{\mathscr{I}}$. By Lemma 4.2, any element of $\bar{R}_{\mathfrak{R}}$ is expressible as a polynomial in elements of the L_{ij} with the help of the operations $\cup$, $\cap$. Therefore every element of $\bar{R}_{\mathfrak{R}}$ is left invariant by $\mathscr{J}^{-1}\bar{\mathscr{I}}$, since its polynomial expression is clearly unchanged by $\mathscr{J}^{-1}\bar{\mathscr{I}}$. This shows that $\mathscr{J}^{-1}\bar{\mathscr{I}}$ is the identity, whence $\mathscr{J} = \bar{\mathscr{I}}$, and the ring-isomorphism $\mathscr{I}$ between $\mathfrak{R}$ and $\mathfrak{R}'$ generates the lattice isomorphism $\mathscr{J}$ between $\bar{R}_{\mathfrak{R}}$ and $\bar{R}_{\mathfrak{R}'}$.

At this point we develop some incidental consequences of the isomorphism theory developed thus far. We shall find an intimate connection between isomorphisms of a lattice $\bar{R}_{\mathfrak{R}}$ ($\mathfrak{R}$ a regular ring) with its dual lattice and anti-automorphisms of $\mathfrak{R}$ which are a generalization of the conjugate-transpose operation for matrices over the real, complex or quaternionic number systems. Throughout the remainder of the chapter, $\mathfrak{R}$ will be a regular ring, and $\mathfrak{R}'$ will be the dual of $\mathfrak{R}$, i.e., the ring obtained from $\mathfrak{R}$ by leaving its elements and the operation of addition unchanged and replacing multiplication xy by the transpose yx.

Definition 4.1: A *dual-automorphism* of a complemented modular lattice L is a one-to-one correspondence $\mathfrak{a} \to \mathfrak{a}'$ of L with itself such that if $\mathfrak{a} \to \mathfrak{a}'$, $\mathfrak{b} \to \mathfrak{b}'$, $\mathfrak{a} < \mathfrak{b}$ then $\mathfrak{a}' > \mathfrak{b}'$.

Note: Clearly the condition

$$\mathfrak{a} \to \mathfrak{a}',\ \mathfrak{b} \to \mathfrak{b}',\ \mathfrak{a} < \mathfrak{b} \text{ implies } \mathfrak{a}' > \mathfrak{b}'$$

is equivalent to the one obtained by replacing in it $<$, $>$ by $\leqq$, $\geqq$, respectively. Moreover, it is equivalent to

$$\mathfrak{a} \to \mathfrak{a}', \ \mathfrak{b} \to \mathfrak{b}' \text{ implies } (\mathfrak{a} \cup \mathfrak{b})' = \mathfrak{a}' \cap \mathfrak{b}',$$

and hence by duality to

$$\mathfrak{a} \to \mathfrak{a}', \ \mathfrak{b} \to \mathfrak{b}' \text{ implies } (\mathfrak{a} \cap \mathfrak{b})' = \mathfrak{a}' \cup \mathfrak{b}',$$

as may be readily verified with the help of the definitions of $\cup$, $\cap$ and the fact that the correspondence is one-to-one.

DEFINITION 4.2: An *anti-automorphism* of $\mathfrak{R}$ is a one-to-one correspondence $x \to x^*$ of $\mathfrak{R}$ with itself such that $x \to x^*$, $y \to y^*$ implies $(x + y)^* = x^* + y^*$, $(xy)^* = y^* x^*$. (I.e., an anti-automorphism of $\mathfrak{R}$ is an isomorphism between $\mathfrak{R}$ and $\mathfrak{R}'$.) An element $x \in \mathfrak{R}$ is *Hermitian* in case $x^* = x$.

COROLLARY: 0, 1 are *Hermitian*.

PROOF: Clearly $0 = 0 + 0$, whence $0^* = 0^* + 0^*$, and $0^* = 0$. Moreover 1 is characterized by $a = a \cdot 1 = 1 \cdot a$ for every $a \in \mathfrak{R}$; hence $a^* = 1^* \cdot a^* = a^* \cdot 1^*$ for every $a \in \mathfrak{R}$, i.e. $b = 1^* \cdot b = b \cdot 1^*$ for every $b \in \mathfrak{R}$. Therefore $1 = 1^*$.

THEOREM 4.3: If $x \to x^*$ is an anti-automorphism of $\mathfrak{R}$, then the transformation

$$\mathscr{I}\colon \quad (a)_r \to (a^*)_l^r$$

(cf. Definition 2.3) is a dual-automorphism of $\bar{R}_{\mathfrak{R}}$. The given anti-automorphism of $\mathfrak{R}$ is said to *generate* $\mathscr{I}$. Conversely, if $\mathfrak{R}$ is of order $n \geqq 3$, every dual-automorphism $\mathscr{I}$ of $\bar{R}_{\mathfrak{R}}$ is generated by a unique anti-automorphism of $\mathfrak{R}$.

PROOF: Clearly $x \to x^*$ defines an isomorphism between $\mathfrak{R}$ and $\mathfrak{R}'$ and hence induces an isomorphism between $\bar{R}_{\mathfrak{R}}$ and $\bar{R}_{\mathfrak{R}'}$. But obviously $\bar{R}_{\mathfrak{R}'} = \bar{L}_{\mathfrak{R}}$, whence $\mathscr{J}\colon (a)_r \to (a^*)_l$ is an isomorphism between $\bar{R}_{\mathfrak{R}}$ and $\bar{L}_{\mathfrak{R}}$. But if $\mathscr{I}_0$ is the anti-isomorphism: $\mathfrak{a} \to \mathfrak{a}^l$ between $\bar{R}_{\mathfrak{R}}$ and $\bar{L}_{\mathfrak{R}}$, then the product $\mathscr{J}\mathscr{I}_0^{-1}$ (in the order written) is an anti-isomorphism between $\bar{R}_{\mathfrak{R}}$ and $\bar{R}_{\mathfrak{R}}$, i.e., a dual-automorphism of $\bar{R}_{\mathfrak{R}}$. But $\mathscr{I} = \mathscr{J}\mathscr{I}_0^{-1}$, whence the desired conclusion follows. To prove the converse, we note that if $\mathscr{J} = \mathscr{I}\mathscr{I}_0$, $\mathscr{I}$ being given and $\mathscr{I}_0$ being defined above, then $\mathscr{J}$ is an isomorphism of $\bar{R}_{\mathfrak{R}}$ with $\bar{R}_{\mathfrak{R}'}$, since $\mathscr{I}$ is an anti-automorphism of $\bar{R}_{\mathfrak{R}}$ and $\mathscr{I}_0$ is an anti-isomorphism between $\bar{R}_{\mathfrak{R}}$ and $\bar{L}_{\mathfrak{R}} = \bar{R}_{\mathfrak{R}'}$. Hence by Theorem 4.2 there exists an isomorphism $x \to x^*$ between $\mathfrak{R}$ and $\mathfrak{R}'$, i.e., an anti-automorphism of $\mathfrak{R}$, which induces the isomorphism $\mathscr{J}$ between $\bar{R}_{\mathfrak{R}}$ and $\bar{R}_{\mathfrak{R}'}$. Let $\mathscr{I}'$ denote the dual-automorphism of $\bar{R}_{\mathfrak{R}}$ generated by the transformation $x \to x^*$. Then

$$\mathscr{I}'\colon \quad (a)_r \to (a^*)_l^r,$$

whence

$$\mathscr{I}'\mathscr{I}_0 : (a)_r \to (a^*)_l^{rl} = (a^*)_l,$$

i.e., $\mathscr{I}'\mathscr{I}_0 = \mathscr{J} = \mathscr{I}\mathscr{I}_0$, and $\mathscr{I}' = \mathscr{I}$. Hence $\mathscr{I}$ is generated by $x \to x^*$. Suppose another anti-automorphism $x \to x'$ generates $\mathscr{I}$. Then $x \to x'$ induces the isomorphism $\mathscr{J}$ between $\bar{R}_{\mathfrak{R}}$ and $\bar{R}_{\mathfrak{R}'}$, whence by Theorem 4.1 it coincides with $x \to x^*$. (It has already been observed that Theorem 4.1 implies the uniqueness of an isomorphism between $\mathfrak{R}$ and $\mathfrak{R}'$ inducing a given isomorphism between $\bar{R}_{\mathfrak{R}}$ and $\bar{R}_{\mathfrak{R}'}$.) This completes the proof.

Corollary: The anti-automorphisms $x \to x^*$ of $\mathfrak{R}$ correspond in one-to-one manner with the dual-automorphisms $\mathscr{I} : \mathfrak{a} \to \mathfrak{a}'$ of $\bar{R}_{\mathfrak{R}}$; if $\mathscr{I}$ is generated by $x \to x^*$, then $\mathscr{I}$ is represented by

$$(18) \qquad \mathscr{I}\colon\ (a)_r \to (a^*)_l^r = (ya^*;\ y \in \mathfrak{R})^r = (x;\ a^*x = 0),$$

and also by

$$(19) \qquad \mathscr{I}\colon\ \mathfrak{a} \to \mathfrak{a}' = (x^*;\ x \in \mathfrak{a})^r.$$

Proof: The first part and the representation of $\mathscr{I}$ by (18) are immediate. We shall establish (19). Clearly $\mathfrak{a} = (a)_r$ yields $(x^*;\ x \in \mathfrak{a}) = ((au)^*;\ u \in \mathfrak{R}) = (u^*a^*;\ u \in \mathfrak{R}) = (ya^*;\ y \in \mathfrak{R}) = (a^*)_l$, whence $\mathfrak{a} \to \mathfrak{a}' = (a^*)_l^r = (x^*;\ x \in \mathfrak{a})^r$.

Lemma 4.3: If $x \to x^*$ is an anti-automorphism of $\mathfrak{R}$ generating the dual-isomorphism $\mathfrak{a} \to \mathfrak{a}'$ of $\bar{R}_{\mathfrak{R}}$, then for every $a \in \mathfrak{R}$, $(a)_r'' = (a^{**})_r$.

Proof: Evidently

$$\begin{aligned}
(a)_r'' &= (y^*;\ y \in (a)_r')^r && \text{(by 19))}\\
&= (y^*;\ a^*y = 0)^r && \text{(by (18))}\\
&= (y^*;\ y^*a^{**} = 0)^r && \text{(since } 0 = 0^*)\\
&= (z;\ za^{**} = 0)^r &&\\
&= (a^{**})_r^{lr} = (a^{**})_r. &&
\end{aligned}$$

Theorem 4.4: Assume that $\mathfrak{R}$ is of order $n \geqq 3$. A dual-automorphism $\mathfrak{a} \to \mathfrak{a}'$ of $\bar{R}_{\mathfrak{R}}$ is *involutoric*, i.e., has the property $\mathfrak{a}'' = \mathfrak{a}$ for every $\mathfrak{a} \in \bar{R}_{\mathfrak{R}}$, if and only if the anti-automorphism $x \to x^*$ which generates it is *involutoric*, i.e. has the property $x = x^{**}$ for every $x \in \mathfrak{R}$.

Proof: Clearly $x \to x^{**}$ is an automorphism of $\mathfrak{R}$, which induces by Lemma 4.3 the automorphism $\mathfrak{a} \to \mathfrak{a}''$ of $\bar{R}_{\mathfrak{R}}$. If $\mathfrak{a} = \mathfrak{a}''$ for every $\mathfrak{a} \in \bar{R}_{\mathfrak{R}}$, then $x \to x^{**}$ is an automorphism of $\mathfrak{R}$ leaving every principal right ideal in $\mathfrak{R}$ invariant, and is therefore the identity by Theorem 4.1. Conversely, of course, $x^{**} = x$ for every $x \in \mathfrak{R}$ implies $\mathfrak{a}'' = \mathfrak{a}$ for every $\mathfrak{a} \in \bar{R}_{\mathfrak{R}}$.

Theorem 4.5: Let the involutoric anti-automorphism $x \to x^*$ of $\mathfrak{R}$ generate the involutoric dual-automorphism $\mathfrak{a} \to \mathfrak{a}'$ of $\bar{R}_{\mathfrak{R}}$. Then the following properties are equivalent:

(a) $\mathfrak{a} \cap \mathfrak{a}' = (0)$ for every $\mathfrak{a} \in \bar{R}_{\mathfrak{R}}$;
(a') $\mathfrak{a} \cup \mathfrak{a}' = \mathfrak{R}$ for every $\mathfrak{a} \in \bar{R}_{\mathfrak{R}}$;
(b) for every $a \in \mathfrak{R}$ there exists $z \in \mathfrak{R}$ with $a = za^*a$;
(c) $a^*a = 0$ implies $a = 0$;
(d) for every $a \in R$ there exists a Hermitian idempotent $e \in \mathfrak{R}$ such that $(a)_r = (e)_r$.

Moreover, in (d) the Hermitian idempotent e is uniquely determined by $(a)_r = (e)_r$.

PROOF: The equivalence of (a) and (a') is immediate since $(\mathfrak{a} \cap \mathfrak{a}')' = \mathfrak{a}' \cup \mathfrak{a}'' = \mathfrak{a} \cup \mathfrak{a}'$, $(\mathfrak{a} \cup \mathfrak{a}')' = \mathfrak{a} \cap \mathfrak{a}'$ (because $\mathfrak{a} \to \mathfrak{a}'$ is involutoric), and since $(0)' = \mathfrak{R}$, $\mathfrak{R}' = (0)$. We shall now establish the equivalences (a) $\rightleftarrows$ (b), (a) $\rightleftarrows$ (c), (b) $\rightleftarrows$ (d). Now $\mathfrak{a} \cap \mathfrak{a}' = (0)$ means that $z \in \mathfrak{a}$, $\mathfrak{a}'$ implies $z = 0$, i.e., $z \in (ax;\ x)$, $(y\colon a^*y = 0)$ implies $z = 0$, i.e., $a^*ax = 0$ implies $ax = 0$. This condition is equivalent to $(x;\ a^*ax = 0) \subset (x;\ ax = 0)$, hence to $(a^*a)_l{}^r \subset (a)_l{}^r$, and therefore to $(a^*a)_l \supset (a)_l$ by Lemma 2.1 (1). But this is manifestly equivalent to $a \in (a^*a)_l$ for every $a \in \mathfrak{R}$, i.e., to (b). Hence (a) $\rightleftarrows$ (b).

Now $\mathfrak{a} \cap \mathfrak{a}' = (0)$ for every $\mathfrak{a} \in \bar{R}_{\mathfrak{R}}$ means that $\mathfrak{a} \subset \mathfrak{a}'$ implies $\mathfrak{a} = (0)$. For, if $\mathfrak{a} \cap \mathfrak{a}' = (0)$, $\mathfrak{a} \subset \mathfrak{a}'$, then $\mathfrak{a} = \mathfrak{a} \cap \mathfrak{a}' \subset (0)$; conversely $\mathfrak{a} \cap \mathfrak{a}' \subset \mathfrak{a} \cup \mathfrak{a}' = (\mathfrak{a} \cap \mathfrak{a}')'$, whence $\mathfrak{a} \cap \mathfrak{a}' = (0)$. Hence (a) means that $(a)_r \subset (y;\ a^*y = 0)$ implies $(a)_r = (0)$, i.e., implies $a = 0$. Hence (a) is equivalent to this property: $a^*ax = 0$ for every $x \in \mathfrak{R}$ implies $a = 0$, i.e., to (c), since $a^*a = 0$ if and only if $a^*ax = 0$ for every $x \in \mathfrak{R}$. Thus (a) $\rightleftarrows$ (c).

Let us assume (b), viz., that every $a \in \mathfrak{R}$ is of the form $a = za^*a$, $z \in \mathfrak{R}$. Define $e \equiv za^*$. Then $e^* = a^{**}z^* = az^*$; hence

$$e^* = a \cdot z^* = za^*a \cdot z^* = za^* \cdot az^* = ee^*,$$

and

$$e = e^{**} = (ee^*)^* = e^{**}e^* = ee^* = e^*;$$

moreover $e^2 = e \cdot e = e \cdot e^* = e$. Thus e is a Hermitian idempotent. Now $a = za^* \cdot a = ea$, whence $a \in (e)_r$, $(a)_r \subset (e)_r$. But $e = az^*$, whence $e \in (a)_r$, $(e)_r \subset (a)_r$. Therefore $(e)_r = (a)_r$, and (d) is established. Conversely, let (d) hold. Then $(a)_r = (e)_r$, with e idempotent and Hermitian. Consequently, e is of the form $e = ax$, $x \in \mathfrak{R}$ and $a = ea$; hence $e = e^* = x^*a^*$, $a = ea = x^*a^*a$, whence $z \equiv x^*$ yields $a = za^*a$. Thus (b) $\rightleftarrows$ (d).

To establish the uniqueness of e in (d), let $(e)_r = (f)_r$, with e, f Hermitian idempotents. Then $fe = e$, $ef = f$, whence $f = f^* = (ef)^* = f^*e^* = fe = e$.

For the purposes of the following theorem, in which we analyze the implications of conditions (b), (c), (d) of Theorem 4.5, we drop the assumption of the regularity of $\mathfrak{R}$.

THEOREM 4.6: Let $\mathfrak{R}$ be a ring (with unit) and let $x \to x^*$ be an involutoric anti-automorphism of $\mathfrak{R}$ (i.e., $x^{**} = x$ for every $x \epsilon \mathfrak{R}$). Then conditions (b) and (d) separately imply that $\mathfrak{R}$ is regular; (c) implies that $\mathfrak{R}$ is semi-simple (cf. Chapter II, Appendix 1 (4)), but not that $\mathfrak{R}$ is regular.

PROOF: Since condition (d) is clearly a strengthened form of Theorem 2.2 (β), it obviously implies regularity. We prove next that (b) implies regularity. Now if $a = za^*a$, then $a^* = a^*a^{**}z^* = a^*az^*$, whence $az^* = za^*a \cdot z^* = z \cdot a^*az^* = za^*$. Therefore $a = za^* \cdot a = az^* \cdot a = axa$ with $x = z^*$, and we have established condition (γ) of Theorem 2.2 and hence regularity.

To prove that (c) implies semi-simplicity, we repeat essentially the argument given in Chapter II, Appendix 2 (I). The contrapositive of (c) is that $a \neq 0$ implies $a^*a \neq 0$. Hence if $a^* = a$, then $a \neq 0$ implies $a^2 \neq 0$, and moreover $(a^2)^* = (a^*)^2 = a^2$. Therefore by iteration, $a \neq 0$, $a^* = a$ implies $a^{(2n)} = (\cdots((a^2)^2)\cdots)^2 \neq 0$ whence $a^m \neq 0$ for every n and $m \leqq 2^n$. Thus $a \neq 0$, $a^* = a$ implies $a^m \neq 0$ for every m. But $(aa^*)^* = a^{**}a^* = aa^*$, whence we see that $a \neq 0$ implies $a^* \neq 0$, $aa^* \neq 0$, $(aa^*)^m \neq 0$. Now if $\mathfrak{a}$ is a right ideal $\neq (0)$, we may select $a \epsilon \mathfrak{a}$ with $a \neq 0$; then $aa^* \epsilon \mathfrak{a}$, $(aa^*)^m \epsilon \mathfrak{a}^m$, $(aa^*)^m \neq 0$, whence $\mathfrak{a}^m \neq (0)$. Thus $\mathfrak{a}$ is not nilpotent, and by definition $\mathfrak{R}$ is semi-simple. To show that (c) does not imply regularity, let $\mathfrak{R}$ be the ring of all bounded operators in a Hilbert space $\mathfrak{H}$ (cf. Chapter II, Appendix 2), and define a^* to be the Hermitian adjoint of a. Then $a^*a = 0$ implies $a = 0$; nevertheless $\mathfrak{R}$ is not regular (cf. loc. cit., (II)).

PART II · CHAPTER V

Projective Isomorphisms in a Complemented Modular Lattice

In what follows, until the contrary is stated, L is supposed to be a given complemented, modular lattice of order n ($n \geqq 4$); for the present we shall assume a homogeneous basis ($\bar{\mathfrak{a}}_i$; $i = 1, \cdots, n$) to be fixed in L. The theory in this chapter is preliminary to the introduction of a regular ring $\mathfrak{R}$ associated with L such that L is isomorphic to the lattice of all principal right ideals in $\mathfrak{R}$; this result has been mentioned at the beginning of Chapter IV. The essential result of this chapter is that contained in Theorem 5.1; Lemmas 5.4—5.10 are preparatory for it. Unless otherwise indicated, indices $i, j, k, \cdots$ vary over the range $(1, \cdots, n)$.

LEMMA 5.1: If $i \neq j \neq k \neq i$, $\mathfrak{b}_{ij} \in L_{ij}$, $\mathfrak{b}_{jk} \in L_{jk}$, and if $\mathfrak{b}_{ik} \equiv (\mathfrak{b}_{ij} \cup \mathfrak{b}_{jk}) \cap (\bar{\mathfrak{a}}_i \cup \bar{\mathfrak{a}}_k)$, then $\mathfrak{b}_{ik} \in L_{ik}$.

PROOF: We have

$$\begin{aligned} \bar{\mathfrak{a}}_k \cup \mathfrak{b}_{ik} &= \bar{\mathfrak{a}}_k \cup ((\mathfrak{b}_{ij} \cup \mathfrak{b}_{jk}) \cap (\bar{\mathfrak{a}}_i \cup \bar{\mathfrak{a}}_k)) \\ &= (\bar{\mathfrak{a}}_k \cup \mathfrak{b}_{ij} \cup \mathfrak{b}_{jk}) \cap (\bar{\mathfrak{a}}_i \cup \bar{\mathfrak{a}}_k) \qquad \text{(by IV)} \\ &= (\bar{\mathfrak{a}}_k \cup \bar{\mathfrak{a}}_j \cup \mathfrak{b}_{ij}) \cap (\bar{\mathfrak{a}}_i \cup \bar{\mathfrak{a}}_k) \\ &= (\bar{\mathfrak{a}}_k \cup \bar{\mathfrak{a}}_j \cup \bar{\mathfrak{a}}_i) \cap (\bar{\mathfrak{a}}_i \cup \bar{\mathfrak{a}}_k) \\ &= \bar{\mathfrak{a}}_i \cup \bar{\mathfrak{a}}_k. \end{aligned}$$

Moreover, since $(\bar{\mathfrak{a}}_i, \bar{\mathfrak{a}}_j, \bar{\mathfrak{a}}_k)\perp$, it follows from Part I, Lemma 2.1, that $(\mathfrak{b}_{ij}, \bar{\mathfrak{a}}_j, \bar{\mathfrak{a}}_k)\perp$, and hence by a second application of this lemma, that $(\mathfrak{b}_{ij}, \mathfrak{b}_{jk}, \bar{\mathfrak{a}}_k)\perp$. Thus $(\mathfrak{b}_{ij} \cup \mathfrak{b}_{jk}) \cap \bar{\mathfrak{a}}_k = 0$, and

$$\bar{\mathfrak{a}}_k \cap \mathfrak{b}_{ik} = (\mathfrak{b}_{ij} \cup \mathfrak{b}_{jk}) \cap (\bar{\mathfrak{a}}_i \cup \bar{\mathfrak{a}}_k) \cap \bar{\mathfrak{a}}_k = (\mathfrak{b}_{ij} \cup \mathfrak{b}_{jk}) \cap \bar{\mathfrak{a}}_k = 0.$$

Consequently $\mathfrak{b}_{ik}$ is an inverse of $\bar{\mathfrak{a}}_k$ in $\bar{\mathfrak{a}}_i \cup \bar{\mathfrak{a}}_k$, i.e., $\mathfrak{b}_{ik} \in L_{ik}$.

DEFINITION 5.1: If $i \neq j \neq j \neq k \neq i$, $\mathfrak{b}_{ij} \in L_{ij}$, $\mathfrak{b}_{jk} \in L_{jk}$, then we define $\mathfrak{b}_{ij} \otimes \mathfrak{b}_{jk} \equiv (\mathfrak{b}_{ij} \cup \mathfrak{b}_{jk}) \cap (\bar{\mathfrak{a}}_i \cup \bar{\mathfrak{a}}_k)$.

COROLLARY: If $\mathfrak{b}_{ij} \in L_{ij}$, $\mathfrak{b}_{jk} \in L_{jk}$, then $\mathfrak{b}_{ij} \otimes \mathfrak{b}_{jk} \in L_{ik}$.

LEMMA 5.2: The operation $\otimes$ is associative, i.e., for i, j, k, l distinct integers, $\mathfrak{b}_{ij} \in L_{ij}$, $\mathfrak{b}_{jk} \in L_{jk}$, $\mathfrak{b}_{kl} \in L_{kl}$ we have

(1) $$(\mathfrak{b}_{ij} \otimes \mathfrak{b}_{jk}) \otimes \mathfrak{b}_{kl} = \mathfrak{b}_{ij} \otimes (\mathfrak{b}_{jk} \otimes \mathfrak{b}_{kl}).$$

PROOF:

$$\text{The left member of (1)} = \{(\mathfrak{b}_{ij} \cup \mathfrak{b}_{jk})(\bar{\mathfrak{a}}_i \cup \bar{\mathfrak{a}}_k \cup \bar{\mathfrak{a}}_l) \cup \mathfrak{b}_{kl}\}\,(\bar{\mathfrak{a}}_i \cup \bar{\mathfrak{a}}_l)$$
$$= (\mathfrak{b}_{ij} \cup \mathfrak{b}_{jk} \cup \mathfrak{b}_{kl})(\bar{\mathfrak{a}}_i \cup \bar{\mathfrak{a}}_l) \qquad \text{(by IV)}$$
$$= \text{the right member of (1).}$$

DEFINITION 5.2: A system $(\bar{\mathfrak{c}}_{ij};\ i, j = 1, \cdots, n,\ i \neq j)$ of axes of perspectivity for the homogeneous basis $(\bar{\mathfrak{a}}_i;\ i = 1, \cdots, n)$ (i.e., $\bar{\mathfrak{c}}_{ij} \in L_{ij}, L_{ji}$) is *normalized* in case

$$\text{(2)} \qquad \bar{\mathfrak{c}}_{ij} = \bar{\mathfrak{c}}_{ji} \qquad (i \neq j)$$

$$\text{(3)} \qquad \bar{\mathfrak{c}}_{ik} = \bar{\mathfrak{c}}_{ij} \otimes \bar{\mathfrak{c}}_{jk} \qquad (i \neq j \neq k \neq i).$$

The combined system $(\bar{\mathfrak{a}}_i, \bar{\mathfrak{c}}_{ij};\ i \neq j)$ is a *normalized frame* for L in case (2) and (3) hold; if (2), (3) are not required to hold, this system is simply called a *frame*.

LEMMA 5.3: There exists a normalized system $(\bar{\mathfrak{c}}_{ij};\ i \neq j)$ for the basis $(\bar{\mathfrak{a}}_i)$; the elements $\bar{\mathfrak{c}}_{i1} \in L_{i1}, L_{1i}$ $(i = 2, \cdots, n)$ may be chosen arbitrarily and the remaining members are then uniquely determined.

PROOF: The existence of the $\bar{\mathfrak{c}}_{i1}$ as common inverses of $\bar{\mathfrak{a}}_1, \bar{\mathfrak{a}}_i$ in $\bar{\mathfrak{a}}_1 \cup \bar{\mathfrak{a}}_i$ $(i = 2, \cdots, n)$ is obvious since $\bar{\mathfrak{a}}_1 \sim \bar{\mathfrak{a}}_1$. If (2) is to hold, we must define $\bar{\mathfrak{c}}_{1j} \equiv \bar{\mathfrak{c}}_{j1}$ $(j = 2, \cdots, n)$, and if (3) is to hold, we must define

$$\text{(4)} \qquad \bar{\mathfrak{c}}_{ij} \equiv \bar{\mathfrak{c}}_{i1} \otimes \bar{\mathfrak{c}}_{1j} = (\bar{\mathfrak{c}}_{i1} \cup \bar{\mathfrak{c}}_{1j}) \cap (\bar{\mathfrak{a}}_i \cup \bar{\mathfrak{a}}_j) \qquad (i, j \neq 1,\ i \neq j).$$

Equation (2) is now evidently satisfied. Since $\bar{\mathfrak{c}}_{i1} \in L_{i1}$ $(i = 2, \cdots, n)$, $\bar{\mathfrak{c}}_{1j} \in L_{1j}$ $(j = 2, \cdots, n)$, it follows from Lemma 5.1 that $\bar{\mathfrak{c}}_{ij} \in L_{ij}$ $(i, j \neq 1, i \neq j)$. Hence $\bar{\mathfrak{c}}_{ij} \in L_{ij}$ for every i, j with $i \neq j$, whence $\bar{\mathfrak{c}}_{ij} = \bar{\mathfrak{c}}_{ji} \in L_{ji}$ also, and $\bar{\mathfrak{c}}_{ij}$ is an axis of perspectivity for $\bar{\mathfrak{a}}_i, \bar{\mathfrak{a}}_j$. It remains to verify (3). If $j = 1$, (3) follows from (4). Suppose then that $j \neq 1$. First, let $i = 1$, whence $k \neq 1, j$. Then

$$\bar{\mathfrak{c}}_{ij} \otimes \bar{\mathfrak{c}}_{jk} = \bar{\mathfrak{c}}_{1j} \otimes (\bar{\mathfrak{c}}_{j1} \otimes \bar{\mathfrak{c}}_{1k}) = \bar{\mathfrak{c}}_{j1} \otimes (\bar{\mathfrak{c}}_{j1} \otimes \bar{\mathfrak{c}}_{1k})$$
$$= \{\bar{\mathfrak{c}}_{j1} \cup [(\bar{\mathfrak{c}}_{j1} \cup \bar{\mathfrak{c}}_{1k}) \cap (\bar{\mathfrak{a}}_j \cup \bar{\mathfrak{a}}_k)]\} \cap (\bar{\mathfrak{a}}_1 \cup \bar{\mathfrak{a}}_k)$$
$$= (\bar{\mathfrak{c}}_{j1} \cup \bar{\mathfrak{c}}_{1k}) \cap (\bar{\mathfrak{c}}_{j1} \cup \bar{\mathfrak{a}}_j \cup \bar{\mathfrak{a}}_k) \cap (\bar{\mathfrak{a}}_1 \cup \bar{\mathfrak{a}}_k) \qquad \text{(by IV)}$$
$$= (\bar{\mathfrak{c}}_{j1} \cup \bar{\mathfrak{c}}_{1k}) \cap (\bar{\mathfrak{a}}_1 \cup \bar{\mathfrak{a}}_j \cup \bar{\mathfrak{a}}_k) \cap (\bar{\mathfrak{a}}_1 \cup \bar{\mathfrak{a}}_k)$$
$$= (\bar{\mathfrak{c}}_{j1} \cup \bar{\mathfrak{c}}_{1k}) \cap (\bar{\mathfrak{a}}_1 \cup \bar{\mathfrak{a}}_k)$$
$$= \bar{\mathfrak{c}}_{1k} \cap \bar{\mathfrak{c}}_{j1}(\bar{\mathfrak{a}}_1 \cup \bar{\mathfrak{a}}_k) \qquad \text{(by IV)}$$
$$= \bar{\mathfrak{c}}_{1k} \cap 0 = \bar{\mathfrak{c}}_{1k}.$$

Thus (3) holds for $i = 1$. Next, let $k = 1$, whence $i \neq 1, j$. Then an analysis similar to the one just made shows that (3) holds for this case also. Finally, let us consider the general case where 1, i, j, k are distinct. Then

$$\begin{aligned} \bar{\mathfrak{c}}_{ij} \otimes \bar{\mathfrak{c}}_{jk} &= (\bar{\mathfrak{c}}_{i1} \otimes \bar{\mathfrak{c}}_{1j}) \otimes \bar{\mathfrak{c}}_{jk} \\ &= \bar{\mathfrak{c}}_{i1} \otimes (\bar{\mathfrak{c}}_{1j} \otimes \bar{\mathfrak{c}}_{jk}) && \text{(by Lemma 5.2)} \\ &= \bar{\mathfrak{c}}_{i1} \otimes \bar{\mathfrak{c}}_{1k} && \text{(by (3) with } i = 1) \\ &= \bar{\mathfrak{c}}_{ik} && \text{(by (4))}, \end{aligned}$$

and (3) is established. This completes the proof.

We shall at this point fix a normalized system $(\bar{\mathfrak{c}}_{ij};\ i \neq j)$ in L so that $(\bar{\mathfrak{a}}_i, \bar{\mathfrak{c}}_{ij};\ i \neq j)$ will be a fixed normalized frame throughout the following discussion until the contrary is asserted.

Throughout the remainder of the chapter we shall be dealing with sequences of m integers $i_\rho, j_\rho, \cdots$ $(\rho = 1, \cdots, m)$, $m < n$; we make the convention that in each of these sequences the m integers shall be distinct and shall belong to the range $(1, \cdots, n)$.

LEMMA 5.4: If $(i_1, \cdots, i_m)$, $(j_1, \cdots, j_m)$ are such that $i_\rho = j_\rho$ except for one value $\bar{\rho}$ of ρ, then

$$\sum_{\rho=1}^{m}{}_{\cup}\, \bar{\mathfrak{a}}_{i_\rho} \sim \sum_{\rho=1}^{m}{}_{\cup}\, \bar{\mathfrak{a}}_{j_\rho} \qquad (\text{mod } \bar{\mathfrak{c}}_{i_{\bar{\rho}} j_{\bar{\rho}}}). \tag{5}$$

PROOF: Since $\mathfrak{a} \equiv \Sigma_{\cup} (\bar{\mathfrak{a}}_{i_\rho};\ \rho = 1, \cdots, m,\ \rho \neq \bar{\rho})$ $= \Sigma_{\cup} (\bar{\mathfrak{a}}_{j_\rho};\ \rho = 1, \cdots, m,\ \rho \neq \bar{\rho})$, and since $(\mathfrak{a}, \bar{\mathfrak{a}}_{i_{\bar{\rho}}}, \bar{\mathfrak{a}}_{j_{\bar{\rho}}}) \perp$, $\bar{\mathfrak{a}}_{i_{\bar{\rho}}} \sim \bar{\mathfrak{a}}_{j_{\bar{\rho}}}$ (mod $\bar{\mathfrak{c}}_{i_{\bar{\rho}} j_{\bar{\rho}}}$) implies by the proof of Theorem 3.5 in Part I, $\mathfrak{a} \cup \bar{\mathfrak{a}}_{i_{\bar{\rho}}} \sim \mathfrak{a} \cup \bar{\mathfrak{a}}_{j_{\bar{\rho}}}$ (mod $\bar{\mathfrak{c}}_{i_{\bar{\rho}} j_{\bar{\rho}}}$). This is precisely (5).

DEFINITION 5.3: Let $(i_1, \cdots, i_m)$, $(j_1, \cdots, j_m)$ be given. If $i_\rho = j_\rho$ $(\rho = 1, \cdots, m)$, then we define $P\begin{pmatrix} i_1 & \cdots & i_m \\ j_1 & \cdots & j_m \end{pmatrix}$ as the identical automorphism of $L(0, \sum_{\rho=1}^{m}{}_{\cup}\, \bar{\mathfrak{a}}_{i_\rho})$. If $i_\rho = j_\rho$ except for one value $\bar{\rho}$ of ρ, then we define $P\begin{pmatrix} i_1 & \cdots & i_m \\ j_1 & \cdots & j_m \end{pmatrix}$ as the perspective isomorphism defined by (5) of the lattices $L(0, \sum_{\rho=1}^{m}{}_{\cup}\, \bar{\mathfrak{a}}_{i_\rho})$, $L(0, \sum_{\rho=1}^{m}{}_{\cup}\, \bar{\mathfrak{a}}_{j_\rho})$ with the axis $\bar{\mathfrak{c}}_{i_{\bar{\rho}} j_{\bar{\rho}}}$. (Cf. Part I, Definition 3.4.)

LEMMA 5.5: Let $(i_1, \cdots, i_m)$, $(j_1, \cdots, j_m)$ be such that $i_\rho = j_\rho$ $(\rho = 1, \cdots, m)$, or such that $i_\rho = j_\rho$ except for $\rho = \bar{\rho}$.

(a) If $l = 1, \cdots, m$ and $\rho_1, \cdots, \rho_l$ are distinct integers in the set $(1, \cdots, m)$, then $P\begin{pmatrix} i_1 & \cdots & i_m \\ j_1 & \cdots & j_m \end{pmatrix}$ coincides with $P\begin{pmatrix} i_{\rho_1} & \cdots & i_{\rho_l} \\ j_{\rho_1} & \cdots & j_{\rho_l} \end{pmatrix}$ for the elements $\mathfrak{a}' \leqq \sum_{\lambda=1}^{l}{}_{\cup}\, \bar{\mathfrak{a}}_{i_{\rho_\lambda}}$.

(b) If $\mu \neq \sigma$, then $P\begin{pmatrix} i_1 \cdots i_m \\ j_1 \cdots j_m \end{pmatrix}$ maps $\bar{\mathfrak{c}}_{i_\mu i_\sigma}$ on $\bar{\mathfrak{c}}_{j_\mu j_\sigma}$.

PROOF: For the case $i_\rho = j_\rho$ $(\rho = 1, \cdots, m)$, both (a) and (b) are obvious. Suppose then $i_{\bar\rho} \neq j_{\bar\rho}$.

(a) Suppose $i_\rho \neq j_\rho$ for $\rho = \bar\rho \neq \tau_\lambda$ $(\lambda = 1, \cdots, l)$. Then $i_{\tau_\lambda} = j_{\tau_\lambda}$ for $\lambda = 1, \cdots, l$, whence $P\begin{pmatrix} i_1 \cdots i_m \\ j_1 \cdots j_m \end{pmatrix}$ is a perspectivity of $\sum_{\rho=1}^{m}{}_\cup\, \bar{\mathfrak{a}}_{i_\rho}$ and $\sum_{\rho=1}^{m}{}_\cup\, \bar{\mathfrak{a}}_{j_\rho}$ which are both $\geqq \sum_{\lambda=1}^{l}{}_\cup\, \bar{\mathfrak{a}}_{i_{\tau_\lambda}} = \sum_{\lambda=1}^{l}{}_\cup\, \bar{\mathfrak{a}}_{j_{\tau_\lambda}}$, and therefore is the identity on $L(0, \sum_{\lambda=1}^{l}{}_\cup\, \bar{\mathfrak{a}}_{i_{\tau_\lambda}})$. Suppose $i_\rho \neq j_\rho$ for $\rho = \bar\rho = \tau_{\bar\lambda}$. Then $i_{\tau_\lambda} = j_{\tau_\lambda}$ except for $\lambda = \bar\lambda$; both our transformations P are perspectivities with the same axes $\bar{\mathfrak{c}}_{i_{\bar\rho} j_{\bar\rho}} = \bar{\mathfrak{c}}_{i_{\tau_{\bar\lambda}} j_{\tau_{\bar\lambda}}}$. Since both domain and range of the second P are contained in those of the first P, respectively, we may conclude that our two transformations P coincide there.

(b) Suppose $i_\rho \neq j_\rho$ for $\rho = \bar\rho = \mu$. We must then prove $\bar{\mathfrak{c}}_{i_\mu i_\sigma} \sim \bar{\mathfrak{c}}_{j_\mu j_\sigma}$ (mod $\bar{\mathfrak{c}}_{i_{\bar\rho} j_{\bar\rho}}$), i.e., $\bar{\mathfrak{c}}_{i_{\bar\rho} i_\sigma} \sim \bar{\mathfrak{c}}_{j_{\bar\rho} j_\sigma}$ (mod $\bar{\mathfrak{c}}_{i_{\bar\rho} j_{\bar\rho}}$). But this holds since

$$\begin{aligned} \bar{\mathfrak{c}}_{i_{\bar\rho} j_{\bar\rho}} \cup \bar{\mathfrak{c}}_{i_{\bar\rho} i_\sigma} &= \bar{\mathfrak{c}}_{i_{\bar\rho} j_{\bar\rho}} \cup [(\bar{\mathfrak{c}}_{i_{\bar\rho} j_{\bar\rho}} \cup \bar{\mathfrak{c}}_{j_{\bar\rho} i_\sigma}) \cap (\bar{\mathfrak{a}}_{i_{\bar\rho}} \cup \bar{\mathfrak{a}}_{i_\sigma})] && \text{(by (3))} \\ &= (\bar{\mathfrak{a}}_{i_{\bar\rho}} \cup \bar{\mathfrak{a}}_{i_\sigma} \cup \bar{\mathfrak{c}}_{i_{\bar\rho} j_{\bar\rho}}) \cap (\mathfrak{c}_{i_{\bar\rho} j_{\bar\rho}} \cup \mathfrak{c}_{j_{\bar\rho} i_\sigma}) && \text{(by IV)} \\ &= (\bar{\mathfrak{a}}_{i_{\bar\rho}} \cup \bar{\mathfrak{a}}_{j_{\bar\rho}} \cup \bar{\mathfrak{a}}_{i_\sigma}) \cap (\bar{\mathfrak{c}}_{i_{\bar\rho} j_{\bar\rho}} \cup \bar{\mathfrak{c}}_{j_{\bar\rho} i_\sigma}) \\ &= \bar{\mathfrak{c}}_{i_{\bar\rho} j_{\bar\rho}} \cup \bar{\mathfrak{c}}_{j_{\bar\rho} i_\sigma} = \bar{\mathfrak{c}}_{i_{\bar\rho} j_{\bar\rho}} \cup \bar{\mathfrak{c}}_{j_{\bar\rho} j_\sigma} && \text{(since } i_\sigma = j_\sigma), \end{aligned}$$

and since $\bar{\mathfrak{c}}_{i_{\bar\rho} i_\sigma} \leqq \sum_{\rho=1}^{m}{}_\cup\, \bar{\mathfrak{a}}_{i_\rho}$, $\bar{\mathfrak{c}}_{j_{\bar\rho} j_\sigma} \leqq \sum_{\rho=1}^{m}{}_\cup\, \bar{\mathfrak{a}}_{j_\rho}$. Suppose $i_\rho \neq j_\rho$ for $\rho = \bar\rho = \sigma$. Then our result follows from the preceding by interchanging the rôles of μ and σ. Suppose finally that $i_\rho \neq j_\rho$ for $\rho = \bar\rho \neq \mu, \sigma$. Then by (a) the effect on $\bar{\mathfrak{c}}_{i_\mu i_\sigma}$ of our perspectivity P is the same as the effect of $P\begin{pmatrix} i_\mu i_\sigma \\ j_\mu j_\sigma \end{pmatrix}$. But this is the identity since $i_\mu = j_\mu$ and $i_\sigma = j_\sigma$, whence $\bar{\mathfrak{c}}_{i_\mu i_\sigma}$ is left invariant, i.e., is transformed into $\bar{\mathfrak{c}}_{j_\mu j_\sigma}$.

LEMMA 5.6: Let i, j, k, h be distinct. Then

$$P\begin{pmatrix} i\ j \\ k\ j \end{pmatrix} P\begin{pmatrix} k\ j \\ k\ h \end{pmatrix} = P\begin{pmatrix} i\ j \\ i\ h \end{pmatrix} P\begin{pmatrix} i\ h \\ k\ h \end{pmatrix} \tag{6}$$

(the operations in a product being applied in the order written).

PROOF: Consider $\mathfrak{u} \leqq \bar{\mathfrak{a}}_i \cup \bar{\mathfrak{a}}_j$. $P\begin{pmatrix} i\ j \\ k\ j \end{pmatrix}$ carries $\mathfrak{u}$ into $\mathfrak{v} \leqq \bar{\mathfrak{a}}_k \cup \bar{\mathfrak{a}}_j$ with $\mathfrak{u} \cup \bar{\mathfrak{c}}_{ik} = \mathfrak{v} \cup \bar{\mathfrak{c}}_{ik}$; $P\begin{pmatrix} k\ j \\ k\ h \end{pmatrix}$ carries $\mathfrak{v}$ into $\mathfrak{w} \leqq \bar{\mathfrak{a}}_k \cup \bar{\mathfrak{a}}_h$ with $\mathfrak{v} \cup \bar{\mathfrak{c}}_{jh} = \mathfrak{w} \cup \bar{\mathfrak{c}}_{jh}$.

On the other hand $P\begin{pmatrix} i & j \\ i & h \end{pmatrix}$ carries $\mathfrak{u}$ into $\mathfrak{v}' \leqq \bar{\mathfrak{a}}_i \cup \bar{\mathfrak{a}}_h$ with $\mathfrak{u} \cup \bar{\mathfrak{c}}_{jh} = \mathfrak{v}' \cup \bar{\mathfrak{c}}_{jh}$; $P\begin{pmatrix} i & h \\ k & h \end{pmatrix}$ carries $\mathfrak{v}'$ into $\mathfrak{w}' \leqq \bar{\mathfrak{a}}_k \cup \bar{\mathfrak{a}}_h$ with $\mathfrak{v}' \cup \bar{\mathfrak{c}}_{ik} = \mathfrak{w}' \cup \bar{\mathfrak{c}}_{ik}$. Hence $\mathfrak{u} \cup \bar{\mathfrak{c}}_{ik} \cup \bar{\mathfrak{c}}_{jh} = \mathfrak{v} \cup \bar{\mathfrak{c}}_{ik} \cup \bar{\mathfrak{c}}_{jh} = \mathfrak{w} \cup \bar{\mathfrak{c}}_{ik} \cup \bar{\mathfrak{c}}_{jh}$; similarly $\mathfrak{u} \cup \bar{\mathfrak{c}}_{ik} \cup \bar{\mathfrak{c}}_{jh} = \mathfrak{w}' \cup \bar{\mathfrak{c}}_{ik} \cup \bar{\mathfrak{c}}_{jh}$, whence $\mathfrak{w} \cup \bar{\mathfrak{c}}_{ik} \cup \bar{\mathfrak{c}}_{jh} = \mathfrak{w}' \cup \bar{\mathfrak{c}}_{ik} \cup \bar{\mathfrak{c}}_{jh}$. Moreover, $\mathfrak{w}, \mathfrak{w}' \leqq \bar{\mathfrak{a}}_k \cup \bar{\mathfrak{a}}_h$. Since $(\bar{\mathfrak{a}}_i, \bar{\mathfrak{a}}_j, \bar{\mathfrak{a}}_k, \bar{\mathfrak{a}}_h) \perp$, and since $\bar{\mathfrak{c}}_{ik} \in L_{ik}$, $\bar{\mathfrak{c}}_{jh} \in L_{jh}$, Part I, Theorem 2.8 yields $(\bar{\mathfrak{c}}_{ik}, \bar{\mathfrak{c}}_{jh}, \bar{\mathfrak{a}}_k, \bar{\mathfrak{a}}_h) \perp$, whence $(\bar{\mathfrak{a}}_k \cup \bar{\mathfrak{a}}_h) \cap (\bar{\mathfrak{c}}_{ik} \cup \bar{\mathfrak{c}}_{jh}) = 0$. Therefore

$$\begin{aligned} \mathfrak{w} = \mathfrak{w} \cup 0 &= \mathfrak{w} \cup [(\bar{\mathfrak{a}}_k \cup \bar{\mathfrak{a}}_h) \cap (\bar{\mathfrak{c}}_{ik} \cup \bar{\mathfrak{c}}_{jh})] \\ &= (\mathfrak{w} \cup \bar{\mathfrak{c}}_{ik} \cup \bar{\mathfrak{c}}_{jh}) \cap (\bar{\mathfrak{a}}_k \cup \bar{\mathfrak{a}}_h) \qquad \text{(by IV)} \\ &= (\mathfrak{w}' \cup \bar{\mathfrak{c}}_{ik} \cup \bar{\mathfrak{c}}_{jh}) \cap (\bar{\mathfrak{a}}_k \cup \bar{\mathfrak{a}}_h) \\ &= \mathfrak{w}' \cup [(\bar{\mathfrak{a}}_k \cup \bar{\mathfrak{a}}_h) \cap (\bar{\mathfrak{c}}_{ik} \cup \bar{\mathfrak{c}}_{jh})] \qquad \text{(by IV)} \\ &= \mathfrak{w}' \cup 0 = \mathfrak{w}', \end{aligned}$$

i.e., $\mathfrak{w} = \mathfrak{w}'$. This establishes (6).

Corollary: If i, j, k, h are distinct, then

$$P\begin{pmatrix} i & j \\ k & j \end{pmatrix} P\begin{pmatrix} k & j \\ k & h \end{pmatrix} P\begin{pmatrix} k & h \\ i & h \end{pmatrix} P\begin{pmatrix} i & h \\ i & j \end{pmatrix} = \text{Identity}.$$

Lemma 5.7: If i, j, k are distinct, and i, h, k are distinct, then

$$P\begin{pmatrix} i & j \\ i & k \end{pmatrix} P\begin{pmatrix} i & k \\ i & h \end{pmatrix} = P\begin{pmatrix} i & j \\ i & h \end{pmatrix}. \tag{7}$$

Proof: If $h = j$, (7) holds since $P\begin{pmatrix} i & j \\ i & k \end{pmatrix}$ and $P\begin{pmatrix} i & k \\ i & j \end{pmatrix}$ are obviously inverse mappings. Suppose $h \neq j$. Consider $\mathfrak{u} \leqq \bar{\mathfrak{a}}_i \cup \bar{\mathfrak{a}}_j$; $\mathfrak{u}$ is transformed by $P\begin{pmatrix} i & j \\ i & k \end{pmatrix}$ into $\mathfrak{v} \leqq \bar{\mathfrak{a}}_i \cup \bar{\mathfrak{a}}_k$, which in turn is transformed by $P\begin{pmatrix} i & k \\ i & h \end{pmatrix}$ into $\mathfrak{w} \leqq \bar{\mathfrak{a}}_i \cup \bar{\mathfrak{a}}_h$. We have

$$\mathfrak{u} \cup \bar{\mathfrak{c}}_{jk} = \mathfrak{v} \cup \bar{\mathfrak{c}}_{jk}, \quad \mathfrak{v} \cup \bar{\mathfrak{c}}_{kh} = \mathfrak{w} \cup \bar{\mathfrak{c}}_{kh}.$$

Now $\mathfrak{u} \cup \bar{\mathfrak{c}}_{jk} \cup \bar{\mathfrak{c}}_{kh} = \mathfrak{v} \cup \bar{\mathfrak{c}}_{jk} \cup \bar{\mathfrak{c}}_{kh} = \mathfrak{w} \cup \bar{\mathfrak{c}}_{jk} \cup \bar{\mathfrak{c}}_{kh}$. Then

$$(\mathfrak{u} \cup \bar{\mathfrak{c}}_{jk} \cup \bar{\mathfrak{c}}_{kh}) \cap (\bar{\mathfrak{a}}_i \cup \bar{\mathfrak{a}}_j \cup \bar{\mathfrak{a}}_h) = (\mathfrak{w} \cup \bar{\mathfrak{c}}_{jk} \cup \bar{\mathfrak{c}}_{kh}) \cap (\bar{\mathfrak{a}}_i \cup \bar{\mathfrak{a}}_j \cup \bar{\mathfrak{a}}_h);$$

since $\mathfrak{u} \leqq \bar{\mathfrak{a}}_i \cup \bar{\mathfrak{a}}_j$, $\mathfrak{w} \leqq \bar{\mathfrak{a}}_i \cup \bar{\mathfrak{a}}_h$, IV applies and we have

$$\mathfrak{u} \cup [(\bar{\mathfrak{c}}_{jk} \cup \bar{\mathfrak{c}}_{kh}) \cap (\bar{\mathfrak{a}}_i \cup \bar{\mathfrak{a}}_j \cup \bar{\mathfrak{a}}_h)] = \mathfrak{w} \cup [(\bar{\mathfrak{c}}_{jk} \cup \bar{\mathfrak{c}}_{kh}) \cap (\bar{\mathfrak{a}}_i \cup \bar{\mathfrak{a}}_j \cup \bar{\mathfrak{a}}_h)].$$

Since $\bar{\mathfrak{c}}_{jk} \cup \bar{\mathfrak{c}}_{kh} \leqq \bar{\mathfrak{a}}_j \cup \bar{\mathfrak{a}}_k \cup \bar{\mathfrak{a}}_h$,

$$\begin{aligned} (\bar{\mathfrak{c}}_{jk} \cup \bar{\mathfrak{c}}_{kh}) \cap (\bar{\mathfrak{a}}_i \cup \bar{\mathfrak{a}}_j \cup \bar{\mathfrak{a}}_h) &= (\bar{\mathfrak{c}}_{jk} \cup \bar{\mathfrak{c}}_{kh}) \cap (\bar{\mathfrak{a}}_j \cup \bar{\mathfrak{a}}_k \cup \bar{\mathfrak{a}}_h) \cap (\bar{\mathfrak{a}}_i \cup \bar{\mathfrak{a}}_j \cup \bar{\mathfrak{a}}_h) \\ &= (\bar{\mathfrak{c}}_{jk} \cup \bar{\mathfrak{c}}_{kh}) \cap (\bar{\mathfrak{a}}_j \cup \bar{\mathfrak{a}}_h) = \bar{\mathfrak{c}}_{jh}, \end{aligned}$$

the last equality but one holding by Part I, Theorem 2.5, since $(\bar{\mathfrak{a}}_i, \bar{\mathfrak{a}}_j, \bar{\mathfrak{a}}_k, \bar{\mathfrak{a}}_h)\perp$, and the last equality holding by (3). Thus $\mathfrak{u} \cup \bar{\mathfrak{c}}_{jh} = \mathfrak{w} \cup \bar{\mathfrak{c}}_{jh}$, and $\mathfrak{w}$ is the transform of $\mathfrak{u}$ under $P\begin{pmatrix} i & j \\ i & h \end{pmatrix}$. This establishes (7).

LEMMA 5.8: If i, j, k are distinct, then

$$\Psi \equiv P\begin{pmatrix} i & j \\ k & j \end{pmatrix} P\begin{pmatrix} k & j \\ k & i \end{pmatrix} P\begin{pmatrix} k & i \\ j & i \end{pmatrix} P\begin{pmatrix} j & i \\ j & k \end{pmatrix} P\begin{pmatrix} j & k \\ i & k \end{pmatrix} P\begin{pmatrix} i & k \\ i & j \end{pmatrix} = \text{Identity}. \tag{8}$$

PROOF: Since $n \geqq 4$, there exists $h = 1, \cdots, n$ such that $h \neq i, j, k$. Then by Lemma 5.7, $P\begin{pmatrix} k & j \\ k & i \end{pmatrix} = P\begin{pmatrix} k & j \\ k & h \end{pmatrix} P\begin{pmatrix} k & h \\ k & i \end{pmatrix}$. Thus

$$\Psi = P\begin{pmatrix} i & j \\ k & j \end{pmatrix} P\begin{pmatrix} k & j \\ k & h \end{pmatrix} P\begin{pmatrix} k & h \\ k & i \end{pmatrix} P\begin{pmatrix} k & i \\ j & i \end{pmatrix} P\begin{pmatrix} j & i \\ j & k \end{pmatrix} P\begin{pmatrix} j & k \\ i & k \end{pmatrix} P\begin{pmatrix} i & k \\ i & j \end{pmatrix}.$$

But by Lemma 5.6

$$P\begin{pmatrix} i & j \\ k & j \end{pmatrix} P\begin{pmatrix} k & j \\ k & h \end{pmatrix} = P\begin{pmatrix} i & j \\ i & h \end{pmatrix} P\begin{pmatrix} i & h \\ k & h \end{pmatrix}, \quad P\begin{pmatrix} k & h \\ k & i \end{pmatrix} P\begin{pmatrix} k & i \\ j & i \end{pmatrix} = P\begin{pmatrix} k & h \\ j & h \end{pmatrix} P\begin{pmatrix} j & h \\ j & i \end{pmatrix},$$

whence

$$\Psi = P\begin{pmatrix} i & j \\ i & h \end{pmatrix} P\begin{pmatrix} i & h \\ k & h \end{pmatrix} P\begin{pmatrix} k & h \\ j & h \end{pmatrix} P\begin{pmatrix} j & h \\ j & i \end{pmatrix} P\begin{pmatrix} j & i \\ j & k \end{pmatrix} P\begin{pmatrix} j & k \\ i & k \end{pmatrix} P\begin{pmatrix} i & k \\ i & j \end{pmatrix}.$$

By Lemma 5.7,

$$P\begin{pmatrix} i & h \\ k & h \end{pmatrix} P\begin{pmatrix} k & h \\ j & h \end{pmatrix} = P\begin{pmatrix} i & h \\ j & h \end{pmatrix}, \qquad P\begin{pmatrix} j & h \\ j & i \end{pmatrix} P\begin{pmatrix} j & i \\ j & k \end{pmatrix} = P\begin{pmatrix} j & h \\ j & k \end{pmatrix},$$

and we have

$$\Psi = P\begin{pmatrix} i & j \\ i & h \end{pmatrix} P\begin{pmatrix} i & h \\ j & h \end{pmatrix} P\begin{pmatrix} j & h \\ j & k \end{pmatrix} P\begin{pmatrix} j & k \\ i & k \end{pmatrix} P\begin{pmatrix} i & k \\ i & j \end{pmatrix}.$$

By Lemma 5.6

$$P\begin{pmatrix} j & h \\ j & k \end{pmatrix} P\begin{pmatrix} j & k \\ i & k \end{pmatrix} = P\begin{pmatrix} j & h \\ i & h \end{pmatrix} P\begin{pmatrix} i & h \\ i & k \end{pmatrix},$$

whence

$$\Psi = P\begin{pmatrix} i & j \\ i & h \end{pmatrix} P\begin{pmatrix} i & h \\ j & h \end{pmatrix} P\begin{pmatrix} j & h \\ i & h \end{pmatrix} P\begin{pmatrix} i & h \\ i & k \end{pmatrix} P\begin{pmatrix} i & k \\ i & j \end{pmatrix};$$

now Lemma 5.7 yields

$$P\begin{pmatrix} i & h \\ j & h \end{pmatrix} P\begin{pmatrix} j & h \\ i & h \end{pmatrix} = \text{Identity},$$

whence

$$\Psi = P\begin{pmatrix} i & j \\ i & h \end{pmatrix} P\begin{pmatrix} i & h \\ i & k \end{pmatrix} P\begin{pmatrix} i & k \\ i & j \end{pmatrix} = P\begin{pmatrix} i & j \\ i & h \end{pmatrix} P\begin{pmatrix} i & h \\ i & j \end{pmatrix} = \text{Identity},$$

the last two equalities holding by Lemma 5.7.

LEMMA 5.9: If

$$(9) \qquad \Delta \equiv \begin{pmatrix} i_1 & i_2 \\ i_1' & i_2' \end{pmatrix} \begin{pmatrix} i_1' & i_2' \\ i_1'' & i_2'' \end{pmatrix} \cdots \begin{pmatrix} i_1^{(\omega-1)} & i_2^{(\omega-1)} \\ i_1^{(\omega)} & i_2^{(\omega)} \end{pmatrix} = \begin{pmatrix} i_1 & i_2 \\ i_1 & i_2 \end{pmatrix} \qquad (i_\nu^{(\omega)} = i_\nu),$$

where Δ represents a product of permutations of pairs of integers in the set $(1, \cdots, n)$ such that for each $\tau = 1, \cdots, n$, $i_\nu^{(\tau-1)} \neq i_\nu^{(\tau)}$ occurs at most once. Then

$$(10) \qquad \Phi \equiv P\begin{pmatrix} i_1 & i_2 \\ i_1' & i_2' \end{pmatrix} \cdots P\begin{pmatrix} i_1^{(\omega-1)} & i_2^{(\omega-1)} \\ i_1^{(\omega)} & i_2^{(\omega)} \end{pmatrix} = \text{Identity}.$$

PROOF: We proceed by induction on ω. For $\omega = 1, 2$ our result is trivial. Suppose $\omega \geqq 3$ and that our result holds for every $\omega' < \omega$. We note first that each of (9) and (10) is equivalent to the statement obtained from it by making a cyclic permutation on the integers $0, 1, \cdots, \omega$; we shall make free use of this fact.

(A) If $i_\nu^{(\tau-1)} = i_\nu^{(\tau)}$ $(\nu = 1, 2)$ occurs for some value of τ, the factor $\begin{pmatrix} i_1^{(\tau-1)} & i_2^{(\tau-1)} \\ i_1^{(\tau)} & i_2^{(\tau)} \end{pmatrix}$ in Δ and the corresponding factor in Φ may be omitted, and our result follows from that for $\omega - 1$ by the induction hypothesis. Hence we may suppose that for each τ, $i_\nu^{(\tau-1)} \neq i_\nu^{(\tau)}$ for exactly one ν, which we denote by ν^τ (we define $\nu^0 \equiv \nu^\omega$). If $\nu^\tau = \nu^{\tau+1}$ ever occurs, we may assume $\nu^\tau = \nu^{\tau+1} = 2$, since the case $\nu^\tau = \nu^{\tau+1} = 1$ may be symmetrically treated. In Δ two consecutive factors of the form

$$\begin{pmatrix} i & i_2^{(\tau-1)} \\ i & i_2^{(\tau)} \end{pmatrix} \begin{pmatrix} i & i_2^{(\tau)} \\ i & i_2^{(\tau+1)} \end{pmatrix}$$

appear and may be replaced by

$$\begin{pmatrix} i & i_2^{(\tau-1)} \\ i & i_2^{(\tau+1)} \end{pmatrix};$$

the corresponding change may be made in Φ by Lemma 5.7. Thus our statement again holds for ω since it holds for $\omega - 1$. Let us then assume $\nu^\tau \neq \nu^{\tau+1}$ for every τ; since each $\nu^\tau = 1, 2$, we see that the ν^τ assume alternatingly the values 1, 2. But $i_\nu^{(\omega)} = i_\nu$ $(\nu = 1, 2)$, whence it is clear that ω must be even. Since $\omega \geqq 3$, we have $\omega = 4, 6, 8, \cdots$.

(B) Suppose now that $i_1^{(\tau)}$, $i_2^{(\tau)}$ are both distinct from $i_1^{(\tau')}$, $i_2^{(\tau')}$ for a pair τ, τ', where $\tau \neq \tau'$. By a cyclic permutation on the numbers $0, \cdots, \omega$ we may reduce the consideration to the case $\tau' = 0$, $0 < \tau < \omega$. Since at least two permutations are required to change (i_1, i_2) into $(i_1^{(\tau)}, i_2^{(\tau)})$, we have $2 \leqq \tau \leqq \omega - 2$. Let $\tau < \omega - 2$. Then we may assume $\nu^1 = 1$, since the case $\nu^1 = 2$ is symmetrically treated. Now the first factors in Δ are

$$\begin{pmatrix} i_1 & i_2 \\ i_1' & i_2' \end{pmatrix} \cdots \begin{pmatrix} i_1^{(\tau-1)} & i_2^{(\tau-1)} \\ i_1^{(\tau)} & i_2^{(\tau)} \end{pmatrix} = \begin{pmatrix} i_1 & i_2 \\ i_1 & i_2^{(\tau)} \end{pmatrix} \begin{pmatrix} i_1 & i_2^{(\tau)} \\ i_1^{(\tau)} & i_2^{(\tau)} \end{pmatrix},$$

and (9) may be written

$$\Delta' \equiv \begin{pmatrix} i_1 & i_2 \\ i_1 & i_2^{(\tau)} \end{pmatrix} \begin{pmatrix} i_1 & i_2^{(\tau)} \\ i_1^{(\tau)} & i_2^{(\tau)} \end{pmatrix} \begin{pmatrix} i_1^{(\tau)} & i_2^{(\tau)} \\ i_1^{(\tau+1)} & i_2^{(\tau+1)} \end{pmatrix} \cdots \begin{pmatrix} i_1^{(\omega-1)} & i_2^{(\omega-1)} \\ i_1^{(\omega)} & i_2^{(\omega)} \end{pmatrix} = \begin{pmatrix} i_1 & i_2 \\ i_1 & i_2 \end{pmatrix}.$$

The length of Δ' is $(\omega - \tau) + 2 \leqq \omega$. If it is $< \omega$, then

$$(11) \quad P\begin{pmatrix} i_1 & i_2 \\ i_1^{(\tau)} & i_2 \end{pmatrix} P\begin{pmatrix} i_1^{(\tau)} & i_2 \\ i_1^{(\tau)} & i_2^{(\tau)} \end{pmatrix} P\begin{pmatrix} i_1^{(\tau)} & i_2^{(\tau)} \\ i_1^{(\tau+1)} & i_2^{(\tau+1)} \end{pmatrix} \cdots P\begin{pmatrix} i_1^{(\omega-1)} & i_2^{(\omega-1)} \\ i_1^{(\omega)} & i_2^{(\omega)} \end{pmatrix} = \text{Identity}$$

by the induction hypothesis. Suppose the length of Δ' to be ω. Now Δ' has $\nu^1 = 2$; but Δ' has $\nu^0 = \nu^\omega = 2$ (since for Δ we assumed $\nu^1 = 1$ and that the numbers $\nu^1, \cdots, \nu^\omega = \nu^0$ assumed alternately the values 1, 2; Δ', Δ have obviously the same $\nu^0 = \nu^\omega$). Hence for Δ', $\nu^0 = \nu^1$ and (11) holds by the result in (A). Thus (11) holds in any case. Moreover,

$$(12) \quad \Delta'' \equiv \begin{pmatrix} i_1 & i_2 \\ i_1' & i_2' \end{pmatrix} \cdots \begin{pmatrix} i_1^{(\tau-1)} & i_2^{(\tau-1)} \\ i_1^{(\tau)} & i_2^{(\tau)} \end{pmatrix} \begin{pmatrix} i_1^{(\tau)} & i_2^{(\tau)} \\ i_1^{(\tau)} & i_2 \end{pmatrix} \begin{pmatrix} i_1^{(\tau)} & i_2 \\ i_1 & i_2 \end{pmatrix} = \begin{pmatrix} i_1 & i_2 \\ i_1 & i_2 \end{pmatrix},$$

and Δ'' has length $\tau + 2 < \omega$. Hence by the induction hypothesis

$$(13) \quad P\begin{pmatrix} i_1 & i_2 \\ i_1' & i_2' \end{pmatrix} \cdots P\begin{pmatrix} i_1^{(\tau-1)} & i_2^{(\tau-1)} \\ i_1^{(\tau)} & i_2^{(\tau)} \end{pmatrix} P\begin{pmatrix} i_1^{(\tau)} & i_2^{(\tau)} \\ i_1^{(\tau)} & i_2 \end{pmatrix} P\begin{pmatrix} i_1^{(\tau)} & i_2 \\ i_1 & i_2 \end{pmatrix} = \text{Identity}.$$

Therefore

$$\begin{aligned} \Phi &= P\begin{pmatrix} i_1 & i_2 \\ i_1' & i_2' \end{pmatrix} \cdots P\begin{pmatrix} i_1^{(\omega-1)} & i_2^{(\omega-1)} \\ i_1^{(\omega)} & i_2^{(\omega)} \end{pmatrix} \\ &= P\begin{pmatrix} i_1 & i_2 \\ i_1' & i_2' \end{pmatrix} \cdots P\begin{pmatrix} i_1^{(\tau-1)} & i_2^{(\tau-1)} \\ i_1^{(\tau)} & i_2^{(\tau)} \end{pmatrix} \cdot P\begin{pmatrix} i_1^{(\tau)} & i_2^{(\tau)} \\ i_1^{(\tau+1)} & i_2^{(\tau+1)} \end{pmatrix} \cdots P\begin{pmatrix} i_1^{(\omega-1)} & i_2^{(\omega-1)} \\ i_1^{(\omega)} & i_2^{(\omega)} \end{pmatrix} \\ &= P\begin{pmatrix} i_1 & i_2 \\ i_1^{(\tau)} & i_2 \end{pmatrix} P\begin{pmatrix} i_1^{(\tau)} & i_2 \\ i_1^{(\tau)} & i_2^{(\tau)} \end{pmatrix} \cdot P\begin{pmatrix} i_1^{(\tau)} & i_2^{(\tau)} \\ i_1^{(\tau+1)} & i_2^{(\tau+1)} \end{pmatrix} \cdots P\begin{pmatrix} i_1^{(\omega-1)} & i_2^{(\omega-1)} \\ i_1^{(\omega)} & i_2^{(\omega)} \end{pmatrix} \quad \text{(by 12))} \\ &= \text{Identity} \qquad \text{(by 11)).} \end{aligned}$$

This completes the proof for the case $\tau < \omega - 2$. We may treat the case $\tau > 2$ in similar manner. There remains then the case $\tau = \omega - 2 = 2$, i.e., $\omega = 4$, $\tau = 2$, $\tau' = 0$. For this case (10) may be written in the form

$$P\begin{pmatrix} i & j \\ k & j \end{pmatrix} P\begin{pmatrix} k & j \\ k & h \end{pmatrix} P\begin{pmatrix} k & h \\ i & h \end{pmatrix} P\begin{pmatrix} i & h \\ i & j \end{pmatrix} = \text{Identity}, \tag{14}$$

in which i, j, k, h are all distinct. Hence (14) holds by the Corollary to Lemma 5.6. This completes the proof for this case.

(C) Suppose now that for every τ, τ': $i_1^{(\tau)}$, $i_2^{(\tau)}$ are not both different from each of $i_1^{(\tau')}$, $i_2^{(\tau')}$. Consider the factors $\begin{pmatrix} i_1^{(\tau'-1)} & i_2^{(\tau'-1)} \\ i_1^{(\tau')} & i_2^{(\tau')} \end{pmatrix}$, $\tau' = \tau + 1$, $\tau + 2$, $\tau + 3$ of Δ, where τ' is taken modulo ω. Since $\nu^\tau = \nu^{\tau+2} \neq \nu^{\tau+1} = \nu^{\tau+3}$, we see that

$$\begin{array}{lll} i_{\nu^{\tau+1}}^{(\tau)} \neq i_{\nu^{\tau+1}}^{(\tau+1)}, & i_{\nu^\tau}^{(\tau)} = i_{\nu^\tau}^{(\tau+1)}, & i_{\nu^\tau}^{(\tau)} \neq i_{\nu^{\tau+1}}^{(\tau)}, \\ i_{\nu^{\tau+1}}^{(\tau+1)} = i_{\nu^{\tau+1}}^{(\tau+2)}, & i_{\nu^\tau}^{(\tau+1)} \neq i_{\nu^\tau}^{(\tau+2)}, & i_{\nu^\tau}^{(\tau+1)} \neq i_{\nu^{\tau+1}}^{(\tau+1)}, \\ i_{\nu^{\tau+1}}^{(\tau+2)} \neq i_{\nu^{\tau+1}}^{(\tau+3)}, & i_{\nu^\tau}^{(\tau+2)} = i_{\nu^\tau}^{(\tau+3)}, & i_{\nu^\tau}^{(\tau+2)} \neq i_{\nu^{\tau+1}}^{(\tau+2)}, \\ & & i_{\nu^\tau}^{(\tau+3)} \neq i_{\nu^{\tau+1}}^{(\tau+3)}, \end{array} \tag{15}$$

and since one of $i_{\nu^\tau}^{(\tau+1)}$, $i_{\nu^{\tau+1}}^{(\tau+1)}$ must be equal to one of $i_{\nu^\tau}^{(\tau+3)}$, $i_{\nu^{\tau+1}}^{(\tau+3)}$, the only possibility is $i_{\nu^\tau}^{(\tau+1)} = i_{\nu^{\tau+1}}^{(\tau+3)}$, i.e.,

$$i_{\nu^\tau}^{(\tau)} = i_{\nu^{\tau+3}}^{(\tau+3)}. \tag{16}$$

By (15) we see that

$$i_{\nu^\tau}^{(\tau)} = i_{\nu^{\tau+3}}^{(\tau+3)} \neq i_{\nu^{\tau+1}}^{(\tau+1)}, \quad i_{\nu^{\tau+2}}^{(\tau+2)}. \tag{17}$$

Hence we have proved that $i_{\nu^\tau}^{(\tau)} = i_{\nu^{\tau'}}^{(\tau')}$ if and only if $\tau \equiv \tau' \pmod 3$. Since $i_{\nu^\omega}^{(\omega)} = i_{\nu^\omega} = i_{\nu^0}$, we have $\omega \equiv 0 \pmod 3$. But we had $\omega = 4, 6, \cdots$; hence $\omega = 6, 12, 18, \cdots$.

Suppose now that $i_1^{(\tau)} = i_1^{(\tau')}$ occurs for a pair τ, τ' such that $\tau' \neq \tau$, $\tau \pm 1 \pmod \omega$. By a cyclic permutation on $0, \cdots, \omega$, we may reduce the consideration to the case $\tau' = 0$, $1 < \tau < \omega$. The first factors in Δ are

$$\begin{pmatrix} i_1 & i_2 \\ i_1' & i_2' \end{pmatrix} \cdots \begin{pmatrix} i^{(\tau-1)} & i_2^{(\tau-1)} \\ i_1 & i_2^{(\tau)} \end{pmatrix} = \begin{pmatrix} i_1 & i_2 \\ i_1 & i_2^{(\tau)} \end{pmatrix},$$

and (9) becomes

$$\Delta' \equiv \begin{pmatrix} i_1 & i_2 \\ i_1 & i_2^{(\tau)} \end{pmatrix} \begin{pmatrix} i_1 & i_2^{(\tau)} \\ i_1^{(\tau+1)} & i_2^{(\tau+1)} \end{pmatrix} \cdots \begin{pmatrix} i_1^{(\omega-1)} & i_2^{(\omega-1)} \\ i_1^{(\omega)} & i_2^{(\omega)} \end{pmatrix} = \begin{pmatrix} i_1 & i_2 \\ i_1 & i_2 \end{pmatrix}.$$

The length of Δ' is $\omega - \tau + 1 < \omega$, whence by the induction hypothesis

$$(18)\qquad P\begin{pmatrix} i_1 & i_2 \\ i_1 & i_2^{(\tau)} \end{pmatrix} P\begin{pmatrix} i_1 & i_2^{(\tau)} \\ i_1^{(\tau+1)} & i_2^{(\tau+1)} \end{pmatrix} \cdots P\begin{pmatrix} i_1^{(\omega-1)} & i_2^{(\omega-1)} \\ i_1^{(\omega)} & i_2^{(\omega)} \end{pmatrix} = \text{Identity.}$$

Moreover,

$$\Delta'' \equiv \begin{pmatrix} i_1 & i_2 \\ i_1' & i_2' \end{pmatrix} \cdots \begin{pmatrix} i_1^{(\tau-1)} & i_2^{(\tau-1)} \\ i_1 & i_2^{(\tau)} \end{pmatrix} \begin{pmatrix} i_1 & i_2^{(\tau)} \\ i_1 & i_2 \end{pmatrix} = \begin{pmatrix} i_1 & i_2 \\ i_1 & i_2 \end{pmatrix},$$

and Δ'' has length $\tau + 1 < \omega$. Hence by the induction hypothesis

$$(19)\qquad P\begin{pmatrix} i_1 & i_2 \\ i_1' & i_2' \end{pmatrix} \cdots P\begin{pmatrix} i_1^{(\tau-1)} & i_2^{(\tau-1)} \\ i_1 & i_2^{(\tau)} \end{pmatrix} P\begin{pmatrix} i_1 & i_2^{(\tau)} \\ i_1 & i_2 \end{pmatrix} = \text{Identity.}$$

By combining (as in the proof for case (B)) (18) and (19) we obtain $\Phi = \text{Identity}$. This discussion shows that the cases remaining to be treated are those in which $\tau' \neq \tau,\ \tau \pm 1 \pmod{\omega}$ implies $i_1^{(\tau)} \neq i_1^{(\tau')}$ and similarly $i_2^{(\tau)} \neq i_2^{(\tau')}$.

Let us assume that $\nu^1 = 1$ (since the case $\nu^1 = 2$ follows by symmetry). By (17) we have $i_1^{(1)} = i_2^{(4)} = i_1^{(7)}$ (since ν^τ is 1 for τ odd and is 2 for τ even). Thus we must have $1 \equiv 7$ or $7 \pm 1 \pmod{\omega}$, i.e., ω must be a divisor of 5 or 6 or 7. But we had $\omega = 6, 12, 18, \cdots$, whence $\omega = 6$. For this case (10) may be written by (15) in the form

$$(20)\quad P\begin{pmatrix} i & j \\ k & j \end{pmatrix} P\begin{pmatrix} k & j \\ k & i \end{pmatrix} P\begin{pmatrix} k & i \\ j & i \end{pmatrix} P\begin{pmatrix} j & i \\ j & k \end{pmatrix} P\begin{pmatrix} j & k \\ i & k \end{pmatrix} P\begin{pmatrix} i & k \\ i & j \end{pmatrix} = \text{Identity,}$$

wherein i, j, k are distinct. Hence (20) holds by Lemma 5.8. This completes the proof.

LEMMA 5.10: If $m \geqq 2$, and if

$$(21)\qquad \begin{pmatrix} i_1 & \cdots & i_m \\ i_1' & \cdots & i_m' \end{pmatrix} \cdots \begin{pmatrix} i_1^{(\omega-1)} & \cdots & i_m^{(\omega-1)} \\ i_1^{(\omega)} & & i_m^{(\omega)} \end{pmatrix} = \begin{pmatrix} i_1 & \cdots & i_m \\ i_1 & \cdots & i_m \end{pmatrix},$$

where for each $\tau = 1, \cdots, \omega$, $i_\nu^{(\tau-1)} \neq i_\nu^{(\tau)}$ occurs at most once, then

$$(22)\qquad \Phi \equiv P\begin{pmatrix} i_1 & \cdots & i_m \\ i_1' & \cdots & i_m' \end{pmatrix} \cdots P\begin{pmatrix} i_1^{(\omega-1)} & \cdots & i_m^{(\omega-1)} \\ i_1^{(\omega)} & & i_m^{(\omega)} \end{pmatrix} = \text{Identity.}$$

Proof: Clearly Φ is an automorphism of $L(0, \sum_{\nu=1}^{m}{}_{\cup}\, \bar{\mathfrak{a}}_{i_\nu})$; this lattice is modular and complemented as is L, and the $\bar{\mathfrak{a}}_{i_\mu}$, $\bar{\mathfrak{c}}_{i_\mu i_\sigma}$ ($\mu, \sigma = 1, \cdots, m$, $\mu \neq \sigma$) constitute in it a normalized frame. Let μ, σ be fixed, and consider an element $\mathfrak{u} \leqq \bar{\mathfrak{a}}_{i_\mu} \cup \bar{\mathfrak{a}}_{i_\sigma}$. The effect of Φ on $\mathfrak{u}$ is the same by Lemma 5.5 (a) as the effect of

$$\Phi' \equiv P\begin{pmatrix} i_\mu & i_\sigma \\ i'_\mu & i'_\sigma \end{pmatrix} \cdots P\begin{pmatrix} i_\mu^{(\omega-1)} & i_\sigma^{(\omega-1)} \\ i_\mu^{(\omega)} & i_\sigma^{(\omega)} \end{pmatrix}.$$

By Lemma 5.9, $\Phi' =$ Identity, whence Φ leaves $\mathfrak{u}$ invariant. Since this holds for all pairs μ, σ, $\mu \neq \sigma$, we have in particular that all elements of each $L_{i_\mu i_\sigma}$ are left invariant by Φ. By Lemma 4.2 every element in $L(0, \sum_{\nu=1}^{m}{}_{\cup}\, \bar{\mathfrak{a}}_{i_\nu})$ may be expressed as a polynomial in elements of the $L_{i_\mu i_\sigma}$ with the help of the operations $\cup$, $\cap$, whence it follows that Φ leaves every element invariant and is therefore the identity.

Corollary: If $m \geqq 2$, and if

$$(23)\quad \begin{pmatrix} i_1 \cdots i_m \\ i'_1 \cdots i'_m \end{pmatrix} \cdots \begin{pmatrix} i_1^{(\omega-1)} \cdots i_m^{(\omega-1)} \\ i_1^{(\omega)} \cdots i_m^{(\omega)} \end{pmatrix} = \begin{pmatrix} j_1 \cdots j_m \\ j'_1 \cdots j'_m \end{pmatrix} \cdots \begin{pmatrix} j_1^{(\omega'-1)} \cdots j_m^{(\omega'-1)} \\ j_1^{(\omega')} \cdots j_m^{(\omega')} \end{pmatrix},$$

where for each $\tau = 1, \cdots, \omega$, $i_\nu^{(\tau-1)} \neq i_\nu^{(\tau)}$ occurs at most once and for each $\tau' = 1, \cdots, \omega'$, $i_\nu^{(\tau'-1)} \neq i_\nu^{(\tau')}$ occurs at most once, then

$$(24)\quad \begin{aligned} &P\begin{pmatrix} i_1 \cdots i_m \\ i'_1 \cdots i'_m \end{pmatrix} \cdots P\begin{pmatrix} i_1^{(\omega-1)} \cdots i_m^{(\omega-1)} \\ i_1^{(\omega)} \cdots i_m^{(\omega)} \end{pmatrix} \\ &\qquad = P\begin{pmatrix} j_1 \cdots j_m \\ j'_1 \cdots j'_m \end{pmatrix} \cdots P\begin{pmatrix} j_1^{(\omega'-1)} \cdots j_m^{(\omega'-1)} \\ j_1^{(\omega')} \cdots j_m^{(\omega')} \end{pmatrix}. \end{aligned}$$

Proof: We may write (23) in the form

$$\begin{aligned} &\begin{pmatrix} i_1 \cdots i_m \\ i'_1 \cdots i'_m \end{pmatrix} \cdots \begin{pmatrix} i_1^{(\omega-1)} \cdots i_m^{(\omega-1)} \\ i_1^{(\omega)} \cdots i_m^{(\omega)} \end{pmatrix} \\ &\qquad \cdot \begin{pmatrix} j_1^{(\omega')} \cdots j_m^{(\omega')} \\ j_1^{(\omega'-1)} \cdots j_m^{(\omega'-1)} \end{pmatrix} \cdots \begin{pmatrix} j'_1 \cdots j'_m \\ j_1 \cdots j_m \end{pmatrix} = \begin{pmatrix} i_1 \cdots i_m \\ i_1 \cdots i_m \end{pmatrix} \end{aligned}$$

since clearly $i_\nu = j_\nu$, $i_\nu^{(\omega)} = j_\nu^{(\omega')}$. Thus Lemma 5.10 applies, yielding

$$\begin{aligned} &P\begin{pmatrix} i_1 \cdots i_m \\ i'_1 \cdots i'_m \end{pmatrix} \cdots P\begin{pmatrix} i_1^{(\omega-1)} \cdots i_m^{(\omega-1)} \\ i_1^{(\omega)} \cdots i_m^{(\omega)} \end{pmatrix} \\ &\qquad \cdot P\begin{pmatrix} j_1^{(\omega')} \cdots j_m^{(\omega')} \\ j_1^{(\omega'-1)} \cdots j_m^{(\omega'-1)} \end{pmatrix} \cdots P\begin{pmatrix} j'_1 \cdots j'_m \\ j_1 \cdots j_m \end{pmatrix} = \text{Identity}, \end{aligned}$$

which may be written also in the form (24).

Theorem 5.1: There exists a system of transformations

$$\left(P\begin{pmatrix} i_1 \cdots i_m \\ j_1 \cdots j_m \end{pmatrix};\ i_\rho, j_\rho = 1, \cdots, n\ (\rho = 1, \cdots, m),\ i_\rho \neq i_\sigma,\ j_\rho \neq j_\sigma \text{ for } \rho \neq \sigma,\ m = 1, \cdots, n-1\right)$$

with the following properties:

(a) $P\begin{pmatrix} i_1 \cdots i_m \\ j_1 \cdots j_m \end{pmatrix}$ is a projective isomorphism of $L(0, \sum_{\cup\, \nu=1}^{m} \bar{\mathfrak{a}}_{i_\nu})$ and $L(0, \sum_{\cup\, \nu=1}^{m} \bar{\mathfrak{a}}_{j_\nu})$.

(b) If $l = 1, \cdots, m$ and $\tau_1, \cdots, \tau_l$ are distinct, then $P\begin{pmatrix} i_1 \cdots i_m \\ j_1 \cdots j_m \end{pmatrix}$ coincides on $L(0, \sum_{\cup\, \lambda=1}^{l} \bar{\mathfrak{a}}_{i_{\tau_\lambda}})$ with $P\begin{pmatrix} i_{\tau_1} \cdots i_{\tau_l} \\ j_{\tau_1} \cdots j_{\tau_l} \end{pmatrix}$.

(c) If $\mu, \sigma = 1, \cdots, m$, $\mu \neq \sigma$ then $P\begin{pmatrix} i_1 \cdots i_m \\ j_1 \cdots j_m \end{pmatrix}$ maps $\bar{\mathfrak{c}}_{i_\mu i_\sigma}$ on $\bar{\mathfrak{c}}_{j_\mu j_\sigma}$.

(d)
$$P\begin{pmatrix} i_1 \cdots i_m \\ j_1 \cdots j_m \end{pmatrix} P\begin{pmatrix} j_1 \cdots j_m \\ k_1 \cdots k_m \end{pmatrix} = P\begin{pmatrix} i_1 \cdots i_m \\ k_1 \cdots k_m \end{pmatrix}.$$

(e) If $i_\rho \neq j_\rho$ for just one value $\bar{\rho}$ of $\rho = 1, \cdots, m$, then $P\begin{pmatrix} i_1 \cdots i_m \\ j_1 \cdots j_m \end{pmatrix}$ is the perspective isomorphism of $L(0, \sum_{\cup\, \rho=1}^{m} \bar{\mathfrak{a}}_{i_\rho})$ and $L(0, \sum_{\cup\, \rho=1}^{m} \bar{\mathfrak{a}}_{j_\rho})$ with the axis $\bar{\mathfrak{c}}_{i_{\bar{\rho}} j_{\bar{\rho}}}$. Properties (a), (b), (c), (d), (e) determine the $P\begin{pmatrix} i_1 \cdots i_m \\ j_1 \cdots j_m \end{pmatrix}$ uniquely.

Proof: If $i_\rho = j_\rho$ for every value of ρ or for every value of ρ but one, then we define $P\begin{pmatrix} i_1 \cdots i_m \\ j_1 \cdots j_m \end{pmatrix}$ as in Definition 5.3. Properties (a), (e) are obvious, and properties (b), (c) are precisely parts (a), (b) of Lemma 5.5. If $m = 1$, this case covers all possibilities, and only property (d) remains to be proved. This holds, since properties (a) $\cdots$ (e) hold for $m = 2$, as we shall see presently, and property (d) for $m = 1$ follows from property (d) for $m = 2$ by (b) with $l = 1$.

Suppose now that $m \neq 1$, i.e., $m = 2, \cdots, n - 1$. Then every permutation $\begin{pmatrix} i_1 \cdots i_m \\ j_1 \cdots j_m \end{pmatrix}$ may be written in the form

$$(25)\quad \begin{pmatrix} i_1 \cdots i_m \\ j_1 \cdots j_m \end{pmatrix} = \begin{pmatrix} i_1 \cdots i_m \\ i_1' \cdots i_m' \end{pmatrix} \begin{pmatrix} i_1' \cdots i_m' \\ i_1'' \cdots i_m'' \end{pmatrix} \cdots \begin{pmatrix} i_1^{(\omega-1)} \cdots i_m^{(\omega-1)} \\ i_1^{(\omega)} \cdots i_m^{(\omega)} \end{pmatrix},$$

where in each factor in (25), $i_\rho^{(\tau-1)} \neq i_\rho^{(\tau)}$ occurs exactly once. We define

$$(26)\quad P\begin{pmatrix} i_1 \cdots i_m \\ j_1 \cdots j_m \end{pmatrix} \equiv P\begin{pmatrix} i_1 \cdots i_m \\ i_1' \cdots i_m' \end{pmatrix} P\begin{pmatrix} i_1' \cdots i_m' \\ i_1'' \cdots i_m'' \end{pmatrix} \cdots P\begin{pmatrix} i_1^{(\omega-1)} \cdots i_m^{(\omega-1)} \\ i_1^{(\omega)} \cdots i_m^{(\omega)} \end{pmatrix};$$

by the Corollary to Lemma 5.10, this definition is independent of the representation (25), and thus gives a unique value to $P\begin{pmatrix} i_1 \cdots i_m \\ j_1 \cdots j_m \end{pmatrix}$ dependent only on $\begin{pmatrix} i_1 \cdots i_m \\ j_1 \cdots j_m \end{pmatrix}$. Thus all properties are obvious.

If two projectivities P, P' exist for the same permutation $\begin{pmatrix} i_1 \cdots i_m \\ j_1 \cdots j_m \end{pmatrix}$ then representation of that permutation in the form (25) shows that $P = P'$. Hence the proof is complete.

COROLLARY: If $m = 1, \cdots, n-1$, $l = 1, \cdots, m$, and if $\tau_1, \cdots, \tau_l$ are distinct, then $P\begin{pmatrix} i_1 \cdots i_m \\ j_1 \cdots j_m \end{pmatrix}$ transforms $\sum_{\lambda=1}^{l} {}_\cup \bar{\mathfrak{a}}_{i_{\tau_\lambda}}$ into $\sum_{\lambda=1}^{l} {}_\cup \bar{\mathfrak{a}}_{j_{\tau_\lambda}}$. In particular, $\bar{\mathfrak{a}}_{i_\nu}$ is transformed into $\bar{\mathfrak{a}}_{j_\nu}$. Moreover if $i_\nu = j_\nu$ for $\nu = \tau_1, \cdots, \tau_l$, then $P\begin{pmatrix} i_1 \cdots i_m \\ j_1 \cdots j_m \end{pmatrix}$ leaves each $\mathfrak{a}' \leqq \sum_{\lambda=1}^{l} {}_\cup \bar{\mathfrak{a}}_{i_{\tau_\lambda}}$ invariant.

PROOF: By (b), $P\begin{pmatrix} i_1 \cdots i_m \\ j_1 \cdots j_m \end{pmatrix}$ has the same effect on $\sum_{\lambda=1}^{l} {}_\cup \bar{\mathfrak{a}}_{i_{\tau_\lambda}}$ as $P\begin{pmatrix} i_{\tau_1} \cdots i_{\tau_l} \\ j_{\tau_1} \cdots j_{\tau_l} \end{pmatrix}$; the latter being a projective isomorphism of $L(0, \sum_{\lambda=1}^{l} {}_\cup \bar{\mathfrak{a}}_{i_{\tau_\lambda}})$ and $L(0, \sum_{\lambda=1}^{l} {}_\cup \bar{\mathfrak{a}}_{j_{\tau_\lambda}})$, the units of these two lattices must correspond under it. Thus $\sum_{\lambda=1}^{l} {}_\cup \bar{\mathfrak{a}}_{i_{\tau_\lambda}}$ is transformed into $\sum_{\lambda=1}^{l} {}_\cup \bar{\mathfrak{a}}_{j_{\tau_\lambda}}$. The second statement follows from the first by putting $l = 1$, $\tau_1 = \nu$. To prove the third part, we note that by (b), $\mathfrak{a}' P\begin{pmatrix} i_1 \cdots i_m \\ j_1 \cdots j_m \end{pmatrix} = \mathfrak{a}' P\begin{pmatrix} i_{\tau_1} \cdots i_{\tau_l} \\ j_{\tau_1} \cdots j_{\tau_l} \end{pmatrix} = \mathfrak{a}'$ (for $\mathfrak{a}' \leqq \sum_{\lambda=1}^{l} {}_\cup \bar{\mathfrak{a}}_{i_{\tau_\lambda}}$) since $P\begin{pmatrix} i_{\tau_1} \cdots i_{\tau_l} \\ i_{\tau_1} \cdots i_{\tau_l} \end{pmatrix}$ is the identity by (d). (The notation $\mathfrak{a}' P\begin{pmatrix} i_1 \cdots i_m \\ j_1 \cdots j_m \end{pmatrix}$ will always be used to represent the transform of $\mathfrak{a}'$ under $P\begin{pmatrix} i_1 \cdots i_m \\ j_1 \cdots j_m \end{pmatrix}$.)

PART II · CHAPTER VI

Definition of L-Numbers; Multiplication

We assume here, as in Chapter V, that L is a complemented modular lattice of order $n \geqq 4$ and that $(\bar{\mathfrak{a}}_i, \bar{\mathfrak{c}}_{ij}; i, j = 1, \ldots, n, i \neq j)$ is a normalized frame in L. The projective isomorphisms $P\begin{pmatrix} i_1 \ldots i_m \\ \vdots \\ j_1 \cdots j_m \end{pmatrix}$ defined in Theorem 5.1 will be used often throughout what follows. We make the convention that whenever referring to elements $\mathfrak{b}_{ij} \in L_{ij}$ we tacitly understand that $i \neq j$. As in Chapter V the indices $i, j, k, \ldots$ run from 1 to n unless the contrary is stated.

Definition 6.1: An *L-number* is a system $(\mathfrak{b}_{ij}; i, j = 1, \cdots, n, i \neq j)$ satisfying

(1) $$\mathfrak{b}_{ij} \in L_{ij},$$

(2) $$\mathfrak{b}_{kh} = \mathfrak{b}_{ij} P\begin{pmatrix} i & j \\ k & h \end{pmatrix}.$$

The set of all L-numbers will be denoted by $\mathfrak{S}$, and the L-numbers will be denoted by $\beta, \gamma, \delta, \ldots$. For a given L-number $\beta = (\mathfrak{b}_{ij}; i, j = 1, \ldots, n)$ and a given pair i, j, the element $\mathfrak{b}_{ij}$ is the i, j-component of β and will be denoted by $(\beta)_{ij}$.

Lemma 6.1: For a given pair i, j, if $\mathfrak{b}_{ij} \in L_{ij}$ then there exists one and only one L-number β such that

(3) $$\mathfrak{b}_{ij} = (\beta)_{ij}.$$

Proof: Define for $k, h = 1, \cdots, n, k \neq h$,

$$\mathfrak{c}_{kh} \equiv \mathfrak{b}_{ij} P\begin{pmatrix} i & j \\ k & h \end{pmatrix}.$$

Then $\mathfrak{c}_{kh} = \mathfrak{b}_{ij} P\begin{pmatrix} i & j \\ k & h \end{pmatrix}$, $\mathfrak{c}_{lp} = \mathfrak{b}_{ij} P\begin{pmatrix} i & j \\ l & p \end{pmatrix}$, whence $\mathfrak{c}_{lp} P\begin{pmatrix} l & p \\ i & j \end{pmatrix} = \mathfrak{b}_{ij}$ and $\mathfrak{c}_{kh} = \mathfrak{c}_{lp} P\begin{pmatrix} l & p \\ i & j \end{pmatrix} P\begin{pmatrix} i & j \\ k & h \end{pmatrix} = \mathfrak{c}_{lp} P\begin{pmatrix} l & p \\ k & h \end{pmatrix}$ by Theorem 5.1 (d). Thus the $\mathfrak{c}_{kh}$ satisfy (2). Since $\mathfrak{b}_{ij} \in L_{ij}$, we have

(4) $$\mathfrak{b}_{ij} \cup \bar{\mathfrak{a}}_j = \bar{\mathfrak{a}}_i \cup \bar{\mathfrak{a}}_j, \quad \mathfrak{b}_{ij} \cap \bar{\mathfrak{a}}_j = 0.$$

Now $P\begin{pmatrix} i & j \\ k & h \end{pmatrix}$ carries $\bar{\mathfrak{a}}_i$ into $\bar{\mathfrak{a}}_k$, and $\bar{\mathfrak{a}}_j$ into $\bar{\mathfrak{a}}_h$ by the Corollary to Theorem 5.1, and is a (projective) isomorphism of $L(0, \bar{\mathfrak{a}}_i \cup \bar{\mathfrak{a}}_j)$, $L(0, \bar{\mathfrak{a}}_k \cup \bar{\mathfrak{a}}_h)$. Thus operation on (4) by $P\begin{pmatrix} i & j \\ k & h \end{pmatrix}$ yields $\mathfrak{c}_{kh} \cup \bar{\mathfrak{a}}_h = \bar{\mathfrak{a}}_k \cup \bar{\mathfrak{a}}_h$, $\mathfrak{c}_{kh} \cap \bar{\mathfrak{a}}_h = 0$, whence $\mathfrak{c}_{kh} \in L_{kh}$; this establishes (1) for the $\mathfrak{c}_{kh}$. Hence the system $(\mathfrak{c}_{kh};\ k, h = 1, \cdots, n)$ is an L-number β. Clearly $(\beta)_{ij} = \mathfrak{c}_{ij} = \mathfrak{b}_{ij} P\begin{pmatrix} i\,j \\ i\,j \end{pmatrix} = \mathfrak{b}_{ij}$, and the proof of the existence is complete. Suppose now that β, δ have the property (3). Then $(\beta)_{ij} = (\delta)_{ij}$, and

$$(\beta)_{kh} = (\beta)_{ij} P\begin{pmatrix} i & j \\ k & h \end{pmatrix} = (\delta)_{ij} P\begin{pmatrix} i & j \\ k & h \end{pmatrix} = (\delta)_{kh},$$

whence $\beta = \delta$.

LEMMA 6.2: If β, γ, $\delta \in \mathfrak{S}$, and if

$$(5) \qquad (\beta)_{ij} \otimes (\gamma)_{jk} = (\delta)_{ik}$$

holds for one triple (i, j, k) of distinct integers, then (5) holds for all such triples.

PROOF: Let (5) hold for the triple (i, j, k), and let (i', j', k') be another triple. Now (5) may be written in the form

$$(6) \qquad [(\beta)_{ij} \cup (\gamma)_{jk}] \cap (\bar{\mathfrak{a}}_i \cup \bar{\mathfrak{a}}_k) = (\delta)_{ik}.$$

Since $P\begin{pmatrix} i\,j\,k \\ i'j'k' \end{pmatrix}$ transforms $\bar{\mathfrak{a}}_i$, $\bar{\mathfrak{a}}_k$ into $\bar{\mathfrak{a}}_{i'}$, $\bar{\mathfrak{a}}_{k'}$, respectively, by the Corollary to Theorem 5.1, operation on (6) with $P\begin{pmatrix} i\,j\,k \\ i'j'k' \end{pmatrix}$ yields

$$\left[(\beta)_{ij} P\begin{pmatrix} i\,j\,k \\ i'j'k' \end{pmatrix} \cup (\gamma)_{jk} P\begin{pmatrix} i\,j\,k \\ i'j'k' \end{pmatrix}\right] \cap (\bar{\mathfrak{a}}_{i'} \cup \bar{\mathfrak{a}}_{k'}) = (\delta)_{ik} P\begin{pmatrix} i\,j\,k \\ i'j'k' \end{pmatrix}.$$

But $(\beta)_{ij} \leqq \bar{\mathfrak{a}}_i \cup \bar{\mathfrak{a}}_j$, whence $(\beta)_{ij} P\begin{pmatrix} i\,j\,k \\ i'j'k' \end{pmatrix} = (\beta)_{ij} P\begin{pmatrix} i\,j \\ i'j' \end{pmatrix} = (\beta)_{i'j'}$ by Theorem 5.1 (b). Similarly, $(\gamma)_{jk} P\begin{pmatrix} i\,j\,k \\ i'j'k' \end{pmatrix} = (\gamma)_{jk} P\begin{pmatrix} j\,k \\ j'k' \end{pmatrix} = (\gamma)_{j'k'}$ and $(\delta)_{ik} P\begin{pmatrix} i\,j\,k \\ i'j'k' \end{pmatrix} = (\delta)_{ik} P\begin{pmatrix} i\,k \\ i'k' \end{pmatrix} = (\delta)_{i'k'}$. Thus

$$[(\beta)_{i'j'} \cup (\gamma)_{j'k'}] \cap (\bar{\mathfrak{a}}_{i'} \cup \bar{\mathfrak{a}}_{k'}) = (\delta)_{i'k'},$$

i.e.,

$$(\beta)_{i'j'} \otimes (\gamma)_{j'k'} = (\delta)_{i'k'}.$$

LEMMA 6.3: If $\beta, \gamma \in \mathfrak{S}$, there exists one and only one element $\delta \in \mathfrak{S}$ such that

$$(7) \qquad (\beta)_{ij} \otimes (\gamma)_{jk} = (\delta)_{ik}$$

for every triple (i, j, k) of distinct integers.

PROOF: Since $n \geqq 4$ there exist three distinct integers i_0, j_0, k_0. Define δ as the unique element of $\mathfrak{S}$ whose i_0, k_0-component is $(\beta)_{i_0 j_0} \otimes (\gamma)_{j_0 k_0}$ (cf. Lemma 6.1). Hence (7) holds for i_0, j_0, k_0, and therefore holds for every triple (i, j, k) of distinct integers by Lemma 6.2. Now if δ' along with δ has the property (7) we have in particular

$$(\delta')_{i_0 k_0} = (\beta)_{i_0 j_0} \otimes (\gamma)_{j_0 k_0} = (\delta)_{i_0 k_0},$$

whence δ', δ have the same i_0, k_0-component. Thus by Lemma 6.1, $\delta = \delta'$.

DEFINITION 6.2: If $\beta, \gamma \in \mathfrak{S}$, we define $\gamma\beta$ as the unique element δ having the property (7) for every triple (i, j, k) of distinct integers (cf. Lemma 6.3).

(A discussion of the motivation of this definition of multiplication is given in the appendix to this chapter.)

THEOREM 6.1: The operation of multiplication in $\mathfrak{S}$ is associative, i.e., if $\beta, \gamma, \delta \in \mathfrak{S}$, then

$$(\beta\gamma)\delta = \beta(\gamma\delta). \tag{8}$$

There exists a unique element $1 \in \mathfrak{S}$ such that for every $\beta \in \mathfrak{S}$

$$\beta 1 = 1 \beta = \beta; \tag{9}$$

there exists a unique element $0 \in \mathfrak{S}$ such that for every $\beta \in \mathfrak{S}$

$$\beta 0 = 0 \beta = 0. \tag{10}$$

Moreover $1 = (\bar{\mathfrak{c}}_{ij};\ i, j = 1, \cdots, n,\ i \neq j)$, $0 = (\bar{\mathfrak{a}}_{ij};\ i, j = 1, \cdots, n,\ i \neq j)$, where $\bar{\mathfrak{a}}_{ij} \equiv \bar{\mathfrak{a}}_i$ $(i, j = 1, \cdots, n,\ i \neq j)$.

PROOF: Since $n \geqq 4$, there exist four distinct integers i, j, k, h. Define $\varepsilon \equiv (\beta\gamma)\delta$, $\varepsilon' \equiv \beta(\gamma\delta)$. Then

$$(\varepsilon)_{ih} = \big((\beta)_{ij} \otimes (\gamma)_{jk}\big) \otimes (\delta)_{kh} = (\beta)_{ij} \otimes \big((\gamma)_{jk} \otimes (\delta)_{kh}\big) = (\varepsilon')_{ih},$$

the last equality but one holding by Lemma 5.2. Hence by Lemma 6.1, $\varepsilon = \varepsilon'$, i.e., (8) holds.

To prove the second part, we observe first that the elements 1, 0 satisfying (9), (10), respectively, are unique if existent. For, if 1, $1'$ satisfy (9), then $1 = 1 \cdot 1' = 1'$; and if 0, $0'$ satisfy (10), then $0 = 0' \cdot 0 = 0'$. Let i, j, k $(= 1, \cdots, n)$ be three fixed distinct integers, and define 1 by $(1)_{ij} = \bar{\mathfrak{c}}_{ij}$ (cf. Lemma 6.1), and 0 by $(0)_{ij} = \bar{\mathfrak{a}}_i$. (Clearly both $\bar{\mathfrak{c}}_{ij}$, $\bar{\mathfrak{a}}_i$ are in L_{ij}.) Then

$$(\beta \cdot 1)_{ik} = (1)_{ij} \otimes (\beta)_{jk} = \big(\bar{\mathfrak{c}}_{ij} \cup (\beta)_{jk}\big) \cap (\bar{\mathfrak{a}}_i \cup \bar{\mathfrak{a}}_k) = (\beta)_{jk} P\begin{pmatrix} j\,k \\ i\,k \end{pmatrix} = (\beta)_{ik},$$

$$(1 \cdot \beta)_{kj} = (\beta)_{ki} \otimes (1)_{ij} = \big((\beta)_{ki} \cup \bar{\mathfrak{c}}_{ij}\big) \cap (\bar{\mathfrak{a}}_k \cup \bar{\mathfrak{a}}_j) = (\beta)_{ki} P\begin{pmatrix} k\,i \\ k\,j \end{pmatrix} = (\beta)_{kj},$$

whence $\beta \cdot 1 = \beta$, $1 \cdot \beta = \beta$ by Lemma 6.1. Moreover,

$$(\beta \cdot 0)_{ik} = (0)_{ij} \otimes (\beta)_{jk} = (\bar{a}_i \cup (\beta)_{jk}) \cap (\bar{a}_i \cup \bar{a}_k) \geqq \bar{a}_i,$$

and since $(\beta \cdot 0)_{ik}$, $\bar{a}_i \in L_{ik}$ we may conclude $(\beta \cdot 0)_{ik} = \bar{a}_i$ by Lemma 3.3; but $(0)_{ik} = (0)_{ij} P\begin{pmatrix} i & j \\ i & k \end{pmatrix} = \bar{a}_i P\begin{pmatrix} i & j \\ i & k \end{pmatrix} = \bar{a}_i$, whence $(\beta \cdot 0)_{ik} = (0)_{ik}$, and $\beta \cdot 0 = 0$ by Lemma 6.1. Also,

$$(0 \cdot \beta)_{kj} = (\beta)_{ki} \otimes (0)_{ij} = ((\beta)_{ki} \cup \bar{a}_i) \cap (\bar{a}_k \cup \bar{a}_j) = (\bar{a}_k \cup \bar{a}_i) \cap (\bar{a}_k \cup \bar{a}_j)$$
$$= \bar{a}_k \ (\text{since } (\bar{a}_i, \bar{a}_j, \bar{a}_k) \perp);$$

but $(0)_{kj} = (0)_{ij} P\begin{pmatrix} i & j \\ k & j \end{pmatrix} = \bar{a}_i P\begin{pmatrix} i & j \\ k & j \end{pmatrix} = \bar{a}_k$, whence $(0 \cdot \beta)_{kj} = (0)_{kj}$, and $0 \cdot \beta = 0$ by Lemma 6.1. This establishes the existence of 1, 0. Since they are unique, a different choice of i, j leads to the same result, i.e., 1, 0, are independent of i, j, whence $(1)_{ij} = \bar{c}_{ij}$, $(0)_{ij} = \bar{a}_i$ for every i, j with $i \neq j$.

APPENDIX

A few illustrative remarks concerning our definition of multiplication (Definition 6.2) will be given here. We shall employ here, as well as in many instances later, two different simplified pictures of the situation in L. (The reader is asked to remember that these pictures are of value only heuristically; the first one refers to a discrete geometry, while the second essentially assumes the situation which our considerations are intended to establish. But they are useful in helping us to visualize the ideas underlying our rather involved and unfamiliar modular-lattice-algebraic expressions and computations.)

PICTURE 1: Assume that $L = L_N$, i.e., that "Case N holds for L" (cf. Part I, Theorem 7.3), and that $N = n$. Then L is the set of all linear subspaces of an $(n-1)$-dimensional projective space (G. Birkhoff, Combinatorial Relations in Projective Geometries, *Annals of Mathematics*, *36* (1935), pages 743—748). The $\bar{a}_i$ $(i = 1, \cdots, n)$ are mutually perspective, and hence all of the same dimension. If their common dimension is $k-1$, then $1 = \bar{a}_1 \cup \cdots \cup \bar{a}_n$ has dimension $nk - 1$; but since its dimension is $n - 1$, we have $k = 1$. Thus $\bar{a}_1, \cdots, \bar{a}_n$ are points and the $\bar{c}_{ij}$ (which are perspective to them) are points also. The position of $\bar{a}_i$, $\bar{a}_j$, $\bar{c}_{ij}$ may be pictured as in Figure 6.1. Thus, if i, j, k are given, we have the situation shown in Figure 6.2. The condition of normalization $\bar{c}_{ik} = (\bar{c}_{ij} \cup \bar{c}_{jk}) \cap (\bar{a}_i \cup \bar{a}_k)$ means that $\bar{c}_{ij}$, $\bar{c}_{jk}$, $\bar{c}_{ik}$ are collinear, as is shown in Figure 6.3.

Figure 6.1

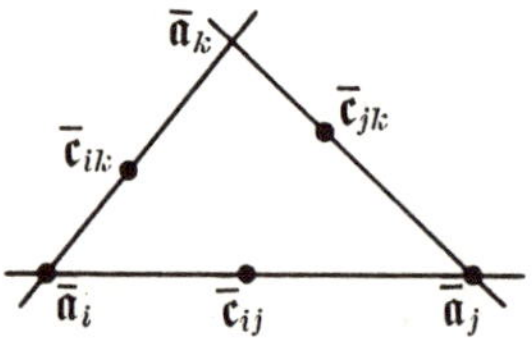

Figure 6.2

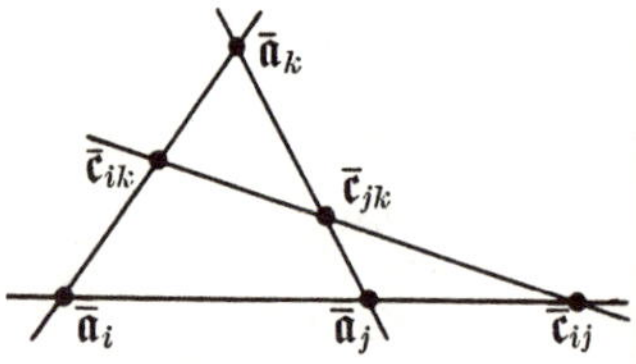

Figure 6.3

The i, j- and i, k-components of an element $\beta \in \mathfrak{S}$ are connected by a perspectivity with axis $\bar{\mathfrak{c}}_{jk}$, as Figure 6.4 illustrates.

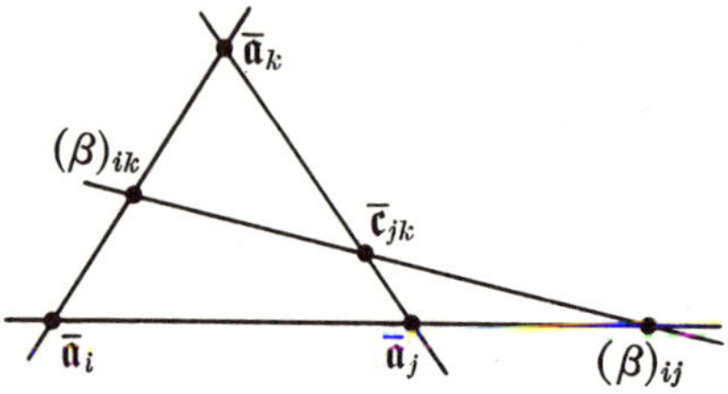

Figure 6.4

Our definition of the product $\gamma\beta$ amounts to $(\gamma\beta)_{ij} = \big((\beta)_{ik} \cup (\gamma)_{kj}\big) \cap (\bar{\mathfrak{a}}_i \cup \bar{\mathfrak{a}}_j)$, as is shown in Figure 6.5.

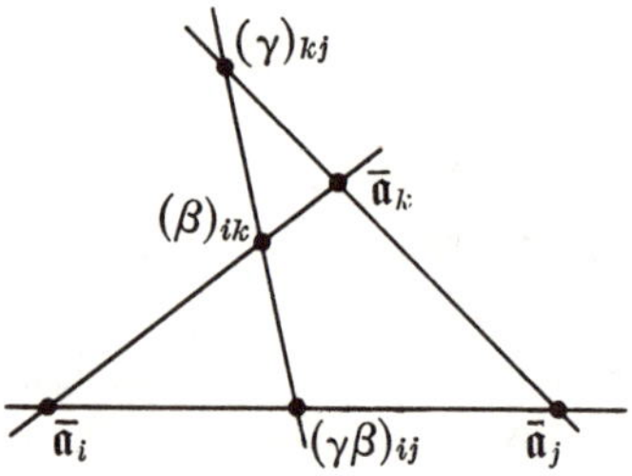

Figure 6.5

If we wish to express all quantities by means of i, j-components, then combining Figure 6.5 with Figure 6.4 (replacing in Figure 6.4, i, j, k by i, j, k for β and by j, i, k for γ) yields the situation in Figure 6.6. This is the well-known definition of multiplication of points (von Staudt) on the line $\bar{\mathfrak{a}}_i \cup \bar{\mathfrak{a}}_j$, where $\bar{\mathfrak{a}}_i$, $\bar{\mathfrak{c}}_{ij}$, $\bar{\mathfrak{a}}_j$ play the roles of 0, 1, ∞, respectively. (Cf. O. Veblen and J. W. Young, *Projective Geometry*, Vol. I, Figure 74 (or 75) on pages 144—145. The points P_0, P_1, P_x, P_y, P_{xy}, P_∞, A, B, X, Y on that figure correspond respectively to $\bar{\mathfrak{a}}_i$, $\bar{\mathfrak{c}}_{ij}$, $(\gamma)_{ij}$, $(\beta)_{ij}$, $(\gamma\beta)_{ij}$, $\bar{\mathfrak{a}}_j$, $\bar{\mathfrak{c}}_{ik}$, $\bar{\mathfrak{c}}_{jk}$, $(\gamma)_{kj}$, $(\beta)_{ik}$ of our Figure 6.6.

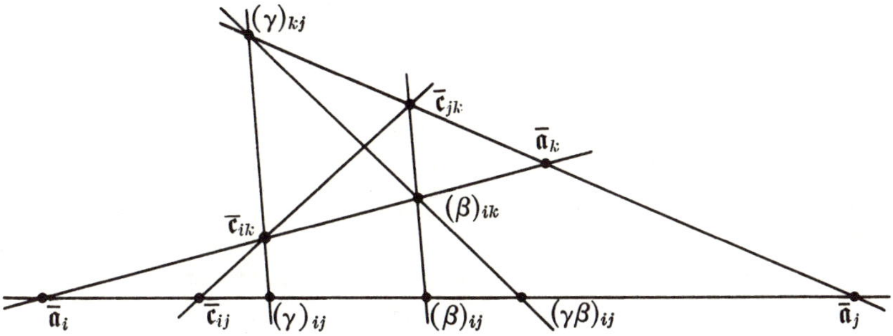

Figure 6.6

PICTURE 2: Let L be the lattice of all principal right ideals of a regular ring $\mathfrak{R} : L = \bar{R}_{\mathfrak{R}}$. Then the existence of the homogeneous basis $\bar{\mathfrak{a}}_1, \cdots, \bar{\mathfrak{a}}_n$ means by Theorem 3.3 that $\mathfrak{R}$ is the matrix ring of order n over another regular ring $\mathfrak{S} : \mathfrak{R} = \mathfrak{S}_n$. In using the vector-set representation (cf. the description at the end of Chapter III, and the conventions at the outset of the proof of Theorem 4.2). We may assume that $\bar{\mathfrak{a}}_i$, $\bar{\mathfrak{c}}_{ij}$, etc., are represented by the following vector sets:

$$\begin{array}{ll} \bar{\mathfrak{a}}_i : (\xi,\ 0,\ 0), & \bar{\mathfrak{a}}_k : (0,\ 0,\ \xi'), \\ \bar{\mathfrak{c}}_{ij} : (\eta,\ -\eta,\ 0), & (\gamma)_{jk} : (0,\ -\zeta',\ -\gamma\zeta'), \\ (\beta)_{ij} : (\zeta,\ -\beta\zeta,\ 0), & (\gamma\beta)_{ik} : (\zeta'',\ 0,\ -\gamma\beta\zeta'') \end{array}$$

(note that only the i, j, k components appear, and in that order). The relation

$$(\gamma\beta)_{ik} = \big((\beta)_{ij} \cup (\gamma)_{jk}\big) \cap (\bar{\mathfrak{a}}_i \cup \bar{\mathfrak{a}}_k)$$

may be immediately verified; the computation will be omitted since it has been made in the proof of Theorem 4.2 (page 108).

PART II · CHAPTER VII

Addition of L-Numbers

We consider here the problem of defining addition of elements in $\mathfrak{S}$. Here again we shall follow the method of von Staudt (cf. Veblen and Young, *Projective Geometry*, Vol. I, pages 141—149); however, in this scheme there are two natural methods of selecting the arbitrary elements, and we shall therefore give first a rather general treatment embodying these as special cases. The specializations will be given in Definition 7.5. For the material preceding this definition $\bar{\mathfrak{a}}_i$, $\bar{\mathfrak{a}}_j$ may be any two elements of L with $\bar{\mathfrak{a}}_i \cap \bar{\mathfrak{a}}_j = 0$.

(The reader is again advised to read the appendix to this chapter together with the definitions and proofs. It should be remembered that the somewhat lengthy computations of our proofs constitute a lattice-algebraic equivalent of what is usually achieved in projective geometry by "inspection" of a figure and visual recognition of the incidence relations between points, lines, planes, etc. [This applies quite particularly to the proof of Theorem 7.2.] The appendix contains some of the familiar projective-geometric figures and states how they are connected with our lattice-algebraic computations.)

LEMMA 7.1: If $(\mathfrak{a} \cup \mathfrak{b}) \cap (\mathfrak{a} \cup \mathfrak{c}) = \mathfrak{a}$ and $\mathfrak{a} \cap \mathfrak{b} = \mathfrak{a} \cap \mathfrak{c} = 0$, then $(\mathfrak{a}, \mathfrak{b}, \mathfrak{c})\perp$.

PROOF: By hypothesis, $\mathfrak{a} \cap \mathfrak{b} = 0$. Moreover,

$$\begin{aligned}(\mathfrak{a} \cup \mathfrak{b}) \cap \mathfrak{c} &= (\mathfrak{a} \cup \mathfrak{b}) \cap [(\mathfrak{a} \cup \mathfrak{c}) \cap \mathfrak{c}] \quad &&(\text{since } \mathfrak{c} \leqq \mathfrak{a} \cup \mathfrak{c})\\ &= [(\mathfrak{a} \cup \mathfrak{b}) \cap (\mathfrak{a} \cup \mathfrak{c})] \cap \mathfrak{c} \\ &= \mathfrak{a} \cap \mathfrak{c} = 0 &&(\text{by hypothesis}).\end{aligned}$$

Thus $(\mathfrak{a}, \mathfrak{b}, \mathfrak{c})\perp$ by Part I, Theorem 2.2.

DEFINITION 7.1: If $\mathfrak{u}, \mathfrak{v} \in L_{ij}$, then we shall call a triple $(\mathfrak{u}', \mathfrak{v}', \mathfrak{w}')$ *permissible for* $\mathfrak{u}, \mathfrak{v}$ $\big((\mathfrak{u}', \mathfrak{v}', \mathfrak{w}')P_{\mathfrak{u}\mathfrak{v}}\big)$ in case

(a) $\bar{\mathfrak{a}}_j \leqq \mathfrak{w}'$, (b) $\bar{\mathfrak{a}}_i \cap \mathfrak{w}' = 0$, (c) $\mathfrak{u}' \geqq \mathfrak{u}$, (d) $\mathfrak{u}'$ is inverse to $\bar{\mathfrak{a}}_j$ in $\bar{\mathfrak{a}}_i \cup \mathfrak{w}'$, (e) $\mathfrak{v}' \geqq \mathfrak{v}$, (f) $\mathfrak{v}'$ is inverse to $\bar{\mathfrak{a}}_j$ in $\bar{\mathfrak{a}}_i \cup \mathfrak{w}'$, (g) $(\mathfrak{u}' \cap \mathfrak{w}') \cup (\mathfrak{v}' \cap \mathfrak{w}') \cup \bar{\mathfrak{a}}_i \geqq \mathfrak{u}$.

COROLLARY: $\mathfrak{u}'(\bar{\mathfrak{a}}_i \cup \bar{\mathfrak{a}}_j) = \mathfrak{u}$.

PROOF: $\mathfrak{u}'(\bar{\mathfrak{a}}_i \cup \bar{\mathfrak{a}}_j) = \mathfrak{u}'(\mathfrak{u} \cup \bar{\mathfrak{a}}_j) = \mathfrak{u} \cup \mathfrak{u}'\bar{\mathfrak{a}}_j = \mathfrak{u} \cup 0 = \mathfrak{u}$.

DEFINITION 7.2: If $\mathfrak{u}, \mathfrak{v} \in L_{ij}$, and $(\mathfrak{u}', \mathfrak{v}', \mathfrak{w}')P_{\mathfrak{u}\mathfrak{v}}$, then we define

$$(1) \quad \mathfrak{u} \oplus \mathfrak{v} \equiv \Big(\Big(\Big(\big(((\mathfrak{v}' \cap \mathfrak{w}') \cup \bar{\mathfrak{a}}_i) \cap \mathfrak{u}'\big) \cup \bar{\mathfrak{a}}_j\Big) \cap \mathfrak{w}'\Big) \cup (\mathfrak{u}' \cap \mathfrak{w}')\Big) \cap (\bar{\mathfrak{a}}_i \cup \bar{\mathfrak{a}}_j);$$

when the dependence of $\oplus$ on $(\mathfrak{u}', \mathfrak{v}', \mathfrak{w}')$ should be indicated, we shall use the notation $\oplus_{(\mathfrak{u}', \mathfrak{v}', \mathfrak{w}')}$.

LEMMA 7.2: If $\mathfrak{u}, \mathfrak{v} \in L_{ij}$, and $(\mathfrak{u}', \mathfrak{v}', \mathfrak{w}')P_{\mathfrak{u}\mathfrak{v}}$, then $\mathfrak{u} \oplus_{(\mathfrak{u}', \mathfrak{v}', \mathfrak{w}')} \mathfrak{v} \in L_{ij}$.

PROOF: (For the purposes of this proof we shall denote the operations $\cup, \cap$ by $+, \cdot$ in order to lessen the apparent complexity of the expressions which we use. This will be done in the future whenever it seems advisable.) We shall prove directly that

$$\text{a)} \quad (\mathfrak{u} \oplus \mathfrak{v})\bar{\mathfrak{a}}_j = 0, \quad \text{and} \quad \text{b)} \quad (\mathfrak{u} \oplus \mathfrak{v}) + \bar{\mathfrak{a}}_j = \bar{\mathfrak{a}}_i + \bar{\mathfrak{a}}_j.$$

a) By (1) we have

$$\begin{aligned}
(\mathfrak{u} \oplus \mathfrak{v})\bar{\mathfrak{a}}_j &= (((\mathfrak{v}'\mathfrak{w}' + \bar{\mathfrak{a}}_i)\mathfrak{u}' + \bar{\mathfrak{a}}_j)\mathfrak{v}' + \mathfrak{u}'\mathfrak{w}')(\bar{\mathfrak{a}}_i + \bar{\mathfrak{a}}_j) \cdot \bar{\mathfrak{a}}_j \\
&= (((\mathfrak{v}'\mathfrak{w}' + \bar{\mathfrak{a}}_i)\mathfrak{u}' + \bar{\mathfrak{a}}_j)\mathfrak{v}' + \mathfrak{u}'\mathfrak{w}') \cdot \bar{\mathfrak{a}}_j \\
&= (((\mathfrak{v}'\mathfrak{w}' + \bar{\mathfrak{a}}_i)\mathfrak{u}' + \bar{\mathfrak{a}}_j)\mathfrak{v}' + \mathfrak{u}'\mathfrak{w}')\mathfrak{w}'\bar{\mathfrak{a}}_j && (\text{since } \bar{\mathfrak{a}}_j \leqq \mathfrak{w}' \text{ by (a)}) \\
&= (((\mathfrak{v}'\mathfrak{w}' + \bar{\mathfrak{a}}_i)\mathfrak{u}' + \bar{\mathfrak{a}}_j)\mathfrak{v}'\mathfrak{w}' + \mathfrak{u}'\mathfrak{w}')\bar{\mathfrak{a}}_j && (\text{by IV since } \mathfrak{u}'\mathfrak{w}' \leqq \mathfrak{w}') \\
&= (((\mathfrak{v}'\mathfrak{w}' + \bar{\mathfrak{a}}_i)\mathfrak{u}' + \bar{\mathfrak{a}}_j]\mathfrak{w}'\mathfrak{v}' + \mathfrak{u}'\mathfrak{w}')\bar{\mathfrak{a}}_j \\
&= (((\mathfrak{v}'\mathfrak{w}' + \bar{\mathfrak{a}}_i)\mathfrak{u}'\mathfrak{w}' + \bar{\mathfrak{a}}_j)\mathfrak{v}' + \mathfrak{u}'\mathfrak{w}')\bar{\mathfrak{a}}_j && (\text{by IV since } \bar{\mathfrak{a}}_j \leqq \mathfrak{w}') \\
&= (((\mathfrak{v}'\mathfrak{w}' + \bar{\mathfrak{a}}_i)\mathfrak{w}'\mathfrak{u}' + \bar{\mathfrak{a}}_j)\mathfrak{v}' + \mathfrak{u}'\mathfrak{w}')\bar{\mathfrak{a}}_j \\
&= (((\mathfrak{v}'\mathfrak{w}' + \bar{\mathfrak{a}}_i\mathfrak{w}')\mathfrak{u}' + \bar{\mathfrak{a}}_j)\mathfrak{v}' + \mathfrak{u}'\mathfrak{w}')\bar{\mathfrak{a}}_j && (\text{by IV since } \mathfrak{v}'\mathfrak{w}' \leqq \mathfrak{w}') \\
&= ((\mathfrak{v}'\mathfrak{w}'\mathfrak{u}' + \bar{\mathfrak{a}}_j)\mathfrak{v}' + \mathfrak{u}'\mathfrak{w}')\bar{\mathfrak{a}}_j && (\text{since } \bar{\mathfrak{a}}_i\mathfrak{w}' = 0 \text{ by (b)}) \\
&= (\mathfrak{v}'\mathfrak{w}'\mathfrak{u}' + \bar{\mathfrak{a}}_j\mathfrak{v}' + \mathfrak{u}'\mathfrak{w}')\bar{\mathfrak{a}}_j && (\text{by IV since } \mathfrak{v}'\mathfrak{w}'\mathfrak{u}' \leqq \mathfrak{v}') \\
&= (\mathfrak{v}'\mathfrak{w}'\mathfrak{u}' + \mathfrak{u}'\mathfrak{w}')\bar{\mathfrak{a}}_j && (\text{since } \bar{\mathfrak{a}}_j\mathfrak{v}' = 0 \text{ by (f)}) \\
&= \mathfrak{u}'\mathfrak{w}'\bar{\mathfrak{a}}_j = \mathfrak{u}'\bar{\mathfrak{a}}_j \cdot \mathfrak{w}' = 0 \cdot \mathfrak{w}' = 0 && (\text{by (d)}).
\end{aligned}$$

b) Moreover,

$$\begin{aligned}
(\mathfrak{u} \oplus \mathfrak{v}) + \bar{\mathfrak{a}}_j &= (((\mathfrak{v}'\mathfrak{w}' + \bar{\mathfrak{a}}_i)\mathfrak{u}' + \bar{\mathfrak{a}}_j)\mathfrak{v}' + \mathfrak{u}'\mathfrak{w}')(\bar{\mathfrak{a}}_i + \bar{\mathfrak{a}}_j) + \bar{\mathfrak{a}}_j \\
&= (((\mathfrak{v}'\mathfrak{w}' + \bar{\mathfrak{a}}_i)\mathfrak{u}' + \bar{\mathfrak{a}}_j)\mathfrak{v}' + \bar{\mathfrak{a}}_j + \mathfrak{u}'\mathfrak{w}')(\bar{\mathfrak{a}}_i + \bar{\mathfrak{a}}_j) && (\text{by IV since } \bar{\mathfrak{a}}_j \leqq \bar{\mathfrak{a}}_i + \bar{\mathfrak{a}}_j) \\
&= (((\mathfrak{v}'\mathfrak{w}' + \bar{\mathfrak{a}}_i)\mathfrak{u}' + \bar{\mathfrak{a}}_j)(\mathfrak{v}' + \bar{\mathfrak{a}}_j) + \mathfrak{u}'\mathfrak{w}')(\bar{\mathfrak{a}}_i + \bar{\mathfrak{a}}_j) && (\text{by IV}) \\
&= (((\mathfrak{v}'\mathfrak{w}' + \bar{\mathfrak{a}}_i)\mathfrak{u}' + \bar{\mathfrak{a}}_j)(\bar{\mathfrak{a}}_i + \mathfrak{w}') + \mathfrak{u}'\mathfrak{w}')(\bar{\mathfrak{a}}_i + \bar{\mathfrak{a}}_j) && (\text{by (f)}).
\end{aligned}$$

Now since $\bar{\mathfrak{a}}_j \leqq (\bar{\mathfrak{a}}_i + \mathfrak{w}')$ (by f), and $(\mathfrak{v}'\mathfrak{w}' + \bar{\mathfrak{a}}_i)\mathfrak{u}' \leqq (\bar{\mathfrak{a}}_i + \mathfrak{w}')$, it is clear that $(\mathfrak{v}'\mathfrak{w}' + \bar{\mathfrak{a}}_i)\mathfrak{u}' + \bar{\mathfrak{a}}_j \leqq (\bar{\mathfrak{a}}_i + \mathfrak{w}')$, whence the factor $(\bar{\mathfrak{a}}_i + \mathfrak{w}')$ may be omitted from our final expression, and we have

$$(\mathfrak{u} \oplus \mathfrak{v}) + \bar{\mathfrak{a}}_j = ((\mathfrak{v}'\mathfrak{w}' + \bar{\mathfrak{a}}_i)\mathfrak{u}' + \bar{\mathfrak{a}}_j + \mathfrak{u}'\mathfrak{w}')(\bar{\mathfrak{a}}_i + \bar{\mathfrak{a}}_j)$$
$$= ((\mathfrak{u}'\mathfrak{w}' + \mathfrak{v}'\mathfrak{w}' + \bar{\mathfrak{a}}_i)\mathfrak{u}' + \bar{\mathfrak{a}}_j)(\bar{\mathfrak{a}}_i + \bar{\mathfrak{a}}_j) \qquad \text{(by IV since } \mathfrak{u}'\mathfrak{w}' \leqq \mathfrak{u}')$$
$$\geqq (\mathfrak{u}\mathfrak{u}' + \bar{\mathfrak{a}}_j)(\bar{\mathfrak{a}}_i + \bar{\mathfrak{a}}_j) \qquad \text{(by (g))}$$
$$= (\mathfrak{u} + \bar{\mathfrak{a}}_j)(\bar{\mathfrak{a}}_i + \bar{\mathfrak{a}}_j) \qquad \text{(by (c))}$$
$$= (\bar{\mathfrak{a}}_i + \bar{\mathfrak{a}}_j)(\bar{\mathfrak{a}}_i + \bar{\mathfrak{a}}_j) \qquad \text{(since } \mathfrak{u} \in L_{ij})$$
$$= \bar{\mathfrak{a}}_i + \bar{\mathfrak{a}}_j.$$

Hence $(\mathfrak{u} \oplus \mathfrak{v}) + \bar{\mathfrak{a}}_j \geqq \bar{\mathfrak{a}}_i + \bar{\mathfrak{a}}_j$. But by (1), $(\mathfrak{u} \oplus \mathfrak{v}) + \bar{\mathfrak{a}}_j \leqq \bar{\mathfrak{a}}_i + \bar{\mathfrak{a}}_j$; hence $(\mathfrak{u} \oplus \mathfrak{v}) + \bar{\mathfrak{a}}_j = \bar{\mathfrak{a}}_i + \bar{\mathfrak{a}}_j$. This completes the proof.

Lemma 7.3: If $\mathfrak{u}, \mathfrak{v} \in L_{ij}$, $(\mathfrak{u}'', \mathfrak{v}'', \mathfrak{w}'')P_{\mathfrak{u}\mathfrak{v}}$, $(\mathfrak{u}''', \mathfrak{v}''', \mathfrak{w}''')P_{\mathfrak{u}\mathfrak{v}}$, and if $(\bar{\mathfrak{a}}_i \cup \mathfrak{w}'') \cap (\bar{\mathfrak{a}}_i \cup \mathfrak{w}''') = \bar{\mathfrak{a}}_i \cup \bar{\mathfrak{a}}_j$, then $(\mathfrak{u}'' \cup \mathfrak{u}''', \mathfrak{v}'' \cup \mathfrak{v}''', \mathfrak{w}'' \cup \mathfrak{w}''')P_{\mathfrak{u}\mathfrak{v}}$.

Proof: It is evident that $\mathfrak{u}'' \cup \mathfrak{u}'''$, $\mathfrak{v}'' \cup \mathfrak{v}'''$, $\mathfrak{w}'' \cup \mathfrak{w}'''$ satisfy conditions (a), (c), (e), (g) of Definition 7.1 as well as the conditions $\mathfrak{u}' \cup \bar{\mathfrak{a}}_j = \bar{\mathfrak{a}}_i \cup \mathfrak{w}'$, $\mathfrak{v}' \cup \bar{\mathfrak{a}}_j = \bar{\mathfrak{a}}_i \cup \mathfrak{w}'$ constituting parts of (d), (f), respectively. It remains then only to verify (b) and the remainders of (d), (f), which are respectively

$$\bar{\mathfrak{a}}_i \cap (\mathfrak{w}'' \cup \mathfrak{w}''') = 0, \tag{2}$$
$$\bar{\mathfrak{a}}_j \cap (\mathfrak{u}'' \cup \mathfrak{u}''') = 0, \tag{3}$$
$$\bar{\mathfrak{a}}_j \cap (\mathfrak{v}'' \cup \mathfrak{v}''') = 0. \tag{4}$$

To prove (2), we note:

$$\begin{aligned}
\bar{\mathfrak{a}}_i(\mathfrak{w}'' \cup \mathfrak{w}''') &= \bar{\mathfrak{a}}_i(\bar{\mathfrak{a}}_i \cup \mathfrak{w}'')(\mathfrak{w}'' \cup \mathfrak{w}''') \\
&\leqq \bar{\mathfrak{a}}_i(\mathfrak{w}'' \cup \mathfrak{w}'''(\bar{\mathfrak{a}}_i \cup \mathfrak{w}'')(\bar{\mathfrak{a}}_i \cup \mathfrak{w}''')) \\
&= \bar{\mathfrak{a}}_i(\mathfrak{w}'' \cup \mathfrak{w}'''(\bar{\mathfrak{a}}_i \cup \bar{\mathfrak{a}}_j)) \\
&= \bar{\mathfrak{a}}_i(\mathfrak{w}'' \cup \bar{\mathfrak{a}}_j) = \bar{\mathfrak{a}}_i\mathfrak{w}'' = 0,
\end{aligned}$$

and (2) is established.

To prove (3), we note:

$$\begin{aligned}
\bar{\mathfrak{a}}_j \cap (\mathfrak{u}'' \cup \mathfrak{u}''') &= \bar{\mathfrak{a}}_j \cap (\bar{\mathfrak{a}}_j \cup \mathfrak{u}''') \cap (\mathfrak{u}'' \cup \mathfrak{u}''') \\
&= \bar{\mathfrak{a}}_j \cap (\mathfrak{u}''' \cup \mathfrak{u}''(\bar{\mathfrak{a}}_j \cup \mathfrak{u}''')) \qquad \text{(by IV)} \\
&= \bar{\mathfrak{a}}_j \cap (\mathfrak{u}''' \cup \mathfrak{u}''(\bar{\mathfrak{a}}_j \cup \mathfrak{u}'')(\bar{\mathfrak{a}}_j \cup \mathfrak{u}''')) \\
&= \bar{\mathfrak{a}}_j \cap (\mathfrak{u}''' \cup \mathfrak{u}''(\bar{\mathfrak{a}}_i \cup \mathfrak{w}'')(\bar{\mathfrak{a}}_i \cup \mathfrak{w}''')) \\
&= \bar{\mathfrak{a}}_j \cap (\mathfrak{u}''' \cup \mathfrak{u}''(\bar{\mathfrak{a}}_i \cup \bar{\mathfrak{a}}_j)) \\
&= \bar{\mathfrak{a}}_j \cap (\mathfrak{u}''' \cup \mathfrak{u}) \\
&= \bar{\mathfrak{a}}_j \cap \mathfrak{u}''' = 0
\end{aligned}$$

(by the Corollary to Definition 7.1)

The proof of (4) is similar to that of (3).

LEMMA 7.4: If $\mathfrak{u}, \mathfrak{v} \in L_{ij}$, $(\mathfrak{u}', \mathfrak{v}', \mathfrak{w}')P_{\mathfrak{uv}}$, $(\mathfrak{u}'', \mathfrak{v}'', \mathfrak{w}'')P_{\mathfrak{uv}}$, and if $\mathfrak{u}' \leqq \mathfrak{u}''$, $\mathfrak{v}' \leqq \mathfrak{v}''$, $\mathfrak{w}' \leqq \mathfrak{w}''$, then

$$\mathfrak{u} \oplus_{(\mathfrak{u}', \mathfrak{v}', \mathfrak{w}')} \mathfrak{v} = \mathfrak{u} \oplus_{(\mathfrak{u}'', \mathfrak{v}'', \mathfrak{w}'')} \mathfrak{v}. \tag{5}$$

PROOF: Let us denote the left and right members of (5) by $\mathfrak{u} \oplus_1 \mathfrak{v}$, $\mathfrak{u} \oplus_2 \mathfrak{v}$, respectively. Then by Lemma 7.2, $\mathfrak{u} \oplus_1 \mathfrak{v}$, $\mathfrak{u} \oplus_2 \mathfrak{v} \in L_{ij}$. But one sees easily from (1) that $\mathfrak{u} \oplus_1 \mathfrak{v} \leqq \mathfrak{u} \oplus_2 \mathfrak{v}$, whence by Lemma 3.3, $\mathfrak{u} \oplus_1 \mathfrak{v} = \mathfrak{u} \oplus_2 \mathfrak{v}$, and (5) is established.

COROLLARY: If $\mathfrak{u}, \mathfrak{v} \in L_{ij}$, $(\mathfrak{u}, \mathfrak{v}', \mathfrak{w}')P_{\mathfrak{uv}}$, $(\mathfrak{u}'', \mathfrak{v}'', \mathfrak{w}'')P_{\mathfrak{uv}}$, $(\mathfrak{u}''', \mathfrak{v}''', \mathfrak{w}''')P_{\mathfrak{uv}}$, and if $\mathfrak{u}', \mathfrak{u}''' \leqq \mathfrak{u}''$, $\mathfrak{v}', \mathfrak{v}''' \leqq \mathfrak{v}''$, $\mathfrak{w}', \mathfrak{w}''' \leqq \mathfrak{w}''$ (or if these relations hold with $\leqq$ replaced by $\geqq$), then $\mathfrak{u} \oplus_{(\mathfrak{u}'\mathfrak{v}'\mathfrak{w}')} \mathfrak{v} = \mathfrak{u} \oplus_{(\mathfrak{u}''', \mathfrak{v}''', \mathfrak{w}''')} \mathfrak{v}$.

PROOF: By Lemma 7.4 we have immediately under either hypothesis

$$\mathfrak{u} \oplus_{(\mathfrak{u}', \mathfrak{v}', \mathfrak{w}')} \mathfrak{v} = \mathfrak{u} \oplus_{(\mathfrak{u}'', \mathfrak{v}'', \mathfrak{w}'')} \mathfrak{v} = \mathfrak{u} \oplus_{(\mathfrak{u}''', \mathfrak{v}''', \mathfrak{w}''')} \mathfrak{v}.$$

LEMMA 7.5: If $\mathfrak{u}, \mathfrak{v} \in L_{ij}$, $(\mathfrak{u}', \mathfrak{v}', \mathfrak{w}')P_{\mathfrak{uv}}$, $(\mathfrak{u}'', \mathfrak{v}'', \mathfrak{w}'')P_{\mathfrak{uv}}$, and if $(\bar{\mathfrak{a}}_i \cup \mathfrak{w}') \cap (\bar{\mathfrak{a}}_i \cup \mathfrak{w}'') = \bar{\mathfrak{a}}_i \cup \bar{\mathfrak{a}}_j$, then $\mathfrak{u} \oplus_{(\mathfrak{u}', \mathfrak{v}', \mathfrak{w}')} \mathfrak{v} = \mathfrak{u} \oplus_{(\mathfrak{u}'', \mathfrak{v}'', \mathfrak{w}'')} \mathfrak{v}$.

PROOF: By Lemma 7.3, $(\mathfrak{u}' \cup \mathfrak{u}'', \mathfrak{v}' \cup \mathfrak{v}'', \mathfrak{w}' \cup \mathfrak{w}'')P_{\mathfrak{uv}}$. Hence since $\mathfrak{u}', \mathfrak{u}'' \leqq \mathfrak{u}' \cup \mathfrak{u}''$, $\mathfrak{v}', \mathfrak{v}'' \leqq \mathfrak{v}' \cup \mathfrak{v}''$, $\mathfrak{w}', \mathfrak{w}'' \leqq \mathfrak{w}' \cup \mathfrak{w}''$, the Corollary to Lemma 7.4 applies and yields the desired result.

We shall now specialize our considerations by giving several examples of permissible triples $(\mathfrak{u}', \mathfrak{v}', \mathfrak{w}')$.

LEMMA 7.6: Let $\mathfrak{u}, \mathfrak{v} \in L_{ij}$. If $(\bar{\mathfrak{a}}_i, \mathfrak{g}, \mathfrak{f})\mathscr{C}$ (cf. Def. 3.4), if $(\bar{\mathfrak{a}}_i, \bar{\mathfrak{a}}_j, \mathfrak{f})\perp$ (or, equivalently, if $(\bar{\mathfrak{a}}_i, \bar{\mathfrak{a}}_j, \mathfrak{g})\perp$), and if

$$\mathfrak{u}' \equiv \mathfrak{u} \cup \mathfrak{g}, \quad \mathfrak{v}' \equiv \mathfrak{v} \cup \mathfrak{f}, \quad \mathfrak{w}' \equiv \bar{\mathfrak{a}}_j \cup \mathfrak{f}, \tag{6}$$

then $(\mathfrak{u}', \mathfrak{v}', \mathfrak{w}')P_{\mathfrak{uv}}$.

PROOF: We first observe that our hypothesis on $\mathfrak{g}, \mathfrak{f}$ is symmetric in $\mathfrak{g}, \mathfrak{f}$, i.e., if $(\bar{\mathfrak{a}}_i, \mathfrak{g}, \mathfrak{f})\mathscr{C}$, then $(\bar{\mathfrak{a}}_i, \bar{\mathfrak{a}}_j, \mathfrak{f})\perp$ is equivalent to $(\bar{\mathfrak{a}}_i, \bar{\mathfrak{a}}_j, \mathfrak{g})\perp$; for, since $\bar{\mathfrak{a}}_i \cup \mathfrak{g} = \bar{\mathfrak{a}}_i \cup \mathfrak{f}$, $\bar{\mathfrak{a}}_i \cap \mathfrak{g} = \bar{\mathfrak{a}}_i \cap \mathfrak{f} = 0$, we may interchange $\mathfrak{g}, \mathfrak{f}$ in any independent system containing $\bar{\mathfrak{a}}_i$ by Part I, Lemma 2.1. We shall now verify that conditions (a)—(g) hold for $\mathfrak{u}', \mathfrak{v}', \mathfrak{w}'$. Indeed, (a), (c), (e) are obvious, and (b) follows from $(\bar{\mathfrak{a}}_i, \bar{\mathfrak{a}}_j, \mathfrak{f})\perp$. To prove (d), we observe first that

$$\begin{aligned}
\mathfrak{u}' \cup \bar{\mathfrak{a}}_j &= \mathfrak{u} \cup \mathfrak{g} \cup \bar{\mathfrak{a}}_j = \mathfrak{u} \cup \bar{\mathfrak{a}}_j \cup \mathfrak{g} = \bar{\mathfrak{a}}_i \cup \bar{\mathfrak{a}}_j \cup \mathfrak{g} && (\text{since } \mathfrak{u} \in L_{ij}) \\
&= \bar{\mathfrak{a}}_i \cup \mathfrak{g} \cup \bar{\mathfrak{a}}_j = \bar{\mathfrak{a}}_i \cup \mathfrak{f} \cup \bar{\mathfrak{a}}_j && (\text{since } (\bar{\mathfrak{a}}_i, \mathfrak{g}, \mathfrak{f})\mathscr{C}) \\
&= \bar{\mathfrak{a}}_i \cup \mathfrak{w}'.
\end{aligned}$$

Now we had $(\bar{\mathfrak{a}}_i, \bar{\mathfrak{a}}_j, \mathfrak{g})\perp$; moreover, $\mathfrak{u} \cup \bar{\mathfrak{a}}_j = \bar{\mathfrak{a}}_i \cup \bar{\mathfrak{a}}_j$, $\mathfrak{u} \cap \bar{\mathfrak{a}}_j = \bar{\mathfrak{a}}_i \cap \bar{\mathfrak{a}}_j = 0$, whence Part I, Lemma 2.1 applies and yields $(\mathfrak{u}, \bar{\mathfrak{a}}_j, \mathfrak{g})\perp$. Hence $\mathfrak{u}' \cap \bar{\mathfrak{a}}_j$

$= (\mathfrak{u} \cup \mathfrak{g}) \cap \bar{\mathfrak{a}}_j = 0$. Thus (d) is established. To prove (f), we note that

$$\mathfrak{v}' \cup \bar{\mathfrak{a}}_j = \mathfrak{v} \cup \mathfrak{f} \cup \bar{\mathfrak{a}}_j = \mathfrak{v} \cup \bar{\mathfrak{a}}_j \cup \mathfrak{f} = \bar{\mathfrak{a}}_i \cup \bar{\mathfrak{a}}_j \cup \mathfrak{f} \quad (\text{since } \mathfrak{v} \in L_{ij}) \\ = \bar{\mathfrak{a}}_i \cup \mathfrak{w}'.$$

Moreover, since $(\bar{\mathfrak{a}}_i, \bar{\mathfrak{a}}_j, \mathfrak{f})\perp$, $\mathfrak{v} \cup \bar{\mathfrak{a}}_j = \bar{\mathfrak{a}}_i \cup \bar{\mathfrak{a}}_j$, $\mathfrak{v} \cap \bar{\mathfrak{a}}_j = \bar{\mathfrak{a}}_i \cap \bar{\mathfrak{a}}_j = 0$ we have by Part I, Lemma 2.1, $(\mathfrak{v}, \bar{\mathfrak{a}}_j, \mathfrak{f})\perp$. Hence $\mathfrak{v}' \cap \bar{\mathfrak{a}}_j = (\mathfrak{v} \cup \mathfrak{f}) \cap \bar{\mathfrak{a}}_j = 0$. Thus (f) is proved. Finally, let us consider (g). We have

$$\begin{aligned}(\mathfrak{u}' \cap \mathfrak{w}') \cup (\mathfrak{v}' \cap \mathfrak{w}') \cup \bar{\mathfrak{a}}_i &= \big((\mathfrak{u} \cup \mathfrak{g}) \cap (\bar{\mathfrak{a}}_j \cup \mathfrak{f})\big) \cup \big((\mathfrak{v} \cup \mathfrak{f}) \cap (\bar{\mathfrak{a}}_j \cup \mathfrak{f})\big) \cup \bar{\mathfrak{a}}_i \\ &\geqq \big((\mathfrak{u} \cup \mathfrak{g}) \cap (\bar{\mathfrak{a}}_j \cup \mathfrak{f})\big) \cup \mathfrak{f} \cup \bar{\mathfrak{a}}_i \\ &= \big((\mathfrak{u} \cup \mathfrak{g}) \cap (\bar{\mathfrak{a}}_j \cup \mathfrak{f})\big) \cup \mathfrak{g} \cup \bar{\mathfrak{a}}_i && (\text{since } (\bar{\mathfrak{a}}_i, \mathfrak{g}, \mathfrak{f})\mathscr{C}) \\ &= \big((\mathfrak{u} \cup \mathfrak{g}) \cap (\bar{\mathfrak{a}}_j \cup \mathfrak{f} \cup \mathfrak{g})\big) \cup \bar{\mathfrak{a}}_i && (\text{by IV}) \\ &= \big((\mathfrak{u} \cup \mathfrak{g}) \cap (\bar{\mathfrak{a}}_j \cup \bar{\mathfrak{a}}_i \cup \mathfrak{g})\big) \cup \bar{\mathfrak{a}}_i && (\text{since } (\bar{\mathfrak{a}}_i, \mathfrak{g}, \mathfrak{f})\mathscr{C}) \\ &\geqq \mathfrak{u} \cap (\bar{\mathfrak{a}}_j \cup \bar{\mathfrak{a}}_i) = \mathfrak{u} && (\text{since } \mathfrak{u} \leqq \bar{\mathfrak{a}}_i \cup \bar{\mathfrak{a}}_j),\end{aligned}$$

whence (g) holds.

DEFINITION 7.3: Let $\mathfrak{u}, \mathfrak{v} \in L_{ij}$. If $(\bar{\mathfrak{a}}_i, \mathfrak{g}, \mathfrak{f})\mathscr{C}$, and if $(\bar{\mathfrak{a}}_i, \bar{\mathfrak{a}}_j, \mathfrak{f})\perp$, then we define

$$(7) \quad \mathfrak{u} \boxplus_{(\mathfrak{g}, \mathfrak{f})} \mathfrak{v} \equiv \Big(\big((\mathfrak{u} \cup \mathfrak{g}) \cap (\bar{\mathfrak{a}}_j \cup \mathfrak{f})\big) \cup \big((\mathfrak{v} \cup \mathfrak{f}) \cap (\bar{\mathfrak{a}}_j \cup \mathfrak{g})\big)\Big) \cap (\bar{\mathfrak{a}}_i \cup \bar{\mathfrak{a}}_j).$$

COROLLARY: If $\mathfrak{u}'$, $\mathfrak{v}'$, $\mathfrak{w}'$ are defined in terms of $\mathfrak{g}$, $\mathfrak{f}$ by (6), then $\mathfrak{u} \boxplus_{(\mathfrak{g}, \mathfrak{f})} \mathfrak{v} = \mathfrak{u} \oplus_{(\mathfrak{u}', \mathfrak{v}', \mathfrak{w}')} \mathfrak{v}$.

PROOF: For the purposes of this proof we replace the symbols $\cup$, $\cap$ by $+$, $\cdot$. We have

$$\begin{aligned}\mathfrak{u} \oplus \mathfrak{v} = \mathfrak{u} \oplus_{(\mathfrak{u}', \mathfrak{v}', \mathfrak{w}')} \mathfrak{v} &= \big(((\mathfrak{v}'\mathfrak{w}' + \bar{\mathfrak{a}}_i)\mathfrak{u}' + \bar{\mathfrak{a}}_j)\mathfrak{v}' + \mathfrak{u}'\mathfrak{w}'\big)(\bar{\mathfrak{a}}_i + \bar{\mathfrak{a}}_j) \\ &= \big((((\mathfrak{v} + \mathfrak{f})(\bar{\mathfrak{a}}_j + \mathfrak{f}) + \bar{\mathfrak{a}}_i)(\mathfrak{u} + \mathfrak{g}) + \bar{\mathfrak{a}}_j)(\mathfrak{v} + \mathfrak{f}) + (\mathfrak{u} + \mathfrak{g})(\bar{\mathfrak{a}}_j + \mathfrak{f})\big)(\bar{\mathfrak{a}}_i + \bar{\mathfrak{a}}_j) \\ &\geqq \big(((\mathfrak{f} + \bar{\mathfrak{a}}_i)(\mathfrak{u} + \mathfrak{g}) + \bar{\mathfrak{a}}_j)(\mathfrak{v} + \mathfrak{f}) + (\mathfrak{u} + \mathfrak{g})(\bar{\mathfrak{a}}_j + \mathfrak{f})\big)(\bar{\mathfrak{a}}_i + \bar{\mathfrak{a}}_j) \\ &= \big(((\mathfrak{g} + \bar{\mathfrak{a}}_i)(\mathfrak{u} + \mathfrak{g}) + \bar{\mathfrak{a}}_j)(\mathfrak{v} + \mathfrak{f}) + (\mathfrak{u} + \mathfrak{g})(\bar{\mathfrak{a}}_j + \mathfrak{f})\big)(\bar{\mathfrak{a}}_i + \bar{\mathfrak{a}}_j) \\ &\qquad (\text{since } (\bar{\mathfrak{a}}_i, \mathfrak{g}, \mathfrak{f})\mathscr{C}) \\ &\geqq \big((\mathfrak{g} + \bar{\mathfrak{a}}_j)(\mathfrak{v} + \mathfrak{f}) + (\mathfrak{u} + \mathfrak{g})(\bar{\mathfrak{a}}_j + \mathfrak{f})\big)(\bar{\mathfrak{a}}_i + \bar{\mathfrak{a}}_j) \\ &= \big((u + \mathfrak{g})(\bar{\mathfrak{a}}_j + \mathfrak{f}) + (\mathfrak{v} + \mathfrak{f})(\bar{\mathfrak{a}}_j + \mathfrak{g})\big)(\bar{\mathfrak{a}}_i + \bar{\mathfrak{a}}_j) \\ &= \mathfrak{u} \boxplus_{(\mathfrak{g}, \mathfrak{f})} \mathfrak{v}.\end{aligned}$$

Hence $\mathfrak{u} \oplus \mathfrak{v} \geqq \mathfrak{u} \boxplus_{(\mathfrak{g}, \mathfrak{f})} \mathfrak{v}$. But $\mathfrak{u} \oplus \mathfrak{v} \in L_{ij}$, whence $(\mathfrak{u} \boxplus_{(\mathfrak{g}, \mathfrak{f})} \mathfrak{v}) \cdot \bar{\mathfrak{a}}_j = 0$. Moreover,

$$\begin{aligned}(\mathfrak{u} \boxplus_{(\mathfrak{g}, \mathfrak{f})} \mathfrak{v}) + \bar{\mathfrak{a}}_j &= \big((\mathfrak{u} + \mathfrak{g})(\bar{\mathfrak{a}}_j + \mathfrak{f}) + (\mathfrak{v} + \mathfrak{f})(\bar{\mathfrak{a}}_j + \mathfrak{g})\big)(\bar{\mathfrak{a}}_i + \bar{\mathfrak{a}}_j) + \bar{\mathfrak{a}}_j \\ &= \big((\mathfrak{u} + \mathfrak{g})(\bar{\mathfrak{a}}_j + \mathfrak{f}) + \bar{\mathfrak{a}}_j + (\mathfrak{v} + \mathfrak{f})(\bar{\mathfrak{a}}_j + \mathfrak{g})\big)(\bar{\mathfrak{a}}_i + \bar{\mathfrak{a}}_j) && (\text{by IV}) \\ &= \big([(\mathfrak{u} + \mathfrak{g})(\bar{\mathfrak{a}}_j + \mathfrak{f}) + \bar{\mathfrak{a}}_j] + [(\mathfrak{v} + \mathfrak{f})(\bar{\mathfrak{a}}_j + \mathfrak{g}) + \bar{\mathfrak{a}}_j]\big)(\bar{\mathfrak{a}}_i + \bar{\mathfrak{a}}_j) \\ &= \big((\mathfrak{u} + \mathfrak{g} + \bar{\mathfrak{a}}_j)(\bar{\mathfrak{a}}_j + \mathfrak{f}) + (\mathfrak{v} + \mathfrak{f} + \bar{\mathfrak{a}}_j)(\bar{\mathfrak{a}}_j + \mathfrak{g})\big)(\bar{\mathfrak{a}}_i + \bar{\mathfrak{a}}_j). && (\text{by IV}).\end{aligned}$$

But $\mathfrak{u} + \mathfrak{g} + \bar{\mathfrak{a}}_j = \mathfrak{u}' + \bar{\mathfrak{a}}_j = \bar{\mathfrak{a}}_i + \mathfrak{w}' = \mathfrak{v}' + \bar{\mathfrak{a}}_j = \mathfrak{v} + \mathfrak{f} + \bar{\mathfrak{a}}_j$. by Def. 7.1 (d), (f) (which hold since $(\mathfrak{u}', \mathfrak{v}', \mathfrak{w}')P_{\mathfrak{u}\mathfrak{v}}$ by Lemma 7.6). Hence

$$\begin{aligned}(\mathfrak{u} \boxplus_{(\mathfrak{g},\mathfrak{f})} \mathfrak{v}) + \bar{\mathfrak{a}}_j &= \big((\mathfrak{v}+\mathfrak{f}+\bar{\mathfrak{a}}_j)(\bar{\mathfrak{a}}_j+\mathfrak{f}) + (\mathfrak{u}+\mathfrak{g}+\bar{\mathfrak{a}}_j)(\bar{\mathfrak{a}}_j+\mathfrak{g})\big)(\bar{\mathfrak{a}}_i+\bar{\mathfrak{a}}_j)\\ &= \big((\bar{\mathfrak{a}}_j+\mathfrak{f}) + (\bar{\mathfrak{a}}_j+\mathfrak{g})\big)(\bar{\mathfrak{a}}_i+\bar{\mathfrak{a}}_j)\\ &= (\bar{\mathfrak{a}}_j+\mathfrak{g}+\mathfrak{f})(\bar{\mathfrak{a}}_i+\bar{\mathfrak{a}}_j)\\ &= (\bar{\mathfrak{a}}_j+\bar{\mathfrak{a}}_i+\mathfrak{f})(\bar{\mathfrak{a}}_i+\bar{\mathfrak{a}}_j) \qquad \text{(since } (\bar{\mathfrak{a}}_i, \mathfrak{g}, \mathfrak{f})\mathscr{C})\\ &= \bar{\mathfrak{a}}_i + \bar{\mathfrak{a}}_j.\end{aligned}$$

This shows that $\mathfrak{u} \boxplus_{(\mathfrak{g},\mathfrak{f})} \mathfrak{v} \in L_{ij}$, and therefore that $\mathfrak{u} \boxplus_{(\mathfrak{g},\mathfrak{f})} \mathfrak{v} = \mathfrak{u} \oplus \mathfrak{v}$ by Lemma 3.3.

Lemma 7.7: Let $\mathfrak{u}, \mathfrak{v} \in L_{ij}$. If $(\bar{\mathfrak{a}}_j, \mathfrak{g}, \mathfrak{f})\mathscr{C}$, if $(\bar{\mathfrak{a}}_i, \bar{\mathfrak{a}}_j, \mathfrak{f})\perp$ (or, equivalently, if $(\bar{\mathfrak{a}}_i, \bar{\mathfrak{a}}_j, \mathfrak{g})\perp$, and if

$$(8) \qquad \mathfrak{u}' \equiv \mathfrak{u} \cup \mathfrak{g}, \quad \mathfrak{v}' \equiv \mathfrak{v} \cup \mathfrak{f}, \quad \mathfrak{w}' \equiv \bar{\mathfrak{a}}_j \cup \mathfrak{g} = \bar{\mathfrak{a}}_j \cup \mathfrak{f},$$

then $(\mathfrak{u}', \mathfrak{v}'\ \mathfrak{w}')P_{\mathfrak{u}\mathfrak{v}}$.

Proof: The observation that our hypothesis is symmetric in $\mathfrak{g}$, $\mathfrak{f}$ follows from the fact that $\mathfrak{g}$, $\mathfrak{f}$ may be interchanged in any independent system containing $\bar{\mathfrak{a}}_i$ by Part I, Lemma 2.1, owing to $\bar{\mathfrak{a}}_j \cup \mathfrak{g} = \bar{\mathfrak{a}}_j \cup \mathfrak{f}$, $\bar{\mathfrak{a}}_j \cap \mathfrak{g} = \bar{\mathfrak{a}}_j \cap \mathfrak{f} = 0$. Conditions (a), (c), (e) of Definition 7.1 are obvious. Also, $\bar{\mathfrak{a}}_i \cap \mathfrak{w}' = \bar{\mathfrak{a}}_i \cap (\bar{\mathfrak{a}}_j \cup \mathfrak{g}) = 0$ since $(\bar{\mathfrak{a}}_i, \bar{\mathfrak{a}}_j, \mathfrak{g})\perp$, and (b) is proved. To establish (d) we note first that

$$\mathfrak{u}' \cup \bar{\mathfrak{a}}_j = \mathfrak{u} \cup \mathfrak{g} \cup \bar{\mathfrak{a}}_j = \mathfrak{u} \cup \bar{\mathfrak{a}}_j \cup \mathfrak{g} = \bar{\mathfrak{a}}_i \cup \bar{\mathfrak{a}}_j \cup \mathfrak{g} = \bar{\mathfrak{a}}_i \cup \mathfrak{w}'.$$

Now we had $(\bar{\mathfrak{a}}_i, \mathfrak{g}, \bar{\mathfrak{a}}_j)\perp$; moreover, $\mathfrak{u} \cup \bar{\mathfrak{a}}_j = \bar{\mathfrak{a}}_i \cup \bar{\mathfrak{a}}_j$, $\mathfrak{u} \cap \bar{\mathfrak{a}}_j = \bar{\mathfrak{a}}_i \cap \bar{\mathfrak{a}}_j = 0$, whence $(\mathfrak{u}, \mathfrak{g}, \bar{\mathfrak{a}}_j)\perp$ by Part I, Lemma 2.1. Hence $\mathfrak{u}' \cap \bar{\mathfrak{a}}_j = (\mathfrak{u} \cup \mathfrak{g}) \cap \bar{\mathfrak{a}}_j = 0$. Thus (d) is proved. In a similar manner (f) is established. Finally, we have

$$\begin{aligned}(\mathfrak{u}' \cap \mathfrak{w}') \cup (\mathfrak{v}' \cap \mathfrak{w}') \cup \bar{\mathfrak{a}}_i &= \big((\mathfrak{u} \cup \mathfrak{g}) \cap (\bar{\mathfrak{a}}_j \cup \mathfrak{g})\big) \cup \big((\mathfrak{v} \cup \mathfrak{f}) \cap (\bar{\mathfrak{a}}_j \cup \mathfrak{f})\big) \cup \bar{\mathfrak{a}}_i\\ &\geqq \mathfrak{g} \cup \mathfrak{f} \cup \bar{\mathfrak{a}}_i = \mathfrak{g} \cup \bar{\mathfrak{a}}_j \cup \bar{\mathfrak{a}}_i \quad \text{(since } (\bar{\mathfrak{a}}_j, \mathfrak{g}, \mathfrak{f})\mathscr{C})\\ &\geqq \bar{\mathfrak{a}}_i \cup \bar{\mathfrak{a}}_j \geqq \mathfrak{u},\end{aligned}$$

whence (g) is proved.

Definition 7.4: Let $\mathfrak{u}, \mathfrak{v} \in L_{ij}$. If $(\bar{\mathfrak{a}}_j, \mathfrak{g}, \mathfrak{f})\mathscr{C}$, and if $(\bar{\mathfrak{a}}_i, \bar{\mathfrak{a}}_j, \mathfrak{f})\perp$, then we define

$$(9) \quad \mathfrak{u} \diamondplus_{(\mathfrak{g},\mathfrak{f})} \mathfrak{v} \equiv \Big(\Big(\big(\big((\mathfrak{u} \cup \mathfrak{g}) \cap (\bar{\mathfrak{a}}_i \cup \mathfrak{f})\big) \cup \bar{\mathfrak{a}}_j\big) \cap (\mathfrak{v} \cap \mathfrak{f})\Big) \cap \mathfrak{g}\Big) \cap (\bar{\mathfrak{a}}_i \cup \bar{\mathfrak{a}}_j).$$

Corollary: If $\mathfrak{u}'$, $\mathfrak{v}'$, $\mathfrak{w}'$ are defined in terms of $\mathfrak{g}$, $\mathfrak{f}$ by (11), then $\mathfrak{u} \diamondplus_{(\mathfrak{g},\mathfrak{f})} \mathfrak{v} = \mathfrak{u} \oplus_{(\mathfrak{u}',\mathfrak{v}',\mathfrak{w}')} \mathfrak{v}$.

Proof: For the purposes of this proof we use the symbols $+$, $\cdot$ in place of $\cup$, $\cap$. Now $(\bar{\mathfrak{a}}_i, \bar{\mathfrak{a}}_j, \mathfrak{f})\perp$, and $\mathfrak{v} + \bar{\mathfrak{a}}_j = \bar{\mathfrak{a}}_i + \bar{\mathfrak{a}}_j$, $\mathfrak{v}\bar{\mathfrak{a}}_j = \bar{\mathfrak{a}}_i\bar{\mathfrak{a}}_j = 0$, whence by Part I, Lemma 2.1, $(\mathfrak{v}, \bar{\mathfrak{a}}_j, \mathfrak{f})\perp$. Hence $(\mathfrak{v}+\mathfrak{f})(\bar{\mathfrak{a}}_j+\mathfrak{f})=\mathfrak{f}$ by Part I, Theorem 2.5. Similarly $(\mathfrak{u}+\mathfrak{g})(\bar{\mathfrak{a}}_j+\mathfrak{g})=\mathfrak{g}$. Therefore

$$\begin{aligned}\mathfrak{u}\oplus\mathfrak{v} &= \mathfrak{u}\oplus_{(\mathfrak{u}',\mathfrak{v}',\mathfrak{w}')}\mathfrak{v} = \big(((\mathfrak{v}'\mathfrak{w}'+\bar{\mathfrak{a}}_i)\mathfrak{u}'+\bar{\mathfrak{a}}_j)\mathfrak{v}'+\mathfrak{u}'\mathfrak{w}'\big)(\bar{\mathfrak{a}}_i\cup\bar{\mathfrak{a}}_j)\\ &= \big((((\mathfrak{v}+\mathfrak{f})(\bar{\mathfrak{a}}_j+\mathfrak{f})+\bar{\mathfrak{a}}_i)(\mathfrak{u}+\mathfrak{g})+\bar{\mathfrak{a}}_j)(\mathfrak{v}+\mathfrak{f})+(\mathfrak{u}+\mathfrak{g})(\bar{\mathfrak{a}}_j+\mathfrak{g})\big)(\bar{\mathfrak{a}}_i+\bar{\mathfrak{a}}_j)\\ &= \big(((\mathfrak{f}+\bar{\mathfrak{a}}_i)(\mathfrak{u}+\mathfrak{g})+\bar{\mathfrak{a}}_j)(\mathfrak{v}+\mathfrak{f})+\mathfrak{g}\big)(\bar{\mathfrak{a}}_i+\bar{\mathfrak{a}}_j)\\ &= \mathfrak{u}\diamond\!\!\!+_{(\mathfrak{g},\mathfrak{f})}\mathfrak{v}.\end{aligned}$$

At this point we assume again that $\bar{\mathfrak{a}}_i$, $\bar{\mathfrak{a}}_j$ are members of a homogeneous basis for L of order $n \geqq 4$.

DEFINITION 7.5: Let $\mathfrak{u}, \mathfrak{v} \in L_{ij}$. We define for $k \neq i, j$

$$(10)\quad \mathfrak{u}\boxplus_k\mathfrak{v} \equiv \big(((\mathfrak{u}\cup\bar{\mathfrak{c}}_{ik})\cap(\bar{\mathfrak{a}}_j\cup\bar{\mathfrak{a}}_k))\cup((\mathfrak{v}\cup\bar{\mathfrak{a}}_k)\cap(\bar{\mathfrak{a}}_j\cup\bar{\mathfrak{c}}_{ik}))\big)\cap(\bar{\mathfrak{a}}_i\cup\bar{\mathfrak{a}}_j),$$

and

$$(11)\quad \mathfrak{u}\diamond\!\!\!+_k\mathfrak{v} \equiv \big((((\mathfrak{u}\cup\bar{\mathfrak{c}}_{jk})\cap(\mathfrak{a}_i\cup\bar{\mathfrak{a}}_k))\cup\bar{\mathfrak{a}}_j)\cap(\mathfrak{v}\cup\bar{\mathfrak{a}}_k)\big)\cup\bar{\mathfrak{c}}_{jk}\big)\cap(\bar{\mathfrak{a}}_i\cup\bar{\mathfrak{a}}_j).$$

COROLLARY: If $\mathfrak{g}\equiv\bar{\mathfrak{c}}_{ik}$, $\mathfrak{f}\equiv\bar{\mathfrak{a}}_k$, then $\mathfrak{g}$, $\mathfrak{f}$ satisfy the hypotheses of Definition 7.3 and $\mathfrak{u}\boxplus_k\mathfrak{v}=\mathfrak{u}\oplus_{(\mathfrak{g},\mathfrak{f})}\mathfrak{v}$. If $\mathfrak{g}'\equiv\bar{\mathfrak{c}}_{jk}$, $\mathfrak{f}'\equiv\bar{\mathfrak{a}}_k$, then $\mathfrak{g}'$, $\mathfrak{f}'$ satisfy the hypothesis of Definition 7.4 and $\mathfrak{u}\diamond\!\!\!+_k\mathfrak{v}=\mathfrak{u}\diamond\!\!\!+_{(\mathfrak{g}',\mathfrak{f}')}\mathfrak{v}$.

PROOF: This is obvious.

LEMMA 7.8: Let $\mathfrak{u}, \mathfrak{v}\in L_{ij}$. Then $\mathfrak{u}\boxplus_k\mathfrak{v}$, $\mathfrak{u}\diamond\!\!\!+_k\mathfrak{v}$ are independent of $k\neq i, j$ and are equal.

PROOF: Let $k, l\neq i, j$ and $k\neq l$. We shall show that $\mathfrak{u}\diamond\!\!\!+_k\mathfrak{v}$, $\mathfrak{u}\diamond\!\!\!+_l\mathfrak{v}$ are both equal to $\mathfrak{u}\boxplus_k\mathfrak{v}$. We have seen that

$$(12)\qquad \mathfrak{u}\boxplus_k\mathfrak{v}=\mathfrak{u}\oplus_{(\mathfrak{u}',\mathfrak{v}',\mathfrak{w}')}\mathfrak{v},$$

with $\mathfrak{u}'=\mathfrak{u}\cup\bar{\mathfrak{c}}_{ik}$, $\mathfrak{v}'=\mathfrak{v}\cup\bar{\mathfrak{a}}_k$, $\mathfrak{w}'=\bar{\mathfrak{a}}_j\cup\bar{\mathfrak{a}}_k$, and

$$(13)\qquad \mathfrak{u}\diamond\!\!\!+_k\mathfrak{v}=\mathfrak{u}\oplus_{(\mathfrak{u}',\mathfrak{v}',\mathfrak{w}')}\mathfrak{v},$$

with $\mathfrak{u}'=\mathfrak{u}\cup\bar{\mathfrak{c}}_{jk}$, $\mathfrak{v}'=\mathfrak{v}\cup\bar{\mathfrak{a}}_k$, $\mathfrak{w}'=\bar{\mathfrak{a}}_j\cup\bar{\mathfrak{a}}_k$ (cf. the Corollaries to Definitions 7.3, 7.4, 7.5), and of course that $(\mathfrak{u}', \mathfrak{v}', \mathfrak{w}')P_{\mathfrak{u}\mathfrak{v}}$ in each of the two cases. These statements are clearly true if k is replaced throughout by l. For $\boxplus_k$ and $\diamond\!\!\!+_k$ we have $\bar{\mathfrak{a}}_i\cup\mathfrak{w}'=\bar{\mathfrak{a}}_i\cup\bar{\mathfrak{a}}_j\cup\bar{\mathfrak{a}}_k$, and for $\boxplus_l$, $\diamond\!\!\!+_l$ we have $\bar{\mathfrak{a}}_i\cup\mathfrak{w}'=\bar{\mathfrak{a}}_i\cup\bar{\mathfrak{a}}_j\cup\bar{\mathfrak{a}}_l$. The product (i.e. lattice meet) of $\bar{\mathfrak{a}}_i\cup\mathfrak{w}'$ for $\boxplus_k$ and $\bar{\mathfrak{a}}_i\cup\mathfrak{w}'$ for either $\boxplus_l$ or $\diamond\!\!\!+_l$ is therefore $(\bar{\mathfrak{a}}_i\cup\bar{\mathfrak{a}}_j\cup\bar{\mathfrak{a}}_k)\cap(\bar{\mathfrak{a}}_i\cup\bar{\mathfrak{a}}_j\cup\bar{\mathfrak{a}}_l)=\bar{\mathfrak{a}}_i\cup\bar{\mathfrak{a}}_j$ by Part I, Theorem 2.5, since $(\bar{\mathfrak{a}}_i, \bar{\mathfrak{a}}_j, \bar{\mathfrak{a}}_k, \bar{\mathfrak{a}}_l)\perp$. Consequently Lemma 7.5 applies yielding $\mathfrak{u}\boxplus_k\mathfrak{v}=\mathfrak{u}\boxplus_l\mathfrak{v}$ and $\mathfrak{u}\boxplus_k\mathfrak{v}=\mathfrak{u}\diamond\!\!\!+_l\mathfrak{v}$. The first equation shows that $\mathfrak{u}\boxplus_k\mathfrak{v}$ does not depend on k. Then the second equation shows that $\mathfrak{u}\diamond\!\!\!+_l\mathfrak{v}$ does not depend on l. And now the second equation shows also that the common value of all $\mathfrak{u}\boxplus_k\mathfrak{v}$ is equal to the common value of all $\mathfrak{u}\diamond\!\!\!+_l\mathfrak{v}$.

DEFINITION 7.6: We define for $\mathfrak{u}, \mathfrak{v} \in L_{ij}$ the element $\mathfrak{u} \oplus \mathfrak{v}$ as the unique common value of all the $\mathfrak{u} \boxplus_k \mathfrak{v}$, $\mathfrak{u} \oplus_k \mathfrak{v}$ with $k \neq i, j$ (cf. Lemma 7.8).

LEMMA 7.9: Let $\mathfrak{u}, \mathfrak{v} \in L_{ij}$. If $(\mathfrak{u}', \mathfrak{v}', \mathfrak{w}')P_{\mathfrak{u}\mathfrak{v}}$, and if there exists $k \neq i$, j such that either

$$(14) \qquad (\mathfrak{a}_i \cup \mathfrak{w}') \cap (\bar{\mathfrak{a}}_i \cup \bar{\mathfrak{a}}_j \cup \bar{\mathfrak{a}}_k) = \bar{\mathfrak{a}}_i \cup \bar{\mathfrak{a}}_j,$$

or

$$(15) \qquad \mathfrak{w}' \leqq \bar{\mathfrak{a}}_i \cup \bar{\mathfrak{a}}_j \cup \bar{\mathfrak{a}}_k,$$

then $\mathfrak{u} \oplus_{(\mathfrak{u}', \mathfrak{v}', \mathfrak{w}')} \mathfrak{v} = \mathfrak{u} \oplus \mathfrak{v}$.

PROOF: In the proof of Lemma 7.8 it was shown that $\bar{\mathfrak{a}}_i \cup \mathfrak{w}'' = \bar{\mathfrak{a}}_i \cup \bar{\mathfrak{a}}_j \cup \bar{\mathfrak{a}}_k$, where $\mathfrak{w}''$ is the element $\mathfrak{w}'$ used in defining $\mathfrak{u} \boxplus_k \mathfrak{v}$. Hence if (14) holds, we have $(\bar{\mathfrak{a}}_i \cup \mathfrak{w}') \cap (\bar{\mathfrak{a}}_i \cup \mathfrak{w}'') = \bar{\mathfrak{a}}_i \cup \bar{\mathfrak{a}}_j$ and by Lemma 7.5, $\mathfrak{u} \oplus \mathfrak{v} = \mathfrak{u} \boxplus_k \mathfrak{v} = \mathfrak{u} \oplus_{(\mathfrak{u}', \mathfrak{v}', \mathfrak{w}')} \mathfrak{v}$. Suppose now that (15) holds, and let l be distinct from i, j, k (existent since $n \geqq 4$). Then since $\mathfrak{w}' \geqq \bar{\mathfrak{a}}_j$ we have

$$(\bar{\mathfrak{a}}_i \cup \mathfrak{w}') \cap (\bar{\mathfrak{a}}_i \cup \bar{\mathfrak{a}}_j \cup \bar{\mathfrak{a}}_l) \geqq (\bar{\mathfrak{a}}_i \cup \bar{\mathfrak{a}}_j) \cap (\bar{\mathfrak{a}}_i \cup \bar{\mathfrak{a}}_j \cup \bar{\mathfrak{a}}_l) = \bar{\mathfrak{a}}_i \cup \bar{\mathfrak{a}}_j;$$

moreover

$$(\bar{\mathfrak{a}}_i \cup \mathfrak{w}') \cap (\bar{\mathfrak{a}}_i \cup \bar{\mathfrak{a}}_j \cup \bar{\mathfrak{a}}_l) \leqq (\bar{\mathfrak{a}}_i \cup \bar{\mathfrak{a}}_j \cup \bar{\mathfrak{a}}_k) \cap (\bar{\mathfrak{a}}_i \cup \bar{\mathfrak{a}}_j \cup \bar{\mathfrak{a}}_l) = \bar{\mathfrak{a}}_i \cup \bar{\mathfrak{a}}_j.$$

Thus (14) holds with k replaced by l, and by the first part of our proof we have that $\mathfrak{u} \oplus \mathfrak{v} = \mathfrak{u} \oplus_{(\mathfrak{u}', \mathfrak{v}', \mathfrak{w}')} \mathfrak{v}$.

LEMMA 7.10: If $\beta, \gamma \in \mathfrak{S}$, there exists a unique element $\delta \in \mathfrak{S}$ such that $(\delta)_{ij} = (\beta)_{ij} \oplus (\gamma)_{ij}$ for every pair i, j with $i \neq j$.

PROOF: Let i, j be fixed with $i \neq j$. Then $(\beta)_{ij}, (\gamma)_{ij} \in L_{ij}$ and hence $(\beta)_{ij} \oplus (\gamma)_{ij} \in L_{ij}$ by Lemma 7.2. Hence there exists an element $\delta \in \mathfrak{S}$ with $(\delta)_{ij} = (\beta)_{ij} \oplus (\gamma)_{ij}$ by Lemma 6.1. Now let i', j' $(= 1, \cdots, n)$ be any two distinct integers. Then

$$(16) \qquad (\delta)_{ij} P\begin{pmatrix} i & j \\ i' & j' \end{pmatrix} = \big((\beta)_{ij} \oplus (\gamma)_{ij}\big) P\begin{pmatrix} i & j \\ i' & j' \end{pmatrix}.$$

Clearly the left member of (16) is $(\delta)_{i'j'}$. We now evaluate the right member of (16). If $i = i'$, $j = j'$, we obtain clearly $(\beta)_{i'j'} \oplus (\gamma)_{i'j'}$. If only two or three of the numbers i, j, i', j' are distinct, then we may choose $k \neq i$, j, i', j', and we have $\mathfrak{u} \oplus \mathfrak{v} = \mathfrak{u} \boxplus_k \mathfrak{v}$. Now $P\begin{pmatrix} i & j \\ i' & j' \end{pmatrix}$ and $P\begin{pmatrix} i & j & k \\ i' & j' & k \end{pmatrix}$ coincide for elements of $L(0, \bar{\mathfrak{a}}_i \cup \bar{\mathfrak{a}}_j)$, in particular for $(\beta)_{ij}$, $(\gamma)_{ij}$ and $(\beta)_{ij} \boxplus_k (\gamma)_{ij}$. Therefore if we operate with $P\begin{pmatrix} i & j \\ i' & j' \end{pmatrix}$ on (10) with $\mathfrak{u} = (\beta)_{ij}$, $\mathfrak{v} = (\gamma)_{ij}$, we see that $\big((\beta)_{ij} \boxplus_k (\gamma)_{ij}\big) P\begin{pmatrix} i & j \\ i' & j' \end{pmatrix} = (\beta)_{i'j'} \boxplus_k (\gamma)_{i'j'} = (\beta)_{i'j'} \oplus (\gamma)_{i'j'}$. If, finally, all the numbers i, j, i', j' are distinct, then the argument just concluded shows that

$$((\beta)_{ij}\oplus(\gamma)_{ij})P\begin{pmatrix} i\ j \\ i'j' \end{pmatrix} = ((\beta)_{ij}\oplus(\gamma)_{ij})\,P\begin{pmatrix} i\ j \\ i'j \end{pmatrix} P\begin{pmatrix} i'j \\ i'j' \end{pmatrix} = ((\beta)_{i'j}\oplus(\gamma)_{i'j})P\begin{pmatrix} i'j \\ i'j' \end{pmatrix}$$
$$= (\beta)_{i'j'} \oplus (\gamma)_{i'j'}.$$

Thus in all cases the right member of (16) is $(\beta)_{i'j'} \oplus (\gamma)_{i'j'}$, whence $(\delta)_{i'j'} = (\beta)_{i'j'} \oplus (\gamma)_{i'j'}$. This proves the existence. To establish the uniqueness of δ, let δ' have the same property and select i, j with $i \neq j$. Then $(\delta)_{ij} = (\beta)_{ij} \oplus (\gamma)_{ij} = (\delta')_{ij}$, from which we may conclude that $\delta = \delta'$ by Lemma 6.1.

Definition 7.7: If $\beta, \gamma \in \mathfrak{S}$, we define $\beta + \gamma$ as the unique element δ such that $(\beta)_{ij} + (\gamma)_{ij} = (\delta)_{ij}$ for every i, j with $i \neq j$ (cf. Lemma 7.10).

Theorem 7.1: Addition in $\mathfrak{S}$ is commutative, i.e., $\beta + \gamma = \gamma + \beta$ for all $\beta, \gamma \in \mathfrak{S}$.

Proof: Let i, j, k be distinct. Then if $\mathfrak{g} = \bar{\mathfrak{c}}_{ik}$, $\mathfrak{f} = \bar{\mathfrak{a}}_k$, we have

$$\begin{aligned}(\beta + \gamma)_{ij} &= (\beta)_{ij} \oplus (\gamma)_{ij} = (\beta)_{ij} \boxplus_k (\gamma)_{ij} \\ &= (\beta)_{ij} \boxplus_{(\mathfrak{g},\mathfrak{f})} (\gamma)_{ij} && \text{(by the Corollary to Definition 7.5)} \\ &= (\gamma)_{ij} \boxplus_{(\mathfrak{f},\mathfrak{g})} (\beta)_{ij} && \text{(by (7))} \\ &= (\gamma)_{ij} \oplus_{(\mathfrak{u}',\mathfrak{v}',\mathfrak{w}')} (\beta)_{ij} && \text{(by the Corollary to Definition 7.3)}\end{aligned}$$

where $\mathfrak{u}'$, $\mathfrak{v}'$, $\mathfrak{w}'$ are defined in terms of $\mathfrak{f}$, $\mathfrak{g}$ by (6) with $(\mathfrak{f}, \mathfrak{g})$ replaced by $(\mathfrak{g}, \mathfrak{f})$. In particular, $\mathfrak{w}' = \bar{\mathfrak{a}}_j \cup \bar{\mathfrak{c}}_{ik} \leqq \bar{\mathfrak{a}}_i \cup \bar{\mathfrak{a}}_j \cup \bar{\mathfrak{a}}_k$. Hence (15) holds and Lemma 7.9 yields $(\beta + \gamma)_{ij} = (\gamma)_{ij} + (\beta)_{ij}$. Thus $(\beta + \gamma)_{ij} = (\gamma + \beta)_{ij}$, and we may conclude $\beta + \gamma = \gamma + \beta$.

Theorem 7.2: Addition in $\mathfrak{S}$ is associative, i.e. $(\beta + \gamma) + \delta = \beta + (\gamma + \delta)$ for all $\beta, \gamma, \delta \in \mathfrak{S}$.

Proof: For the purposes of this proof we replace the symbols $\cup$, $\cap$ by $+$, $\cdot$. It will first be shown that if $\mathfrak{u}, \mathfrak{v}, \mathfrak{w} \in L_{ij}$, then for $\mathfrak{g}$, $\mathfrak{f}$ satisfying the hypotheses of Definition 7.3 and $\mathfrak{g}_1$, $\mathfrak{f}_1$ defined by

$$\mathfrak{g}_1 \equiv (\mathfrak{v} + \mathfrak{f})(\bar{\mathfrak{a}}_j + \mathfrak{g}), \quad \mathfrak{f}_1 \equiv (\bar{\mathfrak{a}}_i + \mathfrak{g}_1)(\bar{\mathfrak{a}}_j + \mathfrak{f}),$$

$(\mathfrak{u} \boxplus \mathfrak{v}) \boxplus_1 \mathfrak{w} = \mathfrak{u} \boxplus (\mathfrak{v} \boxplus_1 \mathfrak{w})$, where $\boxplus$ represents $\boxplus_{(\mathfrak{g},\mathfrak{f})}$, and $\boxplus_1$ represents $\boxplus_{(\mathfrak{g}_1,\mathfrak{f}_1)}$.

We verify first that $\mathfrak{g}_1$, $\mathfrak{f}_1$ satisfy the hypotheses of Definition 7.3. We have

$$\begin{aligned}\mathfrak{f}_1 + \bar{\mathfrak{a}}_i &= (\bar{\mathfrak{a}}_i + \mathfrak{g}_1)(\bar{\mathfrak{a}}_j + \mathfrak{f}) + \bar{\mathfrak{a}}_i \\ &= (\bar{\mathfrak{a}}_i + \mathfrak{g}_1)(\bar{\mathfrak{a}}_j + \mathfrak{f} + \bar{\mathfrak{a}}_i) && \text{(by IV)} \\ &= (\bar{\mathfrak{a}}_i + \mathfrak{g}_1)(\bar{\mathfrak{a}}_j + \mathfrak{g} + \bar{\mathfrak{a}}_i) && \text{(since } (\bar{\mathfrak{a}}_i, \mathfrak{g}, \mathfrak{f})\mathscr{C}) \\ &= \bar{\mathfrak{a}}_i + \mathfrak{g}, && \text{(since } \bar{\mathfrak{a}}_i + \mathfrak{g}_1 \leqq \bar{\mathfrak{a}}_j + \mathfrak{g} + \bar{\mathfrak{a}}_i).\end{aligned}$$

Also,

$$\mathfrak{f}_1 + \mathfrak{g}_1 = (\bar{\mathfrak{a}}_i + \mathfrak{g}_1)(\bar{\mathfrak{a}}_j + \mathfrak{f}) + \mathfrak{g}_1 = (\bar{\mathfrak{a}}_i + \mathfrak{g}_1)(\bar{\mathfrak{a}}_j + \mathfrak{f} + \mathfrak{g}_1) \quad \text{(by IV)}$$
$$= (\bar{\mathfrak{a}}_i + \mathfrak{g}_1)(\bar{\mathfrak{a}}_j + \mathfrak{f} + \mathfrak{g})$$

(the last equality holding since

$$\begin{aligned} \bar{\mathfrak{a}}_j + \mathfrak{g}_1 &= \bar{\mathfrak{a}}_j + (\mathfrak{v} + \mathfrak{f})(\bar{\mathfrak{a}}_j + \mathfrak{g}) \\ &= (\bar{\mathfrak{a}}_j + \mathfrak{v} + \mathfrak{f})(\bar{\mathfrak{a}}_j + \mathfrak{g}) && \text{(by IV)} \\ &= (\bar{\mathfrak{a}}_j + \bar{\mathfrak{a}}_i + \mathfrak{f})(\bar{\mathfrak{a}}_j + \mathfrak{g}) && (\text{since } \mathfrak{v} \in L_{ij}) \\ &= (\bar{\mathfrak{a}}_j + \bar{\mathfrak{a}}_i + \mathfrak{g})(\bar{\mathfrak{a}}_j + \mathfrak{g}) && (\text{since } (\bar{\mathfrak{a}}_i, \mathfrak{g}, \mathfrak{f})\mathscr{C}) \\ &= \bar{\mathfrak{a}}_j + \mathfrak{g}). \end{aligned} \tag{17}$$

Thus

$$\begin{aligned} \mathfrak{f}_1 + \mathfrak{g}_1 &= (\bar{\mathfrak{a}}_i + \mathfrak{g}_1)(\bar{\mathfrak{a}}_j + \mathfrak{f} + \mathfrak{g}) \\ &= (\bar{\mathfrak{a}}_i + \mathfrak{g}_1)(\bar{\mathfrak{a}}_j + \bar{\mathfrak{a}}_i + \mathfrak{g}) && (\text{since } (\bar{\mathfrak{a}}_i, \mathfrak{g}, \mathfrak{f})\mathscr{C}) \\ &= \bar{\mathfrak{a}}_i + \mathfrak{g}_1 && (\text{since } \bar{\mathfrak{a}}_i + \mathfrak{g}_1 \leqq \bar{\mathfrak{a}}_j + \bar{\mathfrak{a}}_i + \mathfrak{g}). \end{aligned}$$

Moreover,

$$\mathfrak{f}_1\bar{\mathfrak{a}}_i \leqq (\bar{\mathfrak{a}}_j + \mathfrak{f})\bar{\mathfrak{a}}_i = 0 \qquad (\text{since } (\bar{\mathfrak{a}}_i, \bar{\mathfrak{a}}_j, \mathfrak{f})\perp),$$
$$\mathfrak{g}_1\bar{\mathfrak{a}}_i \leqq (\bar{\mathfrak{a}}_j + \mathfrak{g})\bar{\mathfrak{a}}_i = 0 \qquad (\text{since } (\bar{\mathfrak{a}}_i, \bar{\mathfrak{a}}_j, \mathfrak{g})\perp).$$

Since $(\bar{\mathfrak{a}}_i, \bar{\mathfrak{a}}_j, \mathfrak{f})\perp$, $\bar{\mathfrak{a}}_i + \mathfrak{f} = \mathfrak{g} + \mathfrak{f}$, $\bar{\mathfrak{a}}_i\mathfrak{f} = \mathfrak{g}\mathfrak{f} = 0$, it follows from Part I, Lemma 2.1 that $(\bar{\mathfrak{a}}_j, \mathfrak{g}, \mathfrak{f})\perp$. Thus $\mathfrak{g}_1\mathfrak{f}_1 \leqq (\bar{\mathfrak{a}}_j + \mathfrak{g})(\bar{\mathfrak{a}}_j + \mathfrak{f}) = \bar{\mathfrak{a}}_j$, and $\mathfrak{g}_1\mathfrak{f}_1 \leqq (\mathfrak{v} + \mathfrak{f})\bar{\mathfrak{a}}_j$. But $(\bar{\mathfrak{a}}_i, \bar{\mathfrak{a}}_j, \mathfrak{f})\perp$, $\mathfrak{v} + \bar{\mathfrak{a}}_j = \bar{\mathfrak{a}}_i + \bar{\mathfrak{a}}_j$, $\mathfrak{v}\bar{\mathfrak{a}}_j = \bar{\mathfrak{a}}_i\bar{\mathfrak{a}}_j = 0$, whence by Part I, Lemma 2.1, $(\mathfrak{v}, \bar{\mathfrak{a}}_j, \mathfrak{f})\perp$. Therefore $\mathfrak{g}_1\mathfrak{f}_1 \leqq (\mathfrak{v} + \mathfrak{f})\bar{\mathfrak{a}}_j = 0$. Thus we have proved

$$\mathfrak{f}_1 + \bar{\mathfrak{a}}_i = \mathfrak{f}_1 + \mathfrak{g}_1 = \mathfrak{g}_1 + \bar{\mathfrak{a}}_i, \ \mathfrak{f}_1\bar{\mathfrak{a}}_i = \mathfrak{f}_1\mathfrak{g}_1 = \mathfrak{g}_1\bar{\mathfrak{a}}_i = 0,$$

which means $(\bar{\mathfrak{a}}_i, \mathfrak{g}_1, \mathfrak{f}_1)\mathscr{C}$. Next, we have

$$\begin{aligned} \mathfrak{g}_1(\bar{\mathfrak{a}}_i + \bar{\mathfrak{a}}_j) &= (\mathfrak{v} + \mathfrak{f})(\bar{\mathfrak{a}}_j + \mathfrak{g})(\bar{\mathfrak{a}}_i + \bar{\mathfrak{a}}_j) \\ &= (\mathfrak{v} + \mathfrak{f})\bar{\mathfrak{a}}_j && (\text{since } (\bar{\mathfrak{a}}_i, \bar{\mathfrak{a}}_j, \mathfrak{f})\perp); \\ &= 0 && (\text{since } (\mathfrak{v}, \bar{\mathfrak{a}}_j, \mathfrak{f})\perp). \end{aligned}$$

Thus we have proved $(\bar{\mathfrak{a}}_i, \bar{\mathfrak{a}}_j, \mathfrak{g}_1)\perp$. Hence $\mathfrak{g}_1$, $\mathfrak{f}_1$ satisfy the hypotheses of Definition 7.3.

We compute next the expressions $(\mathfrak{u} \boxplus \mathfrak{v}) \boxplus_1 \mathfrak{w}$, $\mathfrak{u} \boxplus (\mathfrak{v} \boxplus_1 \mathfrak{w})$. We have

$$\mathfrak{u} \boxplus \mathfrak{v} = \big((\mathfrak{u} + \mathfrak{g})(\bar{\mathfrak{a}}_j + \mathfrak{f}) + (\mathfrak{v} + \mathfrak{f})(\bar{\mathfrak{a}}_j + \mathfrak{g})\big)(\bar{\mathfrak{a}}_i + \bar{\mathfrak{a}}_j),$$
$$(\mathfrak{u} \boxplus \mathfrak{v}) \boxplus_1 \mathfrak{w} = \big(((\mathfrak{u} \boxplus \mathfrak{v}) + \mathfrak{g}_1)(\bar{\mathfrak{a}}_j + \mathfrak{f}_1) + (\mathfrak{w} + \mathfrak{f}_1)(\bar{\mathfrak{a}}_j + \mathfrak{g}_1)\big)(\bar{\mathfrak{a}}_i + \bar{\mathfrak{a}}_j), \tag{18}$$
$$\mathfrak{v} \boxplus_1 \mathfrak{w} = \big((\mathfrak{v} + \mathfrak{g}_1)(\bar{\mathfrak{a}}_j + \mathfrak{f}_1) + (\mathfrak{w} + \mathfrak{f}_1)(\bar{\mathfrak{a}}_j + \mathfrak{g}_1)\big)(\bar{\mathfrak{a}}_i + \bar{\mathfrak{a}}_j),$$
$$\mathfrak{u} \boxplus (\mathfrak{v} \boxplus_1 \mathfrak{w}) = \big((\mathfrak{u} + \mathfrak{g})(\bar{\mathfrak{a}}_j + \mathfrak{f}) + ((\mathfrak{v} \boxplus_1 \mathfrak{w}) + \mathfrak{f})(\bar{\mathfrak{a}}_j + \mathfrak{g})\big)(\bar{\mathfrak{a}}_i + \bar{\mathfrak{a}}_j). \tag{19}$$

We have, moreover,

$$
\begin{aligned}
\bar{\mathfrak{a}}_j + \mathfrak{f}_1 &= \bar{\mathfrak{a}}_j + (\bar{\mathfrak{a}}_i + \mathfrak{g}_1)(\bar{\mathfrak{a}}_j + \mathfrak{f}) \\
&= (\bar{\mathfrak{a}}_i + \bar{\mathfrak{a}}_j + \mathfrak{g}_1)(\bar{\mathfrak{a}}_j + \mathfrak{f}) && \text{(by IV)} \\
&= (\bar{\mathfrak{a}}_i + \bar{\mathfrak{a}}_j + \mathfrak{g})(\bar{\mathfrak{a}}_j + \mathfrak{f}) && (\text{since } \bar{\mathfrak{a}}_j+\mathfrak{g} = \bar{\mathfrak{a}}_j+\mathfrak{g}_1) \\
&= (\bar{\mathfrak{a}}_i + \bar{\mathfrak{a}}_j + \mathfrak{f})(\bar{\mathfrak{a}}_j + \mathfrak{f}) && (\text{since } (\bar{\mathfrak{a}}_i, \mathfrak{g}, \mathfrak{f},)\mathscr{C}) \\
&= \bar{\mathfrak{a}}_j + \mathfrak{f},
\end{aligned} \tag{20}
$$

and also

$$
\begin{aligned}
\mathfrak{v} + \mathfrak{g}_1 &= \mathfrak{v} + (\mathfrak{v} + \mathfrak{f})(\bar{\mathfrak{a}}_j + \mathfrak{g}) \\
&= (\mathfrak{v} + \mathfrak{f})(\mathfrak{v} + \bar{\mathfrak{a}}_j + \mathfrak{g}) && \text{(by IV)} \\
&= (\mathfrak{v} + \mathfrak{f})(\bar{\mathfrak{a}}_i + \bar{\mathfrak{a}}_j + \mathfrak{f}) \\
& && (\text{since } \mathfrak{v} \in L_{ij} \text{ and } (\bar{\mathfrak{a}}_i, \mathfrak{g}, \mathfrak{f})\mathscr{C}) \\
&= \mathfrak{v} + \mathfrak{f} && (\text{since } \mathfrak{v} + \mathfrak{f} \leqq \bar{\mathfrak{a}}_i + \bar{\mathfrak{a}}_j + \mathfrak{f}).
\end{aligned} \tag{21}
$$

Now

$$
\begin{aligned}
&(\mathfrak{u} \boxplus \mathfrak{v}) + \mathfrak{g}_1 \\
&\quad = \big((\mathfrak{u} + \mathfrak{g})(\bar{\mathfrak{a}}_j + \mathfrak{f}) + (\mathfrak{v} + \mathfrak{f})(\bar{\mathfrak{a}}_j + \mathfrak{g})\big)(\bar{\mathfrak{a}}_i + \bar{\mathfrak{a}}_j) + (\mathfrak{v} + \mathfrak{f})(\bar{\mathfrak{a}}_j + \mathfrak{g}) \\
&\quad = \big((\mathfrak{u}+\mathfrak{g})(\bar{\mathfrak{a}}_j+\mathfrak{f}) + (\mathfrak{v}+\mathfrak{f})(\bar{\mathfrak{a}}_j+\mathfrak{g})\big)\big(\bar{\mathfrak{a}}_i+\bar{\mathfrak{a}}_j+(\mathfrak{v}+\mathfrak{f})(\bar{\mathfrak{a}}_j+\mathfrak{g})\big) \quad \text{(by IV)}.
\end{aligned}
$$

But

$$
\begin{aligned}
\bar{\mathfrak{a}}_i + \bar{\mathfrak{a}}_j + (\mathfrak{v} + \mathfrak{f})(\bar{\mathfrak{a}}_j + \mathfrak{g}) &= \bar{\mathfrak{a}}_i + (\bar{\mathfrak{a}}_j + \mathfrak{v} + \mathfrak{f})(\bar{\mathfrak{a}}_j + \mathfrak{g}) && \text{(by IV)} \\
&= \bar{\mathfrak{a}}_i + (\bar{\mathfrak{a}}_j + \bar{\mathfrak{a}}_i + \mathfrak{f})(\bar{\mathfrak{a}}_j + \mathfrak{g}) && (\text{since } \mathfrak{v} \in L_{ij}) \\
&= \bar{\mathfrak{a}}_i + (\bar{\mathfrak{a}}_j + \bar{\mathfrak{a}}_i + \mathfrak{g})(\bar{\mathfrak{a}}_j + \mathfrak{g}) && (\text{since } (\bar{\mathfrak{a}}_i, \mathfrak{g}, \mathfrak{f})\mathscr{C}) \\
&= \bar{\mathfrak{a}}_i + \bar{\mathfrak{a}}_j + \mathfrak{g} \\
&\geqq \begin{cases} \bar{\mathfrak{a}}_i + \bar{\mathfrak{a}}_j \geqq \bar{\mathfrak{a}}_i, \bar{\mathfrak{a}}_j, \mathfrak{u}, \mathfrak{v} \\ \bar{\mathfrak{a}}_i + \mathfrak{g} = \bar{\mathfrak{a}}_i + \mathfrak{f} \geqq \mathfrak{g}, \mathfrak{f}, \end{cases}
\end{aligned}
$$

whence $\bar{\mathfrak{a}}_i + \bar{\mathfrak{a}}_j + (\mathfrak{v} + \mathfrak{f})(\bar{\mathfrak{a}}_j + \mathfrak{g}) \geqq (\mathfrak{u} + \mathfrak{g})(\bar{\mathfrak{a}}_j + \mathfrak{f}) + (\mathfrak{v} + \mathfrak{f})(\bar{\mathfrak{a}}_j + \mathfrak{g})$. Therefore

$$(\mathfrak{u} \boxplus \mathfrak{v}) + \mathfrak{g}_1 = (\mathfrak{u} + \mathfrak{g})(\bar{\mathfrak{a}}_j + \mathfrak{f}) + (\mathfrak{v} + \mathfrak{f})(\bar{\mathfrak{a}}_j + \mathfrak{g}),$$

and by (18)

$$
\begin{aligned}
&(\mathfrak{u} \boxplus \mathfrak{v}) \boxplus_1 \mathfrak{w} \\
&= \big(\big((\mathfrak{u} + \mathfrak{g})(\bar{\mathfrak{a}}_j + \mathfrak{f}) + (\mathfrak{v} + \mathfrak{f})(\bar{\mathfrak{a}}_j + \mathfrak{g})\big)(\bar{\mathfrak{a}}_j + \mathfrak{f}) + (\mathfrak{w} + \mathfrak{f}_1)(\bar{\mathfrak{a}}_j+\mathfrak{g})\big)(\bar{\mathfrak{a}}_i + \bar{\mathfrak{a}}_j) \\
&= \big((\mathfrak{u} + \mathfrak{g})(\bar{\mathfrak{a}}_j + \mathfrak{f}) + (\mathfrak{v} + \mathfrak{f})(\bar{\mathfrak{a}}_j + \mathfrak{g})(\bar{\mathfrak{a}}_j + \mathfrak{f}) + (\mathfrak{w} + \mathfrak{f}_1)(\bar{\mathfrak{a}}_j + \mathfrak{g})\big)(\bar{\mathfrak{a}}_i + \bar{\mathfrak{a}}_j) \\
& \hspace{30em} \text{(by IV)}.
\end{aligned}
$$

But $(\bar{\mathfrak{a}}_i, \bar{\mathfrak{a}}_j, \mathfrak{f})\perp$, $\bar{\mathfrak{a}}_i + \mathfrak{f} = \mathfrak{g} + \mathfrak{f}$, $\bar{\mathfrak{a}}_i\mathfrak{f} = \mathfrak{g}\mathfrak{f} = 0$, whence $(\mathfrak{g}, \bar{\mathfrak{a}}_j, \mathfrak{f})\perp$. Thus

$(\bar{\mathfrak{a}}_j + \mathfrak{g})(\bar{\mathfrak{a}}_j + \mathfrak{f}) = \bar{\mathfrak{a}}_j$, and

$$(22) \quad (\mathfrak{u} \boxplus \mathfrak{v}) \boxplus_1 \mathfrak{w} = \big((\mathfrak{u}+\mathfrak{g})(\bar{\mathfrak{a}}_j+\mathfrak{f}) + (\mathfrak{v}+\mathfrak{f})\bar{\mathfrak{a}}_j + (\mathfrak{w}+\mathfrak{f}_1)(\bar{\mathfrak{a}}_j+\mathfrak{g})\big)(\bar{\mathfrak{a}}_i+\bar{\mathfrak{a}}_j).$$

Now

$$\begin{aligned}
&(\mathfrak{v} \boxplus_1 \mathfrak{w}) + \mathfrak{f} \\
&\quad = \big((\mathfrak{v}+\mathfrak{g}_1)(\bar{\mathfrak{a}}_j+\mathfrak{f}_1) + (\mathfrak{w}+\mathfrak{f}_1)(\bar{\mathfrak{a}}_j+\mathfrak{g}_1)\big)(\bar{\mathfrak{a}}_i+\bar{\mathfrak{a}}_j) + \mathfrak{f} \\
&\quad = \big((\mathfrak{v}+\mathfrak{f})(\bar{\mathfrak{a}}_j+\mathfrak{f}) + (\mathfrak{w}+\mathfrak{f}_1)(\bar{\mathfrak{a}}_j+\mathfrak{g})\big)(\bar{\mathfrak{a}}_i+\bar{\mathfrak{a}}_j) + \mathfrak{f} \quad \text{(by (17), (20), (21))}.
\end{aligned}$$

But $(\mathfrak{v} + \mathfrak{f})(\bar{\mathfrak{a}}_j + \mathfrak{f}) \geqq \mathfrak{f}$ whence IV applies, and

$$\begin{aligned}
&(\mathfrak{v} \boxplus_1 \mathfrak{w}) + \mathfrak{f} \\
&= \big((\mathfrak{v} + \mathfrak{f})(\bar{\mathfrak{a}}_j + \mathfrak{f}) + (\mathfrak{w} + \mathfrak{f}_1)(\bar{\mathfrak{a}}_j + \mathfrak{g})\big)(\bar{\mathfrak{a}}_i + \bar{\mathfrak{a}}_j + \mathfrak{f}) \\
&= \big((\mathfrak{v} + \mathfrak{f})(\bar{\mathfrak{a}}_j + \mathfrak{f}) + (\mathfrak{w} + \mathfrak{f}_1)(\bar{\mathfrak{a}}_j + \mathfrak{g})\big)(\bar{\mathfrak{a}}_i + \bar{\mathfrak{a}}_j + \mathfrak{g}) \quad \text{(since } (\bar{\mathfrak{a}}_i, \mathfrak{g}, \mathfrak{f})\mathscr{C}).
\end{aligned}$$

Thus

$$\begin{aligned}
&\big((\mathfrak{v} \boxplus_1 \mathfrak{w}) + \mathfrak{f}\big)(\bar{\mathfrak{a}}_j + \mathfrak{g}) = \big((\mathfrak{v} + \mathfrak{f})(\bar{\mathfrak{a}}_j + \mathfrak{f}) + (\mathfrak{w} + \mathfrak{f}_1)(\bar{\mathfrak{a}}_j + \mathfrak{g})\big)(\bar{\mathfrak{a}}_j + \mathfrak{g}) \\
&\quad = (\mathfrak{v} + \mathfrak{f})(\bar{\mathfrak{a}}_j + \mathfrak{f})(\bar{\mathfrak{a}}_j + \mathfrak{g}) + (\mathfrak{w} + \mathfrak{f}_1)(\bar{\mathfrak{a}}_j + \mathfrak{g}) \qquad \text{(by IV)} \\
&\quad = (\mathfrak{v} + \mathfrak{f})\bar{\mathfrak{a}}_j + (\mathfrak{w} + \mathfrak{f}_1)(\bar{\mathfrak{a}}_j + \mathfrak{g}) \qquad \text{(since } (\bar{\mathfrak{a}}_j, \mathfrak{g}, \mathfrak{f})\perp).
\end{aligned}$$

Consequently, we have by (19)

$$(23) \quad \mathfrak{u} \boxplus (\mathfrak{v} \boxplus_1 \mathfrak{w}) = \big((\mathfrak{u}+\mathfrak{g})(\bar{\mathfrak{a}}_j+\mathfrak{f}) + (\mathfrak{v}+\mathfrak{f})\bar{\mathfrak{a}}_j + (\mathfrak{w}+\mathfrak{f}_1)(\bar{\mathfrak{a}}_j+\mathfrak{g})\big)(\bar{\mathfrak{a}}_i+\bar{\mathfrak{a}}_j).$$

Comparing (22) and (23), we see that

$$(\mathfrak{u} \boxplus \mathfrak{v}) \boxplus_1 \mathfrak{w} = \mathfrak{u} \boxplus (\mathfrak{v} \boxplus_1 \mathfrak{w}).$$

Now we may select $\mathfrak{g} = \bar{\mathfrak{c}}_{ik}$, $\mathfrak{f} = \bar{\mathfrak{a}}_k$; then $\boxplus$ coincides with $\boxplus_k = \oplus$, and $\boxplus_1$ coincides with $\oplus_{(\mathfrak{u}', \mathfrak{v}', \mathfrak{w}')}$ where

$$\begin{aligned}
\mathfrak{w}' = \bar{\mathfrak{a}}_j + \mathfrak{f}_1 &= \bar{\mathfrak{a}}_i + \mathfrak{f} \qquad \text{(by (20))} \\
&\leqq \bar{\mathfrak{a}}_i + \bar{\mathfrak{a}}_j + \bar{\mathfrak{a}}_k.
\end{aligned}$$

Thus by Lemma 7.9, $\boxplus_1 = \boxplus_k = \oplus$, and we have proved that

$$(\mathfrak{u} \oplus \mathfrak{v}) \oplus \mathfrak{w} = \mathfrak{u} \oplus (\mathfrak{v} \oplus \mathfrak{w}).$$

In particular,

$$\big((\beta)_{ij} \oplus (\gamma)_{ij}\big) \oplus (\delta)_{ij} = (\beta)_{ij} \oplus \big((\gamma)_{ij} \oplus (\delta)_{ij}\big),$$

i.e.,

$$(\beta + \gamma)_{ij} \oplus (\delta)_{ij} = (\beta)_{ij} \oplus (\gamma + \delta)_{ij},$$

whence

$$((\beta+\gamma)+\delta)_{ij} = (\beta+(\gamma+\delta))_{ij},$$

and we may conclude $(\beta+\gamma)+\delta = \beta+(\gamma+\delta)$.

THEOREM 7.3: The element $0 \in \mathfrak{S}$ (cf. Theorem 6.1) has the property

$$0+\beta = \beta+0 = \beta \qquad (\beta \in \mathfrak{S}) \tag{24}$$

and is the only element of $\mathfrak{S}$ with this property.

PROOF: Let i, j, k be three distinct indices (existent since $n \geqq 4$). Then define $\mathfrak{u} \equiv \mathfrak{u}' \equiv (0)_{ij} = \bar{\mathfrak{a}}_i$ (by Theorem 6.1), $\mathfrak{v} \equiv \mathfrak{v}' \equiv (\beta)_{ij}$, $\mathfrak{w}' \equiv \bar{\mathfrak{a}}_j$. Now conditions (a), $\cdots$, (g) of Definition 7.1 are all immediately verified, whence $(\mathfrak{u}', \mathfrak{v}', \mathfrak{w}')P_{\mathfrak{u}\mathfrak{v}}$. Moreover, $\mathfrak{w}' \leqq \bar{\mathfrak{a}}_i \cup \bar{\mathfrak{a}}_j \cup \bar{\mathfrak{a}}_k$. Thus by Lemma 7.9 and Definition 7.2,

$$\begin{aligned}
\mathfrak{u} \oplus \mathfrak{v} &= \Big(\Big(\Big(\Big(\big(\big(((\beta)_{ij} \cap \bar{\mathfrak{a}}_j) \cup \bar{\mathfrak{a}}_i\big) \cap \bar{\mathfrak{a}}_i\big) \cup \bar{\mathfrak{a}}_j\Big) \cap (\beta)_{ij}\Big) \cup (\bar{\mathfrak{a}}_i \cap \bar{\mathfrak{a}}_j)\Big) \cap (\bar{\mathfrak{a}}_i \cup \bar{\mathfrak{a}}_j) \\
&= \big((\bar{\mathfrak{a}}_i \cup \bar{\mathfrak{a}}_j) \cap (\beta)_{ij}\big) \cap (\bar{\mathfrak{a}}_i \cup \bar{\mathfrak{a}}_j) \\
&= (\bar{\mathfrak{a}}_i \cup \bar{\mathfrak{a}}_j) \cap (\beta)_{ij} = (\beta)_{ij},
\end{aligned}$$

i.e., $(0+\beta)_{ij} = (0)_{ij} \oplus (\beta)_{ij} = \mathfrak{u} \oplus \mathfrak{v} = (\beta)_{ij}$, whence $0+\beta = \beta$. By the commutative law for addition, $\beta + 0 = \beta$, and (24) is established. To prove the uniqueness, let $0'$ along with 0 have the property (24). Then $0' = 0 + 0' = 0$, whence 0 is unique.

APPENDIX

We give again, as in the appendix to Chapter VI, two classes of illustrations of our definitions and proofs, first by pointing out the analogy with the usual projective-geometric "figures", and secondly by repeating our computations in the special case of linear sets of n-dimensional vectors. (Cf. the appendix to Chapter VI.)

PICTURE 1: The $\bar{\mathfrak{a}}_i$, $\bar{\mathfrak{c}}_{ij}$ are again points; their relations are described in the appendix to Chapter VI.

Figure 7.1 exhibits the conditions of the relation $(\mathfrak{u}', \mathfrak{v}', \mathfrak{w}')P_{\mathfrak{u}\mathfrak{v}}$ as formulated in Definition 7.1.

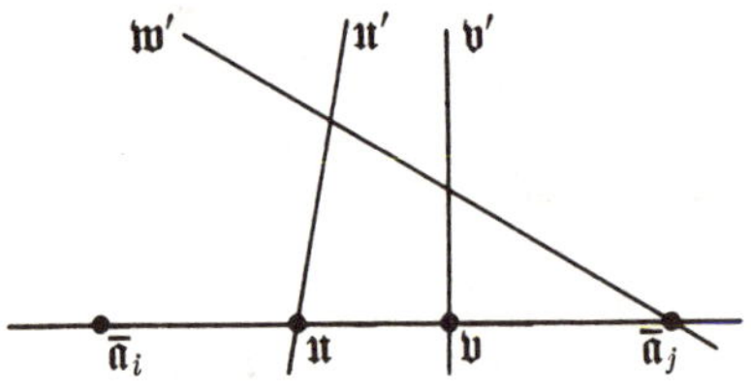

Figure 7.1

It should be observed that $\mathfrak{u}'$, $\mathfrak{v}'$, $\mathfrak{w}'$, which are shown in the figure as lines, might very well have other (generally higher) dimensionality, but they must be equidimensional. For, $\mathfrak{u}'$, $\mathfrak{v}'$ are inverses of $\bar{\mathfrak{a}}_j$ and $\mathfrak{w}'$ is an inverse of $\bar{\mathfrak{a}}_i$ in $\bar{\mathfrak{a}}_j \cup \mathfrak{u}' = \bar{\mathfrak{a}}_j \cup \mathfrak{v}' = \bar{\mathfrak{a}}_i \cup \mathfrak{w}'$, and $\bar{\mathfrak{a}}_i, \bar{\mathfrak{a}}_j$ have the same dimensionality; hence $\mathfrak{u}'$, $\mathfrak{v}'$, $\mathfrak{w}'$ must be equidimensional. Thus the proof of Lemma 7.5 (and the proofs of Lemmas 7.3, 7.4, on which it is based) are definitely dealing with higher dimensional $\mathfrak{u}'$, $\mathfrak{v}'$, $\mathfrak{w}'$.

Definition 7.2, which describes $\mathfrak{u} \oplus_{(\mathfrak{u}',\mathfrak{v}',\mathfrak{w}')} \mathfrak{v}$, is pictured by Figure 7.2.

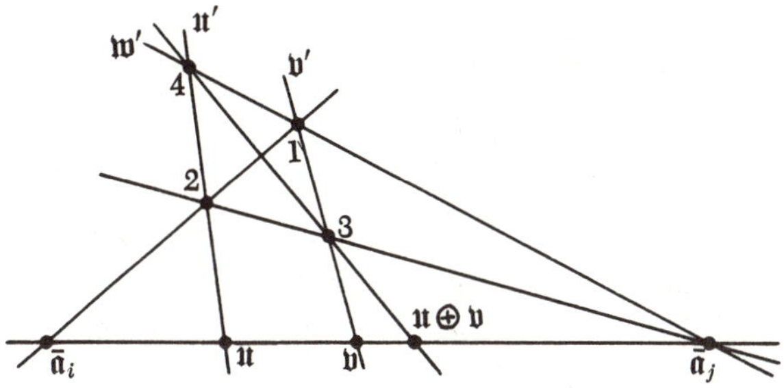

Figure 7.2

This is the well known definition of addition of points (von Staudt) on the line $\bar{\mathfrak{a}}_i \cup \bar{\mathfrak{a}}_j$, where $\bar{\mathfrak{a}}_i, \bar{\mathfrak{a}}_j$ play the roles of 0, ∞, respectively. Differing from multiplication, addition requires no fixed point 1. (Cf. O. Veblen and J. W. Young, *Projective Geometry*, Vol. I, Figure 71 (or 72) on page 142. The points P_0, P_x, P_y, P_{x+y}, P_∞, A, A', X, Y there correspond respectively to $\bar{\mathfrak{a}}_i$, $\mathfrak{u}$, $\mathfrak{v}$, $\mathfrak{u} \oplus \mathfrak{v}$, $\bar{\mathfrak{a}}_j$, 2, 1, 4, 3, on our Figure 7.2).

Lemma 7.6 and Definition 7.3 mean that we choose the point 1 in Figure 7.2 as $\mathfrak{f}$, and the point 2 as $\mathfrak{g}$. Lemma 7.7 and Definition 7.4 mean that we choose the point 1 as $\mathfrak{f}$ and the point 4 as $\mathfrak{g}$. The reader may verify the restrictions on the intersection properties of these points which we imposed in requiring $(\bar{\mathfrak{a}}_i, \mathfrak{g}, \mathfrak{f})\mathscr{C}$, $(\bar{\mathfrak{a}}_i, \bar{\mathfrak{a}}_j, \mathfrak{f})\perp$, and $(\bar{\mathfrak{a}}_j, \mathfrak{g}, \mathfrak{f})\mathscr{C}$, $(\bar{\mathfrak{a}}_i, \bar{\mathfrak{a}}_j, \mathfrak{f})\perp$ in the respective cases. (In the first case, 1 cannot coincide with 2, but it may coincide with 4; in the second case the situation is reversed). In Definition 7.5 and its Corollary we specialized further by taking the point 2 equal to $\mathfrak{g} = \bar{\mathfrak{a}}_k$ and 1 equal to $\mathfrak{f} = \bar{\mathfrak{c}}_{ik}$ in the first case, and 4 equal to $\mathfrak{g} = \bar{\mathfrak{a}}_k$ and 1 equal to $\mathfrak{f} = \bar{\mathfrak{c}}_{jk}$ in the second case. The proof of Theorem 7.1. is sufficiently simple to require no special comment. As to Theorem

7.2, the reader may advantageously compare its proof with the corresponding proof in Veblen and Young, loc. cit., page **143**. There is no difficulty in following each step of our proof on our Figure **7.3**, or on Figure **73**, loc. cit., page **143**.

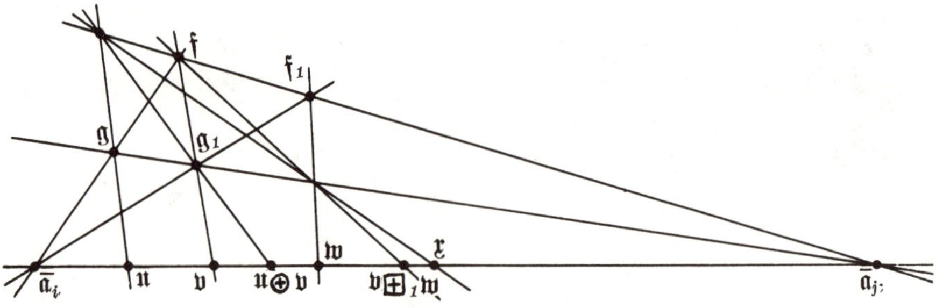

Figure 7.3

Picture 2: As in the Appendix, Chapter VI, we again use the representation by vector sets, displaying only the i, j, k-components in that order as usual. The elements with which we are concerned are represented in the following manner:

$\bar{\mathfrak{a}}_i$: $(\xi, 0, 0)$,		$\bar{\mathfrak{c}}_{jk}$:	$(0, \eta', -\eta')$,
$\bar{\mathfrak{a}}_j$: $(0, \eta, 0)$,		$(\beta)_{ij}$:	$(\zeta', -\beta\zeta', 0)$,
$\bar{\mathfrak{a}}_k$: $(0, 0, \zeta)$,		$(\gamma)_{ij}$:	$(\zeta'', -\gamma\zeta'', 0)$,
$\bar{\mathfrak{c}}_{ik}$: $(\xi', 0, -\zeta')$,		$(\beta+\gamma)_{ij}$:	$(\zeta''', -(\beta+\gamma)\zeta''', 0)$.

Now Definitions **7.5—7.7** state that

$$(\beta+\gamma)_{ij} = \Big(\big(((\beta)_{ij} \cup \bar{\mathfrak{c}}_{ik}) \cap (\bar{\mathfrak{a}}_j \cup \bar{\mathfrak{a}}_k)\big) \cup \big(((\gamma)_{ij} \cup \bar{\mathfrak{a}}_k) \cap (\bar{\mathfrak{a}}_j \cup \bar{\mathfrak{c}}_{ik})\big)\Big) \cap (\bar{\mathfrak{a}}_i \cup \bar{\mathfrak{a}}_j),$$

and

$$(\beta+\gamma)_{ij} = \Big(\Big(\big(\big(((\beta)_{ij} \cup \bar{\mathfrak{c}}_{jk}) \cap (\bar{\mathfrak{a}}_i \cup \bar{\mathfrak{a}}_k)\big) \cup \bar{\mathfrak{a}}_j\big) \cap ((\gamma)_{ij} \cup \bar{\mathfrak{a}}_k)\Big) \cup \bar{\mathfrak{c}}_{jk}\Big) \cap (\bar{\mathfrak{a}}_i \cup \bar{\mathfrak{a}}_j).$$

The reader will find no difficulty in evaluating the vector sets which represent the right members of these equations, and in showing that in both cases the vector set $(\zeta''', -(\beta+\gamma)\zeta''', 0)$ is finally obtained.

PART II · CHAPTER VIII

The Distributive Laws, Subtraction, and Proof that the L-Numbers Form a Ring

In addition to the previous assumptions on L, we suppose that five indices i, j, k, h, l are given, such that i, j, k are distinct and such that $h \neq i, j, k$ and $l \neq i, j, k$. We allow $h = l$. Let also $\mathfrak{b}_{jh} \in L_{jh}$, $\mathfrak{b}_{li} \in L_{li}$ be given. All these are fixed throughout the material preceding Theorem 8.1. We shall prove that multiplication (Chapter VII) is distributive with respect to addition (Chapter VI) on both left and right. A discussion of the distributive laws is given in the appendix.

DEFINITION 8.1: If $\mathfrak{u} \leqq \bar{\mathfrak{a}}_i \cup \bar{\mathfrak{a}}_j \cup \bar{\mathfrak{a}}_k$, we define

$$(1) \qquad \mathfrak{u}^\alpha \equiv (\mathfrak{u} \cup \mathfrak{b}_{jh}) \cap (\bar{\mathfrak{a}}_i \cup \bar{\mathfrak{a}}_h \cup \bar{\mathfrak{a}}_k),$$

$$(2) \qquad \mathfrak{u}^\beta \equiv (\mathfrak{u} \cup \mathfrak{b}_{li}) \cap (\bar{\mathfrak{a}}_l \cup \bar{\mathfrak{a}}_j \cup \bar{\mathfrak{a}}_k).$$

COROLLARY: If $\mathfrak{u} \in L_{ij}$, then $\mathfrak{u}^\alpha = \mathfrak{u} \otimes \mathfrak{b}_{jh} \in L_{ih}$, $\mathfrak{u}^\beta = \mathfrak{b}_{li} \otimes \mathfrak{u} \in L_{lj}$.

PROOF: If $\mathfrak{u} \in L_{ij}$, then $\mathfrak{u} \leqq \bar{\mathfrak{a}}_i \cup \bar{\mathfrak{a}}_j$, whence by (1), $\mathfrak{u} \cup \mathfrak{b}_{jh} \leqq \bar{\mathfrak{a}}_i \cup \bar{\mathfrak{a}}_j \cup \bar{\mathfrak{a}}_h$, and $\mathfrak{u}^\alpha = \mathfrak{u}^\alpha \cap (\bar{\mathfrak{a}}_i \cup \bar{\mathfrak{a}}_j \cup \bar{\mathfrak{a}}_h) = (\mathfrak{u} \cup \mathfrak{b}_{jh}) \cap (\bar{\mathfrak{a}}_i \cup \bar{\mathfrak{a}}_h) = \mathfrak{u} \otimes \mathfrak{b}_{jh} \in L_{ih}$ by Definition 5.1 and its Corollary. Moreover, $\mathfrak{u} \cup \mathfrak{b}_{li} \leqq \bar{\mathfrak{a}}_i \cup \bar{\mathfrak{a}}_j \cup \bar{\mathfrak{a}}_l$, whence $\mathfrak{u}^\beta = \mathfrak{u}^\beta \cap (\bar{\mathfrak{a}}_i \cup \bar{\mathfrak{a}}_j \cup \bar{\mathfrak{a}}_l) = (\mathfrak{u} \cup \mathfrak{b}_{li}) \cap (\bar{\mathfrak{a}}_l \cup \bar{\mathfrak{a}}_j) = \mathfrak{b}_{li} \otimes \mathfrak{u} \in L_{lj}$ by Definition 5.1 and its Corollary.

LEMMA 8.1: If $\mathfrak{u}, \mathfrak{v} \leqq \bar{\mathfrak{a}}_i \cup \bar{\mathfrak{a}}_j \cup \bar{\mathfrak{a}}_k$, then

$$(3) \qquad (\mathfrak{u} \cup \mathfrak{v})^\alpha = \mathfrak{u}^\alpha \cup \mathfrak{v}^\alpha,$$

$$(4) \qquad (\mathfrak{u} \cap \mathfrak{v})^\alpha \leqq \mathfrak{u}^\alpha \cap \mathfrak{v}^\alpha.$$

PROOF: Let $\mathfrak{u}, \mathfrak{v} \leqq \bar{\mathfrak{a}}_i \cup \bar{\mathfrak{a}}_j \cup \bar{\mathfrak{a}}_k$. Then

$$\begin{aligned}
[(\mathfrak{u} \cup \mathfrak{b}_{jh}) \cup (\bar{\mathfrak{a}}_i \cup \bar{\mathfrak{a}}_h \cup \bar{\mathfrak{a}}_k)] \cap [(\mathfrak{v} \cup \mathfrak{b}_{jh}) \cup (\bar{\mathfrak{a}}_i \cup \bar{\mathfrak{a}}_h \cup \bar{\mathfrak{a}}_k)]& \\
\geqq [(\mathfrak{u} \cup \mathfrak{b}_{jh}) \cap (\mathfrak{v} \cup \mathfrak{b}_{jh})] \cup (\bar{\mathfrak{a}}_i \cup \bar{\mathfrak{a}}_h \cup \bar{\mathfrak{a}}_k)& \\
\geqq \mathfrak{b}_{jh} \cup (\bar{\mathfrak{a}}_i \cup \bar{\mathfrak{a}}_h \cup \bar{\mathfrak{a}}_k) = \bar{\mathfrak{a}}_i \cup (\mathfrak{b}_{jh} \cup \bar{\mathfrak{a}}_h) \cup \bar{\mathfrak{a}}_k& \\
= \bar{\mathfrak{a}}_i \cup \bar{\mathfrak{a}}_j \cup \bar{\mathfrak{a}}_h \cup \bar{\mathfrak{a}}_k \quad (\text{since } \mathfrak{b}_{jh} \in L_{jh}).&
\end{aligned}$$

Since all terms in the left member are $\leqq \bar{\mathfrak{a}}_i \cup \bar{\mathfrak{a}}_j \cup \bar{\mathfrak{a}}_h \cup \bar{\mathfrak{a}}_k$, we have also

$[(\mathfrak{u} \cup \mathfrak{b}_{jh}) \cup (\bar{\mathfrak{a}}_i \cup \bar{\mathfrak{a}}_h \cup \bar{\mathfrak{a}}_k)] \cap [(\mathfrak{v} \cup \mathfrak{b}_{jh}) \cup (\bar{\mathfrak{a}}_i \cup \bar{\mathfrak{a}}_h \cup \bar{\mathfrak{a}}_k)] \leqq \bar{\mathfrak{a}}_i \cup \bar{\mathfrak{a}}_j \cup \bar{\mathfrak{a}}_h \cup \bar{\mathfrak{a}}_k$; thus

$$(5) \qquad [(\mathfrak{u} \cup \mathfrak{b}_{jh}) \cup (\bar{\mathfrak{a}}_i \cup \bar{\mathfrak{a}}_h \cup \bar{\mathfrak{a}}_k)] \cap [(\mathfrak{v} \cup \mathfrak{b}_{jh}) \cup (\bar{\mathfrak{a}}_i \cup \bar{\mathfrak{a}}_h \cup \bar{\mathfrak{a}}_k)] = [(\mathfrak{u} \cup \mathfrak{b}_{jh}) \cap (\mathfrak{v} \cup \mathfrak{b}_{jh})] \cup (\bar{\mathfrak{a}}_i \cup \bar{\mathfrak{a}}_h \cup \bar{\mathfrak{a}}_k),$$

whence by Part I, Definition 5.1 and Theorem 5.1,

$$(\mathfrak{u} \cup \mathfrak{b}_{jh}, \mathfrak{v} \cup \mathfrak{b}_{jh}, \bar{\mathfrak{a}}_i \cup \bar{\mathfrak{a}}_h \cup \bar{\mathfrak{a}}_k)D.$$

Thus

$$\begin{aligned}(\mathfrak{u} \cup \mathfrak{v})^\alpha &= (\mathfrak{u} \cup \mathfrak{v} \cup \mathfrak{b}_{jh}) \cap (\bar{\mathfrak{a}}_i \cup \bar{\mathfrak{a}}_h \cup \bar{\mathfrak{a}}_k) \\ &= [(\mathfrak{u} \cup \mathfrak{b}_{jh}) \cup (\mathfrak{v} \cup \mathfrak{b}_{jh})] \cap (\bar{\mathfrak{a}}_i \cup \bar{\mathfrak{a}}_h \cup \bar{\mathfrak{a}}_k) \\ &= [(\mathfrak{u} \cup \mathfrak{b}_{jh}) \cap (\bar{\mathfrak{a}}_i \cup \bar{\mathfrak{a}}_h \cup \bar{\mathfrak{a}}_k)] \cup [(\mathfrak{v} \cup \mathfrak{b}_{jh}) \cap (\bar{\mathfrak{a}}_i \cup \bar{\mathfrak{a}}_h \cup \bar{\mathfrak{a}}_k) \\ &= \mathfrak{u}^\alpha \cup \mathfrak{v}^\alpha,\end{aligned}$$

and (3) is proved. Finally, by (1) since $\mathfrak{u} \cap \mathfrak{v} \leqq \mathfrak{u}$, $(\mathfrak{u} \cap \mathfrak{v})^\alpha \leqq u^\alpha$; similarly $(\mathfrak{u} \cap \mathfrak{v})^\alpha \leqq \mathfrak{v}^\alpha$, whence $(\mathfrak{u} \cap \mathfrak{v})^\alpha \leqq \mathfrak{u}^\alpha \cap \mathfrak{v}^\alpha$, and (4) is proved.

LEMMA 8.2: If $\mathfrak{u}$, $\mathfrak{v} \leqq \bar{\mathfrak{a}}_i \cup \bar{\mathfrak{a}}_j \cup \bar{\mathfrak{a}}_k$, then

$$(6) \qquad (\mathfrak{u} \cap \mathfrak{v})^\beta = \mathfrak{u}^\beta \cap \mathfrak{v}^\beta,$$

$$(7) \qquad (\mathfrak{u} \cup \mathfrak{v})^\beta \geqq \mathfrak{u}^\beta \cup \mathfrak{v}^\beta.$$

PROOF: Let $\mathfrak{u}$, $\mathfrak{v} \leqq \bar{\mathfrak{a}}_i \cup \bar{\mathfrak{a}}_j \cup \bar{\mathfrak{a}}_k$. Then

$$\begin{aligned}(\mathfrak{u} \cup \mathfrak{v}) \cap \mathfrak{b}_{li} &\leqq (\bar{\mathfrak{a}}_i \cup \bar{\mathfrak{a}}_j \cup \bar{\mathfrak{a}}_k) \cap \mathfrak{b}_{li} \\ &= (\bar{\mathfrak{a}}_i \cup \bar{\mathfrak{a}}_j \cup \bar{\mathfrak{a}}_k) \cap (\bar{\mathfrak{a}}_l \cup \bar{\mathfrak{a}}_i) \cap \mathfrak{b}_{li} \\ &= \bar{\mathfrak{a}}_i \cap \mathfrak{b}_{li} = 0,\end{aligned}$$

whence $(\mathfrak{u} \cup \mathfrak{v}) \cap \mathfrak{b}_{li} = 0$, and thus also $\mathfrak{u} \cap \mathfrak{b}_{li}$, $\mathfrak{v} \cap \mathfrak{b}_{li} = 0$. Hence $(\mathfrak{u} \cap \mathfrak{v}) \cap \mathfrak{b}_{li} = (\mathfrak{u} \cap \mathfrak{b}_{li}) \cup (\mathfrak{v} \cap \mathfrak{b}_{li}) = 0$, and $(\mathfrak{u}, \mathfrak{v}, \mathfrak{b}_{li})D$. Consequently,

$$\begin{aligned}(\mathfrak{u} \cap \mathfrak{v})^\beta &= [(\mathfrak{u} \cap \mathfrak{v}) \cup \mathfrak{b}_{li}] \cap (\bar{\mathfrak{a}}_i \cup \bar{\mathfrak{a}}_j \cup \bar{\mathfrak{a}}_k) \\ &= [(\mathfrak{u} \cup \mathfrak{b}_{li}) \cap (\mathfrak{v} \cup \mathfrak{b}_{li})] \cap (\bar{\mathfrak{a}}_i \cup \bar{\mathfrak{a}}_j \cup \bar{\mathfrak{a}}_k) \\ &= [(\mathfrak{u} \cup \mathfrak{b}_{li}) \cap (\bar{\mathfrak{a}}_i \cup \bar{\mathfrak{a}}_j \cup \bar{\mathfrak{a}}_k)] \cap [(\mathfrak{v} \cup \mathfrak{b}_{li}) \cap (\bar{\mathfrak{a}}_i \cup \bar{\mathfrak{a}}_j \cup \bar{\mathfrak{a}}_k)] \\ &= \mathfrak{u}^\beta \cap \mathfrak{v}^\beta,\end{aligned}$$

and (6) is proved. Finally, since $\mathfrak{u} \cup \mathfrak{v} \geqq \mathfrak{u}$ we have by (2), $(\mathfrak{u} \cup \mathfrak{v})^\beta \geqq \mathfrak{u}^\beta$; similarly $(\mathfrak{u} \cup \mathfrak{v})^\beta \geqq \mathfrak{v}^\beta$, whence $(\mathfrak{u} \cup \mathfrak{v})^\beta \geqq \mathfrak{u}^\beta \cup \mathfrak{v}^\beta$, and (7) holds.

LEMMA 8.3:

$$(8) \qquad \bar{\mathfrak{a}}_j^\alpha \leqq \bar{\mathfrak{a}}_h;$$

$$(9) \qquad \bar{\mathfrak{a}}_i^\beta = \bar{\mathfrak{a}}_l;$$

$$(10) \qquad \mathfrak{u} \leqq \bar{\mathfrak{a}}_i \cup \bar{\mathfrak{a}}_k \text{ implies } \mathfrak{u}^\alpha = \mathfrak{u};$$

$$(11) \qquad \mathfrak{u} \leqq \bar{\mathfrak{a}}_j \cup \bar{\mathfrak{a}}_k \text{ implies } \mathfrak{u}^\beta \geqq \mathfrak{u}.$$

PROOF: We have

$$\bar{\mathfrak{a}}_j^\alpha = (\bar{\mathfrak{a}}_j \cup \mathfrak{b}_{jh}) \cap (\bar{\mathfrak{a}}_i \cup \bar{\mathfrak{a}}_h \cup \bar{\mathfrak{a}}_k) \leqq (\bar{\mathfrak{a}}_j \cup \bar{\mathfrak{a}}_h) \cap (\bar{\mathfrak{a}}_i \cup \bar{\mathfrak{a}}_h \cup \bar{\mathfrak{a}}_k) = \bar{\mathfrak{a}}_h,$$

whence (8) holds. Also,

$$\bar{\mathfrak{a}}_i^\beta = (\bar{\mathfrak{a}}_i \cup \mathfrak{b}_{li}) \cap (\bar{\mathfrak{a}}_l \cup \bar{\mathfrak{a}}_j \cup \bar{\mathfrak{a}}_k) = (\bar{\mathfrak{a}}_i \cup \bar{\mathfrak{a}}_l) \cap (\bar{\mathfrak{a}}_l \cup \bar{\mathfrak{a}}_j \cup \bar{\mathfrak{a}}_k) = \bar{\mathfrak{a}}_l,$$

whence (9) holds. Moreover, if $\mathfrak{u} \leqq \bar{\mathfrak{a}}_i \cup \bar{\mathfrak{a}}_k$, then $\mathfrak{u}$, $\bar{\mathfrak{a}}_i \cup \bar{\mathfrak{a}}_k \leqq \bar{\mathfrak{a}}_i \cup \bar{\mathfrak{a}}_k$ and $\mathfrak{b}_{jh}$, $\bar{\mathfrak{a}}_h \leqq \bar{\mathfrak{a}}_j \cup \bar{\mathfrak{a}}_h$ and $(\bar{\mathfrak{a}}_i \cup \bar{\mathfrak{a}}_k) \cap (\bar{\mathfrak{a}}_j \cup \bar{\mathfrak{a}}_h) = 0$. Hence

$$\begin{aligned} \mathfrak{u}^\alpha &= (\mathfrak{u} \cup \mathfrak{b}_{jh}) \cap (\bar{\mathfrak{a}}_i \cup \bar{\mathfrak{a}}_k \cup \bar{\mathfrak{a}}_h) \\ &= [\mathfrak{u} \cap (\bar{\mathfrak{a}}_i \cup \bar{\mathfrak{a}}_k)] \cup [\mathfrak{b}_{jh} \cap \bar{\mathfrak{a}}_h] \quad \text{(by Part I, Theorem 1.2)} \\ &= \mathfrak{u} \cup 0 = \mathfrak{u}, \end{aligned}$$

and (10) holds. Finally, if $\mathfrak{u} \leqq \bar{\mathfrak{a}}_j \cup \bar{\mathfrak{a}}_k$, then $\mathfrak{u}^\beta = (\mathfrak{u} \cup \mathfrak{b}_{li}) \cap (\bar{\mathfrak{a}}_j \cup \bar{\mathfrak{a}}_k \cup \bar{\mathfrak{a}}_l) \geqq \mathfrak{u}$, and (11) is proved.

LEMMA 8.4: Let $\mathfrak{u}, \mathfrak{v} \in L_{ij}$. Then $\mathfrak{u}^\alpha, \mathfrak{v}^\alpha \in L_{ih}$, $\mathfrak{u}^\beta, \mathfrak{v}^\beta \in L_{lj}$, $\mathfrak{u} \oplus \mathfrak{v} \in L_{ij}$, $\mathfrak{u}^\alpha \oplus \mathfrak{v}^\alpha \in L_{ih}$, $\mathfrak{u}^\beta \oplus \mathfrak{v}^\beta \in L_{lj}$,
and

$$(\mathfrak{u} \oplus \mathfrak{v})^\alpha = \mathfrak{u}^\alpha \oplus \mathfrak{v}^\alpha, \tag{12}$$

$$(\mathfrak{u} \oplus \mathfrak{v})^\beta = \mathfrak{u}^\beta \oplus \mathfrak{v}^\beta. \tag{13}$$

PROOF: The first statements are trivial by the Corollary to Definition 8.1 and by Lemma 7.2. Now

$$\begin{aligned} (\mathfrak{u} \oplus \mathfrak{v})^\alpha &= [(((\mathfrak{u} \cup \bar{\mathfrak{c}}_{ik}) \cap (\bar{\mathfrak{a}}_j \cup \bar{\mathfrak{a}}_k)) \cup ((\mathfrak{v} \cup \bar{\mathfrak{a}}_k) \cap (\bar{\mathfrak{a}}_j \cup \bar{\mathfrak{c}}_{ik}))) \cap (\bar{\mathfrak{a}}_i \cup \bar{\mathfrak{a}}_j)]^\alpha \\ &\qquad \text{(by Lemma 7.8, Definition 7.6)} \\ &\leqq (((\mathfrak{u}^\alpha \cup \bar{\mathfrak{c}}_{ik}^\alpha) \cap (\bar{\mathfrak{a}}_j^\alpha \cup \bar{\mathfrak{a}}_k^\alpha)) \cup ((\mathfrak{v}^\alpha \cup \bar{\mathfrak{a}}_k^\alpha) \cap (\bar{\mathfrak{a}}_j^\alpha \cup \bar{\mathfrak{c}}_{ik}^\alpha))) \cap (\bar{\mathfrak{a}}_i^\alpha \cup \bar{\mathfrak{a}}_j^\alpha) \quad \text{(by (3), (4))} \\ &= (((\mathfrak{u}^\alpha \cup \bar{\mathfrak{c}}_{ik}) \cap (\bar{\mathfrak{a}}_j^\alpha \cup \bar{\mathfrak{a}}_k)) \cup ((\mathfrak{v}^\alpha \cup \bar{\mathfrak{a}}_k) \cap (\bar{\mathfrak{a}}_j^\alpha \cup \bar{\mathfrak{c}}_{ik}))) \cap (\bar{\mathfrak{a}}_i \cup \bar{\mathfrak{a}}_j^\alpha) \quad \text{(by (10))} \\ &\leqq (((\mathfrak{u}^\alpha \cup \bar{\mathfrak{c}}_{ik}) \cap (\bar{\mathfrak{a}}_h \cup \bar{\mathfrak{a}}_k)) \cup ((\mathfrak{v}^\alpha \cup \bar{\mathfrak{a}}_k) \cap (\bar{\mathfrak{a}}_h \cup \bar{\mathfrak{c}}_{ik}))) \cap (\bar{\mathfrak{a}}_i \cup \bar{\mathfrak{a}}_h) \quad \text{(by (8))} \\ &= \mathfrak{u}^\alpha \oplus \mathfrak{v}^\alpha \qquad \text{(by Lemma 7.8, Definition 7.6)}, \end{aligned}$$

whence $(\mathfrak{u} \oplus \mathfrak{v})^\alpha \leqq \mathfrak{u}^\alpha \oplus \mathfrak{v}^\alpha$ and since both $(\mathfrak{u} \oplus \mathfrak{v})^\alpha$ and $\mathfrak{u}^\alpha \oplus \mathfrak{v}^\alpha$ are in L_{ih}, we have (12). Moreover,

$$\begin{aligned} (\mathfrak{u} \oplus \mathfrak{v})^\beta &= [(((((\mathfrak{u} \cup \bar{\mathfrak{c}}_{jk}) \cap (\bar{\mathfrak{a}}_i \cup \bar{\mathfrak{a}}_k)) \cup \bar{\mathfrak{a}}_j) \cap (\mathfrak{v} \cup \bar{\mathfrak{a}}_k)) \cup \bar{\mathfrak{c}}_{jk}) \cap (\bar{\mathfrak{a}}_i \cup \bar{\mathfrak{a}}_j)]^\alpha \\ &\qquad \text{(by Lemma 7.8, Definition 7.6)} \\ &\geqq \Big(((((\mathfrak{u}^\beta \cup \bar{\mathfrak{c}}_{jk}^\beta) \cap (\bar{\mathfrak{a}}_i^\beta \cup \bar{\mathfrak{a}}_k^\beta)) \cup \bar{\mathfrak{a}}_j^\beta) \cap (\mathfrak{v}^\beta \cup \bar{\mathfrak{a}}_k^\beta)) \cup \bar{\mathfrak{c}}_{jk}^\beta\Big) \cap (\bar{\mathfrak{a}}_i^\beta \cup \bar{\mathfrak{a}}_j^\beta) \quad \text{(by (6), (7))} \\ &\geqq \Big(((((\mathfrak{u}^\beta \cup \bar{\mathfrak{c}}_{jk}) \cap (\bar{\mathfrak{a}}_i^\beta \cup \bar{\mathfrak{a}}_k)) \cup \bar{\mathfrak{a}}_j) \cap (\mathfrak{v}^\beta \cup \bar{\mathfrak{a}}_k)) \cup \bar{\mathfrak{c}}_{jk}\Big) \cap (\bar{\mathfrak{a}}_i^\beta \cup \bar{\mathfrak{a}}_j) \quad \text{(by (11))} \\ &= \Big(((((\mathfrak{u}^\beta \cup \bar{\mathfrak{c}}_{jk}) \cap (\bar{\mathfrak{a}}_l \cup \bar{\mathfrak{a}}_k)) \cup \bar{\mathfrak{a}}_j) \cap (\mathfrak{v}^\beta \cup \bar{\mathfrak{a}}_k)) \cup \bar{\mathfrak{c}}_{jk}\Big) \cap (\bar{\mathfrak{a}}_l \cup \bar{\mathfrak{a}}_j) \quad \text{(by (9))} \\ &= \mathfrak{u}^\beta \oplus \mathfrak{v}^\beta \qquad \text{(by Lemma 7.8, Definition 7.6)}, \end{aligned}$$

whence $(\mathfrak{u} \oplus \mathfrak{v})^\beta \geqq \mathfrak{u}^\beta \oplus \mathfrak{v}^\beta$, and since both $(\mathfrak{u} \oplus \mathfrak{v})^\beta$ and $\mathfrak{u}^\beta \oplus \mathfrak{v}^\beta$ are in L_{lj}, we have (13).

At this point we drop the assumption that $i, j, k, h, l, \mathfrak{b}_{jh}, \mathfrak{b}_{li}$ are fixed.

THEOREM 8.1: Multiplication in $\mathfrak{S}$ is both right- and left-distributive with respect to addition in $\mathfrak{S}$; i.e., if $\beta, \gamma, \delta \in \mathfrak{S}$, then

$$\beta(\gamma + \delta) = \beta\gamma + \beta\delta, \tag{14}$$

$$(\gamma + \delta)\beta = \gamma\beta + \delta\beta. \tag{15}$$

PROOF: Let i, j, k, h, l be indices selected as indicated at the beginning of the chapter, i.e., i, j, k distinct, $h \neq i, j, k$ and $l \neq i, j, k$. This can be done since $n \geqq 4$. We shall establish (14) by proving that the i, h-components of both members are equal. We have

$$\begin{aligned}
(\beta(\gamma + \delta))_{ih} &= (\gamma + \delta)_{ij} \otimes (\beta)_{jh} && \text{(by Definition 6.2)}\\
&= (\gamma + \delta)_{ij}^\alpha \\
& \qquad \text{(by the Corollary to Definition 8.1 with } \mathfrak{b}_{jh} = (\beta)_{jh})\\
&= ((\gamma)_{ij} \oplus (\delta)_{ij})^\alpha && \text{(by Definition 7.7)}\\
&= (\gamma)_{ij}^\alpha \oplus (\delta)_{ij}^\alpha && \text{(by Lemma 8.4, (12))}\\
&= [(\gamma)_{ij} \otimes (\beta)_{jh}] \oplus [(\delta)_{ij} \otimes (\beta)_{jh}] \\
& \qquad \text{(by the Corollary to Definition 8.1)}\\
&= (\beta\gamma)_{ih} \oplus (\beta\delta)_{ih} && \text{(by Definition 6.2)}\\
&= (\beta\gamma + \beta\delta)_{ih} && \text{(by Definition 7.7),}
\end{aligned}$$

and thus $\beta(\gamma + \delta) = \beta\gamma + \beta\delta$. We establish (15) by proving the equality of the l, j-components of its two members. Indeed,

$$\begin{aligned}
((\gamma + \delta)\beta)_{lj} &= (\beta)_{li} \otimes (\gamma + \delta)_{ij} && \text{(by Definition 6.2)}\\
&= (\gamma + \delta)_{ij}^\beta \\
& \qquad \text{(by the Corollary to Definition 8.1 with } \mathfrak{b}_{li} = (\beta)_{li})\\
&= ((\gamma)_{ij} \oplus (\delta)_{ij})^\beta && \text{(by Definition 7.7)}\\
&= (\gamma)_{ij}^\beta \oplus (\delta)_{ij}^\beta && \text{(by Lemma 8.4, (13))}\\
&= [(\beta)_{li} \otimes (\gamma)_{ij}] \oplus [(\beta)_{li} \otimes (\delta)_{ij}] \\
& \qquad \text{(by the Corollary to Definition 8.1)}\\
&= (\gamma\beta)_{lj} \oplus (\delta\beta)_{lj} && \text{(by Definition 6.2)}\\
&= (\gamma\beta + \delta\beta)_{lj} && \text{(by Definition 7.7),}
\end{aligned}$$

and therefore $(\gamma + \delta)\beta = \gamma\beta + \delta\beta$.

We turn now to a definition of subtraction in $\mathfrak{S}$ and we begin with a construction for $-(\gamma\beta)$ in the next theorem. This construction is discussed in the appendix in a manner similar to that employed in the appendices to the preceding two chapters for multiplication and addition.

THEOREM 8.2: For every $\beta, \gamma \in \mathfrak{S}$ there exists a unique element $\delta \in \mathfrak{S}$ such that $\delta + \gamma\beta = \gamma\beta + \delta = 0$.

PROOF: Let i, j, k be distinct indices (existent since $n \geqq 4$). By Lemma 6.1 and the commutative law for addition it suffices to prove the existence of δ such that $(\delta + \gamma\beta)_{ij} = (0)_{ij} = \bar{\mathfrak{a}}_i$, i.e., such that $(\delta)_{ij} \oplus (\gamma\beta)_{ij} = \bar{\mathfrak{a}}_i$. Define $\mathfrak{b}_{ik} \equiv (\beta)_{ik}$, $\mathfrak{c}_{kj} \equiv (\gamma)_{kj}$,

$$\mathfrak{u} \equiv \big(((\mathfrak{b}_{ik} \cup \bar{\mathfrak{a}}_j) \cap (\mathfrak{c}_{kj} \cup \bar{\mathfrak{a}}_i)) \cup \bar{\mathfrak{a}}_k \big) \cap (\bar{\mathfrak{a}}_i \cup \bar{\mathfrak{a}}_j)$$
$$\mathfrak{v} \equiv (\mathfrak{b}_{ik} \cup \mathfrak{c}_{kj}) \cap (\bar{\mathfrak{a}}_i \cup \bar{\mathfrak{a}}_j) = (\beta)_{ik} \otimes (\gamma)_{kj} = (\gamma\beta)_{ij}.$$

Hence $\mathfrak{v} \in L_{ij}$. We shall show that $\mathfrak{u} \in L_{ij}$ and that $\mathfrak{u} \oplus \mathfrak{v} = \bar{\mathfrak{a}}_i$. Now

$$\begin{aligned}
\mathfrak{u} \cup \bar{\mathfrak{a}}_j &= \big[\big(((\mathfrak{b}_{ik} \cup \bar{\mathfrak{a}}_j) \cap (\mathfrak{c}_{kj} \cup \bar{\mathfrak{a}}_i)) \cup \bar{\mathfrak{a}}_k \big) \cap (\bar{\mathfrak{a}}_i \cup \bar{\mathfrak{a}}_j)\big] \cup \bar{\mathfrak{a}}_j \\
&= \big(((\mathfrak{b}_{ik} \cup \bar{\mathfrak{a}}_j) \cap (\mathfrak{c}_{kj} \cup \bar{\mathfrak{a}}_i)) \cup \bar{\mathfrak{a}}_j \cup \bar{\mathfrak{a}}_k \big) \cap (\bar{\mathfrak{a}}_i \cup \bar{\mathfrak{a}}_j) && \text{(by IV)} \\
&= \big(((\mathfrak{b}_{ik} \cup \bar{\mathfrak{a}}_j) \cap (\mathfrak{c}_{kj} \cup \bar{\mathfrak{a}}_i \cup \bar{\mathfrak{a}}_j)) \cup \bar{\mathfrak{a}}_k \big) \cap (\bar{\mathfrak{a}}_i \cup \bar{\mathfrak{a}}_j) && \text{(by IV)} \\
&= \big(((\mathfrak{b}_{ik} \cup \bar{\mathfrak{a}}_j) \cap (\bar{\mathfrak{a}}_k \cup \bar{\mathfrak{a}}_i \cup \bar{\mathfrak{a}}_j)) \cup \bar{\mathfrak{a}}_k \big) \cap (\bar{\mathfrak{a}}_i \cup \bar{\mathfrak{a}}_j) && \text{(since } \mathfrak{c}_{kj} \in L_{kj}) \\
&= \big((\mathfrak{b}_{ik} \cup \bar{\mathfrak{a}}_j) \cup \bar{\mathfrak{a}}_k \big) \cap (\bar{\mathfrak{a}}_i \cup \bar{\mathfrak{a}}_j) && \text{(since } \mathfrak{b}_{ik} \cup \bar{\mathfrak{a}}_j \leqq \bar{\mathfrak{a}}_i \cup \bar{\mathfrak{a}}_j \cup \bar{\mathfrak{a}}_k) \\
&= (\bar{\mathfrak{a}}_i \cup \bar{\mathfrak{a}}_j \cup \bar{\mathfrak{a}}_k) \cap (\bar{\mathfrak{a}}_i \cup \bar{\mathfrak{a}}_j) && \text{(since } \mathfrak{b}_{ik} \in L_{ik}) \\
&= \bar{\mathfrak{a}}_i \cup \bar{\mathfrak{a}}_j.
\end{aligned}$$

Moreover,

$$\begin{aligned}
\mathfrak{u} \cap \bar{\mathfrak{a}}_j &= \big[\big(((\mathfrak{b}_{ik} \cup \bar{\mathfrak{a}}_j) \cap (\mathfrak{c}_{kj} \cup \bar{\mathfrak{a}}_i)) \cup \bar{\mathfrak{a}}_k \big) \cap (\bar{\mathfrak{a}}_i \cup \bar{\mathfrak{a}}_j)\big] \cap \bar{\mathfrak{a}}_j \\
&= \big(((\mathfrak{b}_{ik} \cup \bar{\mathfrak{a}}_j) \cap (\mathfrak{c}_{kj} \cup \bar{\mathfrak{a}}_i)) \cup \bar{\mathfrak{a}}_k \big) \cap \bar{\mathfrak{a}}_j \\
&\leqq \big(((\mathfrak{b}_{ik} \cup \bar{\mathfrak{a}}_j) \cap (\mathfrak{c}_{kj} \cup \bar{\mathfrak{a}}_i)) \cup \bar{\mathfrak{a}}_k \big) \cap (\mathfrak{b}_{ik} \cup \bar{\mathfrak{a}}_j) \\
&= \big((\mathfrak{b}_{ik} \cup \bar{\mathfrak{a}}_j) \cap (\mathfrak{c}_{kj} \cup \bar{\mathfrak{a}}_i) \big) \cup (\bar{\mathfrak{a}}_k \cap (\mathfrak{b}_{ik} \cup \bar{\mathfrak{a}}_j)) && \text{(by IV).}
\end{aligned}$$

But $(\bar{\mathfrak{a}}_i, \bar{\mathfrak{a}}_j, \bar{\mathfrak{a}}_k) \perp$, $\mathfrak{b}_{ik} \in L_{ik}$, whence by Part I, Lemma 2.1, $(\mathfrak{b}_{ik}, \bar{\mathfrak{a}}_j, \bar{\mathfrak{a}}_k) \perp$ and so $\bar{\mathfrak{a}}_k \cap (\mathfrak{b}_{ik} \cup \bar{\mathfrak{a}}_j) = 0$. Hence

$$\mathfrak{u} \cap \bar{\mathfrak{a}}_j \leqq (\mathfrak{b}_{ik} \cup \bar{\mathfrak{a}}_j) \cap (\mathfrak{c}_{kj} \cup \bar{\mathfrak{a}}_i) \leqq \mathfrak{c}_{kj} \cup \bar{\mathfrak{a}}_i,$$

and since $\mathfrak{u} \cap \bar{\mathfrak{a}}_j \leqq \bar{\mathfrak{a}}_j$ we have $\mathfrak{u} \cap \bar{\mathfrak{a}}_j \leqq (\mathfrak{c}_{kj} \cup \bar{\mathfrak{a}}_i) \cap \bar{\mathfrak{a}}_j$. But since $(\bar{\mathfrak{a}}_i, \bar{\mathfrak{a}}_j, \bar{\mathfrak{a}}_k) \perp$, $\mathfrak{c}_{kj} \in L_{kj}$, we have $(\bar{\mathfrak{a}}_i, \bar{\mathfrak{a}}_j, \mathfrak{c}_{kj}) \perp$ and so $(\mathfrak{c}_{kj} \cup \bar{\mathfrak{a}}_i) \cap \bar{\mathfrak{a}}_j = 0$, whence

$$\mathfrak{u} \cap \bar{\mathfrak{a}}_j = 0.$$

Consequently it follows that $\mathfrak{u} \in L_{ij}$. To compute $\mathfrak{u} \oplus \mathfrak{v}$, we define

(16) $\mathfrak{u}' \equiv \mathfrak{u} \cup \bar{\mathfrak{a}}_k, \quad \mathfrak{v}' \equiv \mathfrak{v} \cup \mathfrak{c}_{kj}, \quad \mathfrak{w}' \equiv \bar{\mathfrak{a}}_j \cup \bar{\mathfrak{a}}_k.$

Then

$$\mathfrak{u}' = [(((\mathfrak{b}_{ik} \cup \bar{\mathfrak{a}}_j) \cap (\mathfrak{c}_{kj} \cup \bar{\mathfrak{a}}_i)) \cup \bar{\mathfrak{a}}_k) \cap (\bar{\mathfrak{a}}_i \cup \bar{\mathfrak{a}}_j)] \cup \bar{\mathfrak{a}}_k$$
$$(17) \quad = (((\mathfrak{b}_{ik} \cup \bar{\mathfrak{a}}_j) \cap (\mathfrak{c}_{kj} \cup \bar{\mathfrak{a}}_i)) \cup \bar{\mathfrak{a}}_k) \cap (\bar{\mathfrak{a}}_i \cup \bar{\mathfrak{a}}_j \cup \bar{\mathfrak{a}}_k) \qquad \text{(by IV)}$$
$$= ((\mathfrak{b}_{ik} \cup \bar{\mathfrak{a}}_j) \cap (\mathfrak{c}_{kj} \cup \bar{\mathfrak{a}}_i)) \cup \bar{\mathfrak{a}}_k \quad (\text{since } \mathfrak{b}_{ik}, \bar{\mathfrak{a}}_j, \mathfrak{c}_{kj}, \bar{\mathfrak{a}}_i \leqq \bar{\mathfrak{a}}_i \cup \bar{\mathfrak{a}}_j \cup \bar{\mathfrak{a}}_k),$$

$$\mathfrak{v}' = ((\mathfrak{b}_{ik} \cup \mathfrak{c}_{kj}) \cap (\bar{\mathfrak{a}}_i \cup \bar{\mathfrak{a}}_j)) \cup \mathfrak{c}_{kj}$$
$$= ((\mathfrak{b}_{ik} \cup \mathfrak{c}_{kj}) \cap (\bar{\mathfrak{a}}_i \cup \bar{\mathfrak{a}}_j \cup \mathfrak{c}_{kj}) \qquad \text{(by IV))}$$
$$(18) \quad = ((\mathfrak{b}_{ik} \cup \mathfrak{c}_{kj}) \cap (\bar{\mathfrak{a}}_i \cup \bar{\mathfrak{a}}_j \cup \bar{\mathfrak{a}}_k) \qquad (\text{since } \mathfrak{c}_{kj} \in L_{kj})$$
$$= \mathfrak{b}_{ik} \cup \mathfrak{c}_{kj} \qquad (\text{since } \mathfrak{b}_{ik}, \mathfrak{c}_{kj} \leqq \bar{\mathfrak{a}}_i \cup \bar{\mathfrak{a}}_j \cup \bar{\mathfrak{a}}_k).$$

We shall show that $(\mathfrak{u}', \mathfrak{v}', \mathfrak{w}')P_{\mathfrak{u}\mathfrak{v}}$ by showing that conditions (a), $\cdots$, (g) of Definition 7.1 are satisfied. Clearly $\mathfrak{u}' \geqq \mathfrak{u}$, $\mathfrak{v}' \geqq \mathfrak{v}$, $\mathfrak{w}' \geqq \bar{\mathfrak{a}}_j$, whence (a), (c), (e) hold. Also, $\bar{\mathfrak{a}}_i \cap \mathfrak{w}' = \bar{\mathfrak{a}}_i \cap (\bar{\mathfrak{a}}_j \cup \bar{\mathfrak{a}}_k) = 0$ since $(\bar{\mathfrak{a}}_i, \bar{\mathfrak{a}}_j, \bar{\mathfrak{a}}_k)\perp$, whence (b) holds. Moreover,

$$\mathfrak{u}' \cup \bar{\mathfrak{a}}_j = \mathfrak{u} \cup \bar{\mathfrak{a}}_k \cup \bar{\mathfrak{a}}_j = \bar{\mathfrak{a}}_i \cup \bar{\mathfrak{a}}_j \cup \bar{\mathfrak{a}}_k = \bar{\mathfrak{a}}_i \cup \mathfrak{w}',$$

and since $(\mathfrak{u}, \bar{\mathfrak{a}}_j, \bar{\mathfrak{a}}_k)\perp$, (because $(\bar{\mathfrak{a}}_i, \bar{\mathfrak{a}}_j, \bar{\mathfrak{a}}_k)\perp$, $\mathfrak{u} \in L_{ij}$), it follows that

$$\mathfrak{u}' \cap \bar{\mathfrak{a}}_j = (\mathfrak{u} \cup \bar{\mathfrak{a}}_k) \cap \bar{\mathfrak{a}}_j = 0,$$

and we have that $\mathfrak{u}'$ is inverse to $\bar{\mathfrak{a}}_j$ in $\bar{\mathfrak{a}}_i \cup \mathfrak{w}'$, viz., that (d) holds. Now

$$\mathfrak{v}' \cup \bar{\mathfrak{a}}_j = \mathfrak{v} \cup \mathfrak{c}_{kj} \cup \bar{\mathfrak{a}}_j = \mathfrak{v} \cup \bar{\mathfrak{a}}_j \cup \bar{\mathfrak{a}}_k = \bar{\mathfrak{a}}_i \cup \bar{\mathfrak{a}}_j \cup \bar{\mathfrak{a}}_k.$$

But since $(\bar{\mathfrak{a}}_i, \bar{\mathfrak{a}}_j, \bar{\mathfrak{a}}_k)\perp$ and $\mathfrak{v} \in L_{ij}$, we have $(\mathfrak{v}, \bar{\mathfrak{a}}_j, \bar{\mathfrak{a}}_k)\perp$, and since $\mathfrak{c}_{kj} \in L_{kj}$ we have $(\mathfrak{v}, \bar{\mathfrak{a}}_j, \mathfrak{c}_{kj})\perp$. Therefore

$$\mathfrak{v}' \cap \bar{\mathfrak{a}}_j = (\mathfrak{v} \cup \mathfrak{c}_{kj}) \cap \bar{\mathfrak{a}}_j = 0,$$

and we have established (f). Finally,

$$(\mathfrak{u}' \cap \mathfrak{w}') \cup (\mathfrak{v}' \cap \mathfrak{w}') \cup \bar{\mathfrak{a}}_i$$
$$= ((\mathfrak{u} \cup \bar{\mathfrak{a}}_k) \cap (\bar{\mathfrak{a}}_j \cup \bar{\mathfrak{a}}_k)) \cup ((\mathfrak{v} \cup \mathfrak{c}_{kj}) \cap (\bar{\mathfrak{a}}_j \cup \bar{\mathfrak{a}}_k)) \cup \bar{\mathfrak{a}}_i$$
$$\geqq \bar{\mathfrak{a}}_k \cup (\mathfrak{c}_{kj} \cap (\bar{\mathfrak{a}}_j \cup \bar{\mathfrak{a}}_k)) \cup \bar{\mathfrak{a}}_i$$
$$= \bar{\mathfrak{a}}_k \cup \mathfrak{c}_{kj} \cup \bar{\mathfrak{a}}_i \qquad (\text{since } \mathfrak{c}_{kj} \leqq \bar{\mathfrak{a}}_j \cup \bar{\mathfrak{a}}_k)$$
$$\geqq ((\mathfrak{b}_{ik} \cup \bar{\mathfrak{a}}_k) \cap (\mathfrak{c}_{kj} \cup \bar{\mathfrak{a}}_i)) \cup \bar{\mathfrak{a}}_k$$
$$\geqq (((\mathfrak{b}_{ik} \cup \bar{\mathfrak{a}}_k) \cap (\mathfrak{c}_{kj} \cup \bar{\mathfrak{a}}_i)) \cup \bar{\mathfrak{a}}_k) \cap (\bar{\mathfrak{a}}_i \cup \bar{\mathfrak{a}}_k) = \mathfrak{u},$$

and (g) is proved. Thus $(\mathfrak{u}', \mathfrak{v}', \mathfrak{w}')P_{\mathfrak{u}\mathfrak{v}}$. Now since $\mathfrak{w}' = \bar{\mathfrak{a}}_j \cup \bar{\mathfrak{a}}_k \leqq \bar{\mathfrak{a}}_i \cup \bar{\mathfrak{a}}_j \cup \bar{\mathfrak{a}}_k$, we have by Lemma 7.9, $\mathfrak{u} \oplus \mathfrak{v} = \mathfrak{u} \oplus_{(\mathfrak{u}', \mathfrak{v}', \mathfrak{w}')} \mathfrak{v}$. Consequently by Definition 7.2 we have

$$\begin{aligned}
\mathfrak{u} \oplus \mathfrak{v} &= ((((((\mathfrak{v}' \cap \mathfrak{w}') \cup \bar{\mathfrak{a}}_i) \cap \mathfrak{u}') \cup \bar{\mathfrak{a}}_j) \cap \mathfrak{v}') \cup (\mathfrak{u}' \cap \mathfrak{w}')) \cap (\bar{\mathfrak{a}}_i \cup \bar{\mathfrak{a}}_j) \\
&\geqq (((((\mathfrak{c}_{kj} \cup \bar{\mathfrak{a}}_i) \cap [((\mathfrak{b}_{ik} \cup \bar{\mathfrak{a}}_j) \cap (\mathfrak{c}_{kj} \cup \bar{\mathfrak{a}}_i) \cup \bar{\mathfrak{a}}_k]) \cup \bar{\mathfrak{a}}_j) \cap \mathfrak{v}') \\
&\qquad \cup (\mathfrak{u}' \cap \mathfrak{w}')) \cap (\bar{\mathfrak{a}}_i \cup \bar{\mathfrak{a}}_j) \quad (\text{since } \mathfrak{v}', \mathfrak{w}' \geqq \mathfrak{c}_{kj}) \\
&\geqq (((((\mathfrak{c}_{kj} \cup \bar{\mathfrak{a}}_i) \cap (\mathfrak{b}_{ik} \cup \bar{\mathfrak{a}}_j)) \cup \bar{\mathfrak{a}}_j) \cap \mathfrak{v}') \cup (\mathfrak{u}' \cap \mathfrak{w}')) \cap (\bar{\mathfrak{a}}_i \cup \bar{\mathfrak{a}}_j) \\
&= ((((\mathfrak{c}_{kj} \cup \bar{\mathfrak{a}}_i \cup \bar{\mathfrak{a}}_j) \cap (\mathfrak{b}_{ik} \cup \bar{\mathfrak{a}}_j)) \cap \mathfrak{v}') \cup (\mathfrak{u}' \cap \mathfrak{w}')) \cap (\bar{\mathfrak{a}}_i \cup \bar{\mathfrak{a}}_j) \quad (\text{by IV}) \\
&= ((((\bar{\mathfrak{a}}_k \cup \bar{\mathfrak{a}}_i \cup \bar{\mathfrak{a}}_j) \cap (\mathfrak{b}_{ik} \cup \bar{\mathfrak{a}}_j)) \cap \mathfrak{v}') \cup (\mathfrak{u}' \cap \mathfrak{w}')) \cap (\bar{\mathfrak{a}}_i \cup \bar{\mathfrak{a}}_j) \\
&\qquad (\text{since } \mathfrak{c}_{kj} \in L_{kj}) \\
&= (((\mathfrak{b}_{ik} \cup \bar{\mathfrak{a}}_j) \cap \mathfrak{v}') \cup (\mathfrak{u}' \cap \mathfrak{w}')) \cap (\bar{\mathfrak{a}}_i \cup \bar{\mathfrak{a}}_j) \\
&\geqq (\mathfrak{b}_{ik} \cup (\mathfrak{u}' \cup \mathfrak{w}')) \cap (\bar{\mathfrak{a}}_i \cup \bar{\mathfrak{a}}_j) \qquad (\text{since } \mathfrak{v}' \geqq \mathfrak{b}_{ik} \text{ by (18)}) \\
&\geqq (\mathfrak{b}_{ik} \cup \bar{\mathfrak{a}}_k) \cap (\bar{\mathfrak{a}}_i \cup \bar{\mathfrak{a}}_j) \\
&= (\bar{\mathfrak{a}}_i \cup \bar{\mathfrak{a}}_k) \cap (\bar{\mathfrak{a}}_i \cup \bar{\mathfrak{a}}_j) = \bar{\mathfrak{a}}_i,
\end{aligned}$$

i.e., $\mathfrak{u} \oplus \mathfrak{v} \geqq \bar{\mathfrak{a}}_i$. But $\bar{\mathfrak{a}}_i \in L_{ij}$, and $\mathfrak{u} \oplus \mathfrak{v} \in L_{ij}$ by Lemma 7.2. Hence $\mathfrak{u} \oplus \mathfrak{v} = \bar{\mathfrak{a}}_i$ by Lemma 3.3. Now since $\mathfrak{u} \in L_{ij}$ we may define δ by $(\delta)_{ij} = \mathfrak{u}$ (cf. Lemma 6.1), and we have

$$(\delta)_{ij} \oplus (\gamma\beta)_{ij} = u \oplus \mathfrak{v} = \bar{\mathfrak{a}}_i,$$

whence δ has the desired property.

To prove the uniqueness of δ let δ' have the property $\delta' + \gamma\beta = \gamma\beta + \delta' = 0$. Then

$$\begin{aligned}
\delta = \delta + 0 = \delta + (\gamma\beta + \delta') &= (\delta + \gamma\beta) + \delta' \quad (\text{by Theorem 7.2}) \\
&= 0 + \delta' = \delta',
\end{aligned}$$

i.e., $\delta = \delta'$, and the uniqueness is established.

THEOREM 8.3: For every $\beta \in \mathfrak{S}$ there exists a unique element $\delta \in \mathfrak{S}$ such that $\delta + \beta = \beta + \delta = 0$.

PROOF: This is an immediate corollary of Theorem 8.2 with $\gamma = 1$.

DEFINITION 8.2: For every $\beta \in \mathfrak{S}$ we define $-\beta$ to be the unique element δ such that $\delta + \beta = \beta + \delta = 0$ (cf. Theorem 8.3).

COROLLARY: The element δ of Theorem 8.2 is $-(\gamma\beta)$.

THEOREM 8.4: The set $\mathfrak{S}$ together with the operations of addition and multiplication is a ring (with unit).

PROOF: We verify (a), $\cdots$, (g) of Definition 1.1 Properties (a), (b), (c), (d), (e) follow respectively from Theorem 7.1, Theorem 7.2, the first part of Theorem 6.1, Theorem 8.1 (14), Theorem 8.1 (15). The second part of Theorem 6.1 is the statement of property (g). To verify (f), let β, γ be given, and define $\delta \equiv (-\beta) + \gamma$. Then

$$\beta + \delta = \beta + ((-\beta) + \gamma) = (\beta + (-\beta)) + \gamma = 0 + \gamma = \gamma,$$

whence (f) holds also.

The ring $\mathfrak{S}$ will be referred to as the auxiliary ring of L. It depends, of course, on the normalized frame in L as well as on L. It should be noted that the symbols 0, 1 are used to denote elements of L and also to denote entirely different elements of $\mathfrak{S}$. This should, however, cause no confusion.

APPENDIX

The proof of the distributive laws

$$\beta(\gamma + \delta) = \beta\gamma + \beta\delta$$
$$(\gamma + \delta)\beta = \gamma\beta + \delta\beta$$

(cf. Theorem 8.1, (14), (15)), in Chapter VIII should be compared with the proof of these laws given in the work of Veblen and Young already referred to (page 147). That proof is based on the fact that (employing our terminology) for an element $\beta \in \mathfrak{R}$ for which a reciprocal β^{-1} exists, both correspondences

$$(\alpha) \qquad (\gamma)_{ij} \to (\beta\gamma)_{ih} = (\gamma)_{ij} \otimes (\beta)_{jh},$$

$$(\beta) \qquad (\gamma)_{ij} \to (\gamma\beta)_{lj} = (\beta)_{li} \otimes (\gamma)_{ij},$$

are isomorphisms of L_{ij}, L_{ih} and of L_{ij}, L_{lj}, respectively. If β^{-1} does not exist, then it follows that $\beta = 0$ for the case considered by Veblen and Young, since the $\bar{\mathfrak{a}}_i$ are all points. (The reader may verify that the existence of β^{-1} means $(\beta)_{ij} \in L_{ji}$ [as well as $(\beta)_{ij} \in L_{ij}$], i.e.,

$$(*) \qquad (\beta)_{ij} \cap \bar{\mathfrak{a}}_i = 0, \quad (\beta)_{ij} \cup \bar{\mathfrak{a}}_i = \bar{\mathfrak{a}}_i \cup \bar{\mathfrak{a}}_j,$$

that $\beta = 0$ means

$$(**) \qquad (\beta)_{ij} = \bar{\mathfrak{a}}_i$$

[as we have already seen], and that if $\bar{\mathfrak{a}}_i$, $\bar{\mathfrak{a}}_j$ are points then the negation of (**) implies (*).) For the case $\beta = 0$, the distributive laws are trivially true.

In our theory the essential complication is due to the following fact. Since $\mathfrak{S}$ is merely a ring (not necessarily a division algebra), the non-existence of β^{-1} does not imply $\beta = 0$ (i.e., $(\beta)_{ij} \cap \bar{\mathfrak{a}}_i \neq 0$ does not imply $(\beta)_{ij} = \bar{\mathfrak{a}}_i$) since $\bar{\mathfrak{a}}_i$ need not be a point. Hence the transformations (α), (β) may fail to be isomorphisms when $\beta \neq 0$. The transformations (α), (β) are precisely $\mathfrak{u} \to \mathfrak{u}^\alpha$, $\mathfrak{u} \to \mathfrak{u}^\beta$, respectively, of Definition 8.1. We overcame the difficulty just described by showing that $\mathfrak{u} \to \mathfrak{u}^\alpha$, $\mathfrak{u} \to \mathfrak{u}^\beta$ are at least "semi-isomorphisms" (cf. Lemmas 8.1, 8.2), and that this suffices to establish the distributive laws (Lemma 8.4). But for these considerations a careful analysis of the definitions of $\mathfrak{u} \oplus \mathfrak{v}$ was needed — in particular we had to use one definition of $\mathfrak{u} \oplus \mathfrak{v}$ for $\mathfrak{u} \to \mathfrak{u}^\alpha$ (viz., that arising from the $\mathfrak{u} \boxplus_k \mathfrak{v}$ of Definition 7.5) and another for $\mathfrak{u} \to \mathfrak{u}^\beta$ (viz., that arising from the $\mathfrak{u} \diamondplus_k \mathfrak{v}$ of Definition 7.5).

We shall give again both illustrative pictures used in the Appendices to the two preceding chapters to clarify the meaning of $-(\gamma\beta)$ (Theorem 8.2 and the Corollary to Definition 8.2).

Picture 1: Figure 8.1 shows the construction of Theorem 8.2 when the $\bar{\mathfrak{a}}_i$ are points.

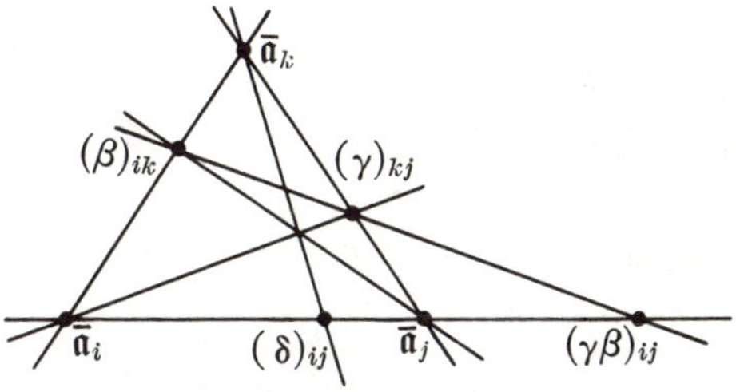

Figure 8.1

One verifies immediately by means of the usual methods of projective geometry that δ is indeed $-(\gamma\beta)$. (Cf. the Appendix to Chapter VI in which the correspondence between our notations and those of Veblen and Young is given.)

Picture 2: We again use the representation of elements of L by means of linear vector-sets (cf. the Appendices to Chapters VI, VII). Then we have

$$\bar{\mathfrak{a}}_i\colon (\xi,\ 0,\ 0), \qquad (\beta)_{ik}\colon (\xi',\ 0,\ -\beta\xi'),$$
$$\bar{\mathfrak{a}}_j\colon (0,\ \eta,\ 0), \qquad (\gamma)_{kj}\colon (0,\ -\gamma\xi'',\ \xi'').$$
$$\bar{\mathfrak{a}}_k\colon (0,\ 0,\ \zeta),$$

It is easily verified that

$$(\delta)_{ij} = \Big(\big(\big((\beta)_{ik} \cup \bar{\mathfrak{a}}_j\big) \cap \big((\gamma)_{kj} \cup \bar{\mathfrak{a}}_i\big)\big) \cup \bar{\mathfrak{a}}_k\Big) \cap (\bar{\mathfrak{a}}_i \cup \bar{\mathfrak{a}}_j)$$

is equal to

$$(-(\gamma\beta))_{ij} \quad : \quad (\xi''',\ \gamma\beta\xi''',\ 0),$$

whence $\delta = -(\gamma\beta)$.

Relations Between the Lattice and its Auxiliary Ring

The preceding chapters have served to define a ring $\mathfrak{S}$ associated with the complemented modular lattice L of order $n \geqq 4$. Our ultimate aim is to prove that the lattice $\bar{R}_{\mathfrak{S}_n}$ is isomorphic to L; in this chapter we shall establish that $\bar{R}_{\mathfrak{S}}$ is isomorphic to $L(0, \bar{\mathfrak{a}}_i)$.

DEFINITION 9.1: For $\beta \in \mathfrak{S}$, and $i, j = 1, \cdots, m$, $i \neq j$, we define $(\beta)_j \equiv ((\beta)_{ij} \cup \bar{\mathfrak{a}}_i) \cap \bar{\mathfrak{a}}_j$.

COROLLARY: The element $(\beta)_j$ is independent of the index i used in its definition, i.e., if $i \neq j$, $k \neq j$, then

$$((\beta)_{ij} \cup \bar{\mathfrak{a}}_i) \cap \bar{\mathfrak{a}}_j = ((\beta)_{kj} \cup \bar{\mathfrak{a}}_k) \cap \bar{\mathfrak{a}}_j.$$

Moreover, $(\beta)_j = (\beta)_h P\begin{pmatrix} h \\ j \end{pmatrix}$.

PROOF: Since $((\beta)_{ij} \cup \bar{\mathfrak{a}}_i) \cap \bar{\mathfrak{a}}_j \leqq \bar{\mathfrak{a}}_j$, the effect on this element of $P\begin{pmatrix} i\ j \\ k\ h \end{pmatrix}$ is the same as that of $P\begin{pmatrix} j \\ h \end{pmatrix}$. Hence

$$(1)\quad [((\beta)_{ij} \cup \bar{\mathfrak{a}}_i) \cap \bar{\mathfrak{a}}_j]\, P\begin{pmatrix} j \\ h \end{pmatrix} = [((\beta)_{ij} \cup \bar{\mathfrak{a}}_i) \cap \bar{\mathfrak{a}}_j]\, P\begin{pmatrix} i\ j \\ k\ h \end{pmatrix}$$

$$= \left((\beta)_{ij} P\begin{pmatrix} i\ j \\ k\ h \end{pmatrix} \cup \bar{\mathfrak{a}}_i P\begin{pmatrix} i\ j \\ k\ h \end{pmatrix}\right) \cap \bar{\mathfrak{a}}_j P\begin{pmatrix} i\ j \\ k\ h \end{pmatrix} = ((\beta)_{kh} \cup \bar{\mathfrak{a}}_k) \cap \bar{\mathfrak{a}}_h.$$

Putting $j = h$ in (1) we obtain the first part of our corollary. Now (1) yields immediately that $(\beta)_j P\begin{pmatrix} j \\ h \end{pmatrix} = (\beta)_h$, i.e., $(\beta)_j = (\beta)_h P\begin{pmatrix} h \\ j \end{pmatrix}$, and the second part is proved.

LEMMA 9.1: Let $\beta, \gamma \in \mathfrak{S}$ and let k be any index. Then $(\gamma)_k \leqq (\beta)_k$ if and only if there exists $\delta \in \mathfrak{S}$ such that $\gamma = \beta\delta$.

PROOF: For the purposes of this proof the symbols $\cup$, $\cap$ will be replaced by $+$, $\cdot$. Let i, j be two distinct indices distinct from k. We prove the converse implication first. Let $\gamma = \beta\delta$. Then

$$\begin{aligned}(\gamma)_k = (\beta\delta)_k &= ((\beta\delta)_{ik} + \bar{\mathfrak{a}}_i) \cdot \bar{\mathfrak{a}}_k \\ &= \big(((\delta)_{ij} + (\beta)_{jk})(\bar{\mathfrak{a}}_i + \bar{\mathfrak{a}}_k) + \bar{\mathfrak{a}}_i\big)\bar{\mathfrak{a}}_k \\ &\leqq \big((\delta)_{ij} + (\beta)_{jk} + \bar{\mathfrak{a}}_i\big)\bar{\mathfrak{a}}_k \\ &\leqq \big(((\beta)_{jk} + \bar{\mathfrak{a}}_j) + \bar{\mathfrak{a}}_i\big)(\bar{\mathfrak{a}}_k + 0) \quad \text{(since } (\delta)_{ij} \leqq \bar{\mathfrak{a}}_i + \bar{\mathfrak{a}}_j).\end{aligned}$$

But

$$\big((\beta)_{jk} + \bar{\mathfrak{a}}_j + \bar{\mathfrak{a}}_k\big)(\bar{\mathfrak{a}}_i + 0) = (\bar{\mathfrak{a}}_j + \bar{\mathfrak{a}}_k)\bar{\mathfrak{a}}_i = 0,$$

whence Part I, Theorem 1.2 applies, yielding

$$\big(((\beta)_{jk} + \bar{\mathfrak{a}}_j) + \bar{\mathfrak{a}}_i\big)(\bar{\mathfrak{a}}_k + 0) = \big((\beta)_{jk} + \bar{\mathfrak{a}}_j\big)\bar{\mathfrak{a}}_k = (\beta)_k.$$

Thus $(\gamma)_k \leqq (\beta)_k$.

Let us suppose now that $(\gamma)_k \leqq (\beta)_k$. We shall prove first that if $\alpha \in \mathfrak{S}$ then $(\alpha)_k + \bar{\mathfrak{a}}_i = (\alpha)_{ik} + \bar{\mathfrak{a}}_i$ for $i \neq k$. In fact,

$$\begin{aligned}(\alpha)_k + \bar{\mathfrak{a}}_i &= ((\alpha)_{ik} + \bar{\mathfrak{a}}_i)\bar{\mathfrak{a}}_k + \bar{\mathfrak{a}}_i \\ &= ((\alpha)_{ik} + \bar{\mathfrak{a}}_i)(\bar{\mathfrak{a}}_k + \bar{\mathfrak{a}}_i) \qquad \text{(by IV)} \\ &= (\alpha)_{ik} + \bar{\mathfrak{a}}_i.\end{aligned}$$

Since $(\gamma)_k \leqq (\beta)_k$, we have then that

(2) $$(\gamma)_{ik} + \bar{\mathfrak{a}}_i \leqq (\beta)_{ik} + \bar{\mathfrak{a}}_i.$$

Now let $\mathfrak{b}$ be inverse to $(\beta)_{ik}\bar{\mathfrak{a}}_i$ in $\bar{\mathfrak{a}}_i$ and define

$$\mathfrak{d} \equiv ((\beta)_{ik} + (\gamma)_{jk})(\mathfrak{b} + \bar{\mathfrak{a}}_j).$$

We shall prove $\mathfrak{d} \in L_{ji}$. Indeed,

$$\begin{aligned}\mathfrak{d}\bar{\mathfrak{a}}_i &= ((\beta)_{ik} + (\gamma)_{jk})(\mathfrak{b} + \bar{\mathfrak{a}}_j)\bar{\mathfrak{a}}_i \\ &= ((\beta)_{ik} + (\gamma)_{jk})(\mathfrak{b} + \bar{\mathfrak{a}}_j\bar{\mathfrak{a}}_i) \qquad \text{(by IV)} \\ &= ((\beta)_{ik} + (\gamma)_{jk})(\mathfrak{b} + 0);\end{aligned}$$

but

$$((\beta)_{ik} + \mathfrak{b})((\gamma)_{jk} + 0) \leqq (\bar{\mathfrak{a}}_i + \bar{\mathfrak{a}}_k)(\gamma)_{jk} = (\bar{\mathfrak{a}}_i + \bar{\mathfrak{a}}_k)(\bar{\mathfrak{a}}_j + \bar{\mathfrak{a}}_k)(\gamma)_{jk} = \bar{\mathfrak{a}}_k(\gamma)_{jk} = 0,$$

whence Part I, Theorem 1.2, applies, yielding

$$\mathfrak{d}\bar{\mathfrak{a}}_i = (\beta)_{ik}\mathfrak{b} = (\beta)_{ik}\bar{\mathfrak{a}}_i\mathfrak{b} = 0.$$

Moreover,

$$\begin{aligned}\mathfrak{d} + \bar{\mathfrak{a}}_i &= ((\beta)_{ik} + (\gamma)_{jk})(\mathfrak{b} + \bar{\mathfrak{a}}_j) + \bar{\mathfrak{a}}_i \\ &= ((\beta)_{ik} + (\gamma)_{jk})(\mathfrak{b} + \bar{\mathfrak{a}}_j) + ((\beta)_{ik}\bar{\mathfrak{a}}_i) + \bar{\mathfrak{a}}_i \\ &= ((\beta)_{ik} + (\gamma)_{jk})(\mathfrak{b} + (\beta)_{ik}\bar{\mathfrak{a}}_i + \bar{\mathfrak{a}}_j) + \bar{\mathfrak{a}}_i \qquad \text{(by IV)} \\ &= ((\beta)_{ik} + (\gamma)_{jk})(\bar{\mathfrak{a}}_i + \bar{\mathfrak{a}}_j) + \bar{\mathfrak{a}}_i \\ &= (\bar{\mathfrak{a}}_i + (\beta)_{ik} + (\gamma)_{jk})(\bar{\mathfrak{a}}_i + \bar{\mathfrak{a}}_j).\end{aligned}$$

Thus $\mathfrak{d} + \bar{\mathfrak{a}}_i \leqq \bar{\mathfrak{a}}_i + \bar{\mathfrak{a}}_j$; but we have also (since $(\gamma)_k \leqq (\beta)_k$)

$$\begin{aligned}
\mathfrak{d} + \bar{\mathfrak{a}}_i &\geqq (\bar{\mathfrak{a}}_i + (\gamma)_{ik} + (\gamma)_{jk})(\bar{\mathfrak{a}}_i + \bar{\mathfrak{a}}_j)\\
&= \left(\bar{\mathfrak{a}}_i + (\gamma)_{jk} P\begin{pmatrix} j\,k \\ i\,k \end{pmatrix} + (\gamma)_{jk}\right)(\bar{\mathfrak{a}}_i + \bar{\mathfrak{a}}_j)\\
&= (\bar{\mathfrak{a}}_i + ((\gamma)_{jk} + \bar{\mathfrak{c}}_{ij})(\bar{\mathfrak{a}}_i + \bar{\mathfrak{a}}_k) + (\gamma)_{jk})(\bar{\mathfrak{a}}_i + \bar{\mathfrak{a}}_j)\\
&= (((\gamma)_{jk} + \bar{\mathfrak{c}}_{ij} + \bar{\mathfrak{a}}_i)(\bar{\mathfrak{a}}_i + \bar{\mathfrak{a}}_k) + (\gamma)_{jk})(\bar{\mathfrak{a}}_i + \bar{\mathfrak{a}}_j) \quad \text{(by IV)}\\
&= ((\gamma)_{jk} + \bar{\mathfrak{c}}_{ij} + \bar{\mathfrak{a}}_i)(\bar{\mathfrak{a}}_i + \bar{\mathfrak{a}}_k + (\gamma)_{jk})(\bar{\mathfrak{a}}_i + \bar{\mathfrak{a}}_j) \quad \text{(by IV)}\\
&= ((\gamma)_{jk} + \bar{\mathfrak{a}}_j + \bar{\mathfrak{a}}_i)(\bar{\mathfrak{a}}_i + \bar{\mathfrak{a}}_k + \bar{\mathfrak{a}}_j)(\bar{\mathfrak{a}}_i + \bar{\mathfrak{a}}_j)\\
&= \bar{\mathfrak{a}}_i + \bar{\mathfrak{a}}_j.
\end{aligned}$$

These considerations show that $\mathfrak{d} + \bar{\mathfrak{a}}_i = \bar{\mathfrak{a}}_i + \bar{\mathfrak{a}}_j$, and when combined with the preceding result, $\mathfrak{d}\bar{\mathfrak{a}}_i = 0$, yield that $\mathfrak{d} \in L_{ji}$. Hence there exists $\delta \in \mathfrak{S}$ such that $(\delta)_{ji} = \mathfrak{d}$. It will be shown that $\gamma = \beta\delta$.

Now

$$\begin{aligned}
(\delta)_{ji} + (\beta)_{ik} &= ((\beta)_{ik} + (\gamma)_{jk})(\mathfrak{b} + \bar{\mathfrak{a}}_j) + (\beta)_{ik}\\
&= ((\beta)_{ik} + (\gamma)_{jk})((\beta)_{ik} + \mathfrak{b} + \bar{\mathfrak{a}}_j) \quad \text{(by IV)}\\
&= ((\beta)_{ik} + (\gamma)_{jk})((\beta)_{ik} + (\beta)_{ik}\bar{\mathfrak{a}}_i + \mathfrak{b} + \bar{\mathfrak{a}}_j)\\
&= ((\beta)_{ik} + (\gamma)_{jk})((\beta)_{ik} + \bar{\mathfrak{a}}_i + \bar{\mathfrak{a}}_j)\\
&\geqq ((\beta)_{ik} + (\gamma)_{jk})((\gamma)_{ik} + \bar{\mathfrak{a}}_i + \bar{\mathfrak{a}}_j) \quad \text{(by (2))}\\
&\geqq ((\beta)_{ik} + (\gamma)_{jk})((\gamma)_{ik} + \bar{\mathfrak{c}}_{ij}).
\end{aligned}$$

But $(\gamma)_{ik} + \bar{\mathfrak{c}}_{ij} \geqq ((\gamma)_{ik} + \bar{\mathfrak{c}}_{ij}) \cdot (\bar{\mathfrak{a}}_j + \bar{\mathfrak{a}}_k) = (\gamma)_{jk}$,
whence $(\delta)_{ji} + (\beta)_{ik} \geqq ((\beta)_{ik} + (\gamma)_{jk})(\gamma)_{jk} = (\gamma)_{jk}$.
Therefore

$$(\beta\delta)_{jk} = ((\delta)_{ji} + (\beta)_{ik})(\bar{\mathfrak{a}}_j + \bar{\mathfrak{a}}_k) \geqq (\gamma)_{jk};$$

but $(\beta\delta)_{jk}, (\gamma)_{jk} \in L_{jk}$, whence $(\beta\delta)_{jk} = (\gamma)_{jk}$ (by Lemma 3.3), and we have $\beta\delta = \gamma$.

Lemma 9.2: For every pair of inverses $\mathfrak{u}, \mathfrak{u}'$ in $\bar{\mathfrak{a}}_k$ there exists $\beta \in \mathfrak{S}$ such that $\mathfrak{u} = (\beta)_k$.

Proof: Let i be an index distinct from k, and define

$$\mathfrak{b}' \equiv ((\mathfrak{u} \cup \bar{\mathfrak{a}}_i) \cap \bar{\mathfrak{c}}_{ik}) \cup \mathfrak{u}'.$$

Now

$$(\mathfrak{u} \cup \bar{\mathfrak{a}}_i) \cap \bar{\mathfrak{c}}_{ik} \cap \bar{\mathfrak{a}}_i \leqq \bar{\mathfrak{c}}_{ik} \cap \bar{\mathfrak{a}}_i = 0;$$

$$(((\mathfrak{u} \cup \bar{\mathfrak{a}}_i) \cap \bar{\mathfrak{c}}_{ik}) \cup \bar{\mathfrak{a}}_i) \cup \mathfrak{u}' \leqq (\mathfrak{u} \cup \bar{\mathfrak{a}}_i) \cap (\mathfrak{u}' \cup 0) = \mathfrak{u} \cap \mathfrak{u}' = 0,$$

the last equality but one holding by Part I, Theorem 1.2, since $(\mathfrak{u} \cup \mathfrak{u}') \cap (\bar{\mathfrak{a}}_i \cup 0) = \bar{\mathfrak{a}}_k \cap \bar{\mathfrak{a}}_i = 0$. Hence $\big((\mathfrak{u} \cup \bar{\mathfrak{a}}_i) \cap \bar{\mathfrak{c}}_{ik}, \bar{\mathfrak{a}}_i, \mathfrak{u}'\big)\perp$, and

$$\mathfrak{b}' \cap \bar{\mathfrak{a}}_i = \big(\big((\mathfrak{u} \cup \bar{\mathfrak{a}}_i) \cap \bar{\mathfrak{c}}_{ik}\big) \cup \mathfrak{u}'\big) \cap \bar{\mathfrak{a}}_i = 0.$$

Moreover,

$$\begin{aligned} \mathfrak{b}' \cup \bar{\mathfrak{a}}_i &= \big((\mathfrak{u} \cup \bar{\mathfrak{a}}_i) \cap \bar{\mathfrak{c}}_{ik}\big) \cup \mathfrak{u}' \cup \bar{\mathfrak{a}}_i \\ &= \big((\mathfrak{u} \cup \bar{\mathfrak{a}}_i) \cap (\bar{\mathfrak{c}}_{ik} \cup \bar{\mathfrak{a}}_i)\big) \cup \mathfrak{u}' && \text{(by IV)} \\ &= \big((\mathfrak{u} \cup \bar{\mathfrak{a}}_i) \cap (\bar{\mathfrak{a}}_i \cup \bar{\mathfrak{a}}_k)\big) \cup \mathfrak{u}' \\ &= \mathfrak{u} \cup \bar{\mathfrak{a}}_i \cup \mathfrak{u}' && \text{(since } \mathfrak{u},\ \bar{\mathfrak{a}}_i \leqq \bar{\mathfrak{a}}_i \cup \bar{\mathfrak{a}}_k) \\ &= \bar{\mathfrak{a}}_k \cup \bar{\mathfrak{a}}_i. \end{aligned}$$

Thus $\mathfrak{b}'$ is inverse to $\bar{\mathfrak{a}}_i$ in $\bar{\mathfrak{a}}_k \cup \bar{\mathfrak{a}}_i$, i.e., $\mathfrak{b}' \in L_{ki}$, and there exists $\beta \in \mathfrak{S}$ with $(\beta)_{ki} = \mathfrak{b}' = \big((\mathfrak{u} \cup \bar{\mathfrak{a}}_i) \cap \bar{\mathfrak{c}}_{ik}\big) \cup \mathfrak{u}'$. Therefore

$$\begin{aligned} (\beta)_i = \big((\beta)_{ki} \cup \bar{\mathfrak{a}}_k\big) \cap \bar{\mathfrak{a}}_i &= \big(\big((\mathfrak{u} \cup \bar{\mathfrak{a}}_i) \cap \bar{\mathfrak{c}}_{ik}\big) \cup \mathfrak{u}' \cup \bar{\mathfrak{a}}_k\big) \cap \bar{\mathfrak{a}}_i \\ &= \big(\big((\mathfrak{u} \cup \bar{\mathfrak{a}}_i) \cap \bar{\mathfrak{c}}_{ik}\big) \cup \bar{\mathfrak{a}}_k\big) \cap \bar{\mathfrak{a}}_i. \end{aligned}$$

Now $(\bar{\mathfrak{a}}_k, \bar{\mathfrak{a}}_i, \bar{\mathfrak{c}}_{ik})\mathscr{C}$, and $\mathfrak{u} \leqq \bar{\mathfrak{a}}_k$, whence Lemma 4.1 applies and yields that $(\beta)_i \sim \mathfrak{u} \pmod{\bar{\mathfrak{c}}_{ik}}$, i.e., $(\beta)_i = \mathfrak{u}P\begin{pmatrix} k \\ i \end{pmatrix}$, whence $\mathfrak{u} = (\beta)_i P\begin{pmatrix} i \\ k \end{pmatrix} = (\beta)_k$ by the Corollary to Definition 9.1.

(Note: It is easy to verify that the β constructed in Lemma 9.2 is idempotent, see Theorem 9.3. - Ed.).

THEOREM 9.1: The transformation $\beta \to \mathfrak{u} = (\beta)_k$ carries elements β of $\mathfrak{S}$ into elements $\mathfrak{u} \leqq \bar{\mathfrak{a}}_k$; its range is the entire set $L(0, \bar{\mathfrak{a}}_k)$. If $\beta \to \mathfrak{u}$, $\gamma \to \mathfrak{v}$, then $\mathfrak{v} \leqq \mathfrak{u}$ is equivalent to the existence of $\delta \in \mathfrak{S}$ with $\gamma = \beta\delta$.

PROOF: By Definition 9.1, $(\beta)_k \in L(0, \bar{\mathfrak{a}}_k)$. Moreover, if $\mathfrak{u} \in L(0, \bar{\mathfrak{a}}_k)$, let $\mathfrak{u}'$ be inverse to $\mathfrak{u}$ in $\bar{\mathfrak{a}}_k$; by Lemma 9.2, there exists $\beta \in \mathfrak{S}$ with $\mathfrak{u} = (\beta)_k$. This proves the first part. The second part is a restatement of Lemma 9.1.

LEMMA 9.3: If $\mathfrak{u}, \mathfrak{u}'$ are inverses in $\bar{\mathfrak{a}}_k$, there exist $\beta, \beta' \in \mathfrak{S}$ with $(\beta)_k = \mathfrak{u}$, $(\beta')_k = \mathfrak{u}'$, $\beta + \beta' = 1$.

PROOF: Let $i \neq k$. The proof of Lemma 9.2 shows that corresponding to the pair $\mathfrak{u}, \mathfrak{u}'$ there exists $\beta \in \mathfrak{S}$ with $(\beta)_k = \mathfrak{u}$ and

$$(3) \qquad (\beta)_{ki} = \big((\mathfrak{u} \cup \bar{\mathfrak{a}}_i) \cap \bar{\mathfrak{c}}_{ik}\big) \cup \mathfrak{u}',$$

and to the pair $\mathfrak{u}', \mathfrak{u}$ there corresponds $\beta' \in \mathfrak{S}$ with $(\beta')_k = \mathfrak{u}'$ and

$$(4) \qquad (\beta')_{ki} = \big((\mathfrak{u}' \cup \bar{\mathfrak{a}}_i) \cap \bar{\mathfrak{c}}_{ik}\big) \cup \mathfrak{u}.$$

We shall prove that $\beta + \beta' = 1$ by showing that $(\beta + \beta')_{ki} = (1)_{ki}$. Let j be an index distinct from k, i (existent since $n \geqq 4$). In order to calculate

$(\beta+\beta')_{ki} = (\beta)_{ki} \oplus (\beta')_{ki}$, we shall use formula (10) of Definition 7.5 with i, j, k there replaced by k, i, j. Thus

$$(\beta+\beta')_{ki} = \Big(\big(\big(((\mathfrak{u} \cup \bar{\mathfrak{a}}_i) \cap \bar{\mathfrak{c}}_{ik}) \cup \mathfrak{u}' \cup \bar{\mathfrak{c}}_{jk}\big) \cap (\bar{\mathfrak{a}}_i \cup \bar{\mathfrak{a}}_j)\big) \cup \big(\big(((\mathfrak{u}' \cup \bar{\mathfrak{a}}_i) \cap \bar{\mathfrak{c}}_{ik}) \cup \mathfrak{u} \cup \bar{\mathfrak{a}}_j\big) \cap (\bar{\mathfrak{a}}_i \cup \bar{\mathfrak{c}}_{jk})\big)\Big) \cap (\bar{\mathfrak{a}}_k \cup \bar{\mathfrak{a}}_i).$$

Now $(\bar{\mathfrak{a}}_i, \bar{\mathfrak{a}}_k, \bar{\mathfrak{c}}_{ik})\mathscr{C}$, whence there exist perspective images $\mathfrak{u}_1, \mathfrak{u}_1' \leqq \bar{\mathfrak{a}}_i$ and $\mathfrak{u}_2, \mathfrak{u}_2' \leqq \bar{\mathfrak{c}}_{ik}$ of $\mathfrak{u}$, $\mathfrak{u}'$, respectively. Hence $\mathfrak{u}_1$, $\mathfrak{u}_1'$ are inverses in $\bar{\mathfrak{a}}_i$, and $\mathfrak{u}_2$, $\mathfrak{u}_2'$ are inverses in $\bar{\mathfrak{c}}_{ik}$. Moreover,

$$\mathfrak{u}_2 = (\mathfrak{u} \cup \bar{\mathfrak{a}}_i) \cap \bar{\mathfrak{c}}_{ik}, \quad \mathfrak{u}_2' = (\mathfrak{u}' \cup \bar{\mathfrak{a}}_i) \cap \bar{\mathfrak{c}}_{ik},$$

whence

$$(\beta+\beta')_{ki} = \Big(\big((\mathfrak{u}_2 \cup \mathfrak{u}' \cup \bar{\mathfrak{c}}_{jk}) \cap (\bar{\mathfrak{a}}_i \cup \bar{\mathfrak{a}}_j)\big) \cup \big((\mathfrak{u}_2' \cup \mathfrak{u} \cup \bar{\mathfrak{a}}_j) \cap (\bar{\mathfrak{a}}_i \cup \bar{\mathfrak{c}}_{jk})\big)\Big) \cap (\bar{\mathfrak{a}}_k \cup \bar{\mathfrak{a}}_i).$$

Thus

$$(\beta+\beta')_{ki} \geqq \Big(\big((\mathfrak{u}_2 \cup \mathfrak{u}' \cup \bar{\mathfrak{c}}_{jk}) \cap (\bar{\mathfrak{a}}_i \cup \bar{\mathfrak{a}}_j)\big) \cup \big((\mathfrak{u} \cup \bar{\mathfrak{a}}_j) \cap \bar{\mathfrak{c}}_{jk}\big)\Big) \cap (\bar{\mathfrak{a}}_k \cup \bar{\mathfrak{a}}_i);$$

but $(\bar{\mathfrak{a}}_j \cup \mathfrak{u}) \cap \bar{\mathfrak{c}}_{jk} \leqq \bar{\mathfrak{c}}_{jk} \leqq \mathfrak{u}_2 \cup \mathfrak{u} \cup \bar{\mathfrak{c}}_{jk}$, whence

$$\begin{aligned}
(\beta+\beta')_{ki} &\geqq (\mathfrak{u}_2 \cup \mathfrak{u}' \cup \bar{\mathfrak{c}}_{jk}) \cap \big(\bar{\mathfrak{a}}_i \cup \bar{\mathfrak{a}}_j \cup ((\bar{\mathfrak{a}}_j \cup \mathfrak{u}) \cap \bar{\mathfrak{c}}_{jk})\big) \cap (\bar{\mathfrak{a}}_i \cup \bar{\mathfrak{a}}_k) \quad \text{(by IV)}\\
&= (\mathfrak{u}_2 \cup \mathfrak{u}' \cup \bar{\mathfrak{c}}_{jk}) \cap \big(\bar{\mathfrak{a}}_i \cup ((\bar{\mathfrak{a}}_j \cup \mathfrak{u}) \cap (\bar{\mathfrak{a}}_j \cup \bar{\mathfrak{c}}_{jk}))\big) \cap (\bar{\mathfrak{a}}_i \cup \bar{\mathfrak{a}}_k) \quad \text{(by IV)}\\
&= (\mathfrak{u}_2 \cup \mathfrak{u}' \cup \bar{\mathfrak{c}}_{jk}) \cap \big(\bar{\mathfrak{a}}_i \cup ((\bar{\mathfrak{a}}_j \cup \mathfrak{u}) \cap (\bar{\mathfrak{a}}_j \cup \bar{\mathfrak{a}}_k))\big) \cap (\bar{\mathfrak{a}}_i \cup \bar{\mathfrak{a}}_k)\\
&= (\mathfrak{u}_2 \cup \mathfrak{u}' \cup \bar{\mathfrak{c}}_{jk}) \cap (\bar{\mathfrak{a}}_i \cup \bar{\mathfrak{a}}_j \cup \mathfrak{u}) \cap (\bar{\mathfrak{a}}_i \cup \bar{\mathfrak{a}}_k)\\
&= (\mathfrak{u}_2 \cup \mathfrak{u}' \cup \bar{\mathfrak{c}}_{jk}) \cap (\mathfrak{u} \cup \bar{\mathfrak{a}}_i \cup \bar{\mathfrak{a}}_j) \cap (\bar{\mathfrak{a}}_i \cup \bar{\mathfrak{a}}_k)\\
&= (\mathfrak{u}_2 \cup \mathfrak{u}' \cup \bar{\mathfrak{c}}_{jk}) \cap (\mathfrak{u}_2 \cup \bar{\mathfrak{a}}_i \cup \bar{\mathfrak{a}}_j) \cap (\bar{\mathfrak{a}}_i \cup \bar{\mathfrak{a}}_k)\\
&\qquad\qquad \text{(since } \mathfrak{u}_2 \sim \mathfrak{u} \ (\text{mod } \bar{\mathfrak{a}}_i) \text{ in } \bar{\mathfrak{c}}_{ik} \cup \bar{\mathfrak{a}}_k)\\
&\geqq \mathfrak{u}_2 \cap \mathfrak{u}_2 \cap (\bar{\mathfrak{a}}_i \cup \bar{\mathfrak{a}}_k) = \mathfrak{u}_2 \qquad \text{(since } \mathfrak{u}_2 \leqq \bar{\mathfrak{c}}_{ik} \leqq \bar{\mathfrak{a}}_i \cup \bar{\mathfrak{a}}_k),
\end{aligned}$$

and we have proved that $(\beta+\beta')_{ki} \geqq \mathfrak{u}_2$. Now in a similar manner we see (by interchanging the roles of $\mathfrak{u}$, $\mathfrak{u}_1$, $\mathfrak{u}_2$, and $\mathfrak{u}'$, $\mathfrak{u}_1'$, $\mathfrak{u}_2'$, and also those of $\bar{\mathfrak{a}}_j$ and $\bar{\mathfrak{c}}_{jk}$) that $(\beta+\beta')_{ki} \geqq \mathfrak{u}_2'$. Consequently $(\beta+\beta')_{ki} \geqq \mathfrak{u}_2 \cup \mathfrak{u}_2' = \bar{\mathfrak{c}}_{ik}$; since $(\beta+\beta')_{ki}$ and $\bar{\mathfrak{c}}_{ik} = \bar{\mathfrak{c}}_{ki}$ are both in L_{ki}, this yields by Lemma 3.3 that $(\beta+\beta')_{ki} = \bar{\mathfrak{c}}_{ki}$. But by Theorem 6.1, $\bar{\mathfrak{c}}_{ki} = (1)_{ki}$, whence $(\beta+\beta')_{ki} = (1)_{ki}$, and our proof is complete.

LEMMA 9.4: Let k be given and $\beta, \gamma \in \mathfrak{S}$. Then

(a) $(\beta)_k = (\gamma)_k$ if and only if $(\beta)_{\underline{r}} = (\gamma)_{\underline{r}}$,
(b) $(\beta)_k \geqq (\gamma)_k$ if and only if $(\beta)_{\underline{r}} \geqq (\gamma)_{\underline{r}}$.

(Note that the symbol $(\beta)_{\underline{r}}$ will denote the principal right ideal generated by β and $(\beta)_{\underline{l}}$ the principal left ideal of β; we underline the r and l to avoid confusion with the symbol $(\beta)_j$ denoting an element of L. We shall use, as in (b) the notation $\geqq$ for the set-inclusion relation $\supset$.)

PROOF: (b) By Lemma 9.1, $(\beta)_k \geqq (\gamma)_k$ is equivalent to the existence of $\delta \in \mathfrak{S}$ with $\gamma = \beta\delta$; this is equivalent to $\gamma \in (\beta)_{\underline{r}}$, i.e., to $(\gamma)_{\underline{r}} \leqq (\beta)_{\underline{r}}$.

(a) By combining (b) and the statement obtained from (b) by interchanging β, γ, we see that the conjunction of $(\beta)_k \geqq (\gamma)_k$ and $(\gamma)_k \geqq (\beta)_k$, i.e., $(\beta)_k = (\gamma)_k$ is equivalent to the conjunction of $(\beta)_{\underline{r}} \geqq (\gamma)_{\underline{r}}$ and $(\gamma)_{\underline{r}} \geqq (\beta)_{\underline{r}}$, i.e., to $(\beta)_{\underline{r}} = (\gamma)_{\underline{r}}$.

THEOREM 9.2: The ring $\mathfrak{S}$ is regular. The set $\bar{R}_{\mathfrak{S}}$ of all principal right ideals in $\mathfrak{S}$ is a lattice-isomorphic to the lattice $L(0, \bar{\mathfrak{a}}_k)$, the correspondence being defined by $(\beta)_{\underline{r}} \rightleftarrows (\beta)_k$.

PROOF: Consider a principal right ideal $\mathfrak{a} \in \bar{R}_{\mathfrak{S}}$. Then $\mathfrak{a}$ is of the form $\mathfrak{a} = (\gamma)_{\underline{r}}$ with $\gamma \in \mathfrak{S}$ and $\mathfrak{u} \equiv (\gamma)_k \leqq \bar{\mathfrak{a}}_k$. Let $\mathfrak{u}'$ be inverse to $\mathfrak{u}$ in $\bar{\mathfrak{a}}_k$. By Lemma 9.3 there exist β, $\beta' \in \mathfrak{S}$ with $(\beta)_k = \mathfrak{u}$, $(\beta')_k = \mathfrak{u}'$, $\beta + \beta' = 1$. We show now that the intersection $(\beta)_{\underline{r}} \cap (\beta')_{\underline{r}}$ is (0). Let $\delta \in (\beta)_{\underline{r}} \cap (\beta')_{\underline{r}}$. Then $\delta \in (\beta)_{\underline{r}}, (\beta')_{\underline{r}}$, i.e., $(\delta)_{\underline{r}} \leqq (\beta)_{\underline{r}}, (\beta')_{\underline{r}}$. By Lemma 9.4 (b), $(\delta)_k \leqq (\beta)_k$, $(\beta')_k$. Thus $(\delta)_k \leqq \mathfrak{u} \cap \mathfrak{u}' = 0$, whence $(\delta)_k = 0$. But $(0)_k = ((0)_{ik} \cup \bar{\mathfrak{a}}_i) \cap \bar{\mathfrak{a}}_k = (\bar{\mathfrak{a}}_i \cup \bar{\mathfrak{a}}_i) \cap \bar{\mathfrak{a}}_k = \bar{\mathfrak{a}}_i \cap \bar{\mathfrak{a}}_k = 0$, whence $(\delta)_k = (0)_k$, i.e., $\delta = 0$. This proves that $(\beta)_{\underline{r}} \cap (\beta')_{\underline{r}} = (0)$. Now since $\beta + \beta' = 1$, we have $1 \in (\beta)_{\underline{r}} \cup (\beta')_{\underline{r}}$, whence $(1)_{\underline{r}} = (\beta)_{\underline{r}} \cup (\beta')_{\underline{r}}$. Consequently $\mathfrak{a} = (\beta)_{\underline{r}}$ and $(\beta')_{\underline{r}}$ are inverse principal right ideals, and $\mathfrak{S}$ is regular by Definition 2.2.

By Theorem 2.4, $\bar{R}_{\mathfrak{S}}$ is a lattice. But by Lemma 9.4 (a) the correspondence $(\beta)_{\underline{r}} \rightleftarrows (\beta)_k$ between $\bar{R}_{\mathfrak{S}}$ and the set of all elements $(\beta)_k$, i.e., $L(0, \bar{\mathfrak{a}}_k)$, is one-to-one, and Lemma 9.4 (b) states that this correspondence is an isomorphism with respect to the relations which partially order these sets; therefore, since $L(0, \bar{\mathfrak{a}}_k)$ is a complemented modular lattice, so is $\bar{R}_{\mathfrak{S}}$, and the correspondence is a lattice-isomorphism.

THEOREM 9.3: Let $\mathfrak{u}$, $\mathfrak{u}'$ be inverses in $\bar{\mathfrak{a}}_k$. Then there exists one and only one idempotent $\beta = \beta(\mathfrak{u}, \mathfrak{u}') \in \mathfrak{S}$ such that

(5) $$(\beta)_k = \mathfrak{u}, \quad (1 - \beta)_k = \mathfrak{u}';$$

moreover,

(6) $$(\beta)_{ki} = ((\mathfrak{u} \cup \bar{\mathfrak{a}}_i) \cap \bar{\mathfrak{c}}_{ik}) \cup \mathfrak{u}', \quad (1 - \beta)_{ki} = ((\mathfrak{u}' \cup \bar{\mathfrak{a}}_i) \cap \bar{\mathfrak{c}}_{ik}) \cup \mathfrak{u}.$$

All idempotents β can be obtained in this manner.

PROOF: Let $\mathfrak{a}$, $\mathfrak{b}$ correspond respectively to $\mathfrak{u}$, $\mathfrak{u}'$ under the transformation $(\gamma)_k \to (\gamma)_{\underline{r}}$. The existence of a unique idempotent β with $(\beta)_{\underline{r}} = \mathfrak{a}$, $(1 - \beta)_{\underline{r}} = \mathfrak{b}$, i.e., such that (5) holds, follows from Theorem 2.1. We now establish (6) by proving that if β has the properties (6), then β has the properties (5) and is idempotent. Let β have the property $(\beta)_{ki} = ((\mathfrak{u} \cup \bar{\mathfrak{a}}_i) \cap \bar{\mathfrak{c}}_{ik}) \cup \mathfrak{u}'$. Then if β' has the property $(\beta')_{ki} = ((\mathfrak{u}' \cup \bar{\mathfrak{a}}_i) \cap \bar{\mathfrak{c}}_{ik}) \cup \mathfrak{u}$ it follows from the proof of Lemma 9.3 that $\beta + \beta' = 1$, i.e., $\beta' = 1 - \beta$

and that $(\beta)_k = \mathfrak{u}$, $(\beta')_k = \mathfrak{u}'$, i.e., that (5) holds. Now since $\mathfrak{u} \cap \mathfrak{u}' = 0$, $(\beta)_{\underline{r}} \cap (\beta')_{\underline{r}} = (0)$. And since $\beta' = 1 - \beta$, $\beta - \beta^2 = (1 - \beta)\beta = \beta(1 - \beta)$, we see that $\beta - \beta^2$ belongs to both $(\beta)_{\underline{r}}$ and $(\beta')_{\underline{r}}$, i.e., that $\beta - \beta^2 = 0$. Hence $\beta = \beta^2$, and β is idempotent. To prove the last statement, let β be idempotent. Then $\mathfrak{u} \equiv (\beta)_k$, $\mathfrak{u}' \equiv (1 - \beta)_k$ are inverses in $\bar{\mathfrak{a}}_k$, and give rise to an idempotent $\beta' \in \mathfrak{S}$ satisfying (5) and (6). Hence $(\beta)_{\underline{r}} = (\beta')_{\underline{r}}$, $(1 - \beta)_{\underline{r}} = (1 - \beta')_{\underline{r}}$, whence $\beta = \beta'$ by Theorem 2.1, and β arises in the manner described.

Corollary: The correspondence $(\mathfrak{u}, \mathfrak{u}') \to \beta(\mathfrak{u}, \mathfrak{u}')$ is a one-to-one correspondence between all pairs of inverses in $\bar{\mathfrak{a}}_k$ and all idempotents $\beta \in \mathfrak{S}$. Moreover, $1 - \beta(\mathfrak{u}, \mathfrak{u}') = \beta(\mathfrak{u}', \mathfrak{u})$.

Proof: This is obvious from Theorem 9.3.

(Note: $(\beta)_{ki} \cap \bar{\mathfrak{a}}_k = \mathfrak{u}' = (1 - \beta)_k$, from (6) and (5). - Ed.)

Lemma 9.5: Let $\mathfrak{u}, \mathfrak{u}'$ be inverses in $\bar{\mathfrak{a}}_k$ and $\beta = \beta(\mathfrak{u}, \mathfrak{u}')$ their corresponding idempotent in $\mathfrak{S}$. Then for every $\xi \in \mathfrak{S}$, $j \neq k$

$$(\beta\xi)_{jk} = ((\xi)_{jk} \cup \mathfrak{u}') \cap (\bar{\mathfrak{a}}_j \cup \mathfrak{u}), \tag{7}$$

$$(\xi\beta)_{kj} = ((\xi)_{kj} \cap (\bar{\mathfrak{a}}_j \cup \mathfrak{u})) \cup \mathfrak{u}'. \tag{8}$$

Proof: For the purposes of this proof the symbols $\cup$, $\cap$ will be replaced by $+$, $\cdot$. Let $i \neq j, k$. Then

$$\begin{aligned}
(\beta\xi)_{jk} &= \big((\beta\xi)_{ji} + \bar{\mathfrak{c}}_{ik}\big)\,(\bar{\mathfrak{a}}_j + \bar{\mathfrak{a}}_k) \\
&= \big(((\xi)_{jk} + (\beta)_{ki})\,(\bar{\mathfrak{a}}_j + \bar{\mathfrak{a}}_i) + \bar{\mathfrak{c}}_{ik}\big)\,(\bar{\mathfrak{a}}_j + \bar{\mathfrak{a}}_k) \\
&= \big(((\xi)_{jk} + (\mathfrak{u} + \bar{\mathfrak{a}}_i)\bar{\mathfrak{c}}_{ik} + \mathfrak{u}')\,(\bar{\mathfrak{a}}_j + \bar{\mathfrak{a}}_i) + \bar{\mathfrak{c}}_{ik}\big)\,(\bar{\mathfrak{a}}_j + \bar{\mathfrak{a}}_k) \\
&= \big(((\xi)_{jk} + (\mathfrak{u} + \bar{\mathfrak{a}}_i)\bar{\mathfrak{c}}_{ik} + \mathfrak{u}')\,(\bar{\mathfrak{a}}_j + \bar{\mathfrak{a}}_i) + (\mathfrak{u} + \bar{\mathfrak{a}}_i)\bar{\mathfrak{c}}_{ik} + \bar{\mathfrak{c}}_{ik}\big)\,(\bar{\mathfrak{a}}_j + \bar{\mathfrak{a}}_k) \\
&= \big(((\xi)_{jk} + (\mathfrak{u} + \bar{\mathfrak{a}}_i)\bar{\mathfrak{c}}_{ik} + \mathfrak{u}')\,(\bar{\mathfrak{a}}_j + \bar{\mathfrak{a}}_i + (\mathfrak{u} + \bar{\mathfrak{a}}_i)\bar{\mathfrak{c}}_{ik}) + \bar{\mathfrak{c}}_{ik}\big)\,(\bar{\mathfrak{a}}_j + \bar{\mathfrak{a}}_k),
\end{aligned}$$

the last equality holding by IV. But

$$\begin{aligned}
\bar{\mathfrak{a}}_j + \bar{\mathfrak{a}}_i + (\mathfrak{u} + \bar{\mathfrak{a}}_i)\bar{\mathfrak{c}}_{ik} &= \bar{\mathfrak{a}}_j + (\bar{\mathfrak{a}}_i + \mathfrak{u})(\bar{\mathfrak{a}}_i + \bar{\mathfrak{c}}_{ik}) && \text{(by IV)} \\
&= \bar{\mathfrak{a}}_j + (\bar{\mathfrak{a}}_i + \mathfrak{u})(\bar{\mathfrak{a}}_i + \bar{\mathfrak{a}}_k) \\
&= \bar{\mathfrak{a}}_j + \bar{\mathfrak{a}}_i + \mathfrak{u} && \text{(since } \bar{\mathfrak{a}}_i,\ \mathfrak{u} \leqq \bar{\mathfrak{a}}_i + \bar{\mathfrak{a}}_k\text{)},
\end{aligned}$$

whence

$$\begin{aligned}
(\beta\xi)_{jk} &= \big(((\xi)_{jk} + (\mathfrak{u} + \bar{\mathfrak{a}}_i)\bar{\mathfrak{c}}_{ik} + \mathfrak{u}')\,(\bar{\mathfrak{a}}_i + \bar{\mathfrak{a}}_j + \mathfrak{u}) + \bar{\mathfrak{c}}_{ik}\big)\,(\bar{\mathfrak{a}}_j + \bar{\mathfrak{a}}_k) \\
&= \big(((\xi)_{jk} + \mathfrak{u}')\,(\bar{\mathfrak{a}}_i + \bar{\mathfrak{a}}_j + \mathfrak{u}) + (\mathfrak{u} + \bar{\mathfrak{a}}_i)\bar{\mathfrak{c}}_{ik} + \bar{\mathfrak{c}}_{ik}\big)\,(\bar{\mathfrak{a}}_j + \bar{\mathfrak{a}}_k) \quad \text{(by IV)} \\
&= \big(((\xi)_{jk} + \mathfrak{u}')\,(\bar{\mathfrak{a}}_i + \bar{\mathfrak{a}}_j + \mathfrak{u}) + \bar{\mathfrak{c}}_{ik}\big)\,(\bar{\mathfrak{a}}_j + \bar{\mathfrak{a}}_k).
\end{aligned}$$

Now $\big(((\xi)_{jk} + \mathfrak{u}') + (\bar{\mathfrak{a}}_j + \mathfrak{u})\big)\,(0 + \bar{\mathfrak{a}}_i) \leqq (\bar{\mathfrak{a}}_j + \bar{\mathfrak{a}}_k)\bar{\mathfrak{a}}_i = 0$, whence by Part I, Theorem 1.2,

$$((\xi)_{jk} + \mathfrak{u}')(\bar{\mathfrak{a}}_i + \bar{\mathfrak{a}}_j + \mathfrak{u}) = \bar{\mathfrak{a}}_i \cdot 0 + ((\xi)_{jk} + \mathfrak{u}')(\bar{\mathfrak{a}}_j + \mathfrak{u})$$
$$= ((\xi)_{jk} + \mathfrak{u}')(\bar{\mathfrak{a}}_j + \mathfrak{u}),$$

and

$$(\beta\xi)_{jk} = \big(((\xi)_{jk} + \mathfrak{u}')(\bar{\mathfrak{a}}_j + \mathfrak{u}) + \bar{\mathfrak{c}}_{ik}\big)(\bar{\mathfrak{a}}_j + \bar{\mathfrak{a}}_k)$$
$$= \big((\xi)_{jk} + \mathfrak{u}'\big)(\bar{\mathfrak{a}}_j + \mathfrak{u}) + \bar{\mathfrak{c}}_{ik}(\bar{\mathfrak{a}}_j + \bar{\mathfrak{a}}_k) \qquad \text{(by IV)}$$
$$= \big((\xi)_{jk} + \mathfrak{u}'\big)(\bar{\mathfrak{a}}_j + \mathfrak{u}),$$

since $\bar{\mathfrak{c}}_{ik}(\bar{\mathfrak{a}}_j + \bar{\mathfrak{a}}_k) = \bar{\mathfrak{c}}_{ik}(\bar{\mathfrak{a}}_i + \bar{\mathfrak{a}}_k)(\bar{\mathfrak{a}}_j + \bar{\mathfrak{a}}_k) = \bar{\mathfrak{c}}_{ik}\bar{\mathfrak{a}}_k = 0$. Thus (7) is proved.

To prove (8), we have

$$(\xi\beta)_{kj} = \big((\beta)_{ki} + (\xi)_{ij}\big)(\bar{\mathfrak{a}}_k + \bar{\mathfrak{a}}_j)$$
$$= \big((\mathfrak{u} + \bar{\mathfrak{a}}_i)\bar{\mathfrak{c}}_{ik} + \mathfrak{u}' + (\xi)_{ij}\big)(\bar{\mathfrak{a}}_k + \bar{\mathfrak{a}}_j)$$
$$= \big((\mathfrak{u} + \bar{\mathfrak{a}}_i)\bar{\mathfrak{c}}_{ik} + \mathfrak{u}' + ((\xi)_{kj} + \bar{\mathfrak{c}}_{ik})(\bar{\mathfrak{a}}_i + \bar{\mathfrak{a}}_j)\big)(\bar{\mathfrak{a}}_k + \bar{\mathfrak{a}}_j)$$
$$= \big(\mathfrak{u}' + ((\xi)_{kj} + \bar{\mathfrak{c}}_{ik})((\mathfrak{u} + \bar{\mathfrak{a}}_i)\bar{\mathfrak{c}}_{ik} + \bar{\mathfrak{a}}_i + \bar{\mathfrak{a}}_j)\big)(\bar{\mathfrak{a}}_k + \bar{\mathfrak{a}}_j) \qquad \text{(by IV)}.$$

Now

$$(\mathfrak{u} + \bar{\mathfrak{a}}_i)\bar{\mathfrak{c}}_{ik} + \bar{\mathfrak{a}}_i + \bar{\mathfrak{a}}_j = (\mathfrak{u} + \bar{\mathfrak{a}}_i)(\bar{\mathfrak{c}}_{ik} + \bar{\mathfrak{a}}_i) + \bar{\mathfrak{a}}_j \qquad \text{(by IV)}$$
$$= (\mathfrak{u} + \bar{\mathfrak{a}}_i)(\bar{\mathfrak{a}}_i + \bar{\mathfrak{a}}_k) + \bar{\mathfrak{a}}_j$$
$$= \mathfrak{u} + \bar{\mathfrak{a}}_i + \bar{\mathfrak{a}}_j \qquad \text{(since } \bar{\mathfrak{a}}_i,\ \mathfrak{u} \leqq \bar{\mathfrak{a}}_i + \bar{\mathfrak{a}}_k).$$

Hence

$$(\xi\beta)_{kj} = \big(\mathfrak{u}' + ((\xi)_{kj} + \bar{\mathfrak{c}}_{ik})(\mathfrak{u} + \bar{\mathfrak{a}}_i + \bar{\mathfrak{a}}_j)\big)(\bar{\mathfrak{a}}_k + \bar{\mathfrak{a}}_j)$$
$$= \mathfrak{u}' + ((\xi)_{kj} + \bar{\mathfrak{c}}_{ik})(\mathfrak{u} + \bar{\mathfrak{a}}_i + \bar{\mathfrak{a}}_j)(\bar{\mathfrak{a}}_k + \bar{\mathfrak{a}}_j)$$
$$= \mathfrak{u}' + ((\xi)_{kj} + \bar{\mathfrak{c}}_{ik})(\mathfrak{u} + \bar{\mathfrak{a}}_j + \bar{\mathfrak{a}}_i(\bar{\mathfrak{a}}_k + \bar{\mathfrak{a}}_j))$$
$$\text{(by IV, since } \mathfrak{u} + \bar{\mathfrak{a}}_j \leqq \bar{\mathfrak{a}}_k + \bar{\mathfrak{a}}_j)$$
$$= \mathfrak{u}' + ((\xi)_{kj} + \bar{\mathfrak{c}}_{ik})(\mathfrak{u} + \bar{\mathfrak{a}}_j).$$

But

$$((\xi)_{kj} + \mathfrak{u} + \bar{\mathfrak{a}}_j)(\bar{\mathfrak{c}}_{ik} + 0) = (\bar{\mathfrak{a}}_k + \bar{\mathfrak{a}}_j + \mathfrak{u})\bar{\mathfrak{c}}_{ik}$$
$$= (\bar{\mathfrak{a}}_k + \bar{\mathfrak{a}}_j)(\bar{\mathfrak{a}}_i + \bar{\mathfrak{a}}_k)\bar{\mathfrak{c}}_{ik} = \bar{\mathfrak{a}}_k \cdot \bar{\mathfrak{c}}_{ik} = 0,$$

whence Part I, Theorem 1.2 applies and yields

$$(\xi\beta)_{kj} = \mathfrak{u}' + (\xi)_{kj}(\mathfrak{u} + \bar{\mathfrak{a}}_j).$$

This completes the proof of (8).

PART II · CHAPTER X

Further Properties of the Auxiliary Ring of the Lattice

This chapter is devoted to a series of lemmas which will aid in establishing the fundamental result of these investigations, viz., that the complemented modular lattice L is isomorphic to the lattice $\bar{R}_{\mathfrak{S}_n}$.

In what follows we shall deal frequently with sums and products of many elements of L. Instead of the symbols $\Sigma_\cup$, $\Pi_\cap$ used heretofore for these sums and products, we shall use the more convenient symbols Σ, Π. It should be remembered, however, that these latter symbols are to be used also for addition and multiplication in $\mathfrak{S}$; however, no confusion should result. The symbol $\sum'_{j \neq i}$ will be used as an abberviation for $\sum_{j=1 (j \neq i)}^{m-1}$.

We shall need a strengthening of Theorem 1.2 of Part I.

LEMMA 10.1: If $k = 1, 2, \cdots$, $\mathfrak{a}$, $\mathfrak{b}_i \in L$ $(i = 1, 2, \cdots, k)$, and if $\mathfrak{a} \cap \sum_{i=1}^k \mathfrak{b}_i = 0$, then

$$\prod_{i=1}^{k} (\mathfrak{a} \cup \mathfrak{b}_i) = \mathfrak{a} \cup \prod_{i=1}^{k} \mathfrak{b}_i.$$

PROOF: The lemma clearly holds for $k = 1$. Suppose that it holds for $k = h$ and let $\mathfrak{a} \cap \sum_{i=1}^{h+1} \mathfrak{b}_i = 0$. Then $\prod_{i=1}^{h+1} (\mathfrak{a} \cup \mathfrak{b}_i) = \prod_{i=1}^{h} (\mathfrak{a} \cup \mathfrak{b}_i) \cap (\mathfrak{a} \cup \mathfrak{b}_{h+1})$. Since $\mathfrak{a} \cap \sum_{i=1}^{h} \mathfrak{b}_i = 0$, the induction hypothesis holds and we have $\prod_{i=1}^{h} (\mathfrak{a} \cup \mathfrak{b}_i) = \mathfrak{a} \cup \prod_{i=1}^{h} \mathfrak{b}_i$. Hence

$$\prod_{i=1}^{h+1} (\mathfrak{a} \cup \mathfrak{b}_i) = \left(\mathfrak{a} \cup \prod_{i=1}^{h} \mathfrak{b}_i\right) \cap (\mathfrak{a} \cup \mathfrak{b}_{h+1}).$$

Since $(\mathfrak{a} \cup \mathfrak{a}) \cap \left(\prod_{i=1}^{h} \mathfrak{b}_i \cup \mathfrak{b}_{h+1}\right) \leqq \mathfrak{a} \cap \sum_{i=1}^{h+1} \mathfrak{b}_i = 0$, Part I, Theorem 1.2 applies, yielding

$$\prod_{i=1}^{k+1} (\mathfrak{a} \cup \mathfrak{b}_i) = \mathfrak{a} \cup \left(\left(\prod_{i=1}^{h} \mathfrak{b}_i\right) \cap \mathfrak{b}_{h+1}\right) = \mathfrak{a} \cup \prod_{i=1}^{h+1} \mathfrak{b}_i.$$

Hence the lemma holds for $k = h + 1$, and thus holds for every k by induction.

DEFINITION 10.1: Let m be an index $\leqq n$. For $\beta, \gamma^1, \cdots, \gamma^{m-1} \in \mathfrak{S}$ we define

(1) $$(\beta; \gamma^1, \cdots, \gamma^{m-1}) \equiv \Big(\sum_{j=1}^{m-1} \bar{\mathfrak{a}}_j \cup (\beta)_m\Big) \cap \prod_{i=1}^{m-1} \Big(\sum_{j \neq i}' \bar{\mathfrak{a}}_j \cup (\gamma^i)_{mi}\Big) \in L.$$

(For the case $m = 1$ we understand that $\sum_{j=1}^{m-1} \bar{\mathfrak{a}}_j = 0$ and $\prod_{i=1}^{m-1} \big(\sum'_{j \neq i} \bar{\mathfrak{a}}_j \cup (\gamma^i)_{mi}\big) = 1$, whence the right member of (1) is $(\beta)_m$.)

COROLLARY: The quantity $(\beta; \gamma^1, \cdots, \gamma^{m-1})$ depends only on $(\beta)_{\underline{r}}$ and $\gamma^1, \cdots, \gamma^{m-1}$.

PROOF: This is clear since $(\beta)_m$ depends only on $(\beta)_{\underline{r}}$.

LEMMA 10.2: If $m = 1, \cdots, n$, and if $\beta \in \mathfrak{S}$ is idempotent, then

(2) $$\bar{\mathfrak{a}}_m \cup (\beta;\ \gamma^1, \cdots, \gamma^{m-1}) = \bar{\mathfrak{a}}_m \cup (1;\ \gamma^1\beta, \cdots, \gamma^{m-1}\beta).$$

PROOF: By Theorem 9.3, β is of the form $\beta = \beta(\mathfrak{u}, \mathfrak{u}')$ with $\mathfrak{u}$, $\mathfrak{u}'$ inverses in $\bar{\mathfrak{a}}_m$. Then since $(\beta)_m = \mathfrak{u}$,

(3) $$\bar{\mathfrak{a}}_m \cup (\beta;\ \gamma^1, \cdots, \gamma^{m-1}) = \bar{\mathfrak{a}}_m \cup \Big(\Big(\sum_{j=1}^{m-1} \bar{\mathfrak{a}}_j \cup \mathfrak{u}\Big) \cap \prod_{i=1}^{m-1} \Big(\sum_{j \neq i}' \bar{\mathfrak{a}}_j \cup (\gamma^i)_{mi}\Big) \Big).$$

Moreover, since $(1)_m = (\bar{\mathfrak{c}}_{im} \cup \bar{\mathfrak{a}}_i) \cap \bar{\mathfrak{a}}_m = (\bar{\mathfrak{a}}_m \cup \bar{\mathfrak{a}}_i) \cap \bar{\mathfrak{a}}_m = \bar{\mathfrak{a}}_m$, we have $\sum_{j=1}^{m-1} \bar{\mathfrak{a}}_j \cup (1)_m = \sum_{j=1}^{m} \bar{\mathfrak{a}}_j$; also, $\sum'_{j \neq i} \bar{\mathfrak{a}}_j \cup (\gamma^i)_{mi} \leqq \sum_{j=1}^{m} \bar{\mathfrak{a}}_j$ for every $i = 1, \cdots, m-1$. Hence

$$\begin{aligned}
\bar{\mathfrak{a}}_m \cup (1;\ \gamma^1\beta, \cdots, \gamma^{m-1}\beta) &= \bar{\mathfrak{a}}_m \cup \Big(\sum_{j=1}^{m} \bar{\mathfrak{a}}_j \cap \prod_{i=1}^{m-1} \Big(\sum_{j \neq i}' \bar{\mathfrak{a}}_j \cup (\gamma^i\beta)_{mi}\Big)\Big) \\
&= \bar{\mathfrak{a}}_m \cup \prod_{i=1}^{m-1} \Big(\sum_{j \neq i}' \bar{\mathfrak{a}}_j \cup (\gamma^i\beta)_{mi}\Big) \\
&= \bar{\mathfrak{a}}_m \cup \prod_{i=1}^{m-1} \Big(\sum_{j \neq i}' \bar{\mathfrak{a}}_j \cup ((\gamma^i)_{mi} \cap (\bar{\mathfrak{a}}_i \cup \mathfrak{u})) \cup \mathfrak{u}'\Big)
\end{aligned}$$

(by Lemma 9.5)

Now $\big(\sum_{i=1}^{m-1} \big(\sum'_{j \neq i} \bar{\mathfrak{a}}_j \cup ((\gamma^i)_{mi} \cap (\bar{\mathfrak{a}}_i \cup \mathfrak{u}))\big)\big) \cap \mathfrak{u}' \leqq \big(\sum_{i=1}^{m-1} \bar{\mathfrak{a}}_j \cup \mathfrak{u}\big) \cap (0 \cup \mathfrak{u}') = 0 \cup (\mathfrak{u} \cap \mathfrak{u}') = 0$ by Part I, Theorem 1.2, since $\big(\sum_{i=1}^{m-1} \bar{\mathfrak{a}}_j \cup 0\big) \cap (\mathfrak{u} \cup \mathfrak{u}') = \sum_{i=1}^{m-1} \bar{\mathfrak{a}}_j \cap \bar{\mathfrak{a}}_m = 0$. Hence Lemma 10.1 applies, yielding

$$\begin{aligned}
\bar{\mathfrak{a}}_m \cup (1;\ \gamma^1\beta, \cdots, \gamma^{m-1}\beta) &= \bar{\mathfrak{a}}_m \cup \prod_{i=1}^{m-1} \Big(\sum_{j \neq i}' \bar{\mathfrak{a}}_j \cup ((\gamma^i)_{mi} \cap (\bar{\mathfrak{a}}_i \cup \mathfrak{u}))\Big) \cup \mathfrak{u}' \\
&= \bar{\mathfrak{a}}_m \cup \prod_{i=1}^{m-1} \Big(\sum_{j \neq i}' \bar{\mathfrak{a}}_j \cup ((\gamma^i)_{mi} \cap (\bar{\mathfrak{a}}_i \cup \mathfrak{u}))\Big).
\end{aligned}$$

But $\big(\sum'_{j \neq i} \bar{\mathfrak{a}}_j\big) \cap (\bar{\mathfrak{a}}_i \cup \mathfrak{u} \cup (\gamma^i)_{mi}) \leqq \big(\sum'_{j \neq i} \bar{\mathfrak{a}}_j\big) \cap (\bar{\mathfrak{a}}_i \cup \bar{\mathfrak{a}}_m) = 0$, whence Part I, Theorem 1.2 applies and yields

$$\bar{\mathfrak{a}}_m \cup (1;\ \gamma^1\beta, \cdots, \gamma^{m-1}\beta) = \bar{\mathfrak{a}}_m \cup \prod_{i=1}^{m-1} \big((\sum_{j\neq i}{}' \bar{\mathfrak{a}}_j \cup \bar{\mathfrak{a}}_i \cup \mathfrak{u}) \cap (\sum_{j\neq i}{}' \bar{\mathfrak{a}}_j \cup (\gamma^i)_{mi}) \big)$$

$$= \bar{\mathfrak{a}}_m \cup \prod_{i=1}^{m-1} \big((\sum_{j=1}^{m-1} \bar{\mathfrak{a}}_j \cup \mathfrak{u}) \cap (\sum_{j\neq i}{}' \bar{\mathfrak{a}}_j \cup (\gamma^i)_{mi}) \big)$$

$$= \bar{\mathfrak{a}}_m \cup \big((\sum_{j=1}^{m-1} \bar{\mathfrak{a}}_j \cup \mathfrak{u}) \cap \prod_{i=1}^{m-1} (\sum_{j\neq i}{}' \bar{\mathfrak{a}}_j \cup (\gamma^i)_{mi}) \big)$$

$$= \bar{\mathfrak{a}}_m \cup (\beta;\ \gamma^1, \cdots, \gamma^{m-1}) \qquad \text{(by (3))}.$$

LEMMA 10.3: If $m = 1, \cdots, n$, and if $\beta \in \mathfrak{S}$ is idempotent, then

$$(4) \qquad (1;\ \gamma^1, \cdots, \gamma^{m-1}) = (\beta;\ \gamma^1, \cdots, \gamma^{m-1}) \cup (1-\beta;\ \gamma^1, \cdots, \gamma^{m-1}).$$

PROOF: Define

$$\mathfrak{u} \equiv \sum_{j=1}^{m-1} \bar{\mathfrak{a}}_j \cup (\beta)_m,\quad \mathfrak{v} \equiv \sum_{j=1}^{m-1} \bar{\mathfrak{a}}_j \cup (1-\beta)_m,\quad \mathfrak{w} \equiv \begin{cases} \prod_{i=1}^{m-1} (\sum_{j\neq i}{}' \bar{\mathfrak{a}}_j \cup (\gamma^i)_{mi}) & (\text{if } m > 1) \\ \bar{\mathfrak{a}}_1 & (\text{if } m = 1). \end{cases}$$

We shall prove $(\mathfrak{u}, \mathfrak{v}, \mathfrak{w})D$. By Theorem 2.1, $(\beta)_{\underline{r}} \cup (1-\beta)_{\underline{r}} = \mathfrak{S}$, whence by Theorem 9.2, $(\beta)_m \cup (1-\beta)_m = \bar{\mathfrak{a}}_m$. Hence $\mathfrak{u} \cup \mathfrak{v} = \sum_{j=1}^{m} \bar{\mathfrak{a}}_j$. Moreover

$$\mathfrak{u} \cap \mathfrak{v} = \big(\sum_{j=1}^{m-1} \bar{\mathfrak{a}}_j \cup (\beta)_m\big) \cap \big(\sum_{j=1}^{m-1} \bar{\mathfrak{a}}_j \cup (1-\beta)_m\big) = \sum_{j=1}^{m-1} \bar{\mathfrak{a}}_j \cup \big((\beta)_m \cap (1-\beta)_m\big)$$

by Part I, Theorem 1.2, since $\big((\beta)_m \cup (1-\beta)_m\big) \cap \sum_{j=1}^{m-1} \bar{\mathfrak{a}}_j \leqq \bar{\mathfrak{a}}_m \cap \sum_{j=1}^{m-1} \bar{\mathfrak{a}}_j = 0$; thus, since $(\beta)_m \cap (1-\beta)_m = 0$ by virtue of the fact β is idempotent (cf. Theorem 2.1), $\mathfrak{u} \cap \mathfrak{v} = \sum_{j=1}^{m-1} \bar{\mathfrak{a}}_j$. Now if $m \neq 1$,

$$\mathfrak{w} \cap \sum_{j=1}^{m-1} \bar{\mathfrak{a}}_j = \prod_{i=1}^{m-1} \big((\sum_{j\neq i}{}' \bar{\mathfrak{a}}_j \cup (\gamma^i)_{mi}) \cap \sum_{j=1}^{m-1} \bar{\mathfrak{a}}_j \big) = \prod_{i=1}^{m-1} \big((\sum_{j\neq i}{}' \bar{\mathfrak{a}}_j \cup (\gamma^i)_{mi}) \cap (\sum_{j\neq i}{}' \bar{\mathfrak{a}}_j \cup \bar{\mathfrak{a}}_i) \big).$$

Since $\sum_{j\neq i}' \bar{\mathfrak{a}}_j \cap \big((\gamma^i)_{mi} \cup \bar{\mathfrak{a}}_i\big) \leqq \sum_{j\neq i}' \bar{\mathfrak{a}}_j \cap (\bar{\mathfrak{a}}_i \cup \bar{\mathfrak{a}}_m) = 0$, we have by Lemma 10.1, $\big(\sum_{j\neq i}' \bar{\mathfrak{a}}_j \cup (\gamma^i)_{mi}\big) \cap \big(\sum_{j\neq i}' \bar{\mathfrak{a}}_j \cup \bar{\mathfrak{a}}_i\big) = \sum_{j\neq i}' \bar{\mathfrak{a}}_j \cup \big((\gamma^i)_{mi} \cap \bar{\mathfrak{a}}_i\big) = \sum_{j\neq i}' \bar{\mathfrak{a}}_j$. Hence

$$\mathfrak{w} \cap \sum_{j=1}^{m-1} \bar{\mathfrak{a}}_j = \prod_{i=1}^{m-1} \sum_{j\neq i}{}' \bar{\mathfrak{a}}_j = 0$$

by Part I, Corollary to Theorem 2.5. If $m = 1$, $\mathfrak{w} \cap \sum_{j=1}^{m-1} \bar{\mathfrak{a}}_j = \mathfrak{w} \cap 0 = 0$. Hence in any case $\mathfrak{w} \cap (\mathfrak{u} \cap \mathfrak{v}) = 0$. Moreover, $\mathfrak{w} \cup \sum_{j=1}^{m-1} \bar{\mathfrak{a}}_j = \prod_{i=1}^{m-1} \big(\sum_{j\neq i}' \bar{\mathfrak{a}}_j \cup (\gamma^i)_{mi}\big) \cup \sum_{j=1}^{m-1} \bar{\mathfrak{a}}_j$. Now we shall show that this expression is equal to

$$\mathfrak{a}(l) \equiv \prod_{i=1}^{l} (\sum_{j\neq i}{}' \bar{\mathfrak{a}}_j \cup (\gamma^i)_{mi}) \cup \sum_{j=1}^{m-1} \bar{\mathfrak{a}}_j$$

for every $l = 1, \cdots, m-1$. It clearly suffices to show that $\mathfrak{a}(l)$ is the same for l as for $l-1$, for each $l = 2, \cdots, m-1$. Indeed,

$$\mathfrak{a}(l) = \prod_{i=1}^{l} \Big({\sum_{j\neq i}}' \bar{\mathfrak{a}}_j \cup (\gamma^i)_{mi} \Big) \cup \sum_{j=1}^{m-1} \bar{\mathfrak{a}}_j$$

$$= \Big(\Big(\prod_{i=1}^{l-1} \Big({\sum_{j\neq i}}' \bar{\mathfrak{a}}_j \cup (\gamma^i)_{mi} \Big) \Big) \cap \Big({\sum_{j\neq l}}' \bar{\mathfrak{a}}_j \cup (\gamma^l)_{ml} \Big) \Big) \cup \bar{\mathfrak{a}}_l \cup \sum_{j=1}^{m-1} \bar{\mathfrak{a}}_j.$$

But $\bar{\mathfrak{a}}_l \leqq \sum'_{j\neq i} \bar{\mathfrak{a}}_j$ for $i = 1, \cdots, l-1$, whence $\bar{\mathfrak{a}}_l \leqq \prod_{i=1}^{l-1} \big(\sum'_{j\neq i} \bar{\mathfrak{a}}_j \cup (\gamma^i)_{mi} \big)$, and by IV,

$$\mathfrak{a}(l) = \Big(\Big(\prod_{i=1}^{l-1} \Big({\sum_{j\neq i}}' \bar{\mathfrak{a}}_j \cup (\gamma^i)_{mi} \Big) \Big) \cap \Big({\sum_{j\neq l}}' \bar{\mathfrak{a}}_j \cup (\gamma^l)_{ml} \cup \bar{\mathfrak{a}}_l \Big) \Big) \cup \sum_{j=1}^{m-1} \bar{\mathfrak{a}}_j$$

$$= \Big(\Big(\prod_{i=1}^{l-1} \Big({\sum_{j\neq i}}' \bar{\mathfrak{a}}_j \cup (\gamma^i)_{mi} \Big) \Big) \cap \Big({\sum_{j\neq l}}' \bar{\mathfrak{a}}_j \cup \bar{\mathfrak{a}}_m \cup \bar{\mathfrak{a}}_l \Big) \Big) \cup \sum_{j=1}^{m-1} \bar{\mathfrak{a}}_j$$

$$= \Big(\Big(\prod_{i=1}^{l-1} \Big({\sum_{j\neq i}}' \bar{\mathfrak{a}}_j \cup (\gamma^i)_{mi} \Big) \Big) \cap \sum_{j=1}^{m} \bar{\mathfrak{a}}_j \Big) \cup \sum_{j=1}^{m-1} \bar{\mathfrak{a}}_j$$

$$= \prod_{i=1}^{l-1} \Big({\sum_{j\neq i}}' \bar{\mathfrak{a}}_j \cup (\gamma^i)_{mi} \Big) \cup \sum_{j=1}^{m-1} \bar{\mathfrak{a}}_j = \mathfrak{a}\,(l-1).$$

Thus $\mathfrak{a}(m-1) = \mathfrak{a}(1)$, and

$$\mathfrak{w} \cup \sum_{j=1}^{m-1} \bar{\mathfrak{a}}_j = \mathfrak{a}(m-1) = \mathfrak{a}(1) = {\sum_{j\neq 1}}' \bar{\mathfrak{a}}_j \cup (\gamma^1)_{m1} \cup \sum_{j=1}^{m-1} \bar{\mathfrak{a}}_j$$

$$= {\sum_{j\neq 1}}' \bar{\mathfrak{a}}_j \cup (\gamma^1)_{m1} \cup \bar{\mathfrak{a}}_1$$

$$= {\sum_{j\neq 1}}' \bar{\mathfrak{a}}_j \cup \bar{\mathfrak{a}}_m \cup \bar{\mathfrak{a}}_1 = \sum_{j=1}^{m} \bar{\mathfrak{a}}_j.$$

Hence $\mathfrak{w} \cup (\mathfrak{u} \cap \mathfrak{v}) = \sum_{j=1}^{m} \bar{\mathfrak{a}}_j = \mathfrak{u} \cup \mathfrak{v}$. These considerations show that $\mathfrak{w}$ is inverse to $\mathfrak{u} \cap \mathfrak{v}$ in $\mathfrak{u} \cup \mathfrak{v}$. Thus $\mathfrak{u} \cup \mathfrak{w} = \mathfrak{u} \cup (\mathfrak{u} \cap \mathfrak{v}) \cup \mathfrak{w} = \mathfrak{u} \cup \mathfrak{u} \cup \mathfrak{v} = \mathfrak{u} \cup \mathfrak{v}$; similarly $\mathfrak{v} \cup \mathfrak{w} = \mathfrak{u} \cup \mathfrak{v}$. Hence $(\mathfrak{u} \cup \mathfrak{w}) \cap (\mathfrak{v} \cup \mathfrak{w}) = (\mathfrak{u} \cup \mathfrak{v}) \cap (\mathfrak{u} \cup \mathfrak{v}) = \mathfrak{u} \cup \mathfrak{v} = (\mathfrak{u} \cap \mathfrak{v}) \cup \mathfrak{w}$. This proves $(\mathfrak{u}, \mathfrak{v}, \mathfrak{w})D$ by Part I, Definition 5.1 and Theorem 5.1. Now

$$(1;\, \gamma^1, \cdots, \gamma^{m-1}) = \Big(\sum_{j=1}^{m-1} \bar{\mathfrak{a}}_j \cup \bar{\mathfrak{a}}_m \Big) \cap \mathfrak{w}$$

$$= \Big(\sum_{j=1}^{m-1} \bar{\mathfrak{a}}_j \cup (\beta)_m \cup (1-\beta)_m \Big) \cap \mathfrak{w} \quad \text{(since } \beta \text{ is idempotent)}$$

$$= \Big(\Big(\sum_{j=1}^{m-1} \bar{\mathfrak{a}}_j \cup (\beta)_m \Big) \cup \Big(\sum_{j=1}^{m-1} \bar{\mathfrak{a}}_j \cup (1-\beta)_m \Big) \Big) \cap \mathfrak{w}$$

$$= (\mathfrak{u} \cup \mathfrak{v}) \cap \mathfrak{w}$$

$$= (\mathfrak{u} \cap \mathfrak{w}) \cup (\mathfrak{v} \cap \mathfrak{w}) \qquad \text{(since } (\mathfrak{u}, \mathfrak{v}, \mathfrak{w})D)$$

$$= (\beta;\, \gamma^1, \cdots, \gamma^{m-1}) \cup (1-\beta;\, \gamma^1, \cdots, \gamma^{m-1}).$$

LEMMA 10.4: If $m = 1, \cdots, n$, and if $\beta \in \mathfrak{S}$ is idempotent, then

$$\bar{\mathfrak{a}}_m \cup (1; \gamma^1, \cdots, \gamma^{m-1}) = \bar{\mathfrak{a}}_m \cup (1; \gamma^1\beta, \cdots, \gamma^{m-1}\beta) \cup (1; \gamma^1(1-\beta), \cdots, \gamma^{m-1}(1-\beta)).$$

PROOF: This is obvious from Lemmas 10.2 and 10.3.

LEMMA 10.5: Let $m = 1, \cdots, n$, and $\gamma^1, \cdots, \gamma^{m-1} \in \mathfrak{S}$. There exists an idempotent $\beta \in \mathfrak{S}$ such that $(\beta)_{\underline{l}} = (\gamma^1)_{\underline{l}}$. For any such idempotent β define

$$\delta^i \equiv \gamma^i\beta, \quad \varepsilon^i \equiv \gamma^i(1 - \beta).$$

Then $\delta^1 = \gamma^1$, and for $i = 2, \cdots, m-1$ there exists η^i such that $\eta^i\gamma^1 = \delta^i$; moreover $\varepsilon^1 = 0$. Finally,

$$(5) \quad \bar{\mathfrak{a}}_m \cup (1; \gamma^1, \cdots, \gamma^{m-1}) = \bar{\mathfrak{a}}_m \cup (1; \gamma^1, \eta^2\gamma^1, \cdots, \eta^{m-1}\gamma^1) \cup (1; 0, \varepsilon^2, \cdots \varepsilon^{m-1}).$$

PROOF: The existence of an idempotent $\beta \in \mathfrak{S}$ such that $(\beta)_{\underline{l}} = (\gamma^1)_{\underline{l}}$ is clear since $\mathfrak{S}$ is regular. Now since $\gamma^1 \in (\gamma^1)_{\underline{l}} = (\beta)_{\underline{l}}$, we have $\gamma^1\beta = \gamma^1$, i.e., $\delta^1 = \gamma^1$. Since $\beta \in (\beta)_{\underline{l}} \leqq (\gamma^1)_{\underline{l}}$, there exists ω such that $\beta = \omega\gamma^1$. Hence if we define for $i = 2, \cdots, m - 1$, $\eta^i \equiv \gamma^i\omega$, we have

$$\delta^i = \gamma^i\beta = \gamma^i\omega\gamma^1 = \eta^i\gamma^1.$$

Furthermore,

$$\varepsilon^1 = \gamma^1(1 - \beta) = \gamma^1 - \gamma^1\beta = \gamma^1 - \gamma^1 = 0.$$

Therefore, by Lemma 10.4,

$$\begin{aligned}
&\bar{\mathfrak{a}}_m \cup (1; \gamma^1, \cdots, \gamma^{m-1}) \\
&\quad = \bar{\mathfrak{a}}_m \cup (1; \gamma^1\beta, \cdots, \gamma^{m-1}\beta) \cup (1; \gamma^1(1 - \beta), \cdots, \gamma^{m-1}(1 - \beta)) \\
&\quad = \bar{\mathfrak{a}}_m \cup (1; \delta^1, \cdots, \delta^{m-1}) \cup (1; \varepsilon^1, \cdots, \varepsilon^{m-1}) \\
&\quad = \bar{\mathfrak{a}}_m \cup (1; \gamma^1, \eta^2\gamma^1, \cdots, \eta^{m-1}\gamma^1) \cup (1; 0, \varepsilon^2, \cdots, \varepsilon^{m-1}),
\end{aligned}$$

and (5) is proved.

LEMMA 10.6: Let i, j, k be three distinct indices, and let $\beta, \gamma \in \mathfrak{S}$. Then

$$(\bar{\mathfrak{a}}_k \cup (\beta)_{ij}) \cap (\bar{\mathfrak{a}}_i \cup (\gamma)_{jk}) = (\bar{\mathfrak{a}}_k \cup (\beta)_{ij}) \cap (\bar{\mathfrak{a}}_j \cup (-\gamma\beta)_{ik}).$$

PROOF: By the proof of Theorem 8.2 (cf. the expression there for $\mathfrak{u}$),

$$\begin{aligned}
&(\bar{\mathfrak{a}}_k \cup (\beta)_{ij}) \cap (\bar{\mathfrak{a}}_j \cup (-\gamma\beta)_{ik}) \\
&\quad = (\bar{\mathfrak{a}}_k \cup (\beta)_{ij}) \cap \big(\bar{\mathfrak{a}}_j \cup [((((\beta)_{ij} \cup \bar{\mathfrak{a}}_k) \cap ((\gamma)_{jk} \cup \bar{\mathfrak{a}}_i)) \cup \bar{\mathfrak{a}}_j) \cap (\bar{\mathfrak{a}}_i \cup \bar{\mathfrak{a}}_k)]\big) \\
&\quad = (\bar{\mathfrak{a}}_k \cup (\beta)_{ij}) \cap \big((((\beta)_{ij} \cup \bar{\mathfrak{a}}_k) \cap ((\gamma)_{jk} \cup \bar{\mathfrak{a}}_i)) \cup \bar{\mathfrak{a}}_j\big) \cap (\bar{\mathfrak{a}}_i \cup \bar{\mathfrak{a}}_j \cup \bar{\mathfrak{a}}_k) \text{ (by IV)} \\
&\quad = (\bar{\mathfrak{a}}_k \cup (\beta)_{ij}) \cap \big((((\beta)_{ij} \cup \bar{\mathfrak{a}}_k) \cap ((\gamma)_{jk} \cup \bar{\mathfrak{a}}_i)) \cup \bar{\mathfrak{a}}_j\big) \\
&\quad = ((\bar{\mathfrak{a}}_k \cup (\beta)_{ij}) \cap \bar{\mathfrak{a}}_j) \cup (((\beta)_{ij} \cup \bar{\mathfrak{a}}_k) \cap ((\gamma)_{jk} \cup \bar{\mathfrak{a}}_i)) \\
&\quad = ((\beta)_{ij} \cup \bar{\mathfrak{a}}_k) \cap ((\gamma)_{jk} \cup \bar{\mathfrak{a}}_i),
\end{aligned}$$

the last equality holding since

$$\begin{aligned}(\bar{\mathfrak{a}}_k \cup (\beta)_{ij}) \cap \bar{\mathfrak{a}}_j &= (\bar{\mathfrak{a}}_k \cup (\beta)_{ij}) \cap (\bar{\mathfrak{a}}_i \cup \bar{\mathfrak{a}}_j) \cap \bar{\mathfrak{a}}_j \\ &= \big((\beta)_{ij} \cup (\bar{\mathfrak{a}}_k \cap (\bar{\mathfrak{a}}_i \cup \bar{\mathfrak{a}}_j))\big) \cap \bar{\mathfrak{a}}_j \qquad \text{(by IV)} \\ &= (\beta)_{ij} \cap \bar{\mathfrak{a}}_j = 0.\end{aligned}$$

LEMMA 10.7: For $m = 2, \cdots, n$, and for every $\gamma^1, \eta^2, \cdots, \eta^{m-1} \in \mathfrak{S}$

(6) $$\bar{\mathfrak{a}}_m \cup (1; \gamma^1, \eta^2\gamma^1, \cdots, \eta^{m-1}\gamma^1)$$
$$= \bar{\mathfrak{a}}_m \cup \Big(\big(\sum_{j=2}^{m-1} \bar{\mathfrak{a}}_j \cup (\gamma^1)_1\big) \cap \prod_{i=2}^{m-1} \big(\sum_{j=2}^{m-1}{}_{j \neq i} \bar{\mathfrak{a}}_j \cup (-\eta^i)_{1i}\big)\Big).$$

(For the case $m = 2$ we understand the second term of the right side of (6) to mean $(\gamma^1)_1$. Cf. the remark after Definition 10.1).

PROOF: We suppose first that $m = 3, \cdots, n$, i.e., $m \neq 2$. Then

$$\begin{aligned}(1; \gamma^1, \eta^2\gamma^1, \cdots, \eta^{m-1}\gamma^1) &= \sum_{j=1}^{m} \bar{\mathfrak{a}}_j \cap \big(\sum_{j=2}^{m-1} \bar{\mathfrak{a}}_j \cup (\gamma^1)_{m1}\big) \cap \prod_{i=2}^{m-1} \big(\sum_{j \neq i}{}' \bar{\mathfrak{a}}_j \cup (\eta^i\gamma^1)_{mi}\big) \\ &= \big(\sum_{j=2}^{m-1} \bar{\mathfrak{a}}_j \cup (\gamma^1)_{m1}\big) \cap \prod_{i=2}^{m-1} \big(\sum_{j \neq i}{}' \bar{\mathfrak{a}}_j \cup (\eta^i\gamma^1)_{mi}\big) \\ &= \prod_{i=2}^{m-1} \big(\sum_{j=2}^{m-1} \bar{\mathfrak{a}}_j \cup (\gamma^1)_{m1}\big) \cap \big(\sum_{j \neq i}{}' \bar{\mathfrak{a}}_j \cup (\eta^i\gamma^1)_{mi}\big) \\ &= \prod_{i=2}^{m-1} \Big[\big(\sum_{j=2}^{m-1}{}_{(j \neq i)} \bar{\mathfrak{a}}_j \cup (\bar{\mathfrak{a}}_i \cup (\gamma^1)_{m1})\big) \cap \big(\sum_{j=2}^{m-1}{}_{(j \neq i)} \bar{\mathfrak{a}}_j \cup (\bar{\mathfrak{a}}_1 \cup (\eta^i\gamma^1)_{mi})\big)\Big].\end{aligned}$$

But $\big(\sum_{j=2}^{m-1}{}_{(j \neq i)} \bar{\mathfrak{a}}_j\big) \cap (\bar{\mathfrak{a}}_i \cup (\gamma^1)_{m1} \cup \bar{\mathfrak{a}}_1 \cup (\eta^i\gamma^1)_{mi}) \leqq \big(\sum_{j=2}^{m-1}{}_{(j \neq i)} \bar{\mathfrak{a}}_j\big) \cap (\bar{\mathfrak{a}}_1 \cup \bar{\mathfrak{a}}_i \cup \bar{\mathfrak{a}}_m) = 0$, whence by Part I, Theorem 1.2,

$$\begin{aligned}(1; \gamma^1, \eta^2\gamma^1, \cdots, \eta^{m-1}\gamma^1) &= \prod_{i=2}^{m-1} \Big[\sum_{j=2}^{m-1}{}_{(j \neq i)} \bar{\mathfrak{a}}_j \cup \big((\bar{\mathfrak{a}}_i \cup (\gamma^1)_{m1}) \cap (\bar{\mathfrak{a}}_1 \cup (\eta^i\gamma^1)_{mi})\big)\Big] \\ &= \prod_{i=2}^{m-1} \Big[\sum_{j=2}^{m-1}{}_{(j \neq i)} \bar{\mathfrak{a}}_j \cup \big((\bar{\mathfrak{a}}_i \cup (\gamma^1)_{m1}) \cap (\bar{\mathfrak{a}}_m \cup (-\eta^i)_{1i})\big)\Big] \\ &\qquad \text{(by Lemma 10.6).}\end{aligned}$$

But $\big(\sum_{j=2}^{m-1}{}_{(j \neq i)} \bar{\mathfrak{a}}_j\big) \cap (\bar{\mathfrak{a}}_i \cup (\gamma^1)_{m1} \cup \bar{\mathfrak{a}}_m \cup (-\eta^i)_{1i}) \leqq \big(\sum_{j=2}^{m-1}{}_{(j \neq i)} \bar{\mathfrak{a}}_j\big) \cap (\bar{\mathfrak{a}}_1 \cup \bar{\mathfrak{a}}_i \cup \bar{\mathfrak{a}}_m) = 0$, whence by Part I, Theorem 1.2,

$(1; \gamma^1, \eta^2\gamma^1, \cdots, \eta^{m-1}\gamma^1)$

$$= \prod_{i=2}^{m-1} [(\textstyle\sum_{j=2 (j\neq i)}^{m-1} \bar{\mathfrak{a}}_j \cup (\bar{\mathfrak{a}}_i \cup (\gamma^1)_{m1})) \cap (\textstyle\sum_{j=2 (j\neq i)}^{m-1} \bar{\mathfrak{a}}_j \cup (\bar{\mathfrak{a}}_m \cup (-\eta^i)_{1i}))]$$

$$= \prod_{i=2}^{m-1} [(\sum_{j=2}^{m-1} \bar{\mathfrak{a}}_j \cup (\gamma^1)_{m1}) \cap (\textstyle\sum_{j=2 (j\neq i)}^{m-1} \bar{\mathfrak{a}}_j \cup \bar{\mathfrak{a}}_m \cup (-\eta^i)_{1i})]$$

$$= (\sum_{j=2}^{m-1} \bar{\mathfrak{a}}_j \cup (\gamma^1)_{m1}) \cap \prod_{i=2}^{m-1} (\textstyle\sum_{j=2 (j\neq i)}^{m-1} \bar{\mathfrak{a}}_j \cup \bar{\mathfrak{a}}_m \cup (-\eta^i)_{1i}).$$

Therefore

$\bar{\mathfrak{a}}_m \cup (1; \gamma^1, \eta^2\gamma^1, \cdots, \eta^{m-1}\gamma^1)$

$$= \bar{\mathfrak{a}}_m \cup (\sum_{j=2}^{m-1} \bar{\mathfrak{a}}_j \cup (\gamma^1)_{m1}) \cap \prod_{i=2}^{m-1} (\textstyle\sum_{j=2 (j\neq i)}^{m-1} \bar{\mathfrak{a}}_j \cup \bar{\mathfrak{a}}_m \cup (-\eta^i)_{1i})$$

$$= (\sum_{j=2}^{m-1} \bar{\mathfrak{a}}_j \cup \bar{\mathfrak{a}}_m \cup (\gamma^1)_{m1}) \cap \prod_{i=2}^{m-1} (\textstyle\sum_{j=2 (j\neq i)}^{m-1} \bar{\mathfrak{a}}_j \cup \bar{\mathfrak{a}}_m \cup (-\eta^i)_{1i}) \quad \text{(by IV)}$$

$$= (\sum_{j=2}^{m-1} \bar{\mathfrak{a}}_j \cup \bar{\mathfrak{a}}_m \cup (\gamma^1)_{1}) \cap \prod_{i=2}^{m-1} (\textstyle\sum_{j=2 (j\neq i)}^{m-1} \bar{\mathfrak{a}}_j \cup \bar{\mathfrak{a}}_m \cup (-\eta^i)_{1i})$$

(since $\bar{\mathfrak{a}}_m \cup (\gamma^1)_{m1} = \bar{\mathfrak{a}}_m \cup (\gamma^1)_1$ by the proof of Lemma 9.1)

$$= \Big(\bar{\mathfrak{a}}_m \cup (\sum_{j=2}^{m-1} \bar{\mathfrak{a}}_j \cup (\gamma^1)_1)\Big) \cap \prod_{i=2}^{m-1} \Big(\bar{\mathfrak{a}}_m \cup (\textstyle\sum_{j=2 (j\neq i)}^{m-1} \bar{\mathfrak{a}}_j \cup (-\eta^i)_{1i})\Big).$$

But $\bar{\mathfrak{a}}_m \cap \Big(\sum_{j=1}^{m-1} \bar{\mathfrak{a}}_j \cup (\gamma^1)_1 \cup \sum_{i=2}^{m-1} (\textstyle\sum_{j=2 (j\neq i)}^{m-1} \bar{\mathfrak{a}}_j \cup (-\eta^i)_{1i})\Big) \leqq \bar{\mathfrak{a}}_m \cap \sum_{j=1}^{m-1} \bar{\mathfrak{a}}_j = 0$, whence by Lemma 10.1

$$\bar{\mathfrak{a}}_m \cup (1; \gamma^1, \eta^2\gamma^1, \cdots, \eta^{m-1}\gamma^1) = \bar{\mathfrak{a}}_m \cup \Big(\sum_{j=2}^{m-1} \bar{\mathfrak{a}}_j \cup (\gamma^1)_1\Big) \cap \Big(\prod_{i=2}^{m-1} (\sum_{j=2}^{m-1} \bar{\mathfrak{a}}_j \cup (-\eta^i)_{1i})\Big).$$

Thus (6) holds for $m \neq 2$. Suppose now that $m = 2$. Then

$$\begin{aligned} \bar{\mathfrak{a}}_m \cup (1;\, \gamma^1) &= \bar{\mathfrak{a}}_m \cup (\gamma^1)_{m1} \quad \text{(cf. Definition 10.1)} \\ &= \bar{\mathfrak{a}}_m \cup (\gamma^1)_1, \end{aligned}$$

whence (6) holds also for this case.

LEMMA 10.8: For $m = 2, \cdots, n$, and for every $\varepsilon^2, \cdots, \varepsilon^{m-1} \in \mathfrak{S}$, we have

$$(7) \quad \bar{\mathfrak{a}}_m \cup (1; 0, \varepsilon^2, \cdots, \varepsilon^{m-1}) = \bar{\mathfrak{a}}_m \cup \Big(\big(\sum_{j=2}^{m-1} \bar{\mathfrak{a}}_j\big) \cap \big(\prod_{i=2}^{m-1} (\textstyle\sum_{j=2 (j\neq i)}^{m-1} \bar{\mathfrak{a}}_j \cup (\varepsilon^i)_{1i}) \cup \bar{\mathfrak{a}}_1\big)\Big).$$

(For the case $m = 2$, $\sum_{j=2}^{m-1} \bar{\mathfrak{a}}_j = 0$ and the right member of (7) is merely $\bar{\mathfrak{a}}_m$.)

PROOF: Suppose that $m \neq 2$. Then

$\bar{\mathfrak{a}}_m \cup (1;\ 0,\ \varepsilon^2, \cdots, \varepsilon^{m-1})$

$$= \bar{\mathfrak{a}}_m \cup \Big(\big(\sum_{j=2}^{m-1} \bar{\mathfrak{a}}_j \cup (0)_{m1}\big) \cap \prod_{i=2}^{m-1} \big(\sum_{j\neq i}' \bar{\mathfrak{a}}_j \cup (\varepsilon^i)_{mi}\big)\Big)$$

$$= \bar{\mathfrak{a}}_m \cup \Big(\big(\sum_{j=2}^{m-1} \bar{\mathfrak{a}}_j \cup \bar{\mathfrak{a}}_m\big) \cap \prod_{i=2}^{m-1} \big(\sum_{j\neq i}' \bar{\mathfrak{a}}_j \cup (\varepsilon^i)_{mi}\big)\Big)$$

$$= \big(\sum_{j=2}^{m-1} \bar{\mathfrak{a}}_j \cup \bar{\mathfrak{a}}_m\big) \cap \big(\prod_{i=2}^{m-1} \big(\sum_{j\neq i}' \bar{\mathfrak{a}}_j \cup (\varepsilon^i)_{mi}\big) \cup \bar{\mathfrak{a}}_m\big) \qquad \text{(by IV)}$$

$$= \bar{\mathfrak{a}}_m \cup \Big(\big(\sum_{j=2}^{m-1} \bar{\mathfrak{a}}_j\big) \cap \big(\prod_{i=2}^{m-1} \big(\sum_{j\neq i}' \bar{\mathfrak{a}}_j \cup (\varepsilon^i)_{mi}\big) \cup \bar{\mathfrak{a}}_m\big)\Big) \qquad \text{(by IV)}$$

$$= \bar{\mathfrak{a}}_m \cup \Big(\big(\sum_{j=2}^{m-1} \bar{\mathfrak{a}}_j\big) \cap \big(\prod_{i=2}^{m-1} \big(\big(\sum_{j=2}^{m-1}{}_{(j\neq i)} \bar{\mathfrak{a}}_j \cup (\varepsilon^i)_{mi}\big) \cup \bar{\mathfrak{a}}_1\big) \cup \bar{\mathfrak{a}}_m\big)\Big).$$

But $\bar{\mathfrak{a}}_1 \cap \sum_{i=2}^{m-1} \big(\sum_{j=2}^{m-1}{}_{(j\neq i)} \bar{\mathfrak{a}}_j \cup (\varepsilon^i)_{mi}\big) \leqq \bar{\mathfrak{a}}_1 \cap \big(\sum_{j=2}^{m-1} \bar{\mathfrak{a}}_j \cup \bar{\mathfrak{a}}_m\big) = 0$, whence by Lemma 10.1,

$\bar{\mathfrak{a}}_m \cup (1;\ 0,\ \varepsilon^2, \cdots, \varepsilon^{m-1})$

$$= \bar{\mathfrak{a}}_m \cup \Big(\big(\sum_{j=2}^{m-1} \bar{\mathfrak{a}}_j\big) \cap \big(\prod_{i=2}^{m-1} \big(\sum_{j=2}^{m-1}{}_{(j\neq i)} \bar{\mathfrak{a}}_j \cup (\varepsilon^i)_{mi}\big) \cup \bar{\mathfrak{a}}_1 \cup \bar{\mathfrak{a}}_m\big)\Big)$$

$$= \bar{\mathfrak{a}}_m \cup \Big(\Big(\big(\sum_{j=2}^{m-1} \bar{\mathfrak{a}}_j\big) \cup 0\Big) \cap \Big(\big(\prod_{i=2}^{m-1} \big(\sum_{j=2}^{m-1}{}_{(j\neq i)} \bar{\mathfrak{a}}_j \cup (\varepsilon^i)_{mi}\big) \cup \bar{\mathfrak{a}}_m\big) \cup \bar{\mathfrak{a}}_1\Big)\Big).$$

Now $(0 \cup \bar{\mathfrak{a}}_1) \cap \big(\sum_{j=2}^{m-1} \bar{\mathfrak{a}}_j \cup \prod_{i=2}^{m-1} \big(\sum_{j=2}^{m-1}{}_{(j\neq i)} \bar{\mathfrak{a}}_j \cup (\varepsilon^i)_{mi}\big) \cup \bar{\mathfrak{a}}_m\big) \leqq \bar{\mathfrak{a}}_1 \cap \big(\sum_{j=2}^{m-1} \bar{\mathfrak{a}}_j \cup \bar{\mathfrak{a}}_m\big) = 0.$ Hence Part I, Theorem 1.2 yields

$\bar{\mathfrak{a}}_m \cup (1;\ 0,\ \varepsilon^2, \cdots, \varepsilon^{m-1})$

$$= \bar{\mathfrak{a}}_m \cup \Big(\Big(\big(\sum_{j=2}^{m-1} \bar{\mathfrak{a}}_j\big) \cap \big(\prod_{i=2}^{m-1} \big(\sum_{j=2}^{m-1}{}_{(j\neq i)} \bar{\mathfrak{a}}_j \cup (\varepsilon^i)_{mi}\big) \cup \bar{\mathfrak{a}}_m\big)\Big) \cup (0 \cap \bar{\mathfrak{a}}_1)\Big)$$

$$= \bar{\mathfrak{a}}_m \cup \Big(\big(\sum_{j=2}^{m-1} \bar{\mathfrak{a}}_j\big) \cap \big(\prod_{i=2}^{m-1} \big(\sum_{j=2}^{m-1}{}_{(j\neq i)} \bar{\mathfrak{a}}_j \cup (\varepsilon^i)_{mi}\big) \cup \bar{\mathfrak{a}}_m\big)\Big).$$

We apply the transformation $P\begin{pmatrix} 2 \cdots m-1 & m \\ 2 \cdots m-1 & 1 \end{pmatrix}$ to

$$\big(\sum_{j=2}^{m-1} \bar{\mathfrak{a}}_j\big) \cap \big(\prod_{i=2}^{m-1} \big(\sum_{j=2}^{m-1}{}_{(j\neq i)} \bar{\mathfrak{a}}_j \cup (\varepsilon^i)_{mi}\big) \cup \bar{\mathfrak{a}}_m\big). \tag{8}$$

Since all elements occurring in this expression are $\leqq \sum_{j=2}^{m} \bar{\mathfrak{a}}_j$, the result is

$$\big(\sum_{j=2}^{m-1} \bar{\mathfrak{a}}_j\big) \cap \big(\prod_{i=2}^{m-1} \big(\sum_{j=2}^{m-1}{}_{(j\neq i)} \bar{\mathfrak{a}}_j \cup (\varepsilon^i)_{1i}\big) \cup \bar{\mathfrak{a}}_1\big). \tag{9}$$

However, the entire expression (8) is $\leqq \sum_{j=2}^{m-1} \bar{\mathfrak{a}}_j$, whence $P\begin{pmatrix} 2\cdots m-1 m \\ 2\cdots m-1\ 1\end{pmatrix}$ coincides for it with $P\begin{pmatrix} 2\cdots m-1 \\ 2\cdots m-1\end{pmatrix}$ = Identity, i.e., the expression (8) is equal to the expression (9). Hence (7) holds. Suppose now that $m = 2$. Then

$$\bar{\mathfrak{a}}_m \cup (1;\ 0) = \bar{\mathfrak{a}}_m \cup (0)_{m1} = \bar{\mathfrak{a}}_m \cup \bar{\mathfrak{a}}_m = \bar{\mathfrak{a}}_m,$$

and (7) holds also for this case.

THEOREM 10.1: Let $m = 2, \cdots, n$ and $\beta, \gamma^1, \cdots, \gamma^{m-1} \in \mathfrak{S}$. There exist two idempotents $\beta', \beta'' \in \mathfrak{S}$ such that $(\beta')_{\underline{r}} = (\beta)_{\underline{r}}$, $(\beta'')_{\underline{l}} = (\gamma^1\beta')_{\underline{l}}$. Define

$$\delta^i \equiv \gamma^i\beta'\beta'', \quad \varepsilon^i \equiv \gamma^i\beta^i(1-\beta'').$$

Then $\delta^1 = \gamma^1\beta'$, $\varepsilon^1 = 0$. There exist $\eta^i \in \mathfrak{S}$ $(i = 2, \cdots, m-1)$ such that $\eta^i\gamma^1\beta' = \delta^i$ and for all such $\eta^i \in \mathfrak{S}$ $(i = 2, \cdots, m-1)$

$$\begin{aligned} &\bar{\mathfrak{a}}_m \cup (\beta;\ \gamma^1, \cdots, \gamma^{m-1}) \\ &\quad = \bar{\mathfrak{a}}_m \cup \Big(\big(\sum_{j=2}^{m-1} \bar{\mathfrak{a}}_j \cup (\gamma^1\beta')_1\big) \cap \prod_{i=2}^{m-1} \big(\sum_{j=2\,(j\neq i)}^{m-1} \bar{\mathfrak{a}}_j \cup (-\eta^i)_{1i}\big)\Big) \\ &\quad\quad \cup \Big(\big(\sum_{j=2}^{m-1} \bar{\mathfrak{a}}_j\big) \cap \Big(\prod_{i=2}^{m-1} \big(\sum_{j=2\,(j\neq i)}^{m-1} \bar{\mathfrak{a}}_j \cup (\varepsilon^i)_{1i}\big) \cup \bar{\mathfrak{a}}_1\Big)\Big). \end{aligned}$$

(For $m = 2$ cf. the explanations at the end of Lemma 10.7 and Lemma 10.8.)

REMARK: It should be observed that β', β'' and the $\delta^i, \varepsilon^i, \eta^i$ are defined in terms of β and the γ^i entirely by means of the structure of the regular ring $\mathfrak{S}$ no direct use being made of the properties of L and of the $\bar{\mathfrak{a}}_i$, $\bar{\mathfrak{c}}_{ij}$.

PROOF: The existence of β', β'' is clear since $\mathfrak{S}$ is regular. By Lemma 10.5, $\delta^1 = \gamma^1\beta'$, $\varepsilon^1 = 0$ and the existence of the η^i follows also from this lemma; in this application of Lemma 10.5 the γ^i are replaced by $\gamma^i\beta'$. Now

$$\begin{aligned} &\bar{\mathfrak{a}}_m \cup (\beta;\ \gamma^1, \cdots, \gamma^{m-1}) \\ &\quad = \bar{\mathfrak{a}}_m \cup (\beta';\ \gamma^1, \cdots, \gamma^{m-1}) \quad \text{(by the Corollary to Definition 10.1)} \\ &\quad = \bar{\mathfrak{a}}_m \cup (1;\ \gamma^1\beta', \cdots, \gamma^{m-1}\beta') \quad \text{(by Lemma 10.2)} \\ &\quad = \bar{\mathfrak{a}}_m \cup (1;\ \gamma^1\beta', \eta^2\gamma^1\beta', \cdots, \eta^{m-1}\gamma^1\beta') \\ &\quad\quad \cup (1;\ 0, \varepsilon^2, \cdots, \varepsilon^{m-1}) \quad \text{(by Lemma 10.5)} \\ &\quad = \bar{\mathfrak{a}}_m \cup \Big(\big(\sum_{j=2}^{m-1} \bar{\mathfrak{a}}_j \cup (\gamma^1\beta')_1\big) \cap \prod_{i=2}^{m-1} \big(\sum_{j=2\,(j\neq i)}^{m-1} \bar{\mathfrak{a}}_j \cup (-\eta^i)_{1i}\big)\Big) \\ &\quad\quad \cup \Big(\big(\sum_{j=2}^{m-1} \bar{\mathfrak{a}}_j\big) \cap \Big(\prod_{i=2}^{m-1} \big(\sum_{j=2\,(j\neq i)}^{m-1} \bar{\mathfrak{a}}_j \cup (\varepsilon^i)_{1i}\big) \cup \bar{\mathfrak{a}}_1\Big)\Big), \end{aligned}$$

the last equality holding by Lemmas 10.7 and 10.8.

PART II · CHAPTER XI

Special Considerations. Statement of the Induction to be Proved

We have seen that if $\mathfrak{S}^\wedge$ is a regular ring, then $L = \bar{R}_{\mathfrak{S}_n^\wedge}$ is a lattice of order n (cf. Theorem 3.3). Moreover, the elements of $\bar{R}_{\mathfrak{S}_n^\wedge}$ may be represented by linear sets of n-dimensional vectors with components in $\mathfrak{S}^\wedge$. If for $i, j = 1, \cdots, n$, $i \neq j$, $\bar{\mathfrak{a}}_i^\wedge$ represents the set of all vectors $(0, \cdots, 0, \xi^\wedge, 0, \cdots, 0)$, $\xi^\wedge \in \mathfrak{S}^\wedge$, in which $\xi^\wedge$ occurs as ith component, and if $\bar{\mathfrak{c}}_{ij}^\wedge$ represents the set of all vectors $(0, \cdots, 0, \xi^\wedge, 0, \cdots, 0, -\xi^\wedge, 0, \cdots, 0)$, $\xi^\wedge \in \mathfrak{S}^\wedge$, in which $\xi^\wedge$ occurs as ith component and $-\xi^\wedge$ as jth component, then the system $(\bar{\mathfrak{a}}_i^\wedge, \bar{\mathfrak{c}}_{ij}^\wedge; i, j = 1, \cdots, n, i \neq j)$ is a frame in $\bar{R}_{\mathfrak{S}_n^\wedge}$ of order n. Throughout the beginning of this chapter $\mathfrak{S}^\wedge$ (and hence also the $\bar{\mathfrak{a}}_i^\wedge$, $\bar{\mathfrak{c}}_{ij}^\wedge$) will be fixed, and it will be shown that the $\bar{\mathfrak{a}}_i^\wedge$, $\bar{\mathfrak{c}}_{ij}^\wedge$ constitute a normalized frame, and that if $\mathfrak{S}^{\wedge\!\wedge}$ is the auxiliary ring associated with $\bar{R}_{\mathfrak{S}_n^\wedge}$ and the $\bar{\mathfrak{a}}_i^\wedge$, $\bar{\mathfrak{c}}_{ij}^\wedge$, then $\mathfrak{S}^{\wedge\!\wedge}$ is isomorphic to $\mathfrak{S}^\wedge$. Of course, we assume $n \geqq 4$.

At the end of the chapter we return to our consideration of the given lattice L and give a statement of the inductive proposition which will establish the desired result that L is isomorphic to $\bar{R}_{\mathfrak{S}_n}$.

LEMMA 11.1: If $\mathfrak{a}^\wedge \leqq \sum_{\tau=1}^{m} \bar{\mathfrak{a}}_{i_\tau}^\wedge$ $(m = 1, \cdots, n-1)$ is represented by a vector set $(\xi_1^\wedge, \cdots, \xi_n^\wedge)$, where the $\xi_i^\wedge$ are restricted in some manner (for the form of these restrictions cf. Appendix 3 to Chapter II), then

(a) for the $(\xi_1^\wedge, \cdots, \xi_n^\wedge)$ of $\mathfrak{a}^\wedge$, $\xi_i^\wedge = 0$ if $i \neq i_1, \cdots, i_m$;

(b) the projective isomorphism $P^\wedge\begin{pmatrix} i_1 & \cdots & i_m \\ j_1 & \cdots & j_m \end{pmatrix}$ $(m = 1, \cdots, n-1)$ transforms $\mathfrak{a}^\wedge$ into an element $\mathfrak{b}^\wedge(\epsilon \bar{R}_{\mathfrak{S}_n^\wedge})$ represented by the set of vectors $(\eta_1^\wedge, \cdots, \eta_n^\wedge)$ in which

$$(1)\qquad \begin{aligned} \eta_{j_\tau}^\wedge &= \xi_{i_\tau}^\wedge & (\tau = 1, \cdots, m) \\ \eta_j^\wedge &= 0 & (j \neq j_1, \cdots, j_m) \end{aligned}$$

$((\xi_1^\wedge, \cdots, \xi_n^\wedge)$ varying over the vector set which represents $\mathfrak{a}^\wedge)$.

(Note: The transformations $P^\wedge\begin{pmatrix} i_1 & \cdots & i_m \\ j_1 & \cdots & j_m \end{pmatrix}$ were defined in Theorem 5.1

under the assumption that the frame is *normalized.* In Lemma 11.1, $P^\wedge$ should be interpreted as *any* product of the form (26) in Chapter V, using Definition 5.3 — Ed.).

PROOF: The statement (a) follows from $\mathfrak{a}^\wedge \leqq \sum_{\tau=1}^{m} \bar{\mathfrak{a}}_{i_\tau}^\wedge$ together with the obvious fact that $\sum_{\tau=1}^{m} \bar{\mathfrak{a}}_{i_\tau}^\wedge$ is represented by the vector set of all $(\xi_1^\wedge, \cdots, \xi_n^\wedge)$ with $\xi_i^\wedge = 0$ if $i \neq i_1, \cdots, i_m$.

As to (b), we note that it suffices to prove it for those $P^\wedge\begin{pmatrix} i_1 \cdots i_m \\ j_1 \cdots j_m \end{pmatrix}$, where $i_\tau \neq j_\tau$ occurs for exactly one value of τ, since the general $P^\wedge\begin{pmatrix} i_1 \cdots i_m \\ j_1 \cdots j_m \end{pmatrix}$ is a product of such special ones, and since (b) (i.e. (1)) holds for a product if it holds for both factors. Let us suppose then, that $i_\tau \neq j_\tau$ occurs only for $\tau = \bar{\tau}$.

Now

$$\mathfrak{a}^\wedge\colon \quad (\xi_1^\wedge, \cdots, \xi_n^\wedge) \qquad \text{(the } \xi_i^\wedge \text{ being restricted as stated above; in particular } \xi_i^\wedge = 0 \text{ if } i \neq i_1, \cdots, i_m)$$

$$\bar{\mathfrak{c}}_{i_{\bar\tau} j_{\bar\tau}}^\wedge\colon \quad (0, \cdots, 0, \underset{i_{\bar\tau}}{\zeta^\wedge}, 0, \cdots, 0, \underset{j_{\bar\tau}}{-\zeta^\wedge}, 0, \cdots, 0)$$

(in which $\zeta^\wedge$ is the $i_{\bar\tau}^{\text{th}}$ component and $-\zeta^\wedge$ the $j_{\bar\tau}^{\text{th}}$ with $\zeta^\wedge$ varying over $\mathfrak{S}^\wedge$). Thus

$$\mathfrak{a}^\wedge \cup \bar{\mathfrak{c}}_{i_{\bar\tau} j_{\bar\tau}}^\wedge\colon (\xi_1^\wedge, \cdots, \xi_{i_{\bar\tau}-1}^\wedge, \xi_{i_{\bar\tau}}^\wedge + \zeta^\wedge, \xi_{i_{\bar\tau}+1}^\wedge, \cdots, \xi_{j_{\bar\tau}-1}^\wedge, \xi_{j_{\bar\tau}}^\wedge - \zeta^\wedge, \xi_{j_{\bar\tau}+1}^\wedge, \cdots, \xi_n^\wedge),$$

and hence $\mathfrak{b}^\wedge = (\mathfrak{a}^\wedge \cup \bar{\mathfrak{c}}_{i_{\bar\tau} j_{\bar\tau}}^\wedge) \cap \sum_{\tau=1}^{m} \bar{\mathfrak{a}}_{j_\tau}^\wedge$ is represented by the same vector set as $\mathfrak{a}^\wedge \cup \mathfrak{c}_{i_{\bar\tau} j_{\bar\tau}}^\wedge$ except that all components $j \neq j_1, \cdots, j_m$ must be 0. Hence all components $j \neq i_1, \cdots, i_m$ except $j = j_{\bar\tau}$ are 0; also the component $j = i_{\bar\tau}$ is 0. Hence $\zeta^\wedge = -\xi_{i_{\bar\tau}}^\wedge$, $\xi_{j_{\bar\tau}}^\wedge - \zeta^\wedge = 0 - (-\xi_{i_{\bar\tau}}^\wedge) = \xi_{i_{\bar\tau}}^\wedge$. Thus $\xi_{i_1}^\wedge, \cdots, \xi_{i_m}^\wedge$ are entirely unrestricted except for their restriction in $\mathfrak{a}^\wedge$, and the components of the general element $(\eta_1^\wedge, \cdots, \eta_n^\wedge)$ of the vector set representing $\mathfrak{b}^\wedge$ are these: $\eta_{j_\tau}^\wedge = \eta_{i_\tau}^\wedge = \xi_{i_\tau}^\wedge$ for $\tau \neq \bar{\tau}$ (for these values of τ: $j_\tau = i_\tau$), $\eta_{j_{\bar\tau}}^\wedge = \xi_{i_{\bar\tau}}^\wedge$, $\eta_j^\wedge = 0$ for $j \neq j_1, \cdots, j_m$ (i.e., for $j \neq i_1, \cdots, i_m$ except $j = j_{\bar\tau}$, and also for $j = i_{\bar\tau}$). (Here $(\xi_1^\wedge, \cdots, \xi_m^\wedge)$ varies over the vector set which represents $\mathfrak{a}^\wedge$.) This is exactly the statement of (b).

LEMMA 11.2: The frame $(\bar{\mathfrak{a}}_i^\wedge, \bar{\mathfrak{c}}_{ij}^\wedge)$ is normalized.

PROOF: It is sufficient to show that

$$\bar{\mathfrak{c}}_{kh}^\wedge = \bar{\mathfrak{c}}_{ij}^\wedge P^\wedge \begin{pmatrix} i & j \\ k & h \end{pmatrix}, \tag{3}$$

since $\bar{\mathfrak{c}}_{ij}^\wedge = \bar{\mathfrak{c}}_{ji}^\wedge$ is obvious from the representation by sets of vectors. Now

$$\bar{\mathfrak{c}}_{ij}^\wedge\colon \; (0, \cdots, 0, \xi^\wedge, 0, \cdots, 0, -\xi^\wedge, 0, \cdots, 0),$$

in which $\xi^\wedge$ is the ith component, $-\xi^\wedge$ is the jth component, and in which $\xi^\wedge$ varies over $\mathfrak{S}^\wedge$. Hence by Lemma 11.1

$$\bar{\mathfrak{c}}_{ij}^\wedge P^\wedge \begin{pmatrix} i & j \\ k & h \end{pmatrix} \colon (0, \cdots, 0, \xi^\wedge, 0, \cdots, 0, -\xi^\wedge, 0, \cdots, 0),$$

in which $\xi^\wedge$ is now the kth component and $-\xi^\wedge$ is the hth component. But this is precisely the representation of $\bar{\mathfrak{c}}_{kh}^\wedge$, whence (3) holds.

THEOREM 11.1: The auxiliary ring $\mathfrak{S}^{\wedge\!\!\wedge}$ of $\bar{R}_{\mathfrak{S}_n^\wedge}$ is isomorphic to $\mathfrak{S}^\wedge$.

PROOF: By Definition 6.1, $\mathfrak{S}^{\wedge\!\!\wedge}$ is the set of all systems $(\mathfrak{b}_{ij}^\wedge; i, j = 1, \cdots, n, i \neq j)$ where $\mathfrak{b}_{ij}^\wedge \in L_{ij}^\wedge$ is inverse to $\bar{\mathfrak{a}}_j$ in $\bar{\mathfrak{a}}_i^\wedge \cup \bar{\mathfrak{a}}_j^\wedge$ and $\mathfrak{b}_{ij}^\wedge = \mathfrak{b}_{kh}^\wedge P^\wedge \begin{pmatrix} k & h \\ i & j \end{pmatrix}$. Now if $\mathfrak{b}_{ij}^\wedge \in L_{ij}^\wedge$, then

$$(4) \qquad \mathfrak{b}_{ij}^\wedge \colon (0, \cdots, 0, \underset{i}{\xi^\wedge}, 0, \cdots, 0, -\underset{j}{\beta^\wedge}\xi^\wedge, 0, \cdots, 0),$$

where $\beta^\wedge$ is fixed in $\mathfrak{S}^\wedge$, and $\xi^\wedge$ varies over $\mathfrak{S}^\wedge$. Moreover, by Lemma 11.1,

$$\mathfrak{b}_{kh}^\wedge \colon (0, \cdots, 0, \underset{k}{\xi^\wedge}, 0, \cdots, 0, -\underset{h}{\beta^\wedge}\xi^\wedge, 0, \cdots, 0),$$

in which $\beta^\wedge$ is the same as in (4). Hence for every $\beta^{\wedge\!\!\wedge} \in \mathfrak{S}^{\wedge\!\!\wedge}$ there exists one and only one element $\beta^\wedge \in \mathfrak{S}^\wedge$ such that

$$(5) \qquad (\beta^{\wedge\!\!\wedge})_{ij} \colon (0, \cdots, 0, \underset{i}{\xi^\wedge}, 0, \cdots, 0, -\underset{j}{\beta^\wedge}\xi^\wedge, 0, \cdots, 0)$$

for every $i, j = 1, \cdots, n$ $(i \neq j)$. Conversely, it is obvious that for every $\beta^\wedge \in \mathfrak{S}^\wedge$ the vector set in (4) represents a unique element $\mathfrak{b}_{ij}^\wedge \in L_{ij}^\wedge$, and hence for every $\beta^\wedge \in \mathfrak{S}^\wedge$ there exists a unique element $\beta^{\wedge\!\!\wedge} \in \mathfrak{S}^{\wedge\!\!\wedge}$ such that (5) holds for every i, j. Therefore (5) defines a one-to-one correspondence between $\mathfrak{S}^\wedge$ and $\mathfrak{S}^{\wedge\!\!\wedge}$. It remains to prove that this correspondence preserves addition and multiplication. That it preserves addition follows from the definitions of addition (cf. Definitions 7.6, 7.7), together with the computations in Appendix, Chapter VI. That it preserves multiplication follows from the definition of multiplication (Definition 6.2) together with the computations in Appendix, Chapter VI.

LEMMA 11.3: Let $m = 1, \cdots, n$. The expression $(\beta^\wedge; \gamma^{\wedge 1}, \cdots, \gamma^{\wedge m-1})$ (cf. Definition 10.1) is represented by the vector set

$$(-\gamma^{\wedge 1}\beta^\wedge\eta^\wedge, -\gamma^{\wedge 2}\beta^\wedge\eta^\wedge, \cdots, -\gamma^{\wedge m-1}\beta^\wedge\eta^\wedge, \underset{m}{\beta^\wedge\eta^\wedge}, 0, \cdots, 0) \quad (\eta^\wedge \text{ arbitrary}).$$

PROOF: By Definition 10.1, we have

$$(\beta^\wedge; \gamma^{\wedge 1}, \cdots, \gamma^{\wedge m-1}) = \Big(\sum_{j=1}^{m-1} \bar{\mathfrak{a}}_j^\wedge \cup (\beta^\wedge)_m\Big) \cap \prod_{i=1}^{m-1} \Big(\sum_{j \neq i}' \bar{\mathfrak{a}}_j^\wedge \cup (\gamma^{\wedge i})_{mi}\Big).$$

Now

$$(\gamma^{\wedge i})_{mi}\colon\ (0,\cdots,0,\ \underset{i}{-\gamma^{\wedge i}\xi_i^{\wedge}},\ 0,\cdots,0,\ \underset{m}{\xi_i^{\wedge}},\ 0,\cdots,0)\qquad(\xi_i^{\wedge}\in\mathfrak{S}^{\wedge})$$

$$\sum_{j\neq i}{}'\ \bar{\mathfrak{a}}_j^{\wedge}\colon\ (\eta_1^{\wedge},\cdots,\eta_{i-1}^{\wedge},\ \underset{i}{0},\ \eta_{i+1}^{\wedge},\cdots,\eta_{m-1}^{\wedge},\ \underset{m}{0},\ 0,\cdots,0),$$

whence

$$(\gamma^{\wedge i})_{mi}\cup\sum_{j\neq i}{}'\ \bar{\mathfrak{a}}_j^{\wedge}\colon\ (\eta_1^{\wedge},\cdots,\eta_{i-1}^{\wedge},\ \underset{i}{-\gamma^{\wedge i}\xi_i^{\wedge}},\ \eta_{i+1}^{\wedge},\cdots,\eta_{m-1}^{\wedge},\ \underset{m}{\xi_i^{\wedge}},\ 0,\cdots,0)$$

and

$$\prod_{i=1}^{m-1}\Big((\gamma^{\wedge i})_{mi}\cup\sum_{j\neq i}{}'\ \bar{\mathfrak{a}}_j^{\wedge}\Big)\colon\ (-\gamma^{\wedge 1}\xi^{\wedge},\ -\gamma^{\wedge 2}\xi^{\wedge},\cdots,-\gamma^{\wedge m-1}\xi^{\wedge},\ \underset{m}{\xi^{\wedge}},\ 0,\cdots,0).$$

Now $(\beta^{\wedge})_m=\big((\beta^{\wedge})_{hm}\cup\bar{\mathfrak{a}}_h^{\wedge}\big)\cap\bar{\mathfrak{a}}_m^{\wedge}$;

$$\bar{\mathfrak{a}}_m^{\wedge}\colon\ (0,\cdots,0,\ \underset{m}{\xi^{\wedge}},\ 0,\cdots,0),$$

$$(\beta^{\wedge})_{hm}\colon\ (0,\cdots,0,\ \underset{h}{\zeta^{\wedge}},\ 0,\cdots,0,\ \underset{m}{-\beta^{\wedge}\zeta^{\wedge}},\ 0,\cdots,0),$$

$$\bar{\mathfrak{a}}_h^{\wedge}\colon\ (0,\cdots,0,\ \underset{h}{\eta^{\wedge}},\ 0,\cdots,0),$$

whence

$$(\beta^{\wedge})_{hm}\cup\mathfrak{a}_h^{\wedge}\colon\quad(0,\cdots,0,\ \underset{h}{\eta^{\wedge}+\zeta^{\wedge}},\ 0,\cdots,0,\ \underset{m}{-\beta^{\wedge}\zeta^{\wedge}},\ 0,\cdots,0),$$

and

$$(\beta^{\wedge})_m\colon\ (0,\cdots,0,\ \beta_m^{\wedge}\eta^{\wedge},\ 0,\cdots,0).$$

Moreover,

$$\sum_{j=1}^{m-1}\bar{\mathfrak{a}}_j^{\wedge}\colon\ (\omega_1^{\wedge},\cdots,\omega_{m-1}^{\wedge},\ 0,\cdots,0),$$

whence

$$\sum_{j=1}^{m-1}\bar{\mathfrak{a}}_j^{\wedge}\cup(\beta^{\wedge})_m\colon\ (\omega_1^{\wedge},\cdots,\omega_{m-1}^{\wedge},\ \underset{m}{\beta^{\wedge}\eta^{\wedge}},\ 0,\cdots,0).$$

Consequently, for vectors in the set representing $(\beta^{\wedge};\gamma^{\wedge 1},\cdots,\gamma^{\wedge m-1})$, $\xi^{\wedge}=\beta^{\wedge}\eta^{\wedge}$, i.e.,

$$(\beta^{\wedge};\gamma^{\wedge 1},\cdots,\gamma^{\wedge m-1})\colon\ (-\gamma^{\wedge 1}\beta^{\wedge}\eta^{\wedge},\ -\gamma^{\wedge 2}\beta^{\wedge}\eta^{\wedge},\cdots,\ -\gamma^{\wedge m-1}\beta^{\wedge}\eta^{\wedge},$$
$$\beta^{\wedge}\eta^{\wedge},\ 0,\cdots,0).$$

We return now to the consideration of the lattice L and our fundamental theorem concerning it, viz., that L is isomorphic to $\bar{R}_{\mathfrak{S}_n}$. Now our desired result is equivalent to the following: If L', L'' are two complemented modular lattices (of order $n\geqq 4$ and possessing normalized frames $(\bar{\mathfrak{a}}_i',\ \bar{\mathfrak{c}}_{ij}')$, $(\bar{\mathfrak{a}}_i'',\ \bar{\mathfrak{c}}_{ij}'')$, respectively), and if L', L'' have the same (i.e. isomorphic) auxiliary ring(s) $\mathfrak{S}$, then L' is isomorphic to L''. For if our desired result holds, then L' is isomorphic to $\bar{R}_{\mathfrak{S}_n}$, so is L'', whence L' is isomorphic to L''; conversely, if we take L'' to be $\bar{R}_{\mathfrak{S}_n}$, then L'' has the auxiliary ring $\mathfrak{S}$ by Theorem 11.1 and if L' and L'' are isomorphic, then L' is isomorphic to $\bar{R}_{\mathfrak{S}_n}$ and our desired result is established.

Now we shall prove our result by establishing the following proposition: Let $n = 4, 5, 6, \cdots$, and let $\mathfrak{S}$ be a given regular ring. Then for each $m = 1, 2, \cdots, n$, the following statement S_m holds:

S_m: Let L', L'' be two complemented modular lattices with normalized frames $(\bar{a}_i', \bar{c}_{ij}'; i, j = 1, \cdots, n, i \neq j)$, $(\bar{a}_i'', \bar{c}_{ij}''; i, j = 1, \cdots, n, i \neq j)$ respectively. Let the auxiliary rings $\mathfrak{S}'$, $\mathfrak{S}''$ of L', L'' respectively be isomorphic to, and henceforth identified with $\mathfrak{S}$. Then there exists an isomorphism $\mathscr{I}_m$ of $L'(0, \sum_{i=1}^m \bar{a}_i')$ and $L''(0, \sum_{i=1}^m \bar{a}_i'')$ with the following properties:

(A_m) For each $\beta \in \mathfrak{S}$ and $i = 1, \cdots, m$, the elements $(\beta)_i'$, $(\beta)_i''$ correspond under $\mathscr{I}_m$;

(B_m) For each $\beta \in \mathfrak{S}$ and $i, j = 1, \cdots, m$, $i > j$, the elements $(\beta)_{ij}'$, $(\beta)_{ij}''$ correspond under $\mathscr{I}_m$.

It should be observed that the case $m = 1$ is already settled. By Theorem 9.2, $L'(0, \bar{a}_1')$, $L'(0, \bar{a}_1'')$ are both isomorphic to $\bar{R}_{\mathfrak{S}'} = \bar{R}_{\mathfrak{S}''}$ and hence to each other, whence the existence of $\mathscr{I}_1$ follows. Property (A_1) is obvious, since by Theorem 9.2, $(\beta)_1' \rightleftarrows (\beta)_{\underline{r}} \rightleftarrows (\beta)_1''$, and property (B_1) is vacuously satisfied.

It remains therefore to prove merely that S_{m-1} implies S_m for $m = 2, \cdots, n$. When this is established, it will follow by induction that S_n holds, i.e., that $L' = L'(0, \sum_{i=1}^n \bar{a}_i')$ and $L'' = L''(0, \sum_{i=1}^n \bar{a}_i'')$ are isomorphic.

Throughout what follows, the (regular) ring $\mathfrak{S}$ and n $(= 4, 5, \cdots)$ will be held fixed. For the purposes of the induction we shall consider m $(= 2, \cdots, n)$, which will also be held fixed, and we shall assume that S_{m-1} holds for these $\mathfrak{S}$, n and m, and for all choices of L', $\bar{a}_i'$, $\bar{c}_{ij}'$ and L'', $\bar{a}_i''$, $\bar{c}_{ij}''$ (corresponding to $\mathfrak{S}$, n). We shall then establish S_m; indeed, it will be found possible to define $\mathscr{I}_m$ as an extension of $\mathscr{I}_{m-1}$. It will be found necessary to distinguish two cases, viz., $m = 2, \cdots, n-1$, and $m = n$. In the second case, it suffices to restrict m to the range $m = 4, \cdots, n$). (Since $n = 4, 5, \cdots$, these two cases exhaust all possibilities for $m = 2, \cdots, n$.) The two cases will be referred to respectively as Case I and Case II.

Apply S_{m-1} to L', $\bar{a}_i'$, $\bar{c}_{ij}'$ and to $L^\wedge = \bar{R}_{\mathfrak{S}_n}$, $\bar{a}_i^\wedge$, $\bar{c}_{ij}^\wedge$, as used before Theorem 11.1 (in place of L'', $\bar{a}_i''$, $\bar{c}_{ij}''$). In view of Theorem 11.1, S_{m-1} implies that $L'(0, \sum_{i=1}^{m-1} \bar{a}_i')$ is isomorphic to $\bar{R}_{\mathfrak{S}_{m-1}}$. Similarly (or by application of

S_{m-1} to L' and L'') $L''(0, \sum_{i=1}^{m-1} \bar{\mathfrak{a}}_i'')$ is isomorphic to $\bar{R}_{\mathfrak{S}_{m-1}}$. In these isomorphisms the correspondences given under (A_{m-1}), (B_{m-1}) above hold. Hence we have the following representations by vector sets:

$$\bar{\mathfrak{a}}_i'(\bar{\mathfrak{a}}_i''):\quad (0, \cdots, 0, \underset{i}{\xi}, 0, \cdots, 0) \qquad (i = 1, \cdots, m-1)$$

$$(\beta)_{ij}'((\beta)_{ij}''):\quad (0, \cdots, 0, \underset{i}{\xi}, 0, \cdots, 0, \underset{j}{-\beta\xi}, 0, \cdots, 0)$$

$$(i, j = 1, \cdots, m-1, i > j).$$

(In those cases where (B_{m-1}) is extended to all $i \neq j$—cf. the discussion below — this holds for all $i \neq j$.) It should be observed also that $\beta = 1$ yields in this relation

$$\bar{\mathfrak{c}}_{ij}'(\bar{\mathfrak{c}}_{ij}''):\quad (0, \cdots, 0, \underset{i}{\xi}, 0, \cdots, 0, \underset{j}{-\xi}, 0, \cdots, 0).$$

This holds for $i > j$, but since both sides are symmetric in i, j, it extends immediately to all $i \neq j$.)

Finally, it should be observed that if $m \geqq 3$, and hence in particular for the $m-1$ when m is in Case II (i.e., $m \geqq 4$, $m - 1 \geqq 3$), the restriction $i > j$ in (B_m) may be replaced by $i \neq j$. Let $i, j = 1, \cdots, m$, $i \neq j$. If $i > j$, then by S_m, $(\beta)_{ij}'$ and $(\beta)_{ij}''$ correspond. Suppose then that $i < j$, and select $l = 1, \cdots, m$ with $l \neq i, j$. This can be done since $m \geqq 3$. Now if $k, h = 1, \cdots, m$, $k \neq h$, then $\bar{\mathfrak{c}}_{kh}' = (1)_{kh}'$ corresponds to $\bar{\mathfrak{c}}_{kh}'' = (1)_{kh}''$ if $k > h$; if $k < h$, then $\bar{\mathfrak{c}}_{kh}'$, $\bar{\mathfrak{c}}_{kh}''$ correspond since $\bar{\mathfrak{c}}_{kh}' = \bar{\mathfrak{c}}_{hk}'$, $\bar{\mathfrak{c}}_{kh}'' = \bar{\mathfrak{c}}_{hk}''$. Now

$$(6') \quad (\beta)_{ij}' = (\beta)_{ji}' P'\begin{pmatrix} j & i \\ j & l \end{pmatrix} P'\begin{pmatrix} j & l \\ i & l \end{pmatrix} P'\begin{pmatrix} i & l \\ i & j \end{pmatrix}$$

$$= (((((((\beta)_{ji}' \cup \bar{\mathfrak{c}}_{il}') \cap (\bar{\mathfrak{a}}_j' \cup \bar{\mathfrak{a}}_l')) \cup \bar{\mathfrak{c}}_{ij}') \cap (\bar{\mathfrak{a}}_i' \cup \bar{\mathfrak{a}}_l')) \cup \bar{\mathfrak{c}}_{jl}') \cap (\bar{\mathfrak{a}}_i' \cup \bar{\mathfrak{a}}_j')$$

$$= (((((((\beta)_{ji}' \cup \bar{\mathfrak{c}}_{il}') \cap ((1)_j' \cup (1)_l')) \cup \bar{\mathfrak{c}}_{ij}') \cap ((1)_i' \cup (1)_l')) \cup \bar{\mathfrak{c}}_{jl}') \cap ((1)_i' \cup (1)_j'),$$

and $(6'')$, obtained from $(6')$ by replacing $'$ by $''$, holds also. Since the right members of $(6')$ and $(6'')$ correspond under the isomorphism $\mathscr{I}_m$, it is clear that $(\beta)_{ij}'$ and $(\beta)_{ij}''$ correspond. Hence S_m with the restriction $i > j$ in (B_m) is equivalent to S_m with the restriction $i \neq j$ in (B_m), provided that $m \geqq 3$.

(Note: When the isomorphism of L and $\bar{R}_{\mathfrak{S}_n}$ has been established it will follow that the restriction $i > j$ in (B_m) could be replaced by $i \neq j$ even for $m = 2$; a direct proof of this can also be given — Ed.)

We shall proceed now to the proof of the implication $S_{m-1} \to S_m$.

First some preliminary considerations will be made, for which a differentiation between Case I and Case II is not necessary. We shall consider the two lattices L', L'' and assume that S_{m-1} holds for them $(m = 2, \cdots, n)$, i.e., that an isomorphism $\mathscr{I}_{m-1}$ exists between $L'(0, \sum_{j=1}^{m-1} \bar{\mathfrak{a}}_j')$ and $L''(0, \sum_{j=1}^{m-1} \bar{\mathfrak{a}}_j'')$ such that

(A_{m-1}) for each $\beta \in \mathfrak{S}$ and $i = 1, \cdots, m-1$, $(\beta)_i'$, $(\beta)_i''$ correspond under $\mathscr{I}_{m-1}$; and

(B_{m-1}) for each $\beta \in \mathfrak{S}$ and $i, j = 1, \cdots, m-1$, $i > j$, $(\beta)_{ij}'$, $(\beta)_{ij}''$ correspond under $\mathscr{I}_{m-1}$.

We shall show how $\mathscr{I}_m$ is defined and indicate what properties will be proved in order to establish that S_m shall hold.

THEOREM 11.2: Every element $\mathfrak{u}' \in L'(0, \sum_{j=1}^{m} \bar{\mathfrak{a}}_j')$ may be represented in the form

$$\text{(I)} \qquad \mathfrak{u}' = \mathfrak{v}' \cup (\beta;\, \gamma^1, \cdots, \gamma^{m-1})',$$

where $\mathfrak{v}' \in L'(0, \sum_{j=1}^{m-1} \bar{\mathfrak{a}}_j')$ and $\beta, \gamma^1, \cdots, \gamma^{m-1} \in \mathfrak{S}$ and $(\beta; \gamma^1, \cdots, \gamma^{m-1})'$ denotes

$$\Big(\sum_{j=1}^{m} \bar{\mathfrak{a}}_j' \cup (\beta)_m'\Big) \cap \prod_{i=1}^{m-1} \Big(\sum_{j \neq 1} \bar{\mathfrak{a}}_j' \cup (\gamma^i)_{mi}'\Big) \in L'.$$

The same statement holds with $'$ replaced by $''$.

PROOF: By the proof of part (d) of Lemma 4.2, we see that $\mathfrak{u}'$ is expressible in the form $\mathfrak{u}' = \mathfrak{v}' \cup \mathfrak{w}'$ where $\mathfrak{v}' \leqq \sum_{j=1}^{m-1} \bar{\mathfrak{a}}_j'$ and $\mathfrak{w}'$ is of the form

$$\mathfrak{w}' = \mathfrak{c}' \cap \Big(\sum_{j=1}^{m-1} \bar{\mathfrak{a}}_j' \cup \mathfrak{b}_m'\Big),$$

where $\mathfrak{b}_m' \leqq \bar{\mathfrak{a}}_m'$ and $\mathfrak{c}'$ is inverse to $\sum_{j=1}^{m-1} \bar{\mathfrak{a}}_j'$ in $\sum_{j=1}^{m} \bar{\mathfrak{a}}_j'$. By part (b) of Lemma 4.2, $\mathfrak{c}'$ is of the form

$$\mathfrak{c}' = \prod_{i=1}^{m-1} \Big(\mathfrak{c}_{mi}' \cup \sum_{j \neq i}' \bar{\mathfrak{a}}_j'\Big),$$

where $\mathfrak{c}_{mi}' \in L_{mi}'$ $(i = 1, \cdots, m-1)$. Now there exist γ^i $(i = 1, \cdots, m-1)$ with $(\gamma^i)_{mi}' = \mathfrak{c}_{mi}'$. Moreover, since $\mathfrak{b}_m' \leqq \bar{\mathfrak{a}}_m'$, there exists $\beta \in \mathfrak{S}$ with $(\beta)_m' = \mathfrak{b}_m'$. Therefore,

$$\begin{aligned} \mathfrak{u}' &= \mathfrak{v}' \cup \Big(\mathfrak{c}' \cap \Big(\sum_{j=1}^{m-1} \bar{\mathfrak{a}}_j' \cup \mathfrak{b}_m'\Big)\Big) \\ &= \mathfrak{v}' \cup \Big(\Big(\sum_{j=1}^{m-1} \bar{\mathfrak{a}}_j' \cup (\beta)_m'\Big) \cap \prod_{i=1}^{m-1} \Big(\sum_{j \neq i}' \bar{\mathfrak{a}}_j' \cup (\gamma^i)_{mi}'\Big)\Big). \end{aligned}$$

The second term in the right member is precisely the element $(\beta; \gamma^1 \cdots, \gamma^{m-1})$ computed in L', and hence (I) holds. Since S_{m-1} is symmetric in L' and L'', replacing $'$ by $''$ obviously leaves the theorem true.

DEFINITION 11.1: Let $\mathfrak{u}' \in L'(0, \sum_{j=1}^{m} \bar{\mathfrak{a}}_j')$, and let $\mathfrak{u}'$ be represented as in (I). Then define

$$\mathscr{I}_m(\mathfrak{u}') \equiv \mathscr{I}_{m-1}(\mathfrak{v}') \cup (\beta;\ \gamma^1, \cdots, \gamma^{m-1})'',$$

where $\mathfrak{v}'$ and $\beta, \gamma^1, \cdots, \gamma^{m-1}$ are the same as those appearing in (I).

REMARK: Clearly $\mathscr{I}_m$ as defined (Definition 11.1) is only a many-to-many correspondence of all elements of $L'(0, \sum_{j=1}^{m} \bar{\mathfrak{a}}_j')$ with all elements of $L''(0, \sum_{j=1}^{m} \bar{\mathfrak{a}}_j'')$. It remains to prove that $\mathscr{I}_m$ is a one-to-one correspondence, that it is a lattice-isomorphism of $L'(0, \sum_{j=1}^{m} \bar{\mathfrak{a}}_j')$ and $L''(0, \sum_{j=1}^{m} \bar{\mathfrak{a}}_j'')$, and that it possesses the properties (A_m), (B_m). We shall show also that $\mathscr{I}_m$ is an extension of $\mathscr{I}_{m-1}$.

In this connection it seems worth while to remark that any lattice-isomorphism $\mathscr{J}_m$ between $L'(0, \sum_{j=1}^{m} \bar{\mathfrak{a}}_j')$ and $L''(0, \sum_{j=1}^{m} \bar{\mathfrak{a}}_j'')$ which is an extension of $\mathscr{I}_{m-1}$ and which has the properties (A_m) and (B_m), must coincide with $\mathscr{I}_m$. Indeed, let $\mathfrak{u}' \in L'(0, \sum_{i=1}^{m} \bar{\mathfrak{a}}_i')$ and $\mathfrak{u}'' \in L''(0, \sum_{i=1}^{m} \bar{\mathfrak{a}}_i'')$ correspond under $\mathscr{J}_m$. Then if $\mathfrak{u}'$ is represented as in (I),

$$\begin{aligned}\mathscr{J}_m(\mathfrak{u}') &= \mathscr{J}_m(\mathfrak{v}') \cup \mathscr{J}_m((\beta;\ \gamma^1, \cdots, \gamma^{m-1})') \\ &= \mathscr{J}_m(\mathfrak{v}') \cup \mathscr{J}_m\Big[\Big(\Big(\sum_{j=1}^{m-1} \bar{\mathfrak{a}}_j' \cup (\beta)_m'\Big) \cap \prod_{i=1}^{m-1} \Big(\sum_{j\neq i}' \bar{\mathfrak{a}}_j' \cup (\gamma^i)_{mi}'\Big)\Big)\Big] \\ &= \mathscr{J}_m(\mathfrak{v}') \cup \Big(\Big(\sum_{j=1}^{m-1} \mathscr{J}_m(\bar{\mathfrak{a}}_j') \cup \mathscr{J}_m((\beta)_m')\Big) \cap \prod_{i=1}^{m-1} \Big(\sum_{j\neq i}' \mathscr{J}_m(\bar{\mathfrak{a}}_j') \cup \mathscr{J}_m((\gamma^i)_{mi}')\Big)\Big)\Big].\end{aligned}$$

But $\bar{\mathfrak{a}}_j' = (1)_j'$, whence $\mathscr{J}_m(\bar{\mathfrak{a}}_j') = \mathscr{J}_m((1)_j') = (1)_j'' = \bar{\mathfrak{a}}_j''$ by (A_m); similarly $\mathscr{J}_m((\beta)_m') = (\beta)_m''$. Moreover $\mathscr{J}_m((\gamma^i)_{mi}') = (\gamma^i)_{mi}''$ $(i = 1, \cdots, m-1)$ by (B_m) since $m > i$. Finally, since $\mathfrak{v}' \in L'(0, \sum_{j=1}^{m-1} \bar{\mathfrak{a}}_j')$, $\mathscr{J}_m(\mathfrak{v}') = \mathscr{I}_{m-1}(\mathfrak{v}')$. Hence

$$\begin{aligned}\mathscr{J}_m(\mathfrak{u}') &= \mathscr{I}_{m-1}(\mathfrak{v}') \cup \Big(\Big(\sum_{j=1}^{m-1} \bar{\mathfrak{a}}_j'' \cup (\beta)_m''\Big) \cap \prod_{i=1}^{m-1} \Big(\sum_{j\neq i}' \bar{\mathfrak{a}}_j'' \cup (\gamma^i)_{mi}''\Big)\Big) \\ &= \mathscr{I}_{m-1}(\mathfrak{v}') \cup (\beta;\ \gamma^1, \cdots, \gamma^{m-1})'' = \mathscr{I}_m(\mathfrak{u}').\end{aligned}$$

We shall prove now, as indicated in the Remark, that $\mathscr{I}_m$ is a one-to-one correspondence, that it is even a lattice-isomorphism of $L'(0, \sum_{j=1}^{m} \bar{\mathfrak{a}}_j')$ and $L''(0, \sum_{j=1}^{m} \bar{\mathfrak{a}}_j'')$, and that it possesses the properties (A_m), (B_m). We shall also show that it is an extension of $\mathscr{I}_{m-1}$.

We prove the last statement first. If $\mathfrak{u}' \leq \sum_{j=1}^{m-1} \bar{\mathfrak{a}}_j'$, we may choose in (I) $\mathfrak{v}' = \mathfrak{u}'$ and $\beta = \gamma^1 = \cdots = \gamma^{m-1} = 0$ (since $(0;\ 0, \cdots, 0)' = \big(\sum_{j=1}^{m-1} \bar{\mathfrak{a}}_j'\big) \cap \prod_{i=1}^{m-1} \big(\sum_{j=1\,(j\neq i)}^{m-1} \bar{\mathfrak{a}}_j' \cup \bar{\mathfrak{a}}_m'\big) = \sum_{j=1}^{m-1} \bar{\mathfrak{a}}_j' \cap \bar{\mathfrak{a}}_m' = 0$, because $(\bar{\mathfrak{a}}', \cdots, \bar{\mathfrak{a}}_m') \perp$); then $\mathfrak{u}'' = \mathscr{I}_{m-1}(\mathfrak{u}')$ (since $(0;\ 0, \cdots, 0)'' = 0$) corresponds to $\mathfrak{u}'$ by $\mathscr{I}_m$.

Let us now consider the three other (preceding) statements. We shall prove that the following statement holds.

(*) Let $\mathfrak{v}', \mathfrak{w}' \leqq \sum_{j=1}^{m-1} \bar{\mathfrak{a}}_j'$, et $\mathfrak{v}'', \mathfrak{w}'' \leqq \sum_{j=1}^{m-1} \bar{\mathfrak{a}}_j''$, and suppose that $\mathfrak{v}', \mathfrak{v}''$ correspond under $\mathscr{I}_{m-1}$ and $\mathfrak{w}', \mathfrak{w}''$ also correspond under $\mathscr{I}_{m-1}$. Then for $\beta, \delta, \gamma^1, \cdots, \gamma^{m-1}, \varepsilon^1, \cdots, \varepsilon^{m-1} \in \mathfrak{S}$,

(7) $$\mathfrak{v}' \cup (\beta;\ \gamma^1, \cdots, \gamma^{m-1})' \leqq \mathfrak{w}' \cup (\delta;\ \varepsilon^1, \cdots, \varepsilon^{m-1})'$$

implies

(8) $$\mathfrak{v}'' \cup (\beta;\ \gamma^1, \cdots, \gamma^{m-1})'' \leqq \mathfrak{w}'' \cup (\delta;\ \varepsilon^1, \cdots, \varepsilon^{m-1})''.$$

If (*) is established, we may argue in the following manner.

By symmetry (in ′ and ″) also (8) implies (7). Hence (7) and (8) are equivalent. Again by symmetry (this time in $\mathfrak{v}', \mathfrak{v}'', \beta, \gamma^1, \cdots, \gamma^{m-1}$ and $\mathfrak{v}', \mathfrak{v}'', \delta, \varepsilon^1, \cdots, \varepsilon^{m-1}$) the same is true if $\leqq$ in (7), (8) is replaced by $\geqq$. These two results give together that (7) and (8) are also equivalent with $=$ in place of $\leqq$. But this means that $\mathscr{I}_m$ is a one-to-one correspondence of $L'(0, \sum_{i=1}^{m} \bar{\mathfrak{a}}_i')$ and $L''(0, \sum_{i=1}^{m} \bar{\mathfrak{a}}_i'')$. Thus the equivalence of the original (7) and (8) implies that $\mathscr{I}_m$ is even a lattice-isomorphism.

Since (A_{m-1}), (B_{m-1}) hold for $\mathscr{I}_{m-1}$, and since $\mathscr{I}_m$ is an extension of $\mathscr{I}_{m-1}$ (cf. above), the statements in (A_m) hold for $i = 1, \cdots, m-1$ and in (B_m) for $i, j = 1, \cdots, m-1$, $i > j$. But we need them for all $i = 1, \cdots, m$ and $i, j = 1, \cdots, m$, $i > j$, respectively; hence it remains for us to consider (A_m) for $i = m$, and (B_m) for $i = m, j = 1, \cdots, m-1$; i.e., we must show that

(9) $$\mathfrak{u}' = (\eta)_m' \text{ and } \mathfrak{u}'' = (\eta)_m'' \text{ correspond under } \mathscr{I}_m,$$

and that

(10) $$\mathfrak{u}' = (\eta)_{mk}' \text{ and } \mathfrak{u}'' = (\eta)_{mk}'' \text{ correspond under } \mathscr{I}_m (k = 1, \cdots, m-1).$$

In order to prove (9), put in (I), $\mathfrak{v}' = \mathfrak{v}'' = 0$, and $\beta = \eta$, $\gamma^1 = \cdots = \gamma^{m-1} = 0$. Then

$$\begin{aligned}(\eta;\ 0, \cdots, 0)' &= \Big(\sum_{j=1}^{m-1} \bar{\mathfrak{a}}_j' \cup (\eta)_m'\Big) \cap \prod_{i=1}^{m-1} \Big(\sum_{j=1\,(j \neq i)}^{m-1} \bar{\mathfrak{a}}_j' \cup \bar{\mathfrak{a}}_m'\Big) \\ &= \Big(\sum_{j=1}^{m-1} \bar{\mathfrak{a}}_j' \cup (\eta)_m'\Big) \cap \bar{\mathfrak{a}}_m' \\ &= (\eta)_m' \qquad \text{(by IV since } (\bar{\mathfrak{a}}_1', \cdots, \bar{\mathfrak{a}}_m') \perp)\end{aligned}$$

and similarly

$$(\eta;\ 0, \cdots, 0)'' = (\eta)_m'',$$

whence (9) holds. In order to prove (10), put in (I) again, $\mathfrak{v}' = \mathfrak{v}'' = 0$, but $\beta = 1$, $\gamma^k = \eta$, $\gamma^i = 0$ for $i \neq k$. Then

$$(1;\ 0, \cdots, 0, \underset{k}{\eta}, 0, \cdots, 0)'$$

$$= \Big(\sum_{j=1}^{m-1} \bar{\mathfrak{a}}_j' \cup \bar{\mathfrak{a}}_m'\Big) \cap \Big(\sum_{j=1}^{m-1}{}_{(j\neq k)} \bar{\mathfrak{a}}_j' \cup (\eta)_{mk}'\Big) \cap \prod_{i=1}^{m-1}{}_{(i\neq k)} \Big(\sum_{j=1}^{m-1}{}_{(j\neq i)} \bar{\mathfrak{a}}_j' \cup \bar{\mathfrak{a}}_m'\Big)$$

$$= (\bar{\mathfrak{a}}_m' \cup \bar{\mathfrak{a}}_k') \cap \Big(\sum_{j=1}^{m-1}{}_{(j\neq k)} \bar{\mathfrak{a}}_j' \cup (\eta)_{mk}'\Big) = (\eta)_{mk}',$$

the last equality holding by IV, since $(\bar{\mathfrak{a}}_1', \cdots, \bar{\mathfrak{a}}_m') \perp$; similarly

$$(1;\ 0, \cdots, 0, \underset{k}{\eta}, 0, \cdots, 0)'' = (\eta)_{mk}'',$$

whence (10) holds.

Thus we see that it remains for us to prove only the statement (*), assuming S_{m-1} and $m = 2, \cdots, n$. Now we shall show that (*) follows readily from a proposition which we shall denote by (**). This proposition is the following:

(**) If L represents either L' or L'', and if $\beta, \gamma^1, \cdots, \gamma^{m-1}, \delta^1, \cdots, \delta^{m-1} \in \mathfrak{S}$, then

$$\Big(\sum_{j=1}^{m-1} \bar{\mathfrak{a}}_j\Big) \cap \big((1;\ \gamma^1, \cdots, \gamma^{m-1}) \cup (\beta;\ \delta^1, \cdots, \delta^{m-1})\big)$$

$$= \Big(\sum_{j=1}^{m-1} \bar{\mathfrak{a}}_j\Big) \cap \big(\bar{\mathfrak{a}}_m \cup (\beta;\ \delta^1 - \gamma^1, \cdots, \delta^{m-1} - \gamma^{m-1})\big).$$

Let us therefore assume (**) and establish (*). Throughout the material preceding Theorem 11.2, L will denote L' or L''.

Lemma 11.4: If $\mathfrak{v}, \mathfrak{w} \leqq \sum_{j=1}^{m-1} \bar{\mathfrak{a}}_j$, and if $\beta, \delta, \gamma^1, \cdots, \gamma^{m-1}, \eta^1, \cdots, \eta^{m-1} \in \mathfrak{S}$, then

$$\mathfrak{v} \cup (\beta;\ \gamma^1, \cdots, \gamma^{m-1}) \leqq \mathfrak{w} \cup (\delta;\ \eta^1, \cdots, \eta^{m-1}) \tag{11}$$

is equivalent to the conjunction of

$$\mathfrak{v} \leqq \mathfrak{w}, \tag{12}$$

$$(\beta)_{\underline{r}} \leqq (\delta)_{\underline{r}}, \tag{13}$$

$$\Big(\sum_{j=1}^{m-1} \bar{\mathfrak{a}}_j\Big) \cap [(\beta;\ \gamma^1, \cdots, \gamma^{m-1}) \cup (1;\ \eta^1, \cdots, \eta^{m-1})] \leqq \mathfrak{w}. \tag{14}$$

Proof: By Theorem 9.2, $(\beta)_{\underline{r}} \leqq (\delta)_{\underline{r}}$ is equivalent to $(\beta)_m \leqq (\delta)_m$, whence (13) may be replaced by this condition. Define

$$\mathfrak{a} \equiv \sum_{j=1}^{m-1} \bar{\mathfrak{a}}_j,\ \mathfrak{b} \equiv \bar{\mathfrak{a}}_m,\ \mathfrak{c} \equiv (\beta)_m,\ \mathfrak{d} \equiv (\delta)_m,\ \mathfrak{g} \equiv (1;\ \gamma^1, \cdots, \gamma^{m-1}),$$
$$\mathfrak{f} \equiv (1;\ \eta^1, \cdots, \eta^{m-1}). \tag{15}$$

Now

$$
\begin{aligned}
(\beta;\, \gamma^1, \cdots, \gamma^{m-1}) &= \Big(\sum_{j=1}^{m-1} \bar{\mathfrak{a}}_j \cup (\beta)_m\Big) \cap \prod_{i=1}^{m-1} \Big(\sum_{j=1}^{m-1}{}_{(j\neq i)}\, \bar{\mathfrak{a}}_j \cup (\gamma^i)_{mi}\Big) \\
&= (\mathfrak{a} \cup \mathfrak{c}) \cap \prod_{i=1}^{m-1} \Big(\sum_{j=1}^{m-1}{}_{(j\neq i)}\, \bar{\mathfrak{a}}_j \cup (\gamma^i)_{mi}\Big) \\
&= (\mathfrak{a} \cup \mathfrak{c}) \cap \Big(\sum_{j=1}^{m-1} \bar{\mathfrak{a}}_j \cup \bar{\mathfrak{a}}_m\Big) \cap \prod_{i=1}^{m-1} \Big(\sum_{j=1}^{m-1}{}_{(j\neq i)}\, \bar{\mathfrak{a}}_j \cup (\gamma^i)_{mi}\Big) \\
&= (\mathfrak{a} \cup \mathfrak{c}) \cap (1;\, \gamma^1, \cdots, \gamma^{m-1}) = (\mathfrak{a} \cup \mathfrak{c}) \cap \mathfrak{g}.
\end{aligned}
$$

Similarly

$$(\delta;\, \eta', \cdots, \eta^{m-1}) = (\mathfrak{a} \cup \mathfrak{d}) \cap \mathfrak{f}.$$

Thus we are to prove that

$$\mathfrak{v} \cup ((\mathfrak{a} \cup \mathfrak{c}) \cap \mathfrak{g}) \leqq \mathfrak{w} \cup ((\mathfrak{a} \cup \mathfrak{d}) \cap \mathfrak{f}) \tag{11'}$$

is equivalent to the conjunction of

$$\mathfrak{v} \leqq \mathfrak{w}, \tag{12'}$$

$$\mathfrak{c} \leqq \mathfrak{d}, \tag{13'}$$

$$\mathfrak{a} \cap (((\mathfrak{a} \cup \mathfrak{c}) \cap \mathfrak{g}) \cup \mathfrak{f}) \leqq \mathfrak{w}. \tag{14'}$$

Let us suppose that (11′) holds. Then

$$\mathfrak{v} \cup ((\mathfrak{a} \cup \mathfrak{c}) \cap \mathfrak{g}) \cup \mathfrak{a} \leqq \mathfrak{w} \cup ((\mathfrak{a} \cup \mathfrak{d}) \cap \mathfrak{f}) \cup \mathfrak{a}. \tag{16}$$

The left member of (16) is

$$
\begin{aligned}
&\mathfrak{v} \cup ((\mathfrak{a} \cup \mathfrak{c}) \cap (\mathfrak{g} \cup \mathfrak{a})) \qquad \text{(by IV)} \\
&= \mathfrak{v} \cup ((\mathfrak{a} \cup \mathfrak{c}) \cap (\mathfrak{a} \cup \mathfrak{b}))
\end{aligned}
$$

since $\mathfrak{g} \cup \mathfrak{a} = \mathfrak{a} \cup \mathfrak{b}$ by the argument on page 171. But $\mathfrak{a} \cup \mathfrak{c} \leqq \mathfrak{a} \cup \mathfrak{b}$, whence the left member of (16) is $\mathfrak{v} \cup \mathfrak{a} \cup \mathfrak{c} = \mathfrak{a} \cup \mathfrak{c}$. Similarly the right member of (16) is $\mathfrak{a} \cup \mathfrak{d}$. Hence $\mathfrak{a} \cup \mathfrak{c} \leqq \mathfrak{a} \cup \mathfrak{d}$, whence $(\mathfrak{a} \cup \mathfrak{c}) \cap \mathfrak{b} \leqq (\mathfrak{a} \cup \mathfrak{d}) \cap \mathfrak{b}$. Since $\mathfrak{c}, \mathfrak{d} \leqq \mathfrak{b}$, IV applies, yielding $\mathfrak{c} \cup (\mathfrak{a} \cap \mathfrak{b}) \leqq \mathfrak{d} \cup (\mathfrak{a} \cap \mathfrak{b})$; but $\mathfrak{a} \cap \mathfrak{b} = 0$, whence $\mathfrak{c} \leqq \mathfrak{d}$, and (13′) holds. Now by (11′),

$$(\mathfrak{v} \cup ((\mathfrak{a} \cup \mathfrak{c}) \cap \mathfrak{g})) \cap \mathfrak{a} \leqq (\mathfrak{w} \cup ((\mathfrak{a} \cup \mathfrak{d}) \cap \mathfrak{f})) \cap \mathfrak{a},$$

whence by IV, since $\mathfrak{v}, \mathfrak{w} \leqq \mathfrak{a}$,

$$\mathfrak{v} \cup ((\mathfrak{a} \cup \mathfrak{c}) \cap \mathfrak{g} \cap \mathfrak{a}) \leqq \mathfrak{w} \cup ((\mathfrak{a} \cup \mathfrak{d}) \cap \mathfrak{f} \cap \mathfrak{a});$$

but $\mathfrak{g} \cap \mathfrak{a} = \mathfrak{f} \cap \mathfrak{a} = 0$, by the argument on page 170. Hence $\mathfrak{v} \leqq \mathfrak{w}$, and (12′) holds. Finally,

$$(\mathfrak{a} \cup \mathfrak{c}) \cap \mathfrak{g} \leqq \mathfrak{v} \cup ((\mathfrak{a} \cup \mathfrak{c}) \cap \mathfrak{g}) \leqq \mathfrak{w} \cup ((\mathfrak{a} \cup \mathfrak{d}) \cap \mathfrak{f}),$$

whence

$$(14'') \qquad (\mathfrak{a} \cup \mathfrak{c}) \cap \mathfrak{g} \leqq \mathfrak{w} \cup ((\mathfrak{a} \cup \mathfrak{d}) \cap \mathfrak{f}).$$

Thus (11′) implies (12′), (13′), (14″). Conversely, if (12′), (13′), (14″) hold, then

$$\mathfrak{v} \cup ((\mathfrak{a} \cup \mathfrak{c}) \cap \mathfrak{g}) \leqq \mathfrak{w} \cup (\mathfrak{w} \cup ((\mathfrak{a} \cup \mathfrak{d}) \cap \mathfrak{f})) = \mathfrak{w} \cup ((\mathfrak{a} \cup \mathfrak{d}) \cap \mathfrak{f}),$$

and (11′) holds. Thus it remains only to show that (12′), (13′), (14′) is equivalent to (12′), (13′), (14″), i.e., that (12′), (13′) implies the equivalence of (14′) and (14″). Suppose then that (14′) holds, i.e.,

$$\mathfrak{a} \cap (((\mathfrak{a} \cup \mathfrak{c}) \cap \mathfrak{g}) \cup \mathfrak{f}) \leqq \mathfrak{w}.$$

Then

$$\begin{aligned} \mathfrak{w} \cup \mathfrak{f} &\geqq (\mathfrak{a} \cap (((\mathfrak{a} \cup \mathfrak{c}) \cap \mathfrak{g}) \cup \mathfrak{f})) \cup \mathfrak{f} = (\mathfrak{a} \cup \mathfrak{f}) \cap (((\mathfrak{a} \cup \mathfrak{c}) \cap \mathfrak{g}) \cup \mathfrak{f}) \qquad \text{(by IV)} \\ &= (\mathfrak{a} \cup \mathfrak{b}) \cap (((\mathfrak{a} \cup \mathfrak{c}) \cap \mathfrak{g}) \cup \mathfrak{f}) = ((\mathfrak{a} \cup \mathfrak{c}) \cap \mathfrak{g}) \cup \mathfrak{f}. \end{aligned}$$

Therefore $(\mathfrak{a} \cup \mathfrak{c}) \cap \mathfrak{g} \leqq \mathfrak{w} \cup \mathfrak{f}$.

But $(\mathfrak{a} \cup \mathfrak{c}) \cap \mathfrak{g} \leqq \mathfrak{a} \cup \mathfrak{c} \leqq \mathfrak{a} \cup \mathfrak{d}$, whence

$$\begin{aligned} (\mathfrak{a} \cup \mathfrak{c}) \cap \mathfrak{g} &\leqq (\mathfrak{a} \cup \mathfrak{d}) \cap (\mathfrak{w} \cup \mathfrak{f}) \\ &= \mathfrak{w} \cup ((\mathfrak{a} \cup \mathfrak{d}) \cap \mathfrak{f}) \qquad \text{(by IV since } \mathfrak{w} \leqq \mathfrak{a}), \end{aligned}$$

and thus (14″) holds. Conversely, suppose (14″) holds. Then $(\mathfrak{a} \cup \mathfrak{c}) \cap \mathfrak{g} \leqq \mathfrak{w} \cup \mathfrak{f}$, and therefore

$$\begin{aligned} &\mathfrak{a} \cap (((\mathfrak{a} \cup \mathfrak{c}) \cap \mathfrak{g}) \cup \mathfrak{f}) \\ &\quad \leqq \mathfrak{a} \cap ((\mathfrak{w} \cup \mathfrak{f}) \cup \mathfrak{f}) \\ &\quad = \mathfrak{a} \cap (\mathfrak{w} \cup \mathfrak{f}) = \mathfrak{w} \cup (\mathfrak{a} \cap \mathfrak{f}) \qquad \text{(by IV)} \\ &\quad = \mathfrak{w} \cup 0 = \mathfrak{w}, \end{aligned}$$

whence (14′) holds. This completes the proof.

Lemma 11.5: If $\mathfrak{v}, \mathfrak{w} \leqq \sum_{j=1}^{m-1} \bar{\mathfrak{a}}_j$, and if $\mu, \rho, \nu^1, \cdots, \nu^{m-1}, \sigma^1, \cdots, \sigma^{m-1} \in \mathfrak{S}$, then

$$(17) \qquad \mathfrak{v} \cup (\mu;\ \nu^1, \cdots, \nu^{m-1}) \leqq \mathfrak{w} \cup (\rho;\ \sigma^1, \cdots, \sigma^{m-1})$$

is equivalent to the conjunction of

$$(18) \qquad \mathfrak{v} \leqq \mathfrak{w},$$

$$(19) \qquad (\mu)_{\underline{r}} \leqq (\rho)_{\underline{r}},$$

$$(20) \quad \left\{ \begin{aligned} &\Big(\sum_{j=2}^{m-1} \bar{\mathfrak{a}}_j \cup (\gamma^1\beta')_1\Big) \cap \prod_{i=2}^{m-1} \Big(\sum_{j=2\,(j \neq i)}^{m-1} \bar{\mathfrak{a}}_j \cup (-\eta^i)_{1i}\Big) \\ &\qquad \cup \Big(\Big(\sum_{j=2}^{m-1} \bar{\mathfrak{a}}_j\Big) \cap \Big(\prod_{i=2}^{m-1} \Big(\sum_{j=2\,(j \neq i)}^{m-1} \bar{\mathfrak{a}}_j \cup (\varepsilon^i)_{1i}\Big) \cup \bar{\mathfrak{a}}_1\Big)\Big) \leqq \mathfrak{w}, \end{aligned} \right.$$

where $\gamma^i = \nu^i - \sigma^i$ $(i = 1, \cdots, m-1)$; β', β'' are any idempotents such that $(\beta')_{\underline{r}} = (\mu)_{\underline{r}}$, $(\beta'')_{\underline{l}} = (\gamma^1\beta')_{\underline{l}}$; $\delta^i = \gamma^i\beta'\beta''$, $\varepsilon^i = \gamma^i\beta'(1-\beta'')$, and η^i is such that $\eta^i\gamma^1\beta' = \delta^i$ (cf. Theorem 10.1.)

PROOF: By Lemma 11.4, (17) is equivalent to (18), (19) and

$$(20)' \qquad \Big(\sum_{j=1}^{m-1} \bar{\mathfrak{a}}_j\Big) \cap [(\mu;\ \nu^1, \cdots, \nu^{m-1}) \cup (1;\ \sigma^1, \cdots, \sigma^{m-1})] \leqq \mathfrak{w}.$$

Now by (**), (20′) is equivalent to

$$\Big(\sum_{j=1}^{m-1} \bar{\mathfrak{a}}_j\Big) \cap \Big(\bar{\mathfrak{a}}_m \cup (\mu;\ \nu^1 - \sigma^1, \cdots, \nu^{m-1} - \sigma^{m-1})\Big) \leqq \mathfrak{w},$$

which in turn is equivalent by Theorem 10.1 to

$$(20)'' \quad \begin{cases} \Big(\sum_{j=1}^{m-1} \bar{\mathfrak{a}}_j\Big) \cap \Big[\bar{\mathfrak{a}}_m \cup \Big(\Big(\sum_{j=2}^{m-1} \bar{\mathfrak{a}}_j \cup (\gamma^1\beta')_1\Big) \cap \prod_{i=2}^{m-1}\Big(\sum_{j=2}^{m-1}{}_{(j\neq i)}\, \bar{\mathfrak{a}}_j \cup (-\eta^i)_{1i}\Big)\Big) \\ \qquad \cup \Big(\Big(\sum_{j=2}^{m-1} \bar{\mathfrak{a}}_j\Big) \cap \Big(\prod_{i=2}^{m-1}\Big(\sum_{j=2}^{m-1}{}_{(j\neq i)}\, \bar{\mathfrak{a}}_j \cup (\varepsilon^i)_{1i}\Big) \cup \bar{\mathfrak{a}}_1\Big)\Big)\Big] \leqq \mathfrak{w}. \end{cases}$$

Now (20″) is of the form $\big(\sum_{j=1}^{m-1} \bar{\mathfrak{a}}_j\big) \cap (\bar{\mathfrak{a}}_m \cup \mathfrak{a})$ with $\mathfrak{a} \leqq \sum_{j=1}^{m-1} \bar{\mathfrak{a}}_j$, whence IV applies and yields that (20″) is equivalent to (20), since $\bar{\mathfrak{a}}_m \cap \sum_{j=1}^{m-1} \bar{\mathfrak{a}}_j = 0$. This completes the proof.

LEMMA 11.6: If $\mathfrak{v}$, $\mathfrak{w} \leqq \sum_{j=1}^{m-1} \bar{\mathfrak{a}}_j$, and if $\mu, \rho, \nu^1, \cdots, \nu^{m-1}, \sigma^1, \cdots, \sigma^{m-1} \in \mathfrak{S}$, then

$$(17) \qquad \mathfrak{v} \cup (\mu;\ \nu^1, \cdots, \nu^{m-1}) \leqq \mathfrak{w} \cup (\rho;\ \sigma^1, \cdots, \sigma^{m-1})$$

is equivalent to the conjunction of

$$(18) \qquad \mathfrak{v} \leqq \mathfrak{w},$$

$$(19) \qquad (\mu)_{\underline{r}} \leqq (\rho)_{\underline{r}},$$

$$(21) \quad \begin{cases} \Big(\Big(\sum_{j=1}^{m-2} \bar{\mathfrak{a}}_j\Big) \cup (\gamma^{m-1}\beta')_{m-1}\Big) \cap \prod_{i=1}^{m-2}\Big(\sum_{j=1}^{m-2}{}_{(j\neq i)}\, \bar{\mathfrak{a}}_j \cup (-\eta^i)_{m-1,\, i}\Big) \\ \qquad \cup \Big(\Big(\sum_{j=1}^{m-2} \bar{\mathfrak{a}}_j\Big) \cap \Big(\prod_{i=1}^{m-2}\Big(\sum_{j=1}^{m-2}{}_{(j\neq i)}\, \bar{\mathfrak{a}}_j \cup (\varepsilon^i)_{m-1,\, i}\Big) \cap \bar{\mathfrak{a}}_{m-1}\Big)\Big) \leqq \mathfrak{w}), \end{cases}$$

where $\gamma^i = \nu^i - \sigma^i$ $(i = 1, \cdots, m-1)$; β', β'' are any idempotents such that $(\beta')_{\underline{r}} = (\mu)_{\underline{r}}$, $(\beta'')_{\underline{l}} = (\gamma^{m-1}\beta')_{\underline{l}}$; $\delta^i = \gamma^i\beta'\beta''$, $\varepsilon^i = \gamma^i\beta'(1-\beta'')$, and η^i is such that $\eta^i\gamma^{m-1}\beta' = \delta^i$.

PROOF: This lemma differs from Lemma 11.5 only in that the indices 1 and $m-1$ have been interchanged.

THEOREM 11.2: The property (*) holds.

PROOF: We must show that if $\mathfrak{v}'$, $\mathfrak{w}' \leqq \sum_{j=1}^{m-1} \bar{\mathfrak{a}}_j$ and $\mathfrak{v}''$, $\mathfrak{w}'' \leqq \sum_{j=1}^{m-1} \bar{\mathfrak{a}}''_j$ and if $\mathfrak{v}'$, $\mathfrak{v}''$ and also $\mathfrak{w}'$, $\mathfrak{w}''$ correspond under $\mathscr{I}_{m-1}$, then for $\mu, \rho, \nu^1, \cdots, \nu^{m-1}, \sigma^1, \cdots, \sigma^{m-1} \in \mathfrak{S}$,

$$(17)' \qquad \mathfrak{v}' \cup (\mu; \nu^1, \cdots, \nu^{m-1})' \leqq \mathfrak{w}' \cup (\rho; \sigma^1, \cdots, \sigma^{m-1})'$$

implies

$$(17)'' \qquad \mathfrak{v}'' \cup (\mu; \nu^1, \cdots, \nu^{m-1})'' \leqq \mathfrak{w}'' \cup (\rho; \sigma^1, \cdots, \sigma^{m-1})''.$$

Now by Lemma 11.6, (17)′ implies the conjunction of (18)′, (19)′, (21)′, obtained from (18), (19), (21) by inserting an accent ′ on every quantity. By the character of $\mathscr{I}_{m-1}$, (18)′, (19)′ imply respectively (18)″, (19)″.

Moreover, if we apply $\mathscr{I}_{m-1}$ to (21)′, since each factor and term in (21)′ is $\leqq \sum_{j=1}^{m-1} \bar{\mathfrak{a}}_j$, we may apply $\mathscr{I}_{m-1}$ separately to these quantities. By properties (A_{m-1}), (B_{m-1}), together with the relation $\bar{\mathfrak{a}}_j = (1)_j$ $(j = 1, \cdots, m-1)$, we obtain (21)″. Hence we have proved (18)″, (19)″, (21)″, and thus by Lemma 11.6, (17)″ holds. This completes the proof.

We have seen that (**) implies (*), whence it remains only to establish (**). In Chapter XII we shall prove (**) in Case I, and in Chapters XIII, XIV we shall establish (**) for Case II.

PART II · CHAPTER XII

Treatment of Case I

The statement (**) (cf. Chapter XI, page 186) will be proved here in Case I. The hypotheses on n, L', L'', $\mathfrak{S}$, m set forth on page 181, in addition to the condition of Case I, viz., $m = 2, \cdots, n-1$, will be assumed. In the material throughout this chapter, L will be used to represent either L' or L''.

LEMMA 12.1: If $\mathfrak{a}, \mathfrak{b}, \mathfrak{c}, \mathfrak{u} \in L$, $\mathfrak{a} \sim \mathfrak{b} \pmod{\mathfrak{u}}$ and $\mathfrak{c} \cap (\mathfrak{a} \cup \mathfrak{u}) = 0$, then $\mathfrak{a} \cup \mathfrak{c} \sim \mathfrak{b} \cup \mathfrak{c} \pmod{\mathfrak{u}}$.

PROOF: Since $\mathfrak{a} \cup \mathfrak{u} = \mathfrak{b} \cup \mathfrak{u}$ we have $(\mathfrak{a} \cup \mathfrak{c}) \cup \mathfrak{u} = (\mathfrak{b} \cup \mathfrak{c}) \cup \mathfrak{u}$. But by hypothesis, $(\mathfrak{c} \cup 0) \cap (\mathfrak{a} \cup \mathfrak{u}) = 0$, $(\mathfrak{c} \cup 0) \cap (\mathfrak{b} \cup \mathfrak{u}) = 0$, whence Part I, Theorem 1.2 applies, yielding $(\mathfrak{a} \cup \mathfrak{c}) \cap \mathfrak{u} = (\mathfrak{b} \cup \mathfrak{c}) \cap \mathfrak{u} = 0$. Thus $\mathfrak{a} \cup \mathfrak{c} \sim \mathfrak{b} \cup \mathfrak{c} \pmod{\mathfrak{u}}$.

THEOREM 12.1: If $\gamma \in \mathfrak{S}$, and $i = 1, \cdots, m-1$, then

$$(1) \qquad \sum_{j=1}^{m} \bar{\mathfrak{a}}_j \sim \sum_{j=1}^{m}{}_{(j \neq i)} \bar{\mathfrak{a}}_j \cup \bar{\mathfrak{a}}_n \qquad \pmod{\bar{\mathfrak{c}}_{ni}},$$

$$(2) \qquad \sum_{j=1}^{m}{}_{(j \neq i)} \bar{\mathfrak{a}}_j \cup \bar{\mathfrak{a}}_n \sim \sum_{j=1}^{m-1}{}_{(j \neq i)} \bar{\mathfrak{a}}_j \cup \bar{\mathfrak{a}}_n \cup (\gamma)_{mi} \qquad \pmod{\bar{\mathfrak{a}}_i},$$

$$(3) \qquad \sum_{j=1}^{m-1}{}_{(j \neq i)} \bar{\mathfrak{a}}_j \cup \bar{\mathfrak{a}}_n \cup (\gamma)_{mi} \sim \sum_{j=1}^{m} \bar{\mathfrak{a}}_j \qquad \pmod{\bar{\mathfrak{c}}_{ni}}.$$

The perspective isomorphisms corresponding to (1), (2), (3) will be denoted by P, Q, R, respectively; P is a perspective isomorphism of $L(0, \sum_{j=1}^{m} \bar{\mathfrak{a}}_j)$ and $L(0, \sum_{j=1}^{m}{}_{(j \neq i)} \bar{\mathfrak{a}}_j \cup \bar{\mathfrak{a}}_n)$ with axis $\bar{\mathfrak{c}}_{ni}$, Q of $L(0, \sum_{j=1}^{m}{}_{(j \neq i)} \bar{\mathfrak{a}}_j \cup \bar{\mathfrak{a}}_n)$ and $L\big(0, \sum_{j=1}^{m-1}{}_{(j \neq i)} \bar{\mathfrak{a}}_j \cup \bar{\mathfrak{a}}_n \cup (\gamma)_{mi}\big)$ with axis $\bar{\mathfrak{a}}_i$, and R of $L\big(0, \sum_{j=1}^{m-1}{}_{(j \neq i)} \bar{\mathfrak{a}}_j \cup \bar{\mathfrak{a}}_n \cup (\gamma)_{mi}\big)$ and $L(0, \sum_{j=1}^{m} \bar{\mathfrak{a}}_j)$ with axis $\bar{\mathfrak{c}}_{ni}$. The product PQR in that order is denoted by $W = W_i(\gamma)$; W is a projective automorphism of $L(0, \sum_{j=1}^{m} \bar{\mathfrak{a}}_j)$.

PROOF: It suffices to prove (1), (2), (3), since the remaining statements are evident when these are established.

(1) Clearly $\bar{\mathfrak{a}}_i \sim \bar{\mathfrak{a}}_n \pmod{\bar{\mathfrak{c}}_{ni}}$; in fact $\bar{\mathfrak{a}}_i \cup \bar{\mathfrak{c}}_{ni} = \bar{\mathfrak{a}}_n \cup \bar{\mathfrak{c}}_{ni} = \bar{\mathfrak{a}}_i \cup \bar{\mathfrak{a}}_n$. Moreover, $(\bar{\mathfrak{a}}_i, \bar{\mathfrak{a}}_n, \sum_{j=1\,(j\neq i)}^{m} \bar{\mathfrak{a}}_j)\perp$, since $m < n$. Hence Part I, Theorem 3.5 applies, and we have

$$\bar{\mathfrak{a}}_i \cup \sum_{j=1\,(j\neq i)}^{m} \bar{\mathfrak{a}}_j \sim \bar{\mathfrak{a}}_n \cup \sum_{j=1\,(j\neq i)}^{m} \bar{\mathfrak{a}}_j \qquad \pmod{\bar{\mathfrak{c}}_{ni}},$$

i.e., (1) holds.

(2) Since $(\gamma)_{mi} \in L_{mi}$, we have $\bar{\mathfrak{a}}_m \sim (\gamma)_{mi} \pmod{\bar{\mathfrak{a}}_i}$. But $(\sum_{j=1\,(j\neq i)}^{m-1} \bar{\mathfrak{a}}_j \cup \bar{\mathfrak{a}}_n) \cap (\bar{\mathfrak{a}}_m \cup \bar{\mathfrak{a}}_i) = 0$, so Lemma 12.1 applies, yielding

$$\bar{\mathfrak{a}}_m \cup \sum_{j=1\,(j\neq i)}^{m-1} \bar{\mathfrak{a}}_j \cup \bar{\mathfrak{a}}_n \sim (\gamma)_{mi} \cup \sum_{j=1\,(j\neq i)}^{m-1} \bar{\mathfrak{a}}_j \cup \bar{\mathfrak{a}}_n \qquad \pmod{\bar{\mathfrak{a}}_i},$$

i.e. (2) holds.

(3) Clearly $\bar{\mathfrak{a}}_n \sim \bar{\mathfrak{a}}_i \pmod{\bar{\mathfrak{c}}_{ni}}$; indeed, $\bar{\mathfrak{a}}_n \cup \bar{\mathfrak{c}}_{ni} = \bar{\mathfrak{a}}_i \cup \bar{\mathfrak{c}}_{ni} = \bar{\mathfrak{a}}_i \cup \bar{\mathfrak{a}}_n$. Moreover, $\bar{\mathfrak{a}}_n \cap \bar{\mathfrak{a}}_i = 0$, and

$$\Big(\sum_{j=1\,(j\neq i)}^{m-1} \bar{\mathfrak{a}}_j \cup (\gamma)_{mi}\Big) \cap (\bar{\mathfrak{a}}_n \cup \bar{\mathfrak{a}}_i) = 0$$

by Part I, Theorem 1.2, since

$$\Big(\sum_{j=1\,(j\neq i)}^{m-1} \bar{\mathfrak{a}}_j \cup \bar{\mathfrak{a}}_n\Big) \cap \big((\gamma)_{mi} \cup \bar{\mathfrak{a}}_i\big) = \Big(\sum_{j=1\,(j\neq i)}^{m-1} \bar{\mathfrak{a}}_j \cup \bar{\mathfrak{a}}_n\Big) \cap (\bar{\mathfrak{a}}_m \cup \bar{\mathfrak{a}}_i) = 0.$$

Thus $\big(\bar{\mathfrak{a}}_n, \bar{\mathfrak{a}}_i, \sum_{j=1\,(j\neq i)}^{m-1} \bar{\mathfrak{a}}_j \cup (\gamma)_{mi}\big)\perp$, and by Part I, Theorem 3.5,

$$\bar{\mathfrak{a}}_n \cup \sum_{j=1\,(j\neq i)}^{m-1} \bar{\mathfrak{a}}_j \cup (\gamma)_{mi} \sim \bar{\mathfrak{a}}_i \cup \sum_{j=1\,(j\neq i)}^{m-1} \bar{\mathfrak{a}}_j \cup (\gamma)_{mi} \qquad \pmod{\bar{\mathfrak{c}}_{ni}},$$

i.e.,

$$\sum_{j=1\,(j\neq i)}^{m-1} \bar{\mathfrak{a}}_j \cup \bar{\mathfrak{a}}_n \cup (\gamma)_{mi} \sim \sum_{j=1\,(j\neq i)}^{m-1} \bar{\mathfrak{a}}_j \cup \bar{\mathfrak{a}}_i \cup (\gamma)_{mi}.$$

But $\bar{\mathfrak{a}}_i \cup (\gamma)_{mi} = \bar{\mathfrak{a}}_i \cup \bar{\mathfrak{a}}_m$, whence the right member is precisely $\sum_{j=1}^{m} \bar{\mathfrak{a}}_j$, and (3) holds.

LEMMA 12.2: Let $\gamma \in \mathfrak{S}$, $i = 1, \cdots, m-1$. Then $W = W_i(\gamma)$ has the property that $\mathfrak{u} \leqq \sum_{j=1}^{m-1} \bar{\mathfrak{a}}_j$ implies $\mathfrak{u}W = \mathfrak{u}$.

PROOF: Since

$$\sum_{j=1}^{m-1} \bar{\mathfrak{a}}_j \cup \bar{\mathfrak{c}}_{ni} = \Big(\sum_{j=1\,(j\neq i)}^{m-1} \bar{\mathfrak{a}}_j \cup \bar{\mathfrak{a}}_n\Big) \cup \bar{\mathfrak{c}}_{ni},$$

it follows that

$$(4)\qquad \sum_{j=1}^{m-1} \bar{\mathfrak{a}}_j \underset{P}{\rightarrow} \sum_{j=1}^{m-1}{}_{(j\neq i)} \bar{\mathfrak{a}}_j \cup \bar{\mathfrak{a}}_n \underset{Q}{\rightarrow} \sum_{j=1}^{m-1}{}_{(j\neq i)} \bar{\mathfrak{a}}_j \cup \bar{\mathfrak{a}}_n \underset{R}{\rightarrow} \sum_{j=1}^{m-1} \bar{\mathfrak{a}}_j,$$

whence if we define $\mathfrak{u}'$, $\mathfrak{u}''$, $\mathfrak{u}'''$ $(= \mathfrak{u}W)$ by

$$\mathfrak{u} \underset{P}{\rightarrow} \mathfrak{u}' \underset{Q}{\rightarrow} \mathfrak{u}'' \underset{R}{\rightarrow} \mathfrak{u}''',$$

we have, since $\mathfrak{u} \leqq \sum_{j=1}^{m-1} \bar{\mathfrak{a}}_j$,

$$\mathfrak{u}',\ \mathfrak{u}'' \leqq \sum_{j=1}^{m-1}{}_{(j\neq i)} \bar{\mathfrak{a}}_j \cup \bar{\mathfrak{a}}_n, \qquad \mathfrak{u}''' \leqq \sum_{j=1}^{m-1} \bar{\mathfrak{a}}_j.$$

Since Q leaves $\sum_{j=1}^{m-1}{}_{(j\neq i)} \bar{\mathfrak{a}}_j \cup \bar{\mathfrak{a}}_n$ invariant by (4), it is clear that Q leaves all elements of $L(0, \sum_{j=1}^{m-1}{}_{(j\neq i)} \bar{\mathfrak{a}}_j \cup \bar{\mathfrak{a}}_n)$ invariant (because Q is a perspectivity; this would not be necessarily true for a projectivity) whence $\mathfrak{u}' = \mathfrak{u}''$. Since P, R have the same axis, viz., $\bar{\mathfrak{c}}_{ni}$, these transformations must be inverse correspondences of $L(0, \sum_{j=1}^{m-1} \bar{\mathfrak{a}}_j)$ and $L(0, \sum_{j=1}^{m-1}{}_{(j\neq i)} \bar{\mathfrak{a}}_j \cup \bar{\mathfrak{a}}_n)$. Thus $\mathfrak{u}''' = \mathfrak{u}$ i.e., $\mathfrak{u}W = \mathfrak{u}$.

LEMMA 12.3: Let $\gamma \in \mathfrak{S}$, $i = 1, \cdots, m-1$. Then $W = W_i(\gamma)$ has the property that $\bar{\mathfrak{a}}_i \leqq \mathfrak{u} \leqq \sum_{j=1}^{m} \bar{\mathfrak{a}}_j$ implies $\mathfrak{u}W = \mathfrak{u}$.

PROOF: Let us define $\mathfrak{u}'$, $\mathfrak{u}''$, $\mathfrak{u}'''(= \mathfrak{u}W)$ by

$$\mathfrak{u} \underset{P}{\rightarrow} \mathfrak{u}' \underset{Q}{\rightarrow} \mathfrak{u}'' \underset{R}{\rightarrow} \mathfrak{u}'''.$$

Since P, Q, R have the axes $\bar{\mathfrak{c}}_{ni}$, $\bar{\mathfrak{a}}_i$, $\bar{\mathfrak{c}}_{ni}$, respectively,

$$\mathfrak{u} \cup \bar{\mathfrak{c}}_{ni} = \mathfrak{u}' \cup \bar{\mathfrak{c}}_{ni},\ \mathfrak{u}' \cup \bar{\mathfrak{a}}_i = \mathfrak{u}'' \cup \bar{\mathfrak{a}}_i,\ \mathfrak{u}'' \cup \bar{\mathfrak{c}}_{ni} = \mathfrak{u}''' \cup \bar{\mathfrak{c}}_{ni}.$$

Thus

$$(5)\qquad \mathfrak{u} \cup \bar{\mathfrak{a}}_i \cup \bar{\mathfrak{a}}_n = \mathfrak{u}' \cup \bar{\mathfrak{a}}_i \cup \bar{\mathfrak{a}}_n = \mathfrak{u}'' \cup \bar{\mathfrak{a}}_i \cup \bar{\mathfrak{a}}_n = \mathfrak{u}''' \cup \bar{\mathfrak{a}}_i \cup \bar{\mathfrak{a}}_n,$$

since $\bar{\mathfrak{a}}_i$, $\bar{\mathfrak{c}}_{ni} \leqq \bar{\mathfrak{a}}_i \cup \bar{\mathfrak{a}}_n$. Now it is clear that

$$\bar{\mathfrak{a}}_i \underset{P}{\rightarrow} \bar{\mathfrak{a}}_n \underset{Q}{\rightarrow} \bar{\mathfrak{a}}_n \underset{R}{\rightarrow} \bar{\mathfrak{a}}_i,$$

whence $\mathfrak{u} \geqq \bar{\mathfrak{a}}_i$ yields $\mathfrak{u}' \geqq \bar{\mathfrak{a}}_n$, $\mathfrak{u}'' \geqq \bar{\mathfrak{a}}_n$, $\mathfrak{u}''' \geqq \bar{\mathfrak{a}}_i$. In particular, $\mathfrak{u} \cup \bar{\mathfrak{a}}_i = \mathfrak{u}$, $\mathfrak{u}''' \cup \bar{\mathfrak{a}}_i = \mathfrak{u}'''$, whence by (5) we have $\mathfrak{u} \cup \bar{\mathfrak{a}}_n = \mathfrak{u}''' \cup \bar{\mathfrak{a}}_n$. Hence

$$(\sum_{j=1}^{m} \bar{\mathfrak{a}}_j) \cap (\mathfrak{u} \cup \bar{\mathfrak{a}}_n) = (\sum_{j=1}^{m} \bar{\mathfrak{a}}_j) \cap (\mathfrak{u}''' \cup \bar{\mathfrak{a}}_n),$$

and

$$\mathfrak{u} \cup ((\sum_{j=1}^{m} \bar{\mathfrak{a}}_j) \cap \bar{\mathfrak{a}}_n) = \mathfrak{u}''' \cup ((\sum_{j=1}^{m} \bar{\mathfrak{a}}_j) \cap \bar{\mathfrak{a}}_n) \qquad \text{(by IV);}$$

since $(\bar{\mathfrak{a}}_1, \cdots, \bar{\mathfrak{a}}_n)\perp$, $m<n$, $(\sum_{j=1}^m \bar{\mathfrak{a}}_j) \cap \bar{\mathfrak{a}}_n = 0$, and we have $\mathfrak{u} = \mathfrak{u}''' = \mathfrak{u}W$.

LEMMA 12.4: Let $\gamma \in \mathfrak{S}$, $i=1, \cdots, m-1$. Then for every $\delta \in \mathfrak{S}$, $W = W_i(\gamma)$ has the property

$$(6) \qquad (\delta)_{mi} W = (\delta + \gamma)_{mi}.$$

PROOF: It is clear that since $(\delta)_{mi} \leqq \bar{\mathfrak{a}}_i \cup \bar{\mathfrak{a}}_m \leqq \sum_{j=1}^m \bar{\mathfrak{a}}_j$, $(\delta)_{mi} W$ is defined. Define $\mathfrak{u} \equiv (\delta)_{mi}$ and $\mathfrak{u}'$, $\mathfrak{u}''$, $\mathfrak{u}'''(=\mathfrak{u}W)$ by

$$\mathfrak{u} \underset{P}{\rightarrow} \mathfrak{u}' \underset{Q}{\rightarrow} \mathfrak{u}'' \underset{R}{\rightarrow} \mathfrak{u}'''.$$

Clearly

$$\bar{\mathfrak{a}}_i \cup \bar{\mathfrak{a}}_m \underset{P}{\rightarrow} \bar{\mathfrak{a}}_n \cup \bar{\mathfrak{a}}_m \underset{Q}{\rightarrow} \bar{\mathfrak{a}}_n \cup (\gamma)_{mi} \underset{R}{\rightarrow} \bar{\mathfrak{a}}_i \cup (\gamma)_{mi} = \bar{\mathfrak{a}}_i \cup \bar{\mathfrak{a}}_m,$$

whence $\mathfrak{u} = (\delta)_{mi} \leqq \bar{\mathfrak{a}}_i \cup \bar{\mathfrak{a}}_m$ yields

$$(7) \qquad \mathfrak{u}' \leqq \bar{\mathfrak{a}}_n \cup \bar{\mathfrak{a}}_m, \quad \mathfrak{u}'' \leqq \bar{\mathfrak{a}}_n \cup (\gamma)_{mi}, \quad \mathfrak{u}''' \leqq \bar{\mathfrak{a}}_i \cup \bar{\mathfrak{a}}_m.$$

The relations (7), together with the fact that P, Q, R are perspective isomorphisms with axes $\bar{\mathfrak{c}}_{ni}$, $\bar{\mathfrak{a}}_i$, $\bar{\mathfrak{c}}_{ni}$ respectively, yield by Part I, Theorem 3.3, Definition 3.4,

$$\mathfrak{u}' = (\mathfrak{u} \cup \bar{\mathfrak{c}}_{ni}) \cap (\bar{\mathfrak{a}}_n \cup \bar{\mathfrak{a}}_m), \quad \mathfrak{u}'' = (\mathfrak{u}' \cup \bar{\mathfrak{a}}_i) \cap (\bar{\mathfrak{a}}_n \cup (\gamma)_{mi}),$$
$$\mathfrak{u}''' = (\mathfrak{u}'' \cup \bar{\mathfrak{c}}_{ni}) \cap (\bar{\mathfrak{a}}_i \cup \bar{\mathfrak{a}}_m).$$

Combining these relations and recalling that $\mathfrak{u} = (\delta)_{mi}$, we obtain

$$(8) \quad \mathfrak{u}''' = ((((((\delta)_{mi} \cup \bar{\mathfrak{c}}_{ni}) \cap (\bar{\mathfrak{a}}_n \cup \bar{\mathfrak{a}}_m)) \cup \bar{\mathfrak{a}}_i) \cap (\bar{\mathfrak{a}}_n \cup (\gamma)_{mi})) \cup \bar{\mathfrak{c}}_{ni}) \cap (\bar{\mathfrak{a}}_i \cup \bar{\mathfrak{a}}_m).$$

By Definition 7.5, the right member of (8) is precisely $(\delta)_{mi} \oplus_n (\gamma)_{mi}$. Thus

$$(\delta)_{mi} W = \mathfrak{u}W = \mathfrak{u}''' = (\delta)_{mi} \oplus (\gamma)_{mi} = (\delta + \gamma)_{mi}.$$

LEMMA 12.5: Let $\gamma \in \mathfrak{S}$, $i = 1, \cdots, m-1$. Then for every $\omega, \psi^1, \cdots, \psi^{m-1} \in \mathfrak{S}$

$$(9) \quad (\omega;\, \psi^1, \cdots, \psi^i, \cdots, \psi^{m-1})W = (\omega;\, \psi^1, \cdots, \psi^i + \gamma, \cdots, \psi^{m-1}).$$

PROOF: Since $(\omega; \psi^1, \cdots, \psi^{m-1}) \leqq \sum_{j=1}^m \mathfrak{a}_j$, the left member of (9) is defined. Now

$$(10) \quad (\omega;\, \psi^1, \cdots, \psi^i, \cdots, \psi^{m-1}) = \Big(\sum_{j=1}^{m-1} \bar{\mathfrak{a}}_j \cup (\omega)_m\Big) \cap \prod_{h=1\,(h\neq i)}^{m-1} \Big(\sum_{j=1\,(j\neq h)}^{m-1} \bar{\mathfrak{a}}_j \cup (\psi^h)_{mh}\Big) \cap \Big(\sum_{j=1\,(j\neq i)}^{m-1} \bar{\mathfrak{a}}_j \cup (\psi^i)_{mi}\Big).$$

Since W is a (projective) automorphism, we may, in applying W to (10), apply it to each factor separately. Each factor with the exception of the last, viz., $\sum_{j=1 (j\neq i)}^{m-1} \bar{\mathfrak{a}}_j \cup (\psi^i)_{mi}$, is $\geqq \bar{\mathfrak{a}}_i$, and is therefore left invariant by Lemma 12.3. In applying W to the last factor, we may apply it to each term separately, again since it is an automorphism. But by Lemma 12.2, $(\sum_{j=1 (j\neq i)}^{m-1} \bar{\mathfrak{a}}_j)W = \sum_{j=1 (j\neq i)}^{m-1} \bar{\mathfrak{a}}_j$, since $\sum_{j=1 (j\neq i)}^{m-1} \bar{\mathfrak{a}}_j \leqq \sum_{j=1}^{m-1} \bar{\mathfrak{a}}_j$. Moreover, by Lemma 12.4, $(\psi^i)_{mi} W = (\psi^i + \gamma)_{mi}$. Hence

$$(\omega;\ \psi^1, \cdots, \psi^i, \cdots, \psi^{m-1})W$$
$$= \Big(\sum_{j=1}^{m-1} \bar{\mathfrak{a}}_j \cup (\omega)_m\Big) \cap \prod_{h=1 (h\neq i)}^{m-1} \Big(\sum_{j=1 (j\neq h)}^{m-1} \bar{\mathfrak{a}}_j \cup (\psi^h)_{mh}\Big) \cap \Big(\sum_{j=1 (j\neq i)}^{m-1} \bar{\mathfrak{a}}_j \cup (\psi^i + \gamma)_{mi}\Big)$$
$$= (\omega;\ \psi^1, \cdots, \psi^i + \gamma, \cdots, \psi^{m-1}).$$

DEFINITION 12.1: If $\gamma^1, \cdots, \gamma^{m-1} \in \mathfrak{S}$, define

$$X \equiv X(\gamma^1, \cdots, \gamma^{m-1}) \equiv W_1(\gamma^1) W_2(\gamma^2) \cdots W_{m-1}(\gamma^{m-1}). \tag{11}$$

LEMMA 12.6: If $\gamma^1, \cdots, \gamma^{m-1} \in \mathfrak{S}$, the transformation X is a projective automorphism of $L(0, \sum_{j=1}^{m} \bar{\mathfrak{a}}_j)$ with the properties

(a) $\mathfrak{u} \leqq \sum_{j=1}^{m-1} \bar{\mathfrak{a}}_j$ or $\mathfrak{u} \geqq \sum_{j=1}^{m-1} \bar{\mathfrak{a}}_j$ (together with $\mathfrak{u} \leqq \sum_{j=1}^{m} \bar{\mathfrak{a}}_j$) implies $\mathfrak{u}X = \mathfrak{u}$;

(b) $(\omega;\ \psi^1, \cdots, \psi^{m-1})X = (\omega;\ \psi^1 + \gamma^1, \cdots, \psi^{m-1} + \gamma^{m-1})$ for every $\omega, \psi^1, \cdots, \psi^{m-1} \in \mathfrak{S}$.

PROOF: Clearly X is a projective automorphism of $\sum_{j=1}^{m} \bar{\mathfrak{a}}_j$ since it is a product of such automorphisms. If $\mathfrak{u} \leqq \sum_{j=1}^{m-1} \bar{\mathfrak{a}}_j$, $\mathfrak{u}$ is left invariant by each $W_i(\gamma^i)$ $(i = 1, \cdots, m-1)$ by Lemma 12.2 and hence by their product X. If $\mathfrak{u} \geqq \sum_{j=1}^{m-1} \bar{\mathfrak{a}}_j$, then $\mathfrak{u} \geqq \bar{\mathfrak{a}}_i$ $(i = 1, \cdots, m-1)$ and $\mathfrak{u}$ is left invariant by each $W_i(\gamma^i)$ $(i = 1, \cdots, m-1)$ by Lemma 12.3 and hence by their product X. This proves (a). Now clearly

$$(\omega;\ \psi^1, \cdots, \psi^{m-1}) W_1(\gamma^1) = (\omega;\ \psi^1 + \gamma^1,\ \psi^2, \cdots, \psi^{m-1})$$

by Lemma 12.5, and also

$$(\omega;\ \psi^1, \cdots, \psi^{m-1}) W_1(\gamma^1) \cdots W_i(\gamma^i)$$
$$= (\omega;\ \psi^1 + \gamma^1, \cdots, \psi^i + \gamma^i, \psi^{i+1}, \cdots, \psi^{m-1})$$

implies

$$(\omega;\ \psi^1, \cdots, \psi^{m-1}) W_1(\gamma^1) \cdots W_{i+1}(\gamma^{i+1})$$
$$= (\omega;\ \psi^1 + \gamma^1, \cdots, \psi^{i+1} + \gamma^{i+1},\ \psi^{i+2}, \cdots, \psi^{m-1})$$

by Lemma 12.5, whence (b) follows by induction on $i = 1, \cdots, m-1$.

LEMMA 12.7: If $\beta, \gamma^1, \cdots, \gamma^{m-1}, \delta^1, \cdots, \delta^{m-1} \in \mathfrak{S}$, then

$$(12)\quad \Big(\sum_{j=1}^{m-1} \bar{\mathfrak{a}}_j\Big) \cap \big((1;\, \gamma^1, \cdots, \gamma^{m-1}) \cup (\beta;\, \delta^1, \cdots, \delta^{m-1})\big)$$
$$= \Big(\sum_{j=1}^{m-1} \bar{\mathfrak{a}}_j\Big) \cap \big(\bar{\mathfrak{a}}_m \cup (\beta;\, \delta^1 - \gamma^1, \cdots, \delta^{m-1} - \gamma^{m-1})\big).$$

PROOF: Denote the left and right members of (12) respectively by $\mathfrak{a}$, $\mathfrak{b}$. Since $\mathfrak{a} \leqq \sum_{j=1}^{m-1} \bar{\mathfrak{a}}_j$, $\mathfrak{a} = \mathfrak{a}X(-\gamma^1, \cdots, -\gamma^{m-1})$ by Lemma 12.6 (a). Since $X = X(-\gamma^1, \cdots, -\gamma^{m-1})$ is an automorphism, we may apply it to the terms and factors in $\mathfrak{a}$ separately. Hence

$$\mathfrak{a} = \mathfrak{a}X = \Big(\Big(\sum_{j=1}^{m-1} \bar{\mathfrak{a}}_j\Big)X\Big) \cap \big((1;\, \gamma^1, \cdots, \gamma^{m-1})X \cup (\beta;\, \delta^1, \cdots, \delta^{m-1})X\big)$$
$$= \Big(\sum_{j=1}^{m-1} \bar{\mathfrak{a}}_j\Big) \cap \big((1;\gamma^1, \cdots, \gamma^{m-1})X \cup (\beta; \delta^1 \cdots, \delta^{m-1})X\big) \quad \text{(by Lemma 12.6(a))}$$
$$= \Big(\sum_{j=1}^{m-1} \bar{\mathfrak{a}}_j\Big) \cap \big((1;\, \gamma^1 - \gamma^1, \cdots, \gamma^{m-1} - \gamma^{m-1}) \cup (\beta; \delta^1 - \gamma^1, \cdots, \delta^{m-1} - \gamma^{m-1})\big)$$
$$\text{(by Lemma 12.6(b))}$$
$$= \Big(\sum_{j=1}^{m-1} \bar{\mathfrak{a}}_j\Big) \cap \big((1;\, 0, \cdots, 0) \cup (\beta;\, \delta^1 - \gamma^1, \cdots, \delta^{m-1} - \gamma^{m-1})\big)$$
$$= \mathfrak{b},$$

since $(1;\, 0, \cdots, 0) = \Big(\sum_{j=1}^{m-1} \bar{\mathfrak{a}}_j \cup (1)_m\Big) \cap \prod_{i=1}^{m-1} \Big(\sum_{j=1\,(j \neq i)}^{m-1} \bar{\mathfrak{a}}_j \cup (0)_{mi}\Big) = \Big(\sum_{j=1}^{m-1} \bar{\mathfrak{a}}_j \cup \bar{\mathfrak{a}}_m\Big) \cap \prod_{i=1}^{m-1} \Big(\sum_{j=1\,(j \neq j)}^{m-1} \bar{\mathfrak{a}}_j \cup \bar{\mathfrak{a}}_m\Big) = \Big(\sum_{j=1}^{m-1} \bar{\mathfrak{a}}_j \cup \bar{\mathfrak{a}}_m\Big) \cap \bar{\mathfrak{a}}_m = \bar{\mathfrak{a}}_m$. Hence (12) is established.

This completes the treatment for Case I, since Lemma 12.7 is precisely (**) for Case I, and since the statement (*) follows now from (**), as was seen at the end of Chapter XI, whence S_{m-1} implies S_m for $m = 2, \cdots, n-1$.

PART II · CHAPTER XIII

Preliminary Lemmas for the Treatment of Case II

The relation (**) (cf. page 186) will be proved in Chapter XIV. In order to carry out the proof, we need two results concerning $\bar{R}_{\mathfrak{S}_{m-1}}$; this chapter is devoted to the proof of these results, Lemma 13.1 and Theorem 13.1, respectively.

First we shall establish the relation (**) with m replaced by $m-1$, for the special lattice $L = \bar{R}_{\mathfrak{S}_{m-1}}$ $(m = 3, 4, \cdots)$.

LEMMA 13.1: If $\beta, \gamma^1, \cdots, \gamma^{m-2}, \delta^1, \cdots, \delta^{m-2} \in \mathfrak{S}$ $(m = 3, 4, \cdots)$, then

$$
\begin{aligned}
(1) \qquad &(\sum_{j=1}^{m-2} \bar{\mathfrak{a}}_j) \cap \left((1;\, \gamma^1, \cdots, \gamma^{m-2}) \cup (\beta;\, \delta^1, \cdots, \delta^{m-2})\right) \\
&\qquad = (\sum_{j=1}^{m-2} \bar{\mathfrak{a}}_j) \cap \left(\bar{\mathfrak{a}}_{m-1} \cup (\beta;\, \delta^1 - \gamma^1, \cdots, \delta^{m-2} - \gamma^{m-2})\right).
\end{aligned}
$$

PROOF: We employ the representation by sets of vectors of order $m-1$ (cf. the Appendix to Chapter II, and page 182):

$$
\begin{aligned}
&\sum_{j=1}^{m-2} \bar{\mathfrak{a}}_j\colon\ (\xi_1, \cdots, \xi_{m-2}, 0), \\
&(1;\, \gamma^1, \cdots, \gamma^{m-2})\colon\ (-\gamma^1\eta, \cdots, -\gamma^{m-2}\eta, \eta), \\
&(\beta;\, \delta^1, \cdots, \delta^{m-2})\colon\ (-\delta^1\beta\zeta, \cdots, -\delta^{m-2}\beta\zeta, \beta\zeta),
\end{aligned}
$$

whence for the vectors representing the left member of (1) we have $\eta + \beta\zeta = 0$, $\eta = -\beta\zeta$, and these vectors therefore constitute the set

$$(2) \qquad (-(\delta^1 - \gamma^1)\beta\zeta, \cdots, -(\delta^{m-2} - \gamma^{m-2})\beta\zeta, 0).$$

Moreover,

$$
\begin{aligned}
&\bar{\mathfrak{a}}_{m-1} \cup (\beta;\, \delta^1 - \gamma^1, \cdots, \delta^{m-2} - \gamma^{m-2})\colon \\
&\qquad\qquad (-(\delta^1 - \gamma^1)\beta\zeta, \cdots, -(\delta^{m-2} - \gamma^{m-2})\beta\zeta, \omega + \beta\zeta),
\end{aligned}
$$

whence for the vectors representing the right member of (1) we have $\omega + \beta\zeta = 0$, $\omega = -\beta\zeta$, and these vectors constitute the set

$$(3) \qquad (-(\delta^1 - \gamma^1)\beta\zeta, \cdots, -(\delta^{m-2} - \gamma^{m-2})\beta\zeta, 0).$$

Comparison of (2) and (3) shows that (1) holds. (Note: this lemma also follows Lemma 12.7 — Ed.)

We shall turn now to the second preliminary result.

LEMMA 13.2: Suppose $\mathfrak{a} \leqq \mathfrak{b}$ in a complemented modular lattice L. Then for any $\mathfrak{x}$ in L there exists an inverse $\mathfrak{u}$ of $\mathfrak{a}$ in $\mathfrak{b}$ such that $\mathfrak{x} = (\mathfrak{x} \cup \mathfrak{u}) \cap (\mathfrak{x} \cup \mathfrak{a})$.

PROOF: Let $\mathfrak{x}_1 \equiv \mathfrak{a} \cap \mathfrak{x}$; let $\mathfrak{y}_1$ be an inverse of $\mathfrak{x}_1$ in $\mathfrak{a}$; let $\mathfrak{x}_2$ be an inverse of $\mathfrak{x}_1$ in $\mathfrak{b} \cap \mathfrak{x}$; let $\mathfrak{y}_2$ be an inverse of $\mathfrak{x}_1 \cup \mathfrak{y}_1 \cup \mathfrak{x}_2$ in $\mathfrak{b}$, and let $\mathfrak{x}_3$ be an inverse of $\mathfrak{x}_1 \cup \mathfrak{x}_2$ in $\mathfrak{x}$. Then $(\mathfrak{x}_1, \mathfrak{y}_1, \mathfrak{x}_2, \mathfrak{y}_2, \mathfrak{x}_3)\perp$ and $\mathfrak{x} = \mathfrak{x}_1 \cup \mathfrak{x}_2 \cup \mathfrak{x}_3$.

Let $\mathfrak{u} = \mathfrak{x}_2 \cup \mathfrak{y}_2$. Then $\mathfrak{u}$ is an inverse of $\mathfrak{a}$ in $\mathfrak{b}$ and $(\mathfrak{x} \cup \mathfrak{u}) \cap (\mathfrak{x} \cup \mathfrak{a}) = (\mathfrak{x} \cup \mathfrak{y}_2) \cap (\mathfrak{x} \cup \mathfrak{y}_1) = \mathfrak{x}$, since $(\mathfrak{x}, \mathfrak{y}_1, \mathfrak{y}_2)\perp$.

THEOREM 13.1: If $m = 3, 4, \cdots, \mathfrak{S}$ is a regular ring, and $\mathfrak{w}, \mathfrak{u} \in \bar{R}_{\mathfrak{S}_{m-1}}$, then $\mathfrak{w} \cup (\gamma)_{ij} = \mathfrak{u} \cup (\gamma)_{ij}$ for every $\gamma \in \mathfrak{S}$ and for every $i, j = 1, \cdots, m-1$, $i \neq j$, implies $\mathfrak{w} = \mathfrak{u}$.

PROOF: It suffices to choose $\mathfrak{b} = \bar{\mathfrak{a}}_1 \cup \bar{\mathfrak{a}}_2$, $\mathfrak{a} = \bar{\mathfrak{a}}_2$ and then to apply Lemma 13.2.

PART II · CHAPTER XIV

Completion of Treatment of Case II. The Fundamental Theorem

The proof of (**) (page 186) will be given here for Case II. The method of proof will be to show that $\mathfrak{w} \cup (\gamma)_{ij} = \mathfrak{u} \cup (\gamma)_{ij}$ for every $\gamma \in \mathfrak{S}$, every $i, j = 1, \cdots, m-1$ $(i \neq j)$, where $\mathfrak{w}$ and $\mathfrak{u}$ are the left and right members respectively in (**); then the result of Chapter XIII, i.e. Theorem 13.1, will be applied to prove $\mathfrak{w} = \mathfrak{u}$. The hypotheses on $L = L'$ or L'', n, m, $\mathfrak{S}$, set forth on page 181 in addition to the condition of Case II, viz, $m = 4, \cdots, n$, will be assumed. It should be recalled that in assuming S_{m-1}, we may replace in property (B_{m-1}) the condition $i > j$ by the condition $i \neq j$ since $m - 1 \geqq 3$ (cf. page 182). The element γ, any element whatever of $\mathfrak{S}$, will be fixed.

DEFINITION 14.1: Define

$$(1) \qquad \bar{\bar{\mathfrak{a}}}_i \equiv \begin{cases} \bar{\mathfrak{a}}_i & (i = 1, \cdots, n,\ i \neq 2) \\ (\gamma)_{21} & (i = 2), \end{cases}$$

$$(2) \qquad \mathfrak{c}^* \equiv (((\bar{\mathfrak{a}}_1 \cup \bar{\mathfrak{c}}_{23}) \cap (\bar{\mathfrak{a}}_3 \cup (\gamma)_{21})) \cup \bar{\mathfrak{c}}_{13}) \cap (\bar{\mathfrak{a}}_1 \cup \bar{\mathfrak{a}}_2),$$

$$(3) \qquad \bar{\bar{\mathfrak{c}}}_{i1} \equiv \bar{\bar{\mathfrak{c}}}_{1i} \equiv \begin{cases} \bar{\mathfrak{c}}_{i1} & (i = 3, \cdots, n) \\ \mathfrak{c}^* & (i = 2). \end{cases}$$

COROLLARY:

$$(4) \qquad \mathfrak{c}^* = (\gamma + 1)_{21}.$$

PROOF: By Definition 7.5,

$$\begin{aligned}(\gamma+1)_{21} &= (1+\gamma)_{21} = (1)_{21} \oplus (\gamma)_{21} = (1)_{21} \boxplus_3 (\gamma)_{21} \\ &= ((((1)_{21} \cup \bar{\mathfrak{c}}_{23}) \cap (\bar{\mathfrak{a}}_1 \cup \bar{\mathfrak{a}}_3)) \cup (((\gamma)_{21} \cup \bar{\mathfrak{a}}_3) \cap (\bar{\mathfrak{a}}_1 \cup \bar{\mathfrak{c}}_{23}))) \cap (\bar{\mathfrak{a}}_1 \cup \bar{\mathfrak{a}}_2) \\ &= (((\bar{\mathfrak{c}}_{12} \cup \bar{\mathfrak{c}}_{23}) \cap (\bar{\mathfrak{a}}_1 \cup \bar{\mathfrak{a}}_3)) \cup (((\gamma)_{21} \cup \bar{\mathfrak{a}}_3) \cap (\bar{\mathfrak{a}}_1 \cup \bar{\mathfrak{c}}_{23}))) \cap (\bar{\mathfrak{a}}_1 \cup \bar{\mathfrak{a}}_2) \\ &= (\bar{\mathfrak{c}}_{13} \cup (((\gamma)_{21} \cup \bar{\mathfrak{a}}_3) \cap (\bar{\mathfrak{a}}_1 \cup \bar{\mathfrak{c}}_{23}))) \cap (\bar{\mathfrak{a}}_1 \cup \bar{\mathfrak{a}}_2) \text{ (by Definition 5.2)} \\ &= \mathfrak{c}^*.\end{aligned}$$

LEMMA 14.1: The system $(\bar{\bar{\mathfrak{a}}}_i;\ i = 1, \cdots, n)$ is a homogeneous basis for L; each $\bar{\bar{\mathfrak{c}}}_{i1}$ is an inverse of $\bar{\bar{\mathfrak{a}}}_i$, $\bar{\bar{\mathfrak{a}}}_1$ in $\bar{\bar{\mathfrak{a}}}_i \cup \bar{\bar{\mathfrak{a}}}_1$. There exists a unique extension $(\bar{\bar{\mathfrak{c}}}_{ij};\ i, j = 1, \cdots, n,\ i \neq j)$ of $(\bar{\bar{\mathfrak{c}}}_{i1};\ i = 2, \cdots, n)$ such that $(\bar{\bar{\mathfrak{a}}}_i, \bar{\bar{\mathfrak{c}}}_{ij};\ i, j = 1, \cdots, n,\ i \neq j)$ is a normalized frame in L.

Proof: The last part follows from the first by Lemma 5.3. Since $(\bar{\mathfrak{a}}_1, \bar{\mathfrak{a}}_2, \cdots, \bar{\mathfrak{a}}_n)\perp$, $(\gamma)_{21} \in L_{21}$, we have $(\bar{\mathfrak{a}}_1, (\gamma)_{21}, \bar{\mathfrak{a}}_3, \cdots, \bar{\mathfrak{a}}_n)\perp$ by Part I, Lemma 2.1; thus $(\bar{\bar{\mathfrak{a}}}_1, \cdots, \bar{\bar{\mathfrak{a}}}_n)\perp$. Since $\bar{\mathfrak{a}}_1 \cup \bar{\mathfrak{a}}_2 \cup \cdots \cup \bar{\mathfrak{a}}_n = 1$, $\bar{\mathfrak{a}}_1 \cup (\gamma)_{21} = \bar{\mathfrak{a}}_1 \cup \bar{\mathfrak{a}}_2$, it is clear that $\bar{\bar{\mathfrak{a}}}_1 \cup \bar{\bar{\mathfrak{a}}}_2 \cup \cdots \cup \bar{\bar{\mathfrak{a}}}_n = 1$. For $i, j = 1, \cdots, n$, $i, j \neq 2$, $\bar{\bar{\mathfrak{a}}}_i = \bar{\mathfrak{a}}_i$, $\bar{\bar{\mathfrak{a}}}_j = \bar{\mathfrak{a}}_j$, whence $\bar{\bar{\mathfrak{a}}}_i \sim \bar{\bar{\mathfrak{a}}}_j$ since $\bar{\mathfrak{a}}_i \sim \bar{\mathfrak{a}}_j$. It remains to show that $(\gamma)_{21} = \bar{\bar{\mathfrak{a}}}_2 \sim \bar{\bar{\mathfrak{a}}}_i = \bar{\mathfrak{a}}_i$ $(i = 1, \cdots, n,\ i \neq 2)$. Now $\bar{\mathfrak{a}}_2 \sim (\gamma)_{21}$ (mod $\bar{\mathfrak{a}}_1$), whence $\bar{\mathfrak{a}}_2 \sim (\gamma)_{21}$; moreover, $\bar{\mathfrak{a}}_2 \sim \bar{\mathfrak{a}}_3$ (mod $\bar{\mathfrak{c}}_{23}$). Since $(\bar{\mathfrak{a}}_2 \cup (\gamma)_{21}) \cap \bar{\mathfrak{c}}_{23} \leqq (\bar{\mathfrak{a}}_2 \cup \bar{\mathfrak{a}}_1) \cap (\bar{\mathfrak{a}}_2 \cup \bar{\mathfrak{a}}_3) \cap \bar{\mathfrak{c}}_{23} = \bar{\mathfrak{a}}_2 \cap \bar{\mathfrak{c}}_{23} = 0$, the proof of Part I, Theorem 3.4 shows that

$$(5) \qquad (\gamma)_{21} \sim \bar{\mathfrak{a}}_3 \qquad \left(\text{mod } \big((\gamma)_{21} \cup \bar{\mathfrak{a}}_3\big) \cap \big(\bar{\mathfrak{c}}_{23} \cup \bar{\mathfrak{a}}_1\big)\right),$$

and that the axis in (5) is an inverse of $(\gamma)_{21}$, $\bar{\mathfrak{a}}_3$ in $(\gamma)_{21} \cup \bar{\mathfrak{a}}_3$. Now $(\bar{\bar{\mathfrak{a}}}_1, \bar{\bar{\mathfrak{a}}}_2, \bar{\bar{\mathfrak{a}}}_3)\perp$, i.e., $(\bar{\mathfrak{a}}_1, (\gamma)_{21}, \bar{\mathfrak{a}}_3)\perp$, whence $\bar{\mathfrak{a}}_1 \sim \bar{\mathfrak{a}}_3$ (mod $\bar{\mathfrak{c}}_{13}$), together with (5) yields by Part I, Theorem 3.4 $(\gamma)_{21} \sim \bar{\mathfrak{a}}_1$ (mod $\mathfrak{c}'$), where

$$(6) \qquad \mathfrak{c}' \equiv \big((\gamma)_{21} \cup \bar{\mathfrak{a}}_1\big) \cap \Big(\bar{\mathfrak{c}}_{13} \cup \Big(\big((\gamma)_{21} \cup \bar{\mathfrak{a}}_3\big) \cap \big(\bar{\mathfrak{c}}_{23} \cup \bar{\mathfrak{a}}_1\big)\Big)\Big),$$

and that $\mathfrak{c}'$ is inverse to $(\gamma)_{21}$, $\bar{\mathfrak{a}}_1$ in $(\gamma)_{21} \cup \bar{\mathfrak{a}}_1$. Since $(\bar{\bar{\mathfrak{a}}}_1, \bar{\bar{\mathfrak{a}}}_2, \bar{\bar{\mathfrak{a}}}_i)\perp$, $\bar{\bar{\mathfrak{a}}}_1 \sim \bar{\bar{\mathfrak{a}}}_2$, $\bar{\bar{\mathfrak{a}}}_1 \sim \bar{\bar{\mathfrak{a}}}_i$, it follows from Part I, Theorem 3.4 that $\bar{\bar{\mathfrak{a}}}_2 \sim \bar{\bar{\mathfrak{a}}}_i$ $(i = 4, \cdots, n)$. Hence we have shown that $\bar{\bar{\mathfrak{a}}}_i \sim \bar{\bar{\mathfrak{a}}}_j$ $(i, j = 1, \cdots, n)$, and $(\bar{\bar{\mathfrak{a}}}_i;\ i = 1, \cdots, n)$ is a homogeneous basis for L.

It will now be shown that $\bar{\bar{\mathfrak{c}}}_{i1}$ is inverse to $\bar{\bar{\mathfrak{a}}}_i$, $\bar{\bar{\mathfrak{a}}}_1$ in $\bar{\bar{\mathfrak{a}}}_i \cup \bar{\bar{\mathfrak{a}}}_1$ $(i = 2, \cdots, n)$. For $i = 3, \cdots, n$ this is obvious since then $\bar{\bar{\mathfrak{c}}}_{i1} = \bar{\mathfrak{c}}_{i1}$, $\bar{\bar{\mathfrak{a}}}_i = \bar{\mathfrak{a}}_i$. Let $i = 2$. Then we shall show that $\mathfrak{c}^*$ is inverse to $\bar{\bar{\mathfrak{a}}}_2$, $\bar{\bar{\mathfrak{a}}}_1$ in $\bar{\bar{\mathfrak{a}}}_2 \cup \bar{\bar{\mathfrak{a}}}_1$, i.e., that $\mathfrak{c}^*$ is inverse to $(\gamma)_{21}$, $\bar{\mathfrak{a}}_1$ in $(\gamma)_{21} \cup \bar{\mathfrak{a}}_1$. It has already been shown that $\mathfrak{c}'$ has this property; hence it suffices to show that $\mathfrak{c}^* = \mathfrak{c}'$. Now since $(\gamma)_{21} \cup \bar{\mathfrak{a}}_1 = \bar{\mathfrak{a}}_2 \cup \bar{\mathfrak{a}}_1$, comparison of (6) and (2) shows that $\mathfrak{c}' = \mathfrak{c}^*$. This completes the proof.

Lemma 14.2:

(a) If $i, j \neq 2$, then $\bar{\bar{\mathfrak{c}}}_{ij} = \bar{\mathfrak{c}}_{ij}$,

(b) $\bar{\bar{\mathfrak{a}}}_1 \cup \bar{\bar{\mathfrak{a}}}_2 = \bar{\mathfrak{a}}_1 \cup \bar{\mathfrak{a}}_2$,

(c) $i \neq 2$ implies $\bar{\bar{\mathfrak{a}}}_i = \bar{\mathfrak{a}}_i$.

Proof: (a) If $i \neq 2$, then $\bar{\bar{\mathfrak{c}}}_{i1} = \bar{\bar{\mathfrak{c}}}_{1i} = \bar{\mathfrak{c}}_{i1} = \bar{\mathfrak{c}}_{1i}$. If $i, j \neq 1, 2$, then $\bar{\bar{\mathfrak{c}}}_{ij} = (\bar{\bar{\mathfrak{c}}}_{i1} \cup \bar{\bar{\mathfrak{c}}}_{j1}) \cap (\bar{\bar{\mathfrak{a}}}_i \cup \bar{\bar{\mathfrak{a}}}_j) = (\bar{\mathfrak{c}}_{i1} \cup \bar{\mathfrak{c}}_{j1}) \cap (\bar{\mathfrak{a}}_i \cup \bar{\mathfrak{a}}_j) = \bar{\mathfrak{c}}_{ij}$.

(b) Clearly $\bar{\bar{\mathfrak{a}}}_1 \cup \bar{\bar{\mathfrak{a}}}_2 = \bar{\mathfrak{a}}_1 \cup (\gamma)_{21} = \bar{\mathfrak{a}}_1 \cup \bar{\mathfrak{a}}_2$.

(c) This is obvious by (1).

Lemma 14.3: As in Theorem 5.1 let $P\begin{pmatrix} i_1 \cdots i_l \\ j_1 \cdots j_l \end{pmatrix}$, $P'\begin{pmatrix} i_1 \cdots i_l \\ j_1 \cdots j_l \end{pmatrix}$ $(l = 1, \cdots, n-1)$ denote the isomorphisms in L associated with the frames $(\bar{\mathfrak{a}}_i, \bar{\mathfrak{c}}_{ij};\ i, j = 1, \cdots, n,\ i \neq j)$, $(\bar{\bar{\mathfrak{a}}}_i, \bar{\bar{\mathfrak{c}}}_{ij};\ i, j = 1, \cdots, n,\ i \neq j)$ respectively. If

(α) $i_\tau, j_\tau \neq 2 \qquad (\tau = 1, \cdots, l)$,

or

(β) there exist $\sigma', \sigma'' = 1, \cdots, l$ with $i_{\sigma'} = j_{\sigma'} = 1$, $i_{\sigma''} = j_{\sigma''} = 2$, then

$$P'\begin{pmatrix} i_1 \cdots i_l \\ j_1 \cdots j_l \end{pmatrix} = P\begin{pmatrix} i_1 \cdots i_l \\ j_1 \cdots j_l \end{pmatrix}.$$

PROOF: If (α) holds and $l < n - 1$ or if (β) holds, the permutation $\begin{pmatrix} i_1 \cdots i_l \\ j_1 \cdots j_l \end{pmatrix}$ can obviously be written as a product of permutations $\begin{pmatrix} i'_1 \cdots i'_l \\ j'_1 \cdots j'_l \end{pmatrix}$ again fulfilling (α) with $l < n - 1$ or (β) respectively, and such that $i'_\tau \neq j'_\tau$ occurs for one τ only. Hence it suffices to prove

$$P'\begin{pmatrix} i'_1 \cdots i'_l \\ j'_1 \cdots j'_l \end{pmatrix} = P\begin{pmatrix} i_1 \cdots i_l \\ j_1 \cdots j_l \end{pmatrix}$$

in this special case. But then it is immediate by Lemma 14.2, (a)—(c), in both cases (α) and $l < n - 1$, (β). If (α) holds and $l = n - 1$ we argue as follows (the argument is valid for all $l \leqq n - 1$): Lemma 14.2, (a)—(c), implies that P' and P have the same effect when applied to any element in $L(0, \bar{\mathfrak{a}}_i \cup \bar{\mathfrak{a}}_j) = L(0, \bar{\bar{\mathfrak{a}}}_i \cup \bar{\bar{\mathfrak{a}}}_j)$ for all $i, j \neq 2$. It follows from Lemma 4.2 (cf. the proof of Lemma 5.10), that $P = P'$.

DEFINITION 14.2: Define $\mathfrak{S}' = (\beta', \gamma', \cdots)$ as the auxiliary ring of L with respect to the normalized frame $(\bar{\bar{\mathfrak{a}}}_i, \bar{\bar{\mathfrak{c}}}_{ij};\ i, j = 1, \cdots, n,\ i \neq j)$.

LEMMA 14.4. The ring $\mathfrak{S}'$ is isomorphic to $\mathfrak{S}$, the isomorphism being defined by

(7) $$(\beta')'_{ij} = (\beta)_{ij} \qquad \text{with, } i, j \neq 2.$$

Any such choice of $i, j \neq 2$ gives the same isomorphism. (In (7) the symbol $(\beta')'_{ij}$ denoted the i, j-component of β' defined in terms of the frame $(\bar{\bar{\mathfrak{a}}}_i, \bar{\bar{\mathfrak{c}}}_{ij};\ i, j = 1, \cdots, n,\ i \neq j)$. A similar remark applies to L'_{ij}, $(\beta')'_i$, and to $(\beta'; \gamma'^1, \cdots, \gamma'^{m-1})')$.

PROOF: $L'_{ij} = L_{ij}$ for $i, j \neq 2$ by Lemma 14.2, (a) and (c), whence (7) defines (for a fixed pair $i, j \neq 2$) a one-to-one correspondence of $\mathfrak{S}'$ and $\mathfrak{S}$. If $i, j, k, l \neq 2$, then application of $P'\begin{pmatrix} i & j \\ k & l \end{pmatrix} = P\begin{pmatrix} i & j \\ k & l \end{pmatrix}$ (cf. Lemma 14.3) shows that (7) gives the same correspondence for i, j and for k, l, i.e., for all pairs $i, j \neq 2$. It remains to prove that this correspondence is a ring-isomorphism. Let $\beta', \gamma' \in \mathfrak{S}'$, $\beta, \gamma \in \mathfrak{S}$ and suppose that β', β as well as γ', γ correspond. Then

$$\begin{aligned}(\beta'\gamma')'_{14} &= (\gamma')'_{13} \otimes (\beta')'_{34}\\ &= \big((\gamma')'_{13} \cup (\beta')'_{34}\big) \cap (\bar{\bar{\mathfrak{a}}}_1 \cup \bar{\bar{\mathfrak{a}}}_4)\\ &= \big((\gamma)_{13} \cup (\beta)_{34}\big) \cap (\bar{\mathfrak{a}}_1 \cup \bar{\mathfrak{a}}_4)\\ &= (\gamma)_{13} \otimes (\beta)_{34} = (\beta\gamma)_{14},\end{aligned}$$

and

$$(\beta' + \gamma')'_{13} = (\beta')'_{13} \boxplus_4 (\gamma')'_{13}$$

$$= \Big(\big(((\beta')'_{13} \cup \bar{\bar{\mathfrak{c}}}_{14}) \cap (\bar{\bar{\mathfrak{a}}}_3 \cup \bar{\bar{\mathfrak{a}}}_4)\big) \cup \big(((\gamma')'_{13} \cup \bar{\bar{\mathfrak{a}}}_4) \cap (\bar{\bar{\mathfrak{a}}}_4 \cup \bar{\bar{\mathfrak{c}}}_{14})\big)\Big) \cap (\bar{\bar{\mathfrak{a}}}_1 \cup \bar{\bar{\mathfrak{a}}}_3)$$

$$= \Big(\big(((\beta)_{13} \cup \bar{\mathfrak{c}}_{14}) \cap (\bar{\mathfrak{a}}_3 \cup \bar{\mathfrak{a}}_4)\big) \cup \big(((\gamma)_{13} \cup \bar{\mathfrak{a}}_4) \cap (\bar{\mathfrak{a}}_4 \cup \bar{\mathfrak{c}}_{14})\big)\Big) \cap (\bar{\mathfrak{a}}_1 \cup \bar{\mathfrak{a}}_3)$$

$$= (\beta)_{13} \boxplus_4 (\gamma)_{13} = (\beta)_{13} \oplus (\gamma)_{13} = (\beta + \gamma)_{13}.$$

Hence $\beta'\gamma'$, $\beta\gamma$ as well as $\beta' + \gamma'$, $\beta + \gamma$ correspond. This completes the proof.

DEFINITION 14.3: We use the notation $\mathfrak{S}$ for both $\mathfrak{S}$ and $\mathfrak{S}'$, which we regard as identified in accordance with the isomorphism (7).

COROLLARY: For every $\beta \in \mathfrak{S}$

(8) $\quad (\beta)'_{ij} = (\beta)_{ij} \qquad (i,\ j \neq 2),$

(9) $\quad (\beta)'_j = (\beta)_j \qquad (j \neq 2).$

PROOF: Clearly (8) is merely (7); to prove (9), we note that we may select $i \neq 2,\ j$ (since $n \geqq 4$), and

$$(\beta)'_j = \big((\beta)'_{ij} \cup \bar{\bar{\mathfrak{a}}}_i\big) \cap \bar{\bar{\mathfrak{a}}}_j = \big((\beta)_{ij} \cup \bar{\mathfrak{a}}_i\big) \cap \bar{\mathfrak{a}}_j = (\beta)_j.$$

LEMMA 14.5: Let $\beta \in \mathfrak{S}$. Then

(10) $$(\beta)'_{m2} = ((\beta)_{m2} \cup \bar{\mathfrak{a}}_1) \cap ((-\gamma\beta)_{m1} \cup \bar{\mathfrak{a}}_2).$$

PROOF: Instead of proving (10), we may prove the equivalent statement

$$(\beta)'_{32} = ((\beta)_{32} \cup \bar{\mathfrak{a}}_1) \cap \big((-\gamma\beta)_{31} \cup \bar{\mathfrak{a}}_2\big)$$

obtained from (10) by applying $P'\begin{pmatrix}1\ 2\ m\\ 1\ 2\ 3\end{pmatrix} = P\begin{pmatrix}1\ 2\ m\\ 1\ 2\ 3\end{pmatrix}$ (by Lemma 14.3 (β)) (note that since $m \geqq 4$, $m - 1 \geqq 3$). But

$$\begin{aligned}(\beta)'_{32} &= \big((\beta)'_{31} \cup \bar{\bar{\mathfrak{c}}}_{12}\big) \cap (\bar{\bar{\mathfrak{a}}}_3 \cup \bar{\bar{\mathfrak{a}}}_2)\\ &= \big((\beta)_{31} \cup (\gamma + 1)_{21}\big) \cap \big(\bar{\mathfrak{a}}_3 \cup (\gamma)_{21}\big),\end{aligned}$$

whence we are to prove

(11) $$\big((\beta)_{31} \cup (\gamma + 1)_{21}\big) \cap \big(\bar{\mathfrak{a}}_3 \cup (\gamma)_{21}\big) = \big((\beta)_{32} \cup \bar{\mathfrak{a}}_1\big) \cap \big((-\gamma\beta)_{31} \cup \bar{\mathfrak{a}}_2\big).$$

Since all quantities in (11) are in the lattice $L(0, \sum_{j=1}^{m-1} \bar{\mathfrak{a}}_j)$, we may prove (11) by means of the representation by vector sets of order $m - 1$ (Cf. page 182; since $m \geqq 4$, $m - 1 \geqq 3$, the unrestricted representation of the

$(\beta)_{ij}$, $i \neq j$ is possible). We shall display only the components 1, 2, 3 since the remaining components will all be 0. Now

$$(\beta)_{31}\colon\ (-\beta\xi,\ 0,\ \xi),\ \ (\gamma+1)_{21}\colon\ (-(\gamma+1)\eta,\ \eta,\ 0),$$
$$(\beta)_{31} \cup (\gamma+1)_{21}\colon\ (-\beta\xi-(\gamma+1)\eta,\ \eta,\ \xi);$$
$$(\gamma)_{21}\colon\ (-\gamma\zeta,\ \zeta,\ 0),\ \bar{\mathfrak{a}}_3\colon\ (0,\ 0,\ \omega),$$
$$(\gamma)_{21} \cup \bar{\mathfrak{a}}_3\colon\ (-\gamma\zeta,\ \zeta,\ \omega).$$

Thus the left member of (11) is represented by $(-\gamma\zeta,\ \zeta,\ \omega)$ with $\omega = \xi$, $\zeta = \eta$, $-\gamma\zeta = -\beta\xi - (\gamma+1)\eta$, i.e., $-\gamma\zeta = -\beta\omega - (\gamma+1)\zeta$, $\zeta = -\beta\omega$. Hence

$$\text{(12)}\qquad ((\beta)_{31} \cup (\gamma+1)_{21}) \cap (\bar{\mathfrak{a}}_3 \cup (\gamma)_{21})\colon\ (\gamma\beta\omega,\ -\beta\omega,\ \omega).$$

Moreover,

$$(\beta)_{32}\colon\ (0,\ -\beta\omega,\ \omega),\ \bar{\mathfrak{a}}_1\colon\ (\psi,\ 0,\ 0),$$
$$(\beta)_{32} \cup \bar{\mathfrak{a}}_1\colon\ (\psi,\ -\beta\omega,\ \omega);$$
$$(-\gamma\beta)_{31}\colon\ (-\gamma\beta\theta,\ 0,\ \theta),\ \bar{\mathfrak{a}}_2\colon\ (0,\ \varphi,\ 0),$$
$$(-\gamma\beta)_{31} \cup \bar{\mathfrak{a}}_2\colon\ (-\gamma\beta\theta,\ \varphi,\ \theta).$$

Thus the right member of (11) is represented by $(-\gamma\beta\theta, \varphi, \theta)$ with $\theta = \omega$, $\varphi = -\beta\omega$. Hence

$$\text{(13)}\qquad ((\beta)_{32} \cup \bar{\mathfrak{a}}_1) \cap ((-\gamma\beta)_{31} \cup \bar{\mathfrak{a}}_2)\colon\ (-\gamma\beta\omega,\ -\beta\omega,\ \omega).$$

Comparison of (12) and (13) shows that (11) holds.

LEMMA 14.6: If $\beta, \gamma^1, \cdots, \gamma^{m-1} \in \mathfrak{S}$, then

$$\text{(14)}\qquad (\beta;\ \gamma^1, \cdots, \gamma^{m-1})' = (\beta;\ \gamma^1 - \gamma\gamma^2,\ \gamma^2, \cdots, \gamma^{m-1}),$$

$$\text{(15)}\qquad (\beta;\ \gamma^1, \cdots, \gamma^{m-1}) = (\beta;\ \gamma^1 + \gamma\gamma^2,\ \gamma^2, \cdots, \gamma^{m-1})'.$$

PROOF: It is clear that (15) follows from (14) by replacing in (14), γ^1 by $\gamma^1 + \gamma\gamma^2$ and leaving $\beta, \gamma^2, \cdots, \gamma^{m-1}$ unchanged. Hence we need prove only (14).

By Definition 10.1

$$\text{(16)}\quad (\beta;\ \gamma^1, \cdots, \gamma^{m-1})' = \Big(\sum_{j=1}^{m-1} \bar{\bar{\mathfrak{a}}}_j \cup (\beta)'_m\Big) \cap \prod_{i=1}^{m-1} \Big(\sum_{j=1\,(j\neq i)}^{m-1} \bar{\bar{\mathfrak{a}}}_j \cup (\gamma^i)'_{mi}\Big).$$

By (9), $(\beta)'_m = (\beta)_m$, and by Lemma 14.2, the first factor is $\sum_{j=1}^{m-1} \bar{\mathfrak{a}}_j \cup (\beta)_m$. Now each factor $\sum_{j=1\,(j\neq i)}^{m-1} \bar{\bar{\mathfrak{a}}}_j \cup (\gamma^i)'_{mi}$ with $i \neq 1,2$, is clearly $\sum_{j=1\,(j\neq i)}^{m-1} \bar{\mathfrak{a}}_j \cup (\gamma^i)_{mi}$. For $i = 1$ the factor is $\sum_{j=3}^{m-1} \bar{\mathfrak{a}}_j \cup (\gamma)_{21} \cup (\gamma^1)_{m1}$. For $i = 2$ we have

$$\sum_{j=1}^{m-1}{}_{(j\neq 2)}\, \bar{\mathfrak{a}}_j \cup (\gamma^2)'_{m2} = \Big(\sum_{j=1}^{m-1}{}_{(j\neq 2)}\, \bar{\mathfrak{a}}_j\Big) \cup \big(((\gamma^2)_{m2} \cup \bar{\mathfrak{a}}_1) \cap ((-\gamma\gamma^2)_{m1} \cup \bar{\mathfrak{a}}_2)\big) \qquad \text{(by (10))}$$

$$= \Big(\sum_{j=3}^{m-1} \bar{\mathfrak{a}}_j \cup \bar{\mathfrak{a}}_1\Big) \cup \big(((\gamma^2)_{m2} \cup \bar{\mathfrak{a}}_1) \cap ((-\gamma\gamma^2)_{m1} \cup \bar{\mathfrak{a}}_2)\big)$$

$$= \Big(\sum_{j=3}^{m-1} \bar{\mathfrak{a}}_j\Big) \cup \big(((\gamma^2)_{m2} \cup \bar{\mathfrak{a}}_1) \cap ((-\gamma\gamma^2)_{m1} \cup \bar{\mathfrak{a}}_1 \cup \bar{\mathfrak{a}}_2)\big) \qquad \text{(by IV)}$$

$$= \Big(\sum_{j=3}^{m-1} \bar{\mathfrak{a}}_j\Big) \cup \big(((\gamma^2)_{m2} \cup \bar{\mathfrak{a}}_1) \cap (\bar{\mathfrak{a}}_m \cup \bar{\mathfrak{a}}_1 \cup \bar{\mathfrak{a}}_2)\big)$$

$$= \Big(\sum_{j=3}^{m-1} \bar{\mathfrak{a}}_j\Big) \cup \big((\gamma^2)_{m2} \cup \bar{\mathfrak{a}}_1\big) \qquad \big(\text{since } (\gamma^2)_{m2}, \bar{\mathfrak{a}}_1 \leqq \bar{\mathfrak{a}}_m \cup \bar{\mathfrak{a}}_1 \cup \bar{\mathfrak{a}}_2\big)$$

$$= \sum_{j=1}^{m-1}{}_{(j\neq 2)}\, \bar{\mathfrak{a}}_j \cup (\gamma^2)_{m2}.$$

Therefore

$$(17)\qquad \begin{aligned} &(\beta;\, \gamma^1, \cdots, \gamma^{m-1})' \\ &\quad = \Big(\sum_{j=1}^{m-1} \bar{\mathfrak{a}}_j \cup (\beta)_m\Big) \cap \prod_{i=3}^{m-1} \Big(\Big(\sum_{j=1}^{m-1}{}_{(j=i)}\, \bar{\mathfrak{a}}_j \cup (\gamma^i)_{mi}\Big) \cap \Big(\sum_{j=1}^{m-1}{}_{(j\neq 2)}\, \bar{\mathfrak{a}}_j \cup (\gamma^2)_{m2}\Big) \\ &\qquad\qquad \cap \Big(\sum_{j=3}^{m-1} \bar{\mathfrak{a}}_j \cup (\gamma)_{21} \cup (\gamma^1)_{m1}\Big)\Big) \\ &\quad = \Big(\sum_{j=1}^{m-1} \bar{\mathfrak{a}}_j \cup (\beta)_m\Big) \cap \prod_{i=3}^{m-1} \Big(\Big(\sum_{j=1}^{m-1}{}_{(j\neq i)}\, \bar{\mathfrak{a}}_j \cup (\gamma^i)_{mi}\Big) \cap \mathfrak{a}\Big) \end{aligned}$$

where

$$(18)\qquad \mathfrak{a} \equiv \Big(\sum_{j=3}^{m-1} \bar{\mathfrak{a}}_j \cup (\bar{\mathfrak{a}}_1 \cup (\gamma^2)_{m2})\Big) \cap \Big(\sum_{j=3}^{m-1} \bar{\mathfrak{a}}_j \cup ((\gamma)_{21} \cup (\gamma^1)_{m1})\Big).$$

Now $(\sum_{j=3}^{m-1} \bar{\mathfrak{a}}_j) \cap (\bar{\mathfrak{a}}_1 \cup (\gamma^2)_{m2} \cup (\gamma)_{21} \cup (\gamma^1)_{m1}) \leqq (\sum_{j=3}^{m-1} \bar{\mathfrak{a}}_j) \cap (\bar{\mathfrak{a}}_1 \cup \bar{\mathfrak{a}}_2 \cup \bar{\mathfrak{a}}_m) = 0$, whence Part I, Theorem 1.2 applies, yielding

$$(19)\qquad \mathfrak{a} = \sum_{j=3}^{m-1} \bar{\mathfrak{a}}_j \cup \mathfrak{b},$$

where

$$(20)\qquad \mathfrak{b} \equiv (\bar{\mathfrak{a}}_1 \cup (\gamma^2)_{m2}) \cap ((\gamma)_{21} \cup (\gamma^1)_{m1}).$$

It will now be shown that

$$(21)\qquad \mathfrak{b} = (\bar{\mathfrak{a}}_1 \cup (\gamma^2)_{m2}) \cap (\bar{\mathfrak{a}}_2 \cup (\gamma^1 - \gamma\gamma^2)_{m1}).$$

Instead of proving (21), we prove the equivalent statement

$$(22)\qquad (\bar{\mathfrak{a}}_1 \cup (\gamma^2)_{32}) \cap ((\gamma)_{21} \cup (\gamma^1)_{31}) = (\bar{\mathfrak{a}}_1 \cup (\gamma^2)_{32}) \cap (\bar{\mathfrak{a}}_2 \cup (\gamma^1 - \gamma\gamma^2)_{31}),$$

obtained from (21) by application of $P\begin{pmatrix}1 & 2 & m\\ 1 & 2 & 3\end{pmatrix}$ (note that $m \geqq 4$, whence $m - 1 \geqq 3$). Since all elements in (22) are in $L(0, \sum_{j=1}^{m-1} \bar{\mathfrak{a}}_j)$, we may employ representation by vector-sets (cf. page 182); all components other than 1, 2, 3 being 0, we display only these. Now

$$\bar{\mathfrak{a}}_1\colon (\eta, 0, 0),\ (\gamma^2)_{32}\colon (0, -\gamma^2\xi, \xi),\ \bar{\mathfrak{a}}_1 \cup (\gamma^2)_{32}\colon (\eta, -\gamma^2\xi, \xi);$$
$$(\gamma)_{21}\colon (-\gamma\zeta, \zeta, 0),\ (\gamma^1)_{31}\colon (-\gamma^1\omega, 0, \omega),\ (\gamma)_{21} \cup (\gamma^1)_{31}\colon (-\gamma\zeta - \gamma^1\omega, \zeta, \omega).$$

Thus the left member of (22) is represented by $(\eta, -\gamma^2\xi, \xi)$, with $\xi = \omega$, $\zeta = -\gamma^2\xi$ and $\eta = -\gamma\zeta - \gamma^1\omega$, i.e., $\eta = \gamma\gamma^2\xi - \gamma^1\xi = -(\gamma^1 - \gamma\gamma^2)\xi$. Hence

$$(23)\qquad (\bar{\mathfrak{a}}_1 \cup (\gamma^2)_{32}) \cap ((\gamma)_{21} \cup (\gamma^1)_{31})\colon (-(\gamma^1 - \gamma\gamma^2)\xi, -\gamma^2\xi, \xi).$$

Moreover,

$$\bar{\mathfrak{a}}_1 \cup (\gamma^2)_{32}\colon (\psi, -\gamma^2\xi, \xi);$$
$$\bar{\mathfrak{a}}_2\colon (0, \varphi, 0),\ (\gamma^1 - \gamma\gamma^2)_{31}\colon (-(\gamma^1 - \gamma\gamma^2)\theta, 0, \theta),$$
$$\bar{\mathfrak{a}}_2 \cup (\gamma^1 - \gamma\gamma^2)_{31}\colon (-(\gamma^1 - \gamma\gamma^2)\theta, \varphi, \theta).$$

Thus the right member of (22) is represented by $(\psi, -\gamma^2\xi, \xi)$ with $\xi = \theta$, $\psi = -(\gamma^1 - \gamma\gamma^2)\theta = -(\gamma^1 - \gamma\gamma^2)\xi$. Hence

$$(24)\quad (\bar{\mathfrak{a}}_1 \cup (\gamma^2)_{32}) \cap (\bar{\mathfrak{a}}_2 \cup (\gamma^1 - \gamma\gamma^2)_{31})\colon (-(\gamma^1 - \gamma\gamma^2)\xi, -\gamma^2\xi, \xi).$$

Comparison of (23) and (24) shows that (22) holds; thus also (21) holds. By (19), (21),

$$\mathfrak{a} = \sum_{j=3}^{m-1} \bar{\mathfrak{a}}_j \cup ((\bar{\mathfrak{a}}_1 \cup (\gamma^2)_{m2}) \cap (\bar{\mathfrak{a}}_2 \cup (\gamma^1 - \gamma\gamma^2)_{m1})).$$

Since $(\sum_{j=3}^{m-1} \bar{\mathfrak{a}}_j) \cap (\bar{\mathfrak{a}}_1 \cup (\gamma^2)_{m2} \cup \bar{\mathfrak{a}}_2 \cup (\gamma^1 - \gamma\gamma^2)_{m1}) \leqq (\sum_{j=3}^{m-1} \bar{\mathfrak{a}}_j) \cap (\bar{\mathfrak{a}}_1 \cup \bar{\mathfrak{a}}_2 \cup \bar{\mathfrak{a}}_m) = 0$, therefore Part I, Theorem 1.2 yields

$$\mathfrak{a} = \Big(\sum_{j=3}^{m-1} \bar{\mathfrak{a}}_j \cup \bar{\mathfrak{a}}_1 \cup (\gamma^2)_{m2}\Big) \cap \Big(\sum_{j=3}^{m-1} \bar{\mathfrak{a}}_j \cup \bar{\mathfrak{a}}_2 \cup (\gamma^1 - \gamma\gamma^2)_{m1}\Big)$$
$$= \Big(\sum_{j=1}^{m-1}{}_{(j\neq 2)}\, \bar{\mathfrak{a}}_j \cup (\gamma^2)_{m2}\Big) \cap \Big(\sum_{j=1}^{m-1}{}_{(j\neq 1)}\, \bar{\mathfrak{a}}_j \cup (\gamma^1 - \gamma\gamma^2)_{m1}\Big).$$

Substitution of this expression for $\mathfrak{a}$ in (17) yields (14).

DEFINITION 14.4: For $\beta, \varepsilon^1, \cdots, \varepsilon^{m-2} \in \mathfrak{S}$ define

$$(\beta; \varepsilon^1, \cdots, \varepsilon^{m-2})'' \equiv (\beta; \varepsilon^1, \cdots, \varepsilon^{m-2})\, P\begin{pmatrix}1 & 2 \cdots m-2 & m-1\\ 1 & 3 \cdots m-1 & 2\end{pmatrix}.$$

LEMMA 14.7: For $\beta, \gamma^1, \cdots, \gamma^{m-1} \in \mathfrak{S}$

$$(\beta;\ \gamma^1, \cdots, \gamma^{m-1})' \cup \bar{\bar{\mathfrak{a}}}_2 = (\beta;\ \gamma^1, \gamma^3, \cdots, \gamma^{m-1})'' \cup \bar{\bar{\mathfrak{a}}}_2. \tag{25}$$

PROOF: We have

$$\begin{aligned}
&(\beta;\ \gamma^1, \cdots, \gamma^{m-1})' \cup \bar{\bar{\mathfrak{a}}}_2\\
&= \Big(\big(\sum_{j=1}^{m-1} \bar{\bar{\mathfrak{a}}}_j \cup (\beta)'_m\big) \cap \big(\prod_{i=1}^{m-1}{}_{(i\neq 2)}\big(\sum_{j=1}^{m-1}{}_{(j\neq i)} \bar{\bar{\mathfrak{a}}}_j \cup (\gamma^i)'_{mi}\big)\big) \cap \big(\sum_{j=1}^{m-1}{}_{(j\neq 2)} \bar{\bar{\mathfrak{a}}}_j \cup (\gamma^2)'_{m2}\big)\Big) \cup \bar{\bar{\mathfrak{a}}}_2\\
&= \big(\sum_{j=1}^{m-1} \bar{\bar{\mathfrak{a}}}_j \cup (\beta)'_m\big) \cap \big(\prod_{i=1}^{m-1}{}_{(i\neq 2)}\big(\sum_{j=1}^{m-1}{}_{(j\neq i)} \bar{\bar{\mathfrak{a}}}_j \cup (\gamma^i)'_{mi}\big)\big) \cap \big(\big(\sum_{j=1}^{m-1}{}_{(j\neq 2)} \bar{\bar{\mathfrak{a}}}_j \cup (\gamma^2)'_{m2}\big) \cup \bar{\bar{\mathfrak{a}}}_2\big)\\
&= \big(\sum_{j=1}^{m-1} \bar{\bar{\mathfrak{a}}}_j \cup (\beta)'_m\big) \cap \big(\prod_{i=1}^{m-1}{}_{(i\neq 2)}\big(\sum_{j=1}^{m-1}{}_{(j\neq i)} \bar{\bar{\mathfrak{a}}}_j \cup (\gamma^i)'_{mi}\big)\big) \cap \big(\sum_{j=1}^{m-1}{}_{(j\neq 2)} \bar{\bar{\mathfrak{a}}}_j \cup (\gamma^2)'_{m2} \cup \bar{\bar{\mathfrak{a}}}_2\big)\\
&= \big(\sum_{j=1}^{m-1} \bar{\bar{\mathfrak{a}}}_j \cup (\beta)'_m\big) \cap \big(\prod_{i=1}^{m-1}{}_{(i\neq 2)}\big(\sum_{j=1}^{m-1}{}_{(j\neq i)} \bar{\bar{\mathfrak{a}}}_j \cup (\gamma^i)_{mi}\big)\big) \cap \big(\sum_{j=1}^{m-1}{}_{(j\neq 2)} \bar{\bar{\mathfrak{a}}}_j \cup \bar{\bar{\mathfrak{a}}}_m \cup \bar{\bar{\mathfrak{a}}}_2\big)\\
&= \big(\sum_{j=1}^{m-1} \bar{\bar{\mathfrak{a}}}_j \cup (\beta)'_m\big) \cap \big(\prod_{i=1}^{m-1}{}_{(i\neq 2)}\big(\sum_{j=1}^{m-1}{}_{(j\neq i)} \bar{\bar{\mathfrak{a}}}_j \cup (\gamma^i)_{mi}\big)\big) \cap \big(\sum_{j=1}^{m} \bar{\bar{\mathfrak{a}}}_j\big)\\
&= \big(\sum_{j=1}^{m-1} \bar{\bar{\mathfrak{a}}}_j \cup (\beta)'_m\big) \cap \prod_{i=1}^{m-1}{}_{(i\neq 2)}\big(\sum_{j=1}^{m-1}{}_{(j\neq i)} \bar{\bar{\mathfrak{a}}}_j \cup (\gamma^i)_{mi}\big)\\
&= \big(\big(\sum_{j=1}^{m-1}{}_{(j\neq 2)} \bar{\bar{\mathfrak{a}}}_j \cup (\beta)'_m\big) \cup \bar{\bar{\mathfrak{a}}}_2\big) \cap \prod_{i=1}^{m-1}{}_{(i\neq 2)}\big(\big(\sum_{j=1}^{m-1}{}_{(j\neq 2,\, i)} \bar{\bar{\mathfrak{a}}}_j \cup (\gamma^i)'_{mi}\big) \cup \bar{\bar{\mathfrak{a}}}_2\big)\\
&= \big(\big(\sum_{j=1}^{m-1}{}_{(j\neq 2)} \bar{\bar{\mathfrak{a}}}_j \cup (\beta)'_m\big) \cap \prod_{i=1}^{m-1}{}_{(i\neq 2)}\big(\sum_{j=1}^{m-1}{}_{(j\neq 2,\, i)} \bar{\bar{\mathfrak{a}}}_j \cup (\gamma^i)'_{mi}\big)\big) \cup \bar{\bar{\mathfrak{a}}}_2\\
&= (\beta;\ \gamma^1, \gamma^3, \cdots, \gamma^{m-1})'' \cup \bar{\bar{\mathfrak{a}}}_2.
\end{aligned}$$

the last equality but one holding by Lemma 10.1, since $\big(\big(\sum_{j=1}^{m-1}{}_{(j\neq 2)} \bar{\bar{\mathfrak{a}}}_j \cup (\beta)'_m\big) \cup \sum_{i=1}^{m-1}{}_{(i\neq 2)}\big(\sum_{j=1}^{m-1}{}_{(j\neq 2,\, i)} \bar{\bar{\mathfrak{a}}}_j \cup (\gamma^i)'_{mi}\big)\big) \cap \bar{\bar{\mathfrak{a}}}_2 \leqq \big(\sum_{j=1}^{m}{}_{(j\neq 2)} \bar{\bar{\mathfrak{a}}}_j\big) \cap \bar{\bar{\mathfrak{a}}}_2 = 0$. Thus (25) holds.

LEMMA 14.8: If $\beta, \gamma^1, \cdots, \gamma^{m-1}, \delta^1, \cdots, \delta^{m-1} \in \mathfrak{S}$, then

$$\begin{aligned}
&\big(\big(\sum_{j=1}^{m-1} \bar{\mathfrak{a}}_j\big) \cap \big((1;\ \gamma^1, \cdots, \gamma^{m-1}) \cup (\beta;\ \delta^1, \cdots, \delta^{m-1})\big)\big) \cup (\gamma)_{21}\\
&\qquad = \big(\big(\sum_{j=1}^{m-1} \bar{\mathfrak{a}}_j\big) \cap \big(\bar{\mathfrak{a}}_m \cup (\beta;\ \delta^1 - \gamma^1, \cdots, \delta^{m-1} - \gamma^{m-1})\big)\big) \cup (\gamma)_{21}.
\end{aligned} \tag{26}$$

PROOF: Define $\overline{\gamma^1} \equiv \gamma^1 + \gamma\gamma^2$, $\overline{\delta^1} \equiv \delta^1 + \gamma\delta^2$. The relation

$$\begin{aligned}
&\big(\sum_{j=1}^{m-1}{}_{(j\neq 2)} \bar{\bar{\mathfrak{a}}}_j\big) \cap \big((1;\ \overline{\gamma^1}, \cdots, \gamma^{m-1})'' \cup (\beta;\ \overline{\delta^1}, \delta^3, \cdots, \delta^{m-1})''\big)\\
&\qquad = \big(\sum_{j=1}^{m-1}{}_{(j\neq 2)} \bar{\bar{\mathfrak{a}}}_j\big) \cap \big(\bar{\bar{\mathfrak{a}}}_m \cup (\beta;\ \overline{\delta^1} - \overline{\gamma^1}, \delta^3 - \gamma^3, \cdots, \delta^{m-1} - \gamma^{m-1})''\big)
\end{aligned} \tag{27}$$

is a statement about elements in $L(0, \bar{\mathfrak{a}}_1 \cup \bar{\mathfrak{a}}_3 \cup \cdots \cup \bar{\mathfrak{a}}_m)$; this lattice is isomorphic with $L_{\mathfrak{S}_{m-1}}$ and the representation by vector-sets of order $m-1$ is applicable (cf. page 182). Hence (27) holds by Lemma 13.1. Since the product of $\bar{\bar{\mathfrak{a}}}_2$ with the sum of all quantities in the left member of (27) is $\leqq \bar{\bar{\mathfrak{a}}}_2 \cap \sum_{j=1 \, (j\neq 2)}^{m} \bar{\bar{\mathfrak{a}}}_j = 0$, and since the same is true of the right member, the equation obtained by adding $\bar{\bar{\mathfrak{a}}}_2$ to both members of (27) becomes, by virtue of Part I, Theorem 1.2,

$$(28) \quad \Big(\sum_{j=1}^{m-1} \bar{\bar{\mathfrak{a}}}_j\Big) \cap \big(\bar{\bar{\mathfrak{a}}}_2 \cup (1; \overline{\gamma^1}, \gamma^3, \cdots, \gamma^{m-1})'' \cup (\beta; \overline{\delta^1}, \delta^3, \cdots, \delta^{m-1})''\big)$$
$$= \Big(\sum_{j=1}^{m-1} \bar{\bar{\mathfrak{a}}}_j\Big) \cap \big(\bar{\bar{\mathfrak{a}}}_m \cup \bar{\bar{\mathfrak{a}}}_2 \cup (\beta; \overline{\delta^1} - \overline{\gamma^1}, \delta^3 - \gamma^3, \cdots, \delta^{m-1} - \gamma^{m-1})''\big).$$

By Lemma 14.7, (28) becomes

$$(29) \quad \Big(\sum_{j=1}^{m-1} \bar{\bar{\mathfrak{a}}}_j\Big) \cap \big(\bar{\bar{\mathfrak{a}}}_2 \cup (1; \overline{\gamma^1}, \gamma^2, \gamma^3, \cdots, \gamma^{m-1})' \cup (\beta; \overline{\delta^1}, \delta^2, \delta^3, \cdots, \delta^{m-1})'\big)$$
$$= \Big(\sum_{j=1}^{m-1} \bar{\bar{\mathfrak{a}}}_j\Big) \cap \big(\bar{\bar{\mathfrak{a}}}_m \cup \bar{\bar{\mathfrak{a}}}_2 \cup (\beta; \overline{\delta^1} - \overline{\gamma^1}, \delta^2 - \gamma^2, \delta^3 - \gamma^3, \cdots, \delta^{m-1} - \gamma^{m-1})'\big).$$

By IV, (29) becomes

$$(30) \quad \Big(\Big(\sum_{j=1}^{m-1} \bar{\mathfrak{a}}_j\Big) \cap \big((1; \overline{\gamma^1}, \gamma^2, \cdots, \gamma^{m-1})' \cup (\beta; \overline{\delta^1}, \delta^2, \cdots, \delta^{m-1})'\big)\Big) \cup \bar{\bar{\mathfrak{a}}}_2$$
$$= \Big(\Big(\sum_{j=1}^{m-1} \bar{\bar{\mathfrak{a}}}_j\Big) \cap \big(\bar{\bar{\mathfrak{a}}}_m \cup (\beta; \overline{\delta^1} - \overline{\gamma^1}, \delta^2 - \gamma^2, \cdots, \delta^{m-1} - \gamma^{m-1})'\big)\Big) \cup \bar{\bar{\mathfrak{a}}}_2.$$

But by (15)

$$(1; \overline{\gamma^1}, \gamma^2, \cdots, \gamma^{m-1})' = (1; \gamma^1 + \gamma\gamma^2, \gamma^2, \cdots, \gamma^{m-1})'$$
$$= (1; \gamma^1, \gamma^2, \cdots, \gamma^{m-1});$$

similarly

$$(1; \overline{\delta^1}, \delta^2, \cdots, \delta^{m-1})' = (1; \delta^1, \delta^2, \cdots, \delta^{m-1}),$$

and

$$(\beta; \overline{\delta^1} - \overline{\gamma^1}, \delta^2 - \gamma^2, \cdots, \delta^{m-1} - \gamma^{m-1})'$$
$$= (\beta; \delta^1 - \gamma^1 + \gamma(\delta^2 - \gamma^2), \delta^2 - \gamma^2, \cdots, \delta^{m-1} - \gamma^{m-1})'$$
$$= (\beta; \delta^1 - \gamma^1, \delta^2 - \gamma^2, \cdots, \delta^{m-1} - \gamma^{m-1}).$$

Moreover, $\bar{\bar{\mathfrak{a}}}_2 = (\gamma)_{21}$, and $\sum_{j=1}^{m-1} \bar{\bar{\mathfrak{a}}}_j = \sum_{j=1}^{m-1} \bar{\mathfrak{a}}_j$, since $\bar{\bar{\mathfrak{a}}}_1 \cup \bar{\bar{\mathfrak{a}}}_2 = \bar{\mathfrak{a}}_1 \cup (\gamma)_{21} = \bar{\mathfrak{a}}_1 \cup \bar{\mathfrak{a}}_2$. Thus (30) becomes (26). This completes the proof.

We now drop the restriction that γ be fixed.

LEMMA 14.9: For every $\beta, \gamma^1, \cdots, \gamma^{m-1}, \delta^1, \cdots, \delta^{m-1} \in \mathfrak{S}$,

$$(31) \qquad (\sum_{j=1}^{m-1} \bar{\mathfrak{a}}_j) \cap ((1;\, \gamma^1, \cdots, \gamma^{m-1}) \cup (\beta;\, \delta^1, \cdots, \delta^{m-1}))$$
$$= (\sum_{j=1}^{m-1} \bar{\mathfrak{a}}_j) \cap (\bar{\mathfrak{a}}_m \cup (\beta;\, \delta^1 - \gamma^1, \cdots, \delta^{m-1} - \gamma^{m-1})).$$

PROOF: Define $\mathfrak{w}$, $\mathfrak{u}$ to be the left and right members of (31) respectively, and let γ be any element of $\mathfrak{S}$. Let i, j be given $(= 1, \cdots, m-1)$, $i \neq j$. In the system $(1, \cdots, m-1)$ interchange $(1, 2)$ and (j, i); the system becomes $(t_1, \cdots, t_{m-1})$. We apply $P\begin{pmatrix} 1 & \cdots & m-1 \\ t_1 & \cdots & t_{m-1} \end{pmatrix}$ to (26). Then γ^1, γ^2, δ^1, δ^2 are interchanged with γ^j, γ^i, δ^j, δ^i, respectively, and $(\gamma)_{21}$ is replaced by $(\gamma)_{ij}$. In the relation thus obtained, let $\gamma^1, \cdots, \gamma^{m-1}$, $\delta^1, \cdots, \delta^{m-1}$ be replaced by $\gamma^{t_1}, \cdots, \gamma^{t_{m-1}}$, $\delta^{t_1}, \cdots, \delta^{t_{m-1}}$, respectively. The result is $\mathfrak{w} \cup (\gamma)_{ij} = \mathfrak{u} \cup (\gamma)_{ij}$. This equation holds for every $\gamma \in \mathfrak{S}$ and every $i, j = 1, \cdots, m-1$. Moreover $\mathfrak{w}, \mathfrak{u} \leqq \sum_{j=1}^{m-1} \bar{\mathfrak{a}}_j$. By virtue of the isomorphism of $L(0, \sum_{j=1}^{m-1} \bar{\mathfrak{a}}_j)$ and $\bar{R}_{\mathfrak{S}_{m-1}}$ (cf. page 181), we may apply Theorem 13.1 and conclude $\mathfrak{w} = \mathfrak{u}$. This completes the proof.

The relation (31) is precisely the statement (**) for Case II, whence we have proved (*) in both Case I and Case II, and therefore the isomorphism of L and $\bar{R}_{\mathfrak{S}_n}$. We now drop our assumptions on L, L', L'', $\mathfrak{S}$, n, m and state our fundamental general result.

THEOREM 14.1: Let L be a complemented modular lattice of order $n \geqq 4$. Then there exists a regular ring $\mathfrak{R}$ such that L is isomorphic to $\bar{R}_{\mathfrak{R}}$. The ring $\mathfrak{R}$ is unique up to isomorphisms.

PROOF: Let $(\bar{\mathfrak{a}}_i, \bar{\mathfrak{c}}_{ij};\, i, j = 1, \cdots, n,\, i \neq j)$ be a normalized frame in L, and let $\mathfrak{S}$ be the auxiliary ring of L relative to this frame. Then define $\mathfrak{R} \equiv \mathfrak{S}_n$. It has been shown that L is isomorphic to $\bar{R}_{\mathfrak{S}_n}$ (Cf. Chapters V—XIV), whence L is isomorphic to $\bar{R}_{\mathfrak{R}}$. Since the lattice $\bar{R}_{\mathfrak{R}}$ is complemented, it follows that $\mathfrak{R}$ is regular by Definition 2.2. (The regularity of $\mathfrak{R}$ follows also from the regularity of $\mathfrak{S}$ by Theorem 2.14.) The uniqueness of $\mathfrak{R}$ is immediate from Theorem 4.2 (Cf. page 108).

PART II · CHAPTER XV

Perspectivities and Projectivities

It has been shown (Chapters V—XIV) that every complemented modular lattice L (of order $n \geqq 4$) may be represented as the lattice $\bar{R}_{\mathfrak{R}}$ of principal right ideals of a regular ring $\mathfrak{R}$. We now return to the theory of lattices $\bar{R}_{\mathfrak{R}}$; in what follows, $\mathfrak{R} = (a, b, c, \cdots)$ is a fixed regular ring and $L \equiv \bar{R}_{\mathfrak{R}}$ is the lattice of all principal right ideals in $\mathfrak{R}$. Elements of L (and occasionally of $\bar{L}_{\mathfrak{R}}$, the lattice of principal left ideals in $\mathfrak{R}$) will be denoted as usual by $\mathfrak{a}, \mathfrak{b}, \mathfrak{c}, \cdots$. The representation in $\mathfrak{R}$ of perspective and projective isomorphisms in L will be discussed. We return to the notation $(a)_r$ (instead of $(a)_{\underline{r}}$) for the principal right ideal of a.

Definition 15.1: If $\mathfrak{a}, \mathfrak{b} \in L$, then a *factor-correspondence* between $\mathfrak{a}, \mathfrak{b}$ is a correspondence of the form

(1) $$b = pa, \quad a = qb \quad (a \in \mathfrak{a},\ b \in \mathfrak{b}),$$

where $p, q \in \mathfrak{R}$ are such that the mappings (1) are mutually inverse. The *factors* of the correspondence are p, q. The *factors* are special in case

(2) $$(q)_r \leqq \mathfrak{a}, \quad (p)_r \leqq \mathfrak{b}.$$

Corollary: If (1) defines a factor correspondence, then

(3) $$a \in \mathfrak{a} \text{ implies } qpa = a; \quad b \in \mathfrak{b} \text{ implies } pqb = b.$$

The conditions (2) of speciality are equivalent to

(4) $$(q)_r = \mathfrak{a}\,,\ (p)_r = \mathfrak{b}.$$

Proof: Since the transformations $b = pa$, $a = qb$ are mutually inverse, (3) is immediate. Now since the correspondence defined by (1) is one-to-one, $\mathfrak{a}$ is the set $(qb;\ b \in \mathfrak{b}) \subset (qu;\ u \in \mathfrak{R}) = (q)_r$, whence $\mathfrak{a} \leqq (q)_r$. Thus $(q)_r \leqq \mathfrak{a}$ is equivalent to $(q)_r = \mathfrak{a}$. Similarly $(p)_r \leqq \mathfrak{b}$ is equivalent to $(p)_r = \mathfrak{b}$.

Lemma 15.1: If $\mathfrak{a}, \mathfrak{b} \in L$, then in every factor correspondence between $\mathfrak{a}, \mathfrak{b}$ the factors may be replaced by special factors.

Proof: Let the given correspondence be defined by (1), and let $\mathfrak{a}$ be generated by the idempotent e. Then $\mathfrak{a} = (e)_r$, whence $\mathfrak{a} = (eu;\ u \in \mathfrak{R})$ and $\mathfrak{b} = (peu;\ u \in \mathfrak{R}) = (pe)_r$. Define $p' \equiv pe$, $q' \equiv eq$, and consider the mappings

$$(5) \qquad b = p'a, \qquad a = q'b \qquad (a \in \mathfrak{a},\ b \in \mathfrak{b}).$$

If $a \in \mathfrak{a}$ then $p'a = pea = pa$; if $b \in \mathfrak{b}$, then $q'b = eqb = qb$ (since $qb \in \mathfrak{a}$). Thus the mappings (5) are the same as the mappings (1). It remains to show that p', q' satisfy (2). Now

$$(p')_r = (pe)_r = \mathfrak{b}, \quad (q')_r = (eq)_r \leqq (e)_r = \mathfrak{a},$$

whence p', q' satisfy (2), and the proof is complete.

LEMMA 15.2: If $\mathfrak{a}, \mathfrak{b} \in L$, and if p, q are special factors defining a factor-correspondence between $\mathfrak{a}$, $\mathfrak{b}$, then

$$(6) \qquad pqp = p, \quad qpq = q,$$

(7) $\qquad pq$, qp are idempotents,

$$(8) \qquad \mathfrak{a} = (q)_r = (qp)_r, \quad \mathfrak{b} = (p)_r = (pq)_r.$$

PROOF: (6) Since p, q satisfy (4), and since $p \in (p)_r = \mathfrak{b}$ we have $pqp = p$ by (3). Similarly $qpq = q$.

(7) Since $pqp = p$, it follows that $pqpq = pq$, i.e., that pq is idempotent. Similarly qp is idempotent.

(8) By (4), $\mathfrak{a} = (q)_r$. It is clear that $(qp)_r \leqq (q)_r$; but since $qpq = q$, $(q)_r \leqq (qp)_r$. Thus $(q)_r = (qp)_r$. Similarly $\mathfrak{b} = (p)_r = (pq)_r$.

LEMMA 15.3: If $a \rightleftarrows b$ is a factor-correspondence between $\mathfrak{a}$, $\mathfrak{b}(\in L)$, and if e and f are idempotents such that $\mathfrak{a} = (e)_r$, $\mathfrak{b} = (f)_r$, then there exists a unique pair p', q' of special factors, such that the mappings $b = p'a$, $a = q'b$ define a factor correspondence which coincides with the given correspondence $a \leftrightarrows b$ and such that

$$(9) \qquad q'p' = e, \quad p'q' = f.$$

PROOF: Let the given correspondence $a \rightleftarrows b$ be defined by the special factors p, q. Define $p' \equiv pe$, $q' \equiv qf$, and consider the mappings $b = p'a$, $a = q'b$. Clearly if $a \in \mathfrak{a}$ then $p'a = pea = pa$, and if $b \in \mathfrak{b}$, then $q'b = qfb = qb$, whence the correspondence defined by p', q' is the same as that defined by p, q, i.e. as $a \rightleftarrows b$. Moreover, the conditions of speciality are preserved, viz.,

$$(p')_r \leqq (p)_r \leqq \mathfrak{b}, \quad (q')_r \leqq (q)_r \leqq \mathfrak{a}.$$

Finally,

$$q'p' = qfpe = qpe \qquad \text{(since } pe \in \mathfrak{b})$$
$$= e \qquad \text{(by (7), (8))},$$

and similarly $p'q' = f$. This completes the proof of the existence. Suppose that p, q along with p', q' have the desired properties. Now if $a \in \mathfrak{a}$, then $p'a = pa$; for $a \equiv e = q'p' = qp$, we obtain $p'q'p' = pqp$. But by (6), $p'q'p' = p'$, $pqp = p$, whence $p' = p$. Similarly $q' = q$, and the uniqueness follows.

LEMMA 15.4: Every factor-correspondence between $\mathfrak{a}$, $\mathfrak{b} (\in L)$ generates a lattice-isomorphism of $L((0), \mathfrak{a})$ and $L((0), \mathfrak{b})$.

PROOF: Let p, q be factors defining the given correspondence (1) of $\mathfrak{a}$, $\mathfrak{b}$. Suppose that a, b as well as a', b' correspond under (1). Then we shall show that $(a)_r \leqq (a')_r$ is equivalent to $(b)_r \leqq (b')_r$. By symmetry it suffices to show that $(a)_r \leqq (a')_r$ implies $(b)_r \leqq (b')_r$. If $(a)_r \leqq (a')_r$, then $a \in (a')_r$, i.e., a is of the form $a = a'u$. Hence $b = pa = pa'u = b'u$, whence $b \in (b')_r$, i.e., $(b)_r \leqq (b')_r$. By interchanging a, b and a', b', we see that $(a)_r \geqq (a')_r$ is equivalent to $(b)_r \geqq (b')_r$. Conjunction of these two results shows that $(a)_r = (a')_r$ is equivalent to $(b)_r = (b')_r$, i.e., that (2) generates a one-to-one correspondence between $L((0), \mathfrak{a})$ and $L((0), \mathfrak{b})$. Now the equivalence of $(a)_r \leqq (a')_r$ and $(b)_r \leqq (b')_r$ shows that this correspondence is a lattice-isomorphism.

DEFINITION 15.2: If $\mathfrak{a}, \mathfrak{b} \in L$, then an isomorphism $\mathscr{I}$ of $L((0), \mathfrak{a})$, $L((0), \mathfrak{b})$ is a *factor-isomorphism* in case it is the isomorphism generated by a factor-correspondence between $\mathfrak{a}$, $\mathfrak{b}$.

THEOREM 15.1: The factor correspondences form a *groupoid*, i.e.,
(a) the identity correspondence between $\mathfrak{a}$ and $\mathfrak{a}$ is a factor-correspondence;
(b) the inverse of a factor-correspondence between $\mathfrak{a}$ and $\mathfrak{b}$ is a factor-correspondence between $\mathfrak{b}$ and $\mathfrak{a}$;
(c) the product of a factor-correspondence between $\mathfrak{a}$ and $\mathfrak{b}$ by one between $\mathfrak{b}$ and $\mathfrak{c}$ is a factor correspondence between $\mathfrak{a}$ and $\mathfrak{c}$.

PROOF: (a) Let $\mathfrak{a} = (e)_r$ (e idempotent), and define $p \equiv q \equiv e$. Then the mappings $a' = ea$, $a = ea'$ ($a, a' \in \mathfrak{a}$) are mutually inverse; indeed, each is the identity. Thus the identity correspondence is a factor-correspondence.

(b) This is obvious, since $\mathfrak{a}$, p and $\mathfrak{b}$, q enter symmetrically into Definition 15.1.

(c) Let

(10) $\quad b = pa, \qquad a = qb \qquad (a \in \mathfrak{a},\ b \in \mathfrak{b})$

(11) $\quad c = rb, \qquad b = sc \qquad (b \in \mathfrak{b},\ c \in \mathfrak{c})$

define respectively factor-correspondences between $\mathfrak{a}$ and $\mathfrak{b}$ and between $\mathfrak{b}$ and $\mathfrak{c}$. Then the product of these correspondences is

(12) $\quad c = rpa, \qquad a = qsc,$

the mappings (12) are mutually inverse (since this holds for (10) and (11)), and the proof is complete.

COROLLARY: In (a), the identity correspondence may be obtained with special factors; in (b), if special factors p, q are used in the correspondence $\mathfrak{a} \rightarrow \mathfrak{b}$ then the inverse correspondence may be represented with special factors q, p; in (c), if the factors in (10) and (11) are special, then those in (12) are special.

PROOF: In (a), since $p = q = e$, we have $(p)_r = (q)_r = \mathfrak{a}$, whence p, q are special. The speciality of p, q in (b) implies the speciality of q, p again by the symmetry in Definition 15.1. Let p, q in (b) and r, s in (c) be special. Then $(q)_r = \mathfrak{a}$, $(p)_r = \mathfrak{b}$, $(s)_r = \mathfrak{b}$, $(r)_r = \mathfrak{c}$. Hence $(rp)_r \leqq (r)_r = \mathfrak{c}$, $(qs)_r \leqq (q)_r = \mathfrak{a}$, and rp, qs are special.

THEOREM 15.2: The factor-isomorphisms form a groupoid.

PROOF: This is immediate from Theorem 15.1, since the factor-isomorphisms are generated by the factor correspondences.

We turn now to a refinement in the representation of factor-isomorphisms (cf. Lemma 15.9). Some preliminary theory of subrings $\mathfrak{S}$ of $\mathfrak{R}$, in particular those of the form $\mathfrak{S} = \mathfrak{R}(e)$ is needed. If $\mathfrak{S}$ is a regular subring of $\mathfrak{R}$ with a unit (which is therefore an idempotent in $\mathfrak{R}$), the theory of Chapters II, III may be applied to $\mathfrak{S}$. In particular, the principal right ideals in $\mathfrak{S}$ form a lattice $\bar{R}_{\mathfrak{S}}$; the principal right ideal in $\mathfrak{S}$ generated by $a \in \mathfrak{S}$ will be denoted by $(a)_{r'}$, and the lattice operations in $\bar{R}_{\mathfrak{S}}$ by $\cup'$, $\cap'$.

DEFINITION 15.3: If $\mathfrak{S}$ is a regular subring of $\mathfrak{R}$, then define $\bar{R}_{\mathfrak{R}}^{(\mathfrak{S})}$ as the set of all $(a)_r$, $a \in \mathfrak{S}$.

LEMMA 15.5: Let $\mathfrak{S}$ be a regular subring of $\mathfrak{R}$. Then the correspondence $(a)_{r'} \rightleftarrows (a)_r$ $(a \in \mathfrak{S})$ is a one-to-one correspondence between $\bar{R}_{\mathfrak{S}}$ and $\bar{R}_{\mathfrak{R}}^{(\mathfrak{S})}$ preserving the inclusion relation $\leqq$.

PROOF: It will be shown first that if a, $b \in \mathfrak{S}$, then

$$(13) \qquad (a)_{r'} \leqq (b)_{r'} \text{ if and only if } (a)_r \leqq (b)_r.$$

Let $(a)_{r'} \leqq (b)_{r'}$. Then $a \in (b)_{r'}$, and a is of the form $a = bu$ $(u \in \mathfrak{S})$. Hence $a = bu$ with $u \in \mathfrak{R}$, and $a \in (b)_r$, i.e., $(a)_r \leqq (b)_r$. Let $(a)_r \leqq (b)_r$, and let $(b)_{r'} = (e)_{r'}$, with $e \in \mathfrak{S}$ idempotent. Then $b = eb$, and $(b)_r = (eb)_r \leqq (e)_r$. But since $(eb)_{r'} = (e)_{r'}$, e is of the form $e = ebu$ $(u \in \mathfrak{S})$. Hence $e = ebu$ with $u \in \mathfrak{R}$, and thus $e \in (eb)_r$, i.e., $(e)_r \leqq (eb)_r$. Therefore $(b)_r = (e)_r$. Since $(a)_r \leqq (b)_r$, $a \in (e)_r$ and $a = ea$; thus $(a)_{r'} \leqq (e)_{r'} = (b)_{r'}$. By interchanging a and b, (13) yields the equivalence of $(a)_{r'} \geqq (b)_{r'}$ and $(a)_r \geqq (b)_r$. Hence by conjunction of these results, there is a one-to-one correspondence between $R_{\mathfrak{S}}$ and $R_{\mathfrak{R}}^{(\mathfrak{S})} = ((a)_r;\ a \in \mathfrak{S})$. Then (13) shows that this correspondence preserves inclusion.

LEMMA 15.6: If $\mathfrak{S}$ is a regular subring of $\mathfrak{R}$ then $\bar{R}_{\mathfrak{R}}^{(\mathfrak{S})}$ is a sublattice of $\bar{R}_{\mathfrak{R}}$.

PROOF: Let $\mathfrak{a}, \mathfrak{b} \in \bar{R}_{\mathfrak{R}}^{(\mathfrak{S})}$. Then $\mathfrak{a} = (a)_r$, $\mathfrak{b} = (b)_r$, with $a, b \in \mathfrak{S}$. We shall show that $\mathfrak{c} \equiv \mathfrak{a} \cup \mathfrak{b}$ and $\mathfrak{d} \equiv \mathfrak{a} \cap \mathfrak{b}$ are in $\bar{R}_{\mathfrak{R}}^{(\mathfrak{S})}$, i.e., that $\mathfrak{c}$ is of the form $\mathfrak{c} = (c)_r$, $c \in \mathfrak{S}$ and $\mathfrak{d}$ is of the form $\mathfrak{d} = (d)_r$, $d \in \mathfrak{S}$. Since $\mathfrak{S}$ is regular, there exist $c, d \in \mathfrak{S}$ such that

$$(a)_{r'} \cup' (b)_{r'} = (c)_{r'}, \quad (a)_{r'} \cap' (b)_{r'} = (d)_{r'}.$$

Let $(g)_{r'}$, $(h)_{r'}$ be inverse to $(a)_{r'} \cap' (b)_{r'}$ in $(a)_{r'}$, $(b)_{r'}$, respectively. Then $((a)_{r'} \cap' (b)_{r'}, (g)_{r'}, (h)_{r'}) \perp$, and by Lemma 3.2 there exist idempotents $e_0, e_1, e_2 \in \mathfrak{S}$ such that

$$(a)_{r'} \cap' (b)_{r'} = (e_0)_{r'}, \quad (g)_{r'} = (e_1)_{r'}, \quad (h)_{r'} = (e_2)_{r'},$$

and such that $e_i e_j = 0$ for $i \neq j$. Then by Lemma 3.1

$$(a)_{r'} = (e_0 + e_1)_{r'}, \quad (b)_{r'} = (e_0 + e_2)_{r'},$$

$$(a)_{r'} \cup' (b)_{r'} = (e_0 + e_1 + e_2)_{r'}.$$

Hence by Lemma 15.5, $(d)_r = (e_0)_r$, $(g)_r = (e_1)_r$, $(h)_r = (e_2)_r$, $(e_0+e_1)_r = (a)_r$, $(e_0+e_2)_r = (b)_r$. Clearly $(e_0)_r + (e_1)_r = (e_0+e_1)_r$, $(e_0)_r + (e_2)_r = (e_0+e_2)_r$ (since $e_i e_j = 0$ for $i \neq j$). Combining these, we obtain $(d)_r \cup (g)_r = (a)_r$ and $(d)_r \cup (h)_r = (b)_r$. But e_0, e_1, e_2 are independent idempotents, whence $((d)_r, (g)_r, (h)_r) \perp$, and $(a)_r \cap (b)_r = (d)_r$. Moreover, $(a)_r \cup (b)_r = (e_0 + e_1 + e_2)_r = (c)_r$. This completes the proof.

COROLLARY: The correspondence of Lemma 15.5 is a lattice isomorphism of $\bar{R}_{\mathfrak{S}}$ and $\bar{R}_{\mathfrak{R}}{}^{(\mathfrak{S})}$.

LEMMA 15.7: Let $e \in \mathfrak{R}$ be idempotent. Then $L((0), (e)_r) = \bar{R}_{\mathfrak{R}}^{(\mathfrak{R}(e))}$.

PROOF: Let $\mathfrak{a} \leqq (e)_r$, $\mathfrak{a} \in \bar{R}_{\mathfrak{R}}$. Now $(1-e)_r$ is inverse to $(e)_r$, whence $(1-e)_r \cap \mathfrak{a} = (0)$. Hence by Part I, Corollary to Theorem 1.4, there exists $\mathfrak{b} \in \bar{R}_{\mathfrak{R}}$ such that $\mathfrak{b} \geqq (1-e)_r$ and $\mathfrak{b}$ is inverse to $\mathfrak{a}$. Then there exists an idempotent $f \in \mathfrak{R}$ such that $\mathfrak{a} = (f)_r$, $\mathfrak{b} = (1-f)_r$, whence $(f)_r \leqq (e)_r$, $(1-f)_r \geqq (1-e)_r$. Hence $(1-f)_r^l \leqq (1-e)_r^l$ (cf. Lemma 2.1 (1), Lemma 2.2 (1)), and $(f)_l \leqq (e)_l$. Thus $f \in (e)_r, (e)_l$, i.e., $f \in \mathfrak{R}(e)$, and $\mathfrak{a} = (f)_r \in \bar{R}_{\mathfrak{R}}^{(\mathfrak{R}(e))}$. This proves that $L((0), (e)_r) \subset \bar{R}_{\mathfrak{R}}^{(\mathfrak{R}(e))}$. Now let $a \in \mathfrak{R}(e)$ and consider $(a)_r \in \bar{R}_{\mathfrak{R}}^{(\mathfrak{R}(e))}$. Then $a = ea$, whence $a \in (e)_r$, $(a)_r \leqq (e)_r$, i.e., $(a)_r \in L((0), (e)_r)$. This proves that $\bar{R}_{\mathfrak{R}}^{(\mathfrak{R}(e))} \subset L((0), (e)_r)$. Consequently $L((0), (e)_r) = \bar{R}_{\mathfrak{R}}^{(\mathfrak{R}(e))}$.

LEMMA 15.8: If $e, f \in \mathfrak{R}$ are idempotents, and if $pq = f$, $qp = e$, then the mappings

$$\text{(14)} \qquad v = puq, \qquad u = qvp \qquad (u \in \mathfrak{R}(e),\ v \in \mathfrak{R}(f))$$

are mutually inverse and define a ring-isomorphism of $\mathfrak{R}(e)$ and $\mathfrak{R}(f)$.

Proof: If $u \in \mathfrak{R}(e)$, then $qpuqp = eue = u$, and if $v \in \mathfrak{R}(f)$, then $pqvpq = fvf = v$; moreover, if $u \in \mathfrak{R}(e)$, then $puq \cdot pq = pueq = puq$, and $pq \cdot puq = peuq = puq$, i.e., $puq \in \mathfrak{R}(f)$, and similarly if $v \in \mathfrak{R}(f)$, then $qvp \in \mathfrak{R}(e)$. Hence the mappings (14) are mutually inverse and therefore define a one-to-one correspondence between $\mathfrak{R}(e)$ and $\mathfrak{R}(f)$. Now let u, v as well as u', v' correspond under (14). Then

$$v + v' = puq + pu'q = p(u + u')q \rightleftarrows u + u',$$

whence (14) preserves addition. Moreover,

$$vv' = puq \cdot pu'q = pueu'q = puu'q \rightleftarrows uu',$$

whence (14) preserves multiplication. Consequently (14) defines a ring-isomorphism of $\mathfrak{R}(e)$ and $\mathfrak{R}(f)$.

Definition 15.4: If $e, f \in \mathfrak{R}$ are idempotents, an isomorphism of $\mathfrak{R}(e)$ and $\mathfrak{R}(f)$ is an *inner-isomorphism* in case it is defined by mappings of the form (14), with $pq = f$, $qp = e$. If $\mathfrak{R}(e) = \mathfrak{R}(f)$, the isomorphism is an *inner-automorphism.*

Note: If $\mathfrak{R}(e) = \mathfrak{R}(f) = \mathfrak{R}$, i.e., if $e = f = 1$, then $pq = qp = 1$, and q is said to be an *inverse* of p. (An inverse of p is unique if existent, since if q' is another inverse, we have $q' = q' \cdot pq = q'p \cdot q = q$.) The notation p^{-1} is used to denote the inverse q of p. The automorphism is then represented by $v = pup^{-1}$, which coincides with the usual concept of inner-automorphism. An element which has an inverse is called *non-singular.*

Lemma 15.9: Let a factor-correspondence be given between $\mathfrak{a}$ and $\mathfrak{b}$, and let the factors p, q be special. Then the factor-isomorphism generated thereby is also generated by the inner-isomorphism

$$(15) \qquad v = puq, \qquad u = qvp \qquad (u \in \mathfrak{R}(qp),\ v \in \mathfrak{R}(pq)).$$

Conversely, any inner-isomorphism of the form (14) generates the same factor-isomorphism as the factor-correspondence defined by p, q.

Proof: Since p, q are special, $e \equiv qp$ and $f \equiv pq$ are idempotent, and $(e)_r = \mathfrak{a}$, $(f)_r = \mathfrak{b}$. By Lemma 15.7 the right ideals $(u)_r$, $u \in \mathfrak{R}(e)$ coincide with those of the set $L((0), \mathfrak{a})$, and the $(v)_r$, $v \in \mathfrak{R}(f)$ constitute $L((0), \mathfrak{b})$. Let $u \in \mathfrak{R}(e)$, let v correspond to u under (15), and let v' correspond to u under the factor-correspondence. Then $(v)_r = (puq)_r \leqq (pu)_r = (v')_r = (pu)_r = (pue)_r = (puqp)_r \leqq (puq)_r = (v)_r$. Thus $(v)_r = (v')_r$. These considerations show that the correspondence generated by (15) between $L((0), \mathfrak{a})$ and $L((0), \mathfrak{b})$ coincides with that generated by the factor-correspondence. Conversely, let (15) be a given inner-isomorphism of $\mathfrak{R}(e)$, $\mathfrak{R}(f)$, where $\mathfrak{a} = (e)_r$, $\mathfrak{b} = (f)_r$. Then $a \in \mathfrak{a}$ implies $qpa = ea = a$, and $b \in \mathfrak{b}$ implies $pqb = fb = b$, whence p, q define a factor-correspondence of $\mathfrak{a}, \mathfrak{b}$ and the argument given above (which depends only on the relations $pq = f$, $qp = e$ and not on the speciality of p, q) shows that the generated factor-isomorphism is the same as that generated by (15).

COROLLARY: Every factor-isomorphism of $L((0), \mathfrak{a})$, $L((0), \mathfrak{b})$ may be generated by an inner-isomorphism of the form (15) where

$$pqp = q, \quad qpq = q. \tag{16}$$

PROOF: This is clear from Lemma 15.9, since the relations (16) follow from the speciality of p, q (cf. Lemma 15.2).

THEOREM 15.3:
(a) Every perspective isomorphism of $L((0), \mathfrak{a})$, $L((0), \mathfrak{b})$ is a factor-isomorphism.
(b) Every projective isomorphism of $L((0), \mathfrak{a})$, $L((0), \mathfrak{b})$ is a factor-isomorphism.
(c) If $\mathfrak{a} \cap \mathfrak{b} = (0)$, then every factor-isomorphism of $L((0), \mathfrak{a})$, $L((0), \mathfrak{b})$ is a perspective isomorphism.

PROOF: (a) Let $\mathfrak{c}$ be a common inverse of $\mathfrak{a}$ and $\mathfrak{b}$. Then there exist idempotents e, f such that

$$\mathfrak{a} = (e)_r, \ \mathfrak{c} = (1 - e)_r = (1 - f)_r, \ \mathfrak{b} = (f)_r.$$

Now since $(1 - e)_r = (1 - f)_r$, we have also $(1 - e)_r^l = (1 - f)_r^l$, i.e. $(e)_l = (f)_l$. This last condition is equivalent to $ef = e$, $fe = f$. Define $p \equiv f$, $q \equiv e$. Then $pq = f$, $qp = e$, whence $a \in \mathfrak{a}$ implies $qpa = ea = a$, and $b \in \mathfrak{b}$ implies $pqb = fb = b$. Therefore p, q define a factor-correspondence between $\mathfrak{a}$ and $\mathfrak{b}$. (Indeed, p, q are special, since $(q)_r = (e)_r = \mathfrak{a}$, $(p)_r = (f)_r = \mathfrak{b}$). We shall show that the factor-isomorphism generated by this correspondence is the perspective isomorphism of $L((0), \mathfrak{a})$, $L((0), \mathfrak{b})$ with the axis $\mathfrak{c}$. It suffices to show that if $a \in \mathfrak{a}$ and $b \in \mathfrak{b}$ correspond, i.e., if $b = pa$, $a = qb$, then

$$(a)_r \cup \mathfrak{c} = (b)_r \cup \mathfrak{c}. \tag{17}$$

Now $a - b = a - pa = (1-p)a = (1-f)a$, whence $(a)_r \leqq (b)_r \cup (1-f)_r = (b)_r \cup \mathfrak{c}$ and $(b)_r \leqq (a)_r \cup (1 - f)_r = (a)_r \cup \mathfrak{c}$. Thus $(a)_r \cup \mathfrak{c} \leqq (b)_r \cup \mathfrak{c}$, $(b)_r \cup \mathfrak{c} \leqq (a)_r \cup \mathfrak{c}$ and (17) holds. This completes the proof.

(b) This is obvious since every projective isomorphism is a product of a finite number of perspective-isomorphisms and the factor-isomorphisms form a groupoid (by Theorem 15.2).

(c) Let $\mathfrak{a} \cap \mathfrak{b} = (0)$ and let p, q define a factor-correspondence between $\mathfrak{a}$ and $\mathfrak{b}$ with p, q special. Then $e \equiv qp$ and $f \equiv pq$ are idempotents such that $(e)_r = \mathfrak{a}$, $(f)_r = \mathfrak{b}$. Moreover, $p = pqp$, $q = qpq$ by Lemma 15.2, whence

$$fp = pe = p, \qquad eq = qf = q.$$

The factor-isomorphism of $L((0), \mathfrak{a})$ and $L((0), \mathfrak{b})$ is generated by the mappings $b = pa$, $a = qb$ $(a \in \mathfrak{a},\ b \in \mathfrak{b})$. Now

$$(e - p)(-q) = -eq + pq = -q + f = f - q,$$

whence $(e - p)_r \leqq (f - q)_r$; also

$$(f - q)(-p) = -fp + qp = -p + e = e - p,$$

whence $(f-q)_r \leqq (e-p)_r$. Thus $(e-p)_r = (f-q)_r$ and we may define $\mathfrak{c} \equiv (e-p)_r = (f - q)_r$. It will be shown that $\mathfrak{c}$ is inverse to $\mathfrak{a}$, $\mathfrak{b}$ in $\mathfrak{a} \cup \mathfrak{b}$, and that the given factor-isomorphism is the perspective isomorphism of $L((0), \mathfrak{a})$, $L((0), \mathfrak{b})$ with the axis $\mathfrak{c}$. Suppose $x \in \mathfrak{a}, \mathfrak{c}$. Then $x = (e - p)u (u \in \mathfrak{R})$, and $x = ex$, whence $x = e(e - p)u$. Hence $eu - pu = eu - epu$, i.e., $pu = epu$. Hence $pu \in (e)_r$. But $pu = fpu$, whence $pu \in (f)_r$, and since $(e)_r \cap (f)_r = \mathfrak{a} \cap \mathfrak{b} = (0)$, we have $pu = 0$. But then $x = (e - p)u = (qp - p)u = (q - 1)pu = 0$. Therefore $\mathfrak{a} \cap \mathfrak{c} = (0)$, and similarly $\mathfrak{b} \cap \mathfrak{c} = (0)$. Now if $a \in \mathfrak{a}$ and $b = pa$, $a - b = a - pa = ea - pa = (e-p)a$. Thus $(a)_r \leqq (b)_r \cup (e-p)_r = (b)_r \cup \mathfrak{c}$, and $(b)_r \leqq (a)_r \cup (e-p)_r = (a)_r \cup \mathfrak{c}$, whence

$$(a)_r \cup \mathfrak{c} = (b)_r \cup \mathfrak{c}. \tag{18}$$

Now $a \equiv e$ yields $b = pe = p$, whence $(b)_r = (p)_r = \mathfrak{b}$; thus (18) for $a = e$ states that $\mathfrak{a} \cup \mathfrak{c} = \mathfrak{b} \cup \mathfrak{c}$. Now $\mathfrak{a} \cup \mathfrak{c} = \mathfrak{b} \cup \mathfrak{c} \geqq \mathfrak{a} \cup \mathfrak{b}$; but since $\mathfrak{c} = (e - p)_r$, $\mathfrak{c} \leqq (e)_r \cup (p)_r = \mathfrak{a} \cup \mathfrak{b}$, whence $\mathfrak{a} \cup \mathfrak{c} = \mathfrak{b} \cup \mathfrak{c} \leqq \mathfrak{a} \cup \mathfrak{b}$. Hence we have shown that $\mathfrak{c}$ is inverse to $\mathfrak{a}$, $\mathfrak{b}$ in $\mathfrak{a} \cup \mathfrak{b}$. Now (18) shows that if $(a)_r \in L((0), \mathfrak{a})$, then its correspondent under the perspective isomorphism with axis $\mathfrak{c}$ is $(b)_r$ and conversely, if $(b)_r \in L((0), \mathfrak{b})$, then its correspondent under this isomorphism is $(a)_r$. Hence the perspective isomorphism coincides with the factor-isomorphism.

COROLLARY: If $\mathfrak{a} \cap \mathfrak{b} = (0)$, the set (a) of all perspective isomorphisms of $L((0), \mathfrak{a})$, $L((0), \mathfrak{b})$, the set (b) of all projective isomorphisms of $L((0), \mathfrak{a})$, $L((0), \mathfrak{b})$, and the set (c) of all factor-isomorphisms of $L((0), \mathfrak{a})$, $L((0), \mathfrak{b})$ all coincide.

PROOF: Clearly the set (a) is contained in the set (b); the set (b) is contained in the set (c) by Theorem 15.3 (b). Finally, the set (c) is contained in the set (a) by Theorem 15.3 (c). Hence the three sets all coincide.

Inner Automorphisms

This chapter is devoted to a study of inner-automorphisms of $\mathfrak{R}$ and the lattice-automorphisms of $L = \bar{R}_{\mathfrak{R}}$ that they generate. These lattice-automorphisms wil be referred to as *inner-automorphisms* of L. It will be supposed throughout the chapter that $\mathfrak{R}$ (i.e., L) is of order n $(= 1, 2, \cdots)$.

LEMMA 16.1: If $(s_{ij};\ i, j = 1, \cdots, n)$ and $(s'_{ij};\ i, j = 1, \cdots, n)$ are two systems of matrix-units in $\mathfrak{R}$ such that there exists a factor-correspondence between $(s_{11})_r$ and $(s'_{11})_r$, then for $i = 1, \cdots, n$, there exists a factor correspondence between $(s_{ii})_r$ and $(s'_{ii})_r$, and there exists an inner-automorphism of $\mathfrak{R}$ which carries each s_{ii} into s'_{ii}.

PROOF: Since the $(s_{ii})_r$ form a homogeneous basis for L by Lemma 3.6, it follows that $(s_{ii})_r \sim (s_{11})_r$. Hence there exists, by Theorem 15.3 (a), a factor correspondence between $(s_{ii})_r$ and $(s_{11})_r$. Similarly there is a factor-correspondence between $(s'_{ii})_r$ and $(s'_{11})_r$. But the product of any number of factor-correspondences is again a factor-correspondence, by Theorem 15.1, whence there is a factor-correspondence between $(s_{ii})_r$ and $(s'_{ii})_r$. Now by Lemma 15.3 there exist special factors $p_i, q_i (i = 1, \cdots, n)$ such that $q_i p_i = s_{ii}$, $p_i q_i = s'_{ii}$, $p_i \epsilon (s'_{ii})_r$, $q_i \epsilon (s_{ii})_r$, and such that the correspondence between $(s_{ii})_r$ and $(s'_{ii})_r$ is defined by the mappings

$$y = p_i x, \quad x = q_i y \qquad (x \epsilon (s_{ii})_r,\ y \epsilon (s'_{ii})_r).$$

Hence

$$q_i = s_{ii} q_i = q_i p_i q_i = q_i s'_{ii},$$
$$p_i = s'_{ii} p_i = p_i q_i p_i = p_i s_{ii},$$

and for $i \neq j$

$$q_i p_j = q_i s'_{ii} s'_{jj} p_j = 0,$$
$$p_j q_i = p_j s_{jj} s_{ii} q_i = 0.$$

Define $u \equiv \sum_{i=1}^{n} p_i$, $v \equiv \sum_{i=1}^{n} q_i$. Then

$$uv = \sum_{i,j=1}^{n} p_j q_i = \sum_{i=1}^{n} p_i q_i = \sum_{i=1}^{n} s'_{ii} = 1,$$

$$vu = \sum_{i,j=1}^{n} q_i p_j = \sum_{i=1}^{n} q_i p_i = \sum_{i=1}^{n} s_{ii} = 1,$$

whence $v = u^{-1}$. It will be shown that the inner automorphism $x \to uxu^{-1}$ has the desired property. We have

$$us_{ii}u^{-1} = us_{ii}v = \sum_{k,h=1}^{n} p_k s_{ii} q_h = \sum_{k,h=1}^{n} p_k q_i p_i q_h = p_i q_i p_i q_i = s'_{ii} s'_{ii} = s'_{ii}.$$

Hence the automorphism carries each s_{ii} into s'_{ii}, and the proof is complete.

LEMMA 16.2: Let $n \geqq 2$, let $(s_{ij};\, i, j = 1, \cdots, n)$ be a system of matrix-units for $\mathfrak{R}$ and let A be a given ring-automorphism of $\mathfrak{R}$. Then A carries the s_{ij} into matrix-units s'_{ij}, and generates a lattice-isomorphism of $(s_{11} + s_{22})_r$ and $(s'_{11} + s'_{22})_r$. If this lattice-isomorphism is a factor-isomorphism, then A is an inner-automorphism.

PROOF: Clearly if s'_{ij} is the transform of s_{ij} under A $(i, j = 1, \cdots, n)$, then the properties

$$s_{ij} s_{kl} = \begin{cases} s_{il} & (j = k) \\ 0 & (j \neq k), \end{cases}$$

$$\sum_{i=1}^{n} s_{ii} = 1$$

are possessed also by the s'_{ij} since A is an automorphism; hence the s'_{ij} are matrix-units.

Now suppose A generates a factor-isomorphism between $(s_{11} + s_{22})_r$ and $(s'_{11} + s'_{22})_r$. Then A generates a factor-isomorphism between $(s_{11})_r$ and $(s'_{11})_r$ (since s_{11} and s'_{11} correspond under A). By Lemma 16.1 there exists an inner-automorphism B of $\mathfrak{R}$ which carries each s'_{ii} into s_{ii}, whence AB carries each s_{ii} into itself. Since B is inner, it suffices to show that AB is inner, since then $A = AB \cdot B^{-1}$ is also inner. Now if p, q are factors defining the factor-correspondence between $(s_{11} + s_{22})_r$ and $(s'_{11} + s'_{22})_r$, and if B is defined by $y = wxw^{-1}$ $(x \in \mathfrak{R})$, then AB generates in $L((0), (s_{11} + s_{22})_r)$ the same lattice-automorphism as the correspondence defined by

$$(1) \qquad z = wpxw^{-1}, \quad x = qw^{-1}zw \qquad (x, z \in (s_{11} + s_{22})_r).$$

Note that the mappings (1) are mutually inverse, since $pqw^{-1}zw = w^{-1}zw$ because $w^{-1}zw \in (s_{11} + s_{22})_r$, and so $wpqw^{-1}zww^{-1} = ww^{-1}zww^{-1} = z$; and $qw^{-1}wpxw^{-1}w = qpx = x$. Hence the mappings:

$$z = wpx, \quad x = qw^{-1}z \qquad (x, z \in (s_{11} + s_{22})_r)$$

are also mutually inverse, since $wpqw^{-1}z = wpqw^{-1}zww^{-1} = z$ for $z \in (s_{11} + s_{22})_r$, and $qw^{-1}wpx = qpx = x$ for $x \in (s_{11} + s_{22})_r$. Hence they define a factor-correspondence between $(s_{11} + s_{22})_r$ and $(s_{11} + s_{22})_r$. This correspondence generates the same lattice-automorphism of $(s_{11} + s_{22})_r$ as (1), since if $x \in (s_{11} + s_{22})_r$, $(wpx)_r = (wpxu;\ u \in \mathfrak{R}) = (wpx \cdot w^{-1}u;\ u \in \mathfrak{R}) = (wpxw^{-1})_r$ and similarly $(qw^{-1}z)_r = (qw^{-1}zw)_r$ for $z \in (s_{11} + s_{22})_r$. Hence there exists, by Lemma 15.9, an inner-automorphism of $\mathfrak{R}(s_{11} + s_{22})$ which generates the same lattice-automorphism of $(s_{11} + s_{22})_r$ as does AB. Let this inner-automorphism be defined by

$$(2) \qquad y = u_0xv_0, \quad x = v_0yu_0 \qquad (x, y \in \mathfrak{R}(s_{11} + s_{22}))$$

where u_0, $v_0 \in \mathfrak{R}(s_{11} + s_{22})$, and $u_0v_0 = v_0u_0 = s_{11} + s_{22}$. Define $u \equiv u_0 + \sum_{i=3}^n s_{ii}$, $v \equiv v_0 + \sum_{i=3}^n s_{ii}$. Now $uv = u_0v_0 + u_0\sum_{i=3}^n s_{ii} + (\sum_{i=3}^n s_{ii})v_0 + (\sum_{i=3}^n s_{ii})(\sum_{i=3}^n s_{ii})$. Now since u_0, $v_0 \in \mathfrak{R}(s_{11}+s_{22})$, we have $u_0 = u_0(s_{11} + s_{22})$, $v_0 = (s_{11} + s_{22})v_0$; moreover, $\sum_{i=3}^n s_{ii}$ is clearly idempotent. Hence $uv = u_0v_0 + \sum_{i=3}^n s_{ii} = s_{11} + s_{22} + \sum_{i=3}^n s_{ii} = 1$. Similarly $vu = 1$. Hence $v = u^{-1}$, and $x \to uxu^{-1} (x \in \mathfrak{R})$ defines an inner-automorphism C of $\mathfrak{R}$. But if $x \in \mathfrak{R}(s_{11} + s_{22})$, $uxu^{-1} = (u_0 + \sum_{i=3}^n s_{ii})x(v_0 + \sum_{i=3}^n s_{ii}) = u_0xv_0 + \sum_{i=3}^n s_{ii}xv_0 + u_0x\sum_{i=3}^n s_{ii} + \sum_{i=3}^n s_{ii}x\sum_{i=3}^n s_{ii} = u_0xv_0$ since $\sum_{i=3}^n s_{ii}x = x\sum_{i=3}^n s_{ii} = 0$. Hence C coincides with the isomorphism (2). Hence C and AB, as ring-automorphisms of $\mathfrak{R}(s_{11} + s_{22})$, generate the same lattice-automorphism of $(s_{11} + s_{22})_r$, whence C and AB agree on $\mathfrak{R}(s_{11} + s_{22})$ by Theorem 4.1 (cf. remarkson page 103). Now $D \equiv ABC^{-1}$ leaves invariant every $x \in \mathfrak{R}(s_{11} + s_{22})$; moreover C leaves invariant every s_{kk}, $(k = 3, \cdots, n)$ since

$$\begin{aligned} us_{kk}u^{-1} &= (u_0 + \sum_{i=3}^n s_{ii})s_{kk}(v_0 + \sum_{i=3}^n s_{ii}) \\ &= u_0s_{kk}v_0 + \sum_{i=3}^n s_{ii}s_{kk}v_0 + u_0s_{kk}\sum_{i=3}^n s_{ii} + \sum_{i=3}^n s_{ii}s_{kk}\sum_{i=3}^n s_{ii} \\ &= 0 + 0 + 0 + s_{kk}s_{kk}s_{kk} = s_{kk}, \end{aligned}$$

whence D also leaves each s_{ii} invariant. Now since C is an inner-automorphism of $\mathfrak{R}$, C^{-1} is also inner, and AB will be inner if D is inner since $AB = DC$.

Now we shall show that D is inner. Let s''_{ij} be the transform of s_{ij} under D $(i, j = 1, \cdots, n)$. Then the s''_{ij} are matrix-units with $s''_{ii} = s_{ii}$ $(i = 1, \cdots, n)$. Define

$$y \equiv \sum_{i=1}^n s_{i1}s''_{1i}, \qquad z \equiv \sum_{i=1}^n s''_{i1}s_{1i}.$$

Then

$$yz = \sum_{k,h=1}^{n} s_{k1}s''_{1k}s''_{h1}s_{1h} = \sum_{k=1}^{n} s_{k1}s''_{11}s_{1k} = \sum_{k=1}^{n} s_{k1}s_{11}s_{1k} = \sum_{k=1}^{n} s_{k1}s_{1k} = \sum_{k=1}^{n} s_{kk} = 1,$$

$$zy = \sum_{k,h=1}^{n} s''_{k1}s_{1k}s_{h1}s''_{1h} = \sum_{k=1}^{n} s''_{k1}s_{11}s''_{1k} = \sum_{k=1}^{n} s''_{k1}s''_{11}s''_{1k} = \sum_{k=1}^{n} s''_{k1}s''_{1k} = \sum_{k=1}^{n} s''_{kk} = 1,$$

whence $y = z^{-1}$. Let E denote the inner automorphism of $\mathfrak{R}$ defined by $x \to zxz^{-1}$ $(x \in \mathfrak{R})$. Now

$$s_{ij} \to zs_{ij}z^{-1} = zs_{ij}y = \sum_{k,h=1}^{n} s''_{k1}s_{1k}s_{ij}s_{h1}s''_{1h} = s''_{i1}s_{1i}s_{ij}s_{j1}s''_{1j} = s''_{i1}s_{11}s''_{1j}$$

$$= s''_{i1}s''_{11}s''_{1j} = s''_{i1}s''_{1j} = s''_{ij},$$

i.e., E has the same effect on the s_{ij} as D. Now

$$s_{11}y = s_{11}\sum_{i=1}^{n} s_{i1}s''_{1i} = s_{11}s_{11}s''_{11} = s_{11}s_{11}s_{11} = s_{11},$$

$$zs_{11} = \sum_{i=1}^{n} s''_{i1}s_{1i}s_{11} = s''_{11}s_{11}s_{11} = s_{11}s_{11}s_{11} = s_{11},$$

whence for every $x \in \mathfrak{R}(s_{11})$,

$$zxz^{-1} = zxy = zs_{11}xs_{11}y = s_{11}xs_{11} = x.$$

Hence E (as well as D) leaves each $x \in \mathfrak{R}(s_{11})$ invariant. Thus D and E agree for the s_{ij} and the elements $x \in \mathfrak{R}(s_{11})$, hence also for all elements x of the form

$$x = \sum_{i,j=1}^{n} s_{i1}x_{ij}s_{1j} \qquad (x_{ij} \in \mathfrak{R}(s_{11})),$$

i.e., for all elements $x \in \mathfrak{R}$ (by the proof of Theorem 3.3 — cf. equation (11)). Hence $D = E$. But E is an inner automorphism, whence D is also an inner automorphism.

THEOREM 16.1: If $n \geqq 3$, and $(s_{ij};\, i, j = 1, \cdots, n)$ is a system of matrix-units for $\mathfrak{R}$, then any lattice-automorphism $\mathscr{I}$ of L such that $(s_{11} + s_{22})_r \to (s_{11} + s_{22})_r\mathscr{I}$ is a factor-isomorphism may be generated by an inner-automorphism of $\mathfrak{R}$.

PROOF: By Theorem 4.2, there exists a ring-automorphism A which generates $\mathscr{I}$ since $n \geqq 3$. Then A is inner by Lemma 16.2.

COROLLARY: If the correspondence $(s_{11} + s_{22})_r \to (s_{11} + s_{22})_r\mathscr{I}$ is a projective isomorphism or if it is the identity (i.e., if it leaves every

$\mathfrak{a} \leqq (s_{11} + s_{22})_r$ invariant), then $\mathscr{I}$ may be generated by an inner-automorphism of $\mathfrak{R}$.

PROOF: This is evident since the identity is a projective isomorphism, and since every projective isomorphism is a factor-isomorphism by Theorem 15.3 (b).

PART II · CHAPTER XVII

Properties of Continuous Rings

Since every complemented modular lattice L of order $\geqq 4$ is isomorphic to the lattice $\bar{R}_{\mathfrak{R}}$ of principal right ideals of a regular ring $\mathfrak{R}$ by Theorem 14.1, and since the finite-dimensional projective geometries L_n $(n \geqq 4)$ and the continuous geometries L_∞ (cf. Part I, pages 57—58) are such lattices (by Axioms I—VI), it follows that each L_n $(n \geqq 4)$ as well as each L_∞ is isomorphic to a lattice $\bar{R}_{\mathfrak{R}}$ with $\mathfrak{R}$ regular. In this chapter we seek the properties of such rings $\mathfrak{R}$ for which $\bar{R}_{\mathfrak{R}}$ satisfies Axioms I—VI, i.e., for which $\bar{R}_{\mathfrak{R}}$ is a finite-dimensional projective geometry L_n $(n = 1, 2, \cdots)$ or a continuous geometry L_∞. We shall call these rings $\mathfrak{R}$ *discrete* or *continuous* rings in the two respective cases. The properties to admit a numerical-valued rank-function, to be complete with respect to the rank-metric and to be irreducible, are shown in Chapter XVIII of Part II to characterize such rings. Hence $\mathfrak{R}$ will be supposed regular, and $L = \bar{R}_{\mathfrak{R}}$ will be assumed to be finite-dimensional projective geometry or a continuous geometry i.e., to satisfy Axioms I—VI in Part I. We may make free use of the unique dimension-function $D(\mathfrak{a})$, $\mathfrak{a} \in L$, whose range Δ is either $\Delta_n = (0, 1/n, 2/n, \cdots, n-1/n, 1)$ for $n = 1, 2, \cdots$ or the set Δ_∞ of all real numbers x, $0 \leqq x \leqq 1$. (Cf. Part I, Theorem 7.3.) We denote by L' the lattice dual to L (which also satisfies Axioms I—VI) and by D' the unique dimension function associated with L'. Free use will be made of the results of Chapters II, III, XV, XVI of Part II, and also of the results right-left symmetric (i.e., dual) to them. For example, we shall need to employ frequently the theory of factor-correspondences between principal left ideals as well as between principal right ideals. The theory of dimensionality (Part I, Chapters VI, VII) will also be used freely.

Lemma 17.1: If there is a factor-correspondence between $(a)_r$ and $(b)_r$, then $D((a)_r) = D((b)_r)$.

Proof: By Lemma 15.4 the given factor-correspondence generates a lattice-isomorphism of $L((0), (a)_r)$ and $L((0), (b)_r)$. Suppose first that $(a)_r = (b)_r$. Then obviously $D((a)_r) = D((b)_r)$. Hence we may assume $(a)_r \neq (b)_r$. Moreover each of the conditions $a = 0$ and $b = 0$ implies the other, and thus implies $(a)_r = (b)_r$; hence we may also assume a, $b \neq 0$. Clearly the lattices $L((0), (a)_r)$, $L((0), (b)_r)$ satisfy Axioms I—VI (by Part I, Theorem 5.12); moreover $D(\mathfrak{a})/D((a)_r)$, $\mathfrak{a} \leqq (a)_r$, and $D(\mathfrak{b})/D((b)_r)$, $\mathfrak{b} \leqq (b)_r$, are respective dimension-functions for them. If $\mathfrak{a}$, $\mathfrak{b}$ correspond under the factor-isomorphism, then $D(\mathfrak{a})/D((b)_r) = D(\mathfrak{b})/D((a)_r)$, since both members are dimension-functions for $L((0), (b)_r)$, and are equal by the uniqueness of such a dimension function (Part I, Corollary 1 to Theorem 7.4). Thus there exists a real constant C such that $D(\mathfrak{a}) = CD(\mathfrak{b})$ for every $\mathfrak{a}$, $\mathfrak{b}$ which correspond. Suppose now that $(a)_r \nleqq (b)_r$. Then $(a)_r \cap (b)_r < (a)_r$, and there exists $\mathfrak{a}'$ inverse to $(a)_r \cap (b)_r$ in $(a)_r$, whence $\mathfrak{a}' \cap (b)_r = \mathfrak{a}' \cap (a)_r \cap (b)_r = 0$, $\mathfrak{a}' \neq 0$. Let $\mathfrak{b}'$ correspond to $\mathfrak{a}'$ then $\mathfrak{b}' \neq 0$, and the factor correspondence of $(a)_r$, $(b)_r$ generates a lattice-isomorphism of $L((0), \mathfrak{a}')$, $L((0), \mathfrak{b}')$. Since $\mathfrak{a}' \cap \mathfrak{b}' \leqq \mathfrak{a}' \cap (b)_r = (0)$, this isomorphism is a perspective isomorphism by Theorem 15.3 (c). Therefore $\mathfrak{a}' \sim \mathfrak{b}'$, whence $D(\mathfrak{a}') = D(\mathfrak{b}') = CD(\mathfrak{a}')$, and $C = 1$ since $D(\mathfrak{a}') \neq 0$. Thus $(a)_r \nleqq (b)_r$ implies $D((b)_r) = D((a)_r)$. Similarly $(a)_r \ngeqq (b)_r$ implies $D((b)_r) = D((a)_r)$. Finally if neither $(a)_r \nleqq (b)_r$ nor $(a)_r \ngeqq (b)_r$ holds, then $(a)_r = (b)_r$, contrary to the assumption $(a)_r \neq (b)_r$. This completes the proof.

LEMMA 17.2: The lattice L' is isomorphic to $\bar{L}_{\mathfrak{R}}$ and for $a \in \mathfrak{R}$,

$$D((a)_r) = D'((a)_l) = 1 - D((a)_l^r) = 1 - D'((a)_r^l).$$

PROOF: By the Corollary 2 to Lemma 2.2, $\bar{L}_{\mathfrak{R}}$ is anti-isomorphic to $\bar{R}_{\mathfrak{R}}$, but L' is anti-isomorphic to L, whence L' is isomorphic to $\bar{L}_{\mathfrak{R}}$, the isomorphism being defined by $\mathfrak{a} \rightleftarrows \mathfrak{a}' = \mathfrak{a}^l$ $(\mathfrak{a} \in L')$. Hence we may identify L' and $\bar{L}_{\mathfrak{R}}$. The equation $D((a)_r) = D'((a)_l)$ will first be proved, the method being to show that $D((a)_r)$ is a function φ of $D'((a)_l)$, and then to prove that φ is the identity function. In order to show that such a function φ exists, it suffices to prove that $D'((a)_l) = D'((b)_l)$ if and only if $D((a)_r) = D((b)_r)$. We prove first that

$$(1) \qquad (a)_l = (b)_l \text{ implies } D((a)_r) = D((b)_r).$$

Since $(a)_l = (b)_l$, there exist p, q with $b = pa$, $a = qb$, whence $a = qpa$, $b = pqb$, and p, q are factors of a factor corrrespondence between $(a)_r$ and $(b)_r$. But then Lemma 17.1 applies, and we have $D((a)_r) = D((b)_r)$.

Suppose now that $D'((a)_l) = D'((b)_l)$. Then $(a)_l \sim (b)_l$ by Part I, Theorem 6.9 (iii)''. There exist idempotents e, f such that $(e)_l = (a)_l$, $(f)_l = (b)_l$, $(e)_r = (f)_r$ (cf. the proof of Theorem 15.3 (a)). Hence by (1), $D((e)_r) = D((a)_r)$, $D((f)_r) = D((b)_r)$; but it is obvious that $D((e)_r) = D((f)_r)$, whence $D((a)_r) = D((b)_r)$. Hence we have shown that $D'((a)_l) =$

$D'((b)_l)$ implies $D((a)_r) = D((b)_r)$. Similarly the converse is proved. This shows that there exist functions φ, ψ such that

$$(2) \qquad \varphi(D'((a)_l)) = D((a)_r), \quad \psi(D((a)_r)) = D'((a)_l).$$

Since the range Δ of D coincides with the range of D', φ and ψ are mutually inverse functions whose range and domain are both Δ; therefore φ is a one-to-one mapping of Δ on itself.

Let $x, y \in \Delta$, so that also $x + y \in \Delta$. Then there exist $\mathfrak{a}', \mathfrak{b}' \in L'$ with $\mathfrak{a}' \cap \mathfrak{b}' = (0)$, $D'(\mathfrak{a}') = x$, $D'(\mathfrak{b}') = y$ (cf. Part I, Proof of Theorem 7.4). Since $(\mathfrak{a}', \mathfrak{b}') \perp$, there are, by Lemma 3.2, two idempotents e, f, with $ef = fe = 0$, such that $\mathfrak{a}' = (e)_l, \mathfrak{b}' = (f)_l$ whence $\mathfrak{a}' \cup \mathfrak{b}' = (e + f)_l$ by Lemma 3.1. Now $D'((e+f)_l) = D'(\mathfrak{a}' \cup \mathfrak{b}') = D'(\mathfrak{a}') + D'(\mathfrak{b}') = x + y$, whence

$$\begin{aligned} \varphi(x+y) &= \varphi(D'((e+f)_l)) = D((e+f)_r) = D((e)_r \cup (f)_r) \\ &= D((e)_r) + D((f)_r) = \varphi(D'((e)_l)) + \varphi(D'((f)_l)) \\ &= \varphi(D'(\mathfrak{a}')) + \varphi(D'(\mathfrak{b}')) = \varphi(x) + \varphi(y). \end{aligned}$$

Now it is well known that a one-to-one mapping φ of Δ on itself with the property $\varphi(x+y) = \varphi(x) + \varphi(y)$ is the identity. (Cf. Part I, Proof of Theorem 7.4.) Thus $D((a)_r) = D'((a)_l)$.

Now let e be an idempotent generating $(a)_r$. Then $(a)_r = (e)_r$, whence $(1-e)_r$ is inverse to $(a)_r$, and

$$\begin{aligned} D((a)_r) &= D((e)_r) = 1 - D((1-e)_r) = 1 - D'((1-e)_l) \\ &= 1 - D'((e)_r^l) \qquad \text{(by Lemma 2.2 (i))}. \\ &= 1 - D'((a)_r^l). \end{aligned}$$

Similarly $D'((a)_l) = 1 - D((a)_l^r)$, and the proof is complete.

Definition 17.1: For every $a \in \mathfrak{R}$ we define the rank of a $(R(a))$ by

$$R(a) \equiv D((a)_r) = D'((a)_l) = 1 - D((a)_l^r) = 1 - D'((a)_r^l).$$

Corollary: The range of $R(a)$ is Δ, the range of $D(\mathfrak{a})$, $\mathfrak{a} \in L$.

Theorem 17.1:

(a) $0 \leqq R(a) \leqq 1$ for every $a \in \mathfrak{R}$;
(b) $R(a) = 0$ if and only if $a = 0$;
(c) $R(a) = 1$ if and only if a is non-singular;
(d) $R(a) = R(b)$ if and only if a is of the form $a = vbu$ where u, v are non-singular;
(e) $R(ab) \leqq R(a), R(b)$;
(f) $R(a+b) \leqq R(a) + R(b)$;
(g) if e, f are independent idempotents, i.e., if $e^2 = e$, $f^2 = f$, $ef = fe = 0$, then $R(e+f) = R(e) + R(f)$.

Proof: (a) Since $x \in \Delta$ implies $0 \leqq x \leqq 1$, and since $R(a) \in \Delta$ for $a \in \mathfrak{R}$, (a) holds.

(b) Let $a = 0$; then $R(a) = D((0)_r) = D((0)) = 0$. Let $R(a) = 0$; then $D((a)_r) = 0$, and $(a)_r = (0)$, whence $a = 0$.

(c) Let $R(a) = 1$. Then $1 = D((a)_r) = D'((a)_l)$, whence $(a)_r = (1)_r$, $(a)_l = (1)_l$, and there exist $u, v \in \mathfrak{R}$ with $1 = au = va$. Hence $u = vau = v$, and $1 = au = ua$, whence $u = a^{-1}$, and a is non-singular. Conversely, let a be non-singular. Then $1 = aa^{-1}$, and $(a)_r = (1)_r$, whence $R(a) = D((a)_r) = 1$.

(e) Clearly $ab \in (a)_r$, whence $(ab)_r \leqq (a)_r$, and $R(ab) = D((ab)_r) \leqq D((a)_r) = R(a)$. By right-left duality, $R(ab) \leqq R(b)$.

(d) Let $a = ubv$ with u, v non-singular. Then $R(a) = R(ubv) \leqq R(bv) \leqq R(b)$ by (e). But $b = u^{-1}av^{-1}$ so $R(b) \leqq R(a)$. Thus $R(a) = R(b)$.

Conversely, let $R(a) = R(b)$. We shall first prove that $(c)_r = (d)_r$ implies the existence of u (non-singular) with $d = cu$. Since $(c)_r = (d)_r$, there exist p, q such that $d = cp$, $c = dq$. Now $d = dqp$, $c = cpq$, whence p, q are factors of a factor-correspondence between $(c)_l$ and $(d)_l$. If e and f are idempotents with $(e)_l = (c)_l$, $(f)_l = (d)_l$, then p, q may be replaced by special factors p', q' with $p'q' = e$, $q'p' = f$, $p'q'p' = p'$, $q'p'q' = q'$, without altering the correspondence. Since c and d correspond, we have $d = cp'$, $c = dq'$. Since $R(c) = R(d)$, $D((c)_l) = D((d)_l)$, whence by Part I, Theorem 6.9 (iii)″, $(c)_l \sim (d)_l$, and $(e)_l \sim (f)_l$. Thus also $(1-e)_l \sim (1-f)_l$. By Theorem **15.3** (a) there exists a factor-correspondence between $(1-e)_l$ and $(1-f)_l$ with special factors, $\bar{p}$, $\bar{q}$ such that $\bar{p}\bar{q} = 1 - e$, $\bar{q}\bar{p} = 1 - f$, $\bar{p}\bar{q}\bar{p} = \bar{p}$, $\bar{q}\bar{p}\bar{q} = \bar{q}$. Hence $e\bar{p} = e\bar{p}\bar{q}\bar{p} = e(1-e)\bar{p} = 0$. Similarly, $\bar{p}f = f\bar{q} = \bar{q}e = 0$. Define $u \equiv p' + \bar{p}$, $v \equiv q' + \bar{q}$. Then

$$uv = p'q' + \bar{p}q' + p'\bar{q} + \bar{p}\bar{q} = e + \bar{p}q'p'q' + p'q'p'\bar{q} + 1 - e = 1 + \bar{p}fq' + p'f\bar{q} = 1,$$

$$vu = q'p' + q'\bar{p} + \bar{q}p' + \bar{q}\bar{p} = f + q'p'q'\bar{p} + \bar{q}p'q'p' + 1 - f = 1 + q'e\bar{p} + \bar{q}ep' = 1.$$

Hence $v = u^{-1}$, and u is non-singular. Finally,

$$d = cp' = cp' + ce\bar{p} = cp' + c\bar{p} = c(p' + \bar{p}) = cu.$$

Consequently our result, viz.,

(3) $\qquad (c)_r = (d)_r$ implies $d = cu'$, with u' non-singular,

follows. The dual statement,

(4) $\qquad (c)_l = (d)_l$ implies $d = v'c$, with v' non-singular,

follows similarly.

Let us now return to our proposition concerning a and b. Since $R(a) = R(b)$ we have $(a)_r \sim (b)_r$. Then there exist idempotents e, f

with $(a)_r = (e)_r$, $(b)_r = (f)_r$, $(e)_l = (f)_l$ (cf. the proof of Theorem 15.3 (a)). By (3) there exist u_1, u_2 (non-singular) with $e = au_1$, $b = fu_2$, and by (4) there exists v (non-singular) with $f = ve$. Hence

$$b = fu_2 = veu_2 = vau_1u_2.$$

Define $u \equiv u_1u_2$. Since u_1, u_2 are non-singular, u is also non-singular (its inverse being $u_2^{-1}u_1^{-1}$), whence $b = vau$ with u, v non-singular.

(f) Since $(a+b)_r \leqq (a_r) \cup (b)_r$, $D((a+b)_r) \leqq D((a)_r \cup (b)_r) \leqq D((a)_r) + D((b)_r)$, whence $R(a+b) \leqq R(a) + R(b)$.

(g) If $e^2 = e$, $f^2 = f$, $ef = fe = 0$, then by Lemma 3.1, $(e+f)_r = (e)_r \cup (f)_r$, and $D((e)_r \cup (f)_r) = D((e)_r) + D((f)_r)$, since $(e)_r \cap (f)_r = (0)$. Thus

$$R(e+f) = D((e+f)_r) = D((e)_r \cup (f)_r) = D((e)_r) + D((f)_r) = R(e) + R(f).$$

THEOREM 17.2: If $\bar{R}(a)$ is a real-valued function of $a \in \mathfrak{R}$ such that

(a′) $0 \leqq \bar{R}(a) \leqq 1$ for every $a \in \mathfrak{R}$,

(b′) $\bar{R}(0) = 0$,

(c′) $\bar{R}(1) = 1$,

(e′) $\bar{R}(ab) \leqq \bar{R}(a)$ for every $a, b \in \mathfrak{R}$, or $\bar{R}(ab) \leqq \bar{R}(b)$ for every $a, b \in \mathfrak{R}$,

(g′) $e^2 = e$, $f^2 = f$, $ef = fe = 0$ implies $\bar{R}(e+f) = \bar{R}(e) + \bar{R}(f)$,

then $\bar{R}(a) = R(a)$ for every $a \in \mathfrak{R}$.

PROOF: Suppose that (e') the first condition holds (the case when the second holds is treated similarly, by employing right-left symmetry). Suppose $(a)_r = (b)_r$. Then a is of the form $a = bx$, and by (e'), $\bar{R}(a) \leqq \bar{R}(b)$. Similarly $\bar{R}(a) \geqq \bar{R}(b)$, whence $\bar{R}(a) = \bar{R}(b)$. Hence $\bar{R}(a)$ depends on $(a)_r$, and we may define $\bar{D}(\mathfrak{a}) \equiv \bar{R}(a)$ for every $(a)_r = \mathfrak{a}$ and obtain a single-valued function $\bar{D}(\mathfrak{a})$, $\mathfrak{a} \in \bar{R}_{\mathfrak{R}}$. Clearly

(5) $0 \leqq \bar{D}(\mathfrak{a}) \leqq 1$ for every $\mathfrak{a} \in \bar{R}_{\mathfrak{R}}$ (by (a')),

(6) $\bar{D}((0)) = 0$ (by (b')),

(7) $\bar{D}(\mathfrak{R}) = 1$ (by (e′)).

Let $\mathfrak{a}, \mathfrak{b} \in \bar{R}_{\mathfrak{R}}$. Then if $\mathfrak{a}_1$, $\mathfrak{b}_1$ are inverses of $\mathfrak{a} \cap \mathfrak{b}$ in $\mathfrak{a}$, $\mathfrak{b}$ respectively, $(\mathfrak{a}_1, \mathfrak{b}_1, \mathfrak{a} \cap \mathfrak{b}) \perp$, and there exist independent idempotents e_1, e_2, e_3 with $(e_1)_r = \mathfrak{a}_1$, $(e_2)_r = \mathfrak{b}_1$, $(e_3)_r = \mathfrak{a} \cap \mathfrak{b}$. Now

$$\begin{aligned}\bar{D}(\mathfrak{a} \cup \mathfrak{b}) = \bar{D}(\mathfrak{a}_1 \cup \mathfrak{b}_1 \cup (\mathfrak{a} \cap \mathfrak{b})) &= \bar{D}((e_1 + e_2 + e_3)_r) = \bar{R}(e_1 + e_2 + e_3)\\ &= \bar{R}(e_1 + e_2) + \bar{R}(e_3) \quad \text{(by (g′))}\\ &= \bar{R}(e_1) + \bar{R}(e_2) + \bar{R}(e_3) \quad \text{(by (g′))}\\ &= (\bar{R}(e_1) + \bar{R}(e_3)) + (\bar{R}(e_2) + \bar{R}(e_3)) - \bar{R}(e_3)\\ &= \bar{R}(e_1 + e_3) + \bar{R}(e_2 + e_3) - \bar{R}(e_3) \quad \text{(by (g′))}\\ &= \bar{D}(\mathfrak{a}) + \bar{D}(\mathfrak{b}) - \bar{D}(\mathfrak{a} \cap \mathfrak{b}),\end{aligned}$$

whence

$$\bar{D}(\mathfrak{a} \cup \mathfrak{b}) + \bar{D}(\mathfrak{a} \cap \mathfrak{b}) = \bar{D}(\mathfrak{a}) + \bar{D}(\mathfrak{b}). \tag{8}$$

Now it has been shown that if $\bar{D}$ has properties (5), (6), (7), (8), then $\bar{D} = D$ (cf. Part I, Corollary 1, Theorem 7.4). Thus $\bar{R}(a) = \bar{D}((a)_r) = D((a)_r) = R(a)$ for every $a \in \mathfrak{R}$.

COROLLARY 1: There exists a unique real-valued function $\bar{R}(a)$, $a \in \mathfrak{R}$ satisfying properties (a′), (b′), (c′), (e′), (g′), and $\bar{R}(a) = R(a)$ for every $a \in \mathfrak{R}$.

PROOF: This is clear from Theorems 17.1, 17.2, since properties (a'), (b′), (c′), (e′), (g′) are statements implied respectively by properties (a), (b), (c), (e), (g).

COROLLARY 2: Every automorphism or anti-automorphism of $\mathfrak{R}$ leaves $R(a)$ invariant.

PROOF: Let $a \rightarrow a'$ be the given automorphism or anti-automorphism of $\mathfrak{R}$, and define $R'(a) \equiv R(a')$. Since all statements (a′), (b′), (c′), (e′), (g′) are invariant under an automorphism or anti-automorphism (in (e′) the two alternatives would be interchanged by an anti-automorphism), $R'(a)$ satisfies these conditions along with $R(a)$. Hence by Corollary 1, $R'(a) = R(a)$, i.e., $R(a') = R(a)$ for every $a \in \mathfrak{R}$, and $R(a)$ is invariant under $a \rightarrow a'$.

DEFINITION 17.2: We define the *rank-distance* $\delta(a, b)$ *between* a *and* b by

$$\delta(a, b) \equiv R(a - b).$$

THEOREM 17.3: The ring $\mathfrak{R}$ is a metric space relative to the distance-function $\delta(a, b)$, i.e.,

(a) $\delta(a, b) = 0$ if and only if $a = b$, $\delta(a, b) > 0$ if $a \neq b$,

(b) $\delta(a, b) = \delta(b, a)$,

(c) $\delta(a, c) \leqq \delta(a, b) + \delta(b, c)$.

Moreover,

(d) $\delta(a + b, c + d) \leqq \delta(a, c) + \delta(b, d)$,

(e) $\delta(ab, cd) \leqq \delta(a, c) + \delta(b, d)$,

(f) $|R(a) - R(b)| \leqq \delta(a, b)$.

PROOF: (a) Clearly $\delta(a, b) = 0$ if and only if $R(a - b) = 0$, i.e., if and only if $a = b$, by Theorem 17.1 (b). Let $a \neq b$. Then $\delta(a, b) = R(a - b) \geqq 0$ by Theorem 17.1 (a). If $R(a - b) = 0$, then $a - b = 0$, $a = b$, contrary to $a \neq b$.

(b) It is obvious from Theorem 17.1 (d) that $R(-c) = R(c)$, since $-c = (-1) \cdot c \cdot 1$ and since 1, -1 are non-singular, having 1, -1, respectively, as inverses. Hence $R(a-b) = R(b-a)$, i.e., $\delta(a, b) = \delta(b, a)$.

(c) Obviously $R(a - c) = R\big((a - b) + (b - c)\big) \leqq R(a - b) + R(b - c)$ by Theorem 17.1 (f). Hence $\delta(a, c) \leqq \delta(a, b) + \delta(b, c)$.

(d) Since $\delta(a + b, c + b) = R(a - c) = \delta(a, c)$, and similarly $\delta(c + b, c + d) = \delta(b, d)$, it follows from (c) that $\delta(a + b, c + d) \leqq \delta(a + b, c + b) + \delta(c + b, c + d) = \delta(a, c) + \delta(b, d)$.

(e) We have

$$\begin{aligned}\delta(ab, cd) = R(ab-cd) = R(ab - cb + cb - cd) &= R\big((a - c)b + c(b - d)\big) \\ &\leqq R\big((a - c)b\big) + R\big(c(b - d)\big) \quad \text{(by Theorem 17.1 (f))} \\ &\leqq R(a - c) + R(b - d) \quad \text{(by Theorem 17.1 (e))} \\ &= \delta(a, c) + \delta(b, d).\end{aligned}$$

(f) It is clear that $R(a) = R\big((a - b) + b\big) \leqq R(a - b) + R(b)$ (by Theorem 17.1 (f)), whence $R(a) - R(b) \leqq R(a - b)$. Similarly $R(b) - R(a) \leqq R(b - a) = R(a - b)$, whence $|R(a) - R(b)| \leqq R(a - b) = \delta(a, b)$.

By Theorem 17.3, the distance function $\delta(a, b)$ defines a topology in $\mathfrak{R}$, in which the operations $a + b$, ab are continuous, and even satisfy Lipschitz conditions. The usual topological theory of metric spaces will be assumed. For a complete treatment of this subject, cf. Hausdorff, *Mengenlehre*, Berlin, 1927, particularly Chapter VI, Kuratowski, *Topologie I*, Warsaw-Lwow, 1933, particularly Chapters II, III. The definitions of neighborhoods, open and closed sets, limit points, etc., are the ones commonly given and will therefore be omitted. We give here a few of the salient ideas. A sequence $(a_i;\ i = 1, 2, \cdots)$ is *convergent to the limit* a in case $\lim_{i\to\infty} \delta(a_i, a_j) = 0$. In this case we write $a = \lim^{(R)}_{i\to\infty} a_i$, since if a sequence converges, it has just one limit. It is known that every subsequence of a convergent sequence converges to the same limit. A sequence (a_i) is *fundamental* in case $\lim_{i, j\to\infty} \delta(a_i, a_j) = 0$. It is known that every subsequence of a fundamental sequence is also fundamental, and that every convergent sequence is fundamental. The space $\mathfrak{R}$ is called *complete* in case every fundamental sequence is convergent. It will be shown (Theorem 17.4) that $\mathfrak{R}$ is complete.

Lemma 17.3: Let $e \in \mathfrak{R}$ be idempotent, and let $(x_i;\ i = 1, 2, \cdots)$ be a fundamental sequence such that $ex_i = 0$, $x_i e = x_i$ (i.e., $x_i = (1 - e)x_i e$) $(i = 1, 2, \cdots)$. Then there exists x such that $ex = 0$, $xe = x$, and such that $x = \lim^{(R)}_{i\to\infty} x_i$.

Proof: For every $k = 1, 2, \cdots$ there exists an integer m_k such that $i, j \geqq m_k$ implies $R(x_i - x_j) \leqq 1/2^k$, since (x_i) is fundamental. Hence there exists a sequence of integers $n_1, n_2, \cdots$ such that $n_1 < n_2 < \cdots$, and $n_k \geqq m_k$ $(k = 1, 2, \cdots)$. Then $h, l \geqq k$ implies $n_h, n_l \geqq m_k$, which in turn implies $R(x_{n_h} - x_{n_l}) \leqq 1/2^k$. Define $y_i \equiv x_{n_i}$ $(i = 1, 2, \cdots)$. The sequence (y_i) is fundamental (being a subsequence of (x_i)); moreover, if it can be shown that (y_i) has a limit x with the desired properties, then $\lim_{i\to\infty} R(x_{n_i} - x) = 0$ and $\lim_{i\to\infty} R(x_{n_i} - x_i) = 0$ (because (x_i) is fundamental) together yield $\lim_{i\to\infty} R(x_i - x) = 0$. Let us then prove that (y_i) converges to x such that $ex = 0$, $xe = x$ under the assumptions $ey_i = 0$, $ye_i = y$ $(i = 1, 2, \cdots)$ and $R(y_h - y_l) \leqq 1/2^k$ for $k \leqq h, l$.

Define $\mathfrak{a}_i \equiv (e + y_i)_r$, $\mathfrak{b}_j \equiv \sum_{i=j}^{\infty} \mathfrak{a}_i$, $\mathfrak{c} \equiv \prod_{j=1}^{\infty} \mathfrak{b}_j$. Clearly $\mathfrak{b}_i \geqq \mathfrak{a}_i$, $\mathfrak{b}_1 \geqq \mathfrak{b}_2 \geqq \cdots$. Now $1 = (e + y_i) + (1 - e - y_i) = (e + y_i) + (1 - e)(1 - e - y_i)$ (since $ey_i = 0$), whence $1 \in (e + y_i)_r \cup (1 - e)_r = \mathfrak{a}_i \cup (1 - e)_r$, and $\mathfrak{R} = \mathfrak{a}_i \cup (1 - e)_r$. Now $\mathfrak{b}_i \geqq \mathfrak{a}_i$ yields $\mathfrak{R} = \mathfrak{b}_i \cup (1 - e)_r$, and we have also $\mathfrak{R} = \prod_{i=1}^{\infty} (\mathfrak{b}_i \cup (1 - e)_r)$. Since $\mathfrak{b}_1 \geqq \mathfrak{b}_2 \geqq \cdots$, we may apply Axion III_1; thus $\mathfrak{R} = (\prod_{i=1}^{\infty} \mathfrak{b}_i) \cup (1 - e)_r = \mathfrak{c} \cup (1 - e)_r$. Hence there exists (cf. Part I, Theorem 1.4, Corollary 2) $\mathfrak{c}' \leqq \mathfrak{c}$ such that $\mathfrak{c}'$ is inverse to $(1 - e)_r$ (in $\mathfrak{R}$). (It might be shown that $\mathfrak{c} = \mathfrak{c}'$, but this is unessential.) Now there exists an idempotent $f \in \mathfrak{R}$ such that $(f)_r = \mathfrak{c}'$, $(1 - f)_r = (1 - e)_r$. Hence $f \in (f)_r = \mathfrak{c}' \leqq \mathfrak{c}$, whence $f \in \mathfrak{c}$ and $(1 - f)_r^l = (1 - e)_r^l$ whence $(f)_l = (e)_l$ by Lemma 2.2 (i). Define $x \equiv f - e$. Then $ex = e(f - e) = ef - e = e - e = 0$, $xe = (f - e)e = fe - e = f - e = x$. Moreover, $e + x = f \in \mathfrak{c}$. Now for $i \geqq j$

$$\mathfrak{a}_i = (e + y_i)_r = \Big((e + y_j) + \sum_{k=j+1}^{i} (y_k - y_{k-1})\Big)_r \leqq (e + y_j)_r \cup \sum_{k=j+1}^{\infty} (y_k - y_{k-1})_r$$
$$= \mathfrak{a}_j \cup \sum_{k=j+1}^{\infty} (y_k - y_{k-1})_r,$$

$$\mathfrak{b}_j = \sum_{k=j}^{\infty} \mathfrak{a}_k \leqq \mathfrak{a}_j \cup \sum_{k=j+1}^{\infty} (y_k - y_{k-1})_r.$$

Hence

$$(9)\quad D(\mathfrak{b}_j) \leqq D(\mathfrak{a}_j) + \sum_{k=j+1}^{\infty} D\big((y_k - y_{k-1})_r\big) = D(\mathfrak{a}_j) + \sum_{k=j+1}^{\infty} R(y_k - y_{k-1})$$
$$\leqq D(\mathfrak{a}_i) + \sum_{k=j+1}^{\infty} \frac{1}{2^{k-1}} = D(\mathfrak{a}_j) + \frac{1}{2^{j-1}}.$$

Let $\mathfrak{d}_j = (u_j)_r$ be an inverse of $\mathfrak{a}_j$ in $\mathfrak{b}_j$. Then (9) yields $D(\mathfrak{d}_j) = D(\mathfrak{b}_j) - D(\mathfrak{a}_j) \leqq 1/2^{j-1}$ i.e., $R(u_j) \leqq 1/2^{j-1}$. Now $e + x \in \mathfrak{c} \leqq \mathfrak{b}_j = \mathfrak{a}_j \cup \mathfrak{d}_j = (e + y_j)_r = (u_j)_r$, whence $e + x$ is of the form $e + x = (e + y_j)y + u_j z$. Hence

$$(e + y_j)(e + x) = (e + y_j)(e + y_j)y + (e + y_j)u_j z;$$

but $(e+y_j)(e+x) = e + y_je + ex + y_jx = e + y_j + 0 + y_jex = e + y_j$, and similarly $(e+y_j)(e+y_j) = e + y_j$. Hence

$$e + y_j = (e+y_j)y + (e+y_j)u_jz,$$

and $y_j - x = (e+y_j) - (e+x) = (e+y_j-1)u_jz$. Thus $R(y_j - x) \leqq R(u_j) \leqq 1/2^{j-1}$ (by Theorem 17.1 (e)), and $x = \lim^{(R)}_{j\to\infty} y_j$. This completes the proof.

LEMMA 17.4: Let $(y_i;\ i = 1, 2, \cdots)$ be a fundamental sequence. Then there exists $y \in \mathfrak{R}$ with $y = \lim^{(R)}_{i\to\infty} y_i$.

PROOF: This is trivial if $\mathfrak{R}$ is discrete, since then the range of $R(a)$ is a discrete set of numbers in the interval (0, 1), whence any fundamental sequence is ultimately constant and so converges. Hence we may assume that $\mathfrak{R}$ is continuous. Thus $\mathfrak{R}$ has order 2, whence by Theorem 3.2 there exist matrix units s_{hl} $(h, l = 1, 2)$ in $\mathfrak{R}$. By the proof of Theorem 3.3 (cf. page 99, equations (10), (11)), each y_i is of the form

$$y_i = \sum_{h,\, l=1}^{2} s_{h1} y_{i,hl} s_{1l}\,,$$

where

$$y_{i,hl} = s_{1h} y_i s_{l1}\,,$$

whence

$$s_{11} y_{i,hl} = y_{i,hl} s_{11} = y_{i,hl}\,.$$

Now $R(y_{i,hl} - y_{j,hl}) = R\big(s_{1h}(y_i - y_j)s_{l1}\big) \leqq R(y_i - y_j)$, whence $(y_{i,hl};\ i = 1, 2, \cdots)$ is fundamental for each $k, l = 1, 2$. We shall show that these sequences $(y_{i,hl})$ converge. Define $y'_{i,hl} \equiv s_{21} y_{i,hl}$. Then $s_{11} y'_{i,hl} = s_{11}s_{21}y_{i,hl} = 0$, $y'_{i,hl}s_{11} = s_{21}y_{i,hl}s_{11} = s_{21}y_{i,hl} = y'_{i,hl}$. Moreover, $R(y'_{i,hl} - y'_{j,hl}) = R\big(s_{21}(y_{i,hl} - y_{j,hl})\big) \leqq R(y_{i,hl} - y_{j,hl})$, whence the sequence $(y'_{i,hl})$ is fundamental for every h, l. Hence we may apply Lemma 17.3 and obtain a limit y'_{hl} satisfying $s_{11}y'_{hl} = 0$, $y'_{hl}s_{11} = y'_{hl}$. Define $y_{hl} \equiv s_{12}y'_{hl}$. Since

$$s_{12} y'_{i,hl} = s_{12}s_{21}y_{i,hl} = s_{11}y_{i,hl} = y_{i,hl}\,,$$

it is clear that $(y_{i,hl})$ converges to y_{hl}. Define $y \equiv \sum^2_{h,\,l=1} s_{h1}y_{hl}s_{1l}$. Since y_{hl} is the limit of $(y_{i,hl})$, it is clear that y is the limit of (y_i).

THEOREM 17.4: The ring $\mathfrak{R}$ is complete.

PROOF: This is a restatement of Lemma 17.4.

THEOREM 17.5: The ring $\mathfrak{R}$ is irreducible (cf. Definition 2.5). Its center is a division algebra.

PROOF: By Axiom VI, L is irreducible. Hence by Theorem 2.9, $\mathfrak{R}$ is irreducible; by Theorem 2.7 the center of $\mathfrak{R}$ is a division algebra.

PART II · CHAPTER XVIII

Rank-Rings and Characterization of Continuous Rings

In Chapter XVII it was shown that the ring $\mathfrak{R}$ associated with a finite-dimensional or continuous geometry admits a numerical-valued rank-function (Definition 17.1, Theorem 17.1), is complete in the "rank-metric", and is irreducible. We shall see that these properties characterize all discrete and continuous rings.

DEFINITION 18.1: A regular ring $\mathfrak{R}$ is a *rank-ring* in case there exists a real-valued *rank-function* $\bar{R}(a)$, $a \in \mathfrak{R}$, i.e., such that

($\bar{\text{a}}$) $0 \leqq \bar{R}(a) \leqq 1$ for every $a \in \mathfrak{R}$;

($\bar{\text{b}}$) $\bar{R}(a) = 0$ if and only if $a = 0$;

($\bar{\text{c}}$) $\bar{R}(1) = 1$;

($\bar{\text{e}}$) $\bar{R}(ab) \leqq \bar{R}(a),\ \bar{R}(b)$;

($\bar{\text{g}}$) $e^2 = e,\ f^2 = f,\ ef = ef = 0$ implies $\bar{R}(e + f) = \bar{R}(e) + \bar{R}(f)$. (Cf. Theorem 17.1.)

COROLLARY: Conditions ($\bar{\text{e}}$), ($\bar{\text{g}}$) imply

($\bar{\text{f}}$) $\bar{R}(a + b) \leqq \bar{R}(a) + \bar{R}(b)$.

PROOF: Let $\mathfrak{a}'$, $\mathfrak{b}'$ be inverses of $(a)_r \cap (b)_r$ in $(a)_r$, $(b)_r$, respectively. Then $\big(\mathfrak{a}', \mathfrak{b}', (a)_r \cap (b)_r\big) \perp$, and there exist by Lemma 3.2 independent idempotents e_1, e_2, e_3 such that $\mathfrak{a}' = (e_1)_r$, $\mathfrak{b}' = (e_2)_r$, $(a)_r \cap (b)_r = (e_3)_r$. Then, by Lemma 3.1, $(a)_r = (e_1 + e_3)_r$, $(b)_r = (e_2 + e_3)_r$, $(a)_r \cup (b)_r = (e_1 + e_2 + e_3)_r$, whence $a + b$ is of the form $(e_1 + e_2 + e_3) \cdot u$, and

$$\begin{aligned}\bar{R}(a+b) \leqq \bar{R}(e_1+e_2+e_3) = \bar{R}(e_1) + \bar{R}(e_2) + \bar{R}(e_3) \leqq \bar{R}(e_1) + \bar{R}(e_3) + \bar{R}(e_2)\\ + \bar{R}(e_3) = \bar{R}(e_1 + e_3) + \bar{R}(e_2 + e_3) \leqq \bar{R}(a) + \bar{R}(b),\end{aligned}$$

the last inequality holding since $e_1 + e_3 = av$, $e_2 + e_3 = bw$, with $v, w \in \mathfrak{R}$.

Let us fix attention on a given rank-ring $\mathfrak{R}$ and let $\bar{R}(a)$, $a \in \mathfrak{R}$ be a given rank-function. We shall suppose, moreover, that $\mathfrak{R}$ is irreducible. Denote by L the lattice $\bar{R}_{\mathfrak{R}}$.

DEFINITION 18.2: For $a, b \in \mathfrak{R}$, define the *rank-distance* $\bar{\delta}(a, b)$ by

$$\bar{\delta}(a, b) \equiv \bar{R}(a - b).$$

LEMMA 18.1: The function $\bar{\delta}(a, b)$ defines a metric in $\mathfrak{R}$ i.e.,

(a) $\bar{\delta}(a, b) = 0$ if and only if $a = b$, $\bar{\delta}(a, b) > 0$ if $a \neq b$,

(b) $\bar{\delta}(a, b) = \bar{\delta}(b, a)$,

(c) $\bar{\delta}(a, c) \leqq \bar{\delta}(a, b) + \bar{\delta}(b, c)$.

Moreover,

(e) $\bar{\delta}(a + b, c + d) \leqq \bar{\delta}(a, c) + \bar{\delta}(b, d)$,

(f) $\bar{\delta}(ab, cd) \leqq \bar{\delta}(a, c) + \bar{\delta}(b, d)$,

(g) $|\bar{R}(a) - \bar{R}(b)| \leqq \bar{\delta}(a, b)$.

PROOF: The proof is precisely the same as that of Theorem 17.3, and will not be repeated here. Our assumptions on $\bar{R}(a)$ include those properties of $R(a)$ used in the proof.

It should be observed that (e), (f) are Lipschitz conditions for addition and multiplication, which therefore imply the continuity of these operations. Moreover, the function $\bar{R}(a)$ is continuous by (g).

Since $\mathfrak{R}$ is regular, L is a complemented modular lattice by Theorem 2.4. Since $\mathfrak{R}$ is irreducible, L is irreducible by Theorem 2.9. Hence L satisfies Axioms, I, IV, V, VI for continuous geometry. We shall proceed to prove that the completeness of $\mathfrak{R}$ in the rank-metric implies Axioms II, III. Let us then suppose that $\mathfrak{R}$ is complete. Since the conditions $(\bar{\text{a}})$, $(\bar{\text{b}})$, $(\bar{\text{c}})$, $(\bar{\text{e}})$, $(\bar{\text{g}})$ are right-left symmetric, any proposition concerning $\bar{R}(a)$, L when proved holds also for $\bar{R}(a)$, $L' = \bar{L}_{\mathfrak{R}}$.

LEMMA 18.2: If $(a)_r = (b)_r$, then $\bar{R}(a) = \bar{R}(b)$ (in particular $\bar{R}(a) = \bar{R}(-a)$ for all $a \in \mathfrak{R}$).

PROOF: Since there exist $u, v \in \mathfrak{R}$ with $a = bu$, $b = av$, we have $\bar{R}(a) \leqq \bar{R}(b) \leqq \bar{R}(a)$ by $(\bar{\text{e}})$. Hence $\bar{R}(a) = \bar{R}(b)$.

DEFINITION 18.3: For $\mathfrak{a} \in L$ define $\bar{D}(\mathfrak{a})$ as the unique common value of all $\bar{R}(a)$, with $(a)_r = \mathfrak{a}$ (existent by Lemma 18.2).

COROLLARY: The function $\bar{D}(\mathfrak{a})$ is a dimension function in L, i.e.,

(a) $0 \leqq \bar{D}(\mathfrak{a}) \leqq 1$,

(b) $\bar{D}((0)) = 0$, $\bar{D}(\mathfrak{R}) = 1$,

(c) $\bar{D}(\mathfrak{a} \cup \mathfrak{b}) + \bar{D}(\mathfrak{a} \cap \mathfrak{b}) = \bar{D}(\mathfrak{a}) + \bar{D}(\mathfrak{b})$,

(d) $\mathfrak{a} < \mathfrak{b}$ implies $\bar{D}(\mathfrak{a}) < \bar{D}(\mathfrak{b})$.

PROOF: Parts (a), (b) are trivial. To prove (c), let $\mathfrak{a}'$, $\mathfrak{b}'$ be inverse to $\mathfrak{a} \cap \mathfrak{b}$ in $\mathfrak{a}$, $\mathfrak{b}$ respectively. Then $(\mathfrak{a}', \mathfrak{b}', \mathfrak{a} \cap \mathfrak{b})\perp$, and by Lemma 3.2 there exist independent idempotents e_1, e_2, e_3 such that $(e_1)_r = \mathfrak{a}'$, $(e_2)_r = \mathfrak{b}'$, $(e_3)_r = \mathfrak{a} \cap \mathfrak{b}$. Then by Lemma 3.1, $\mathfrak{a} = (e_1 + e_3)_r$, $\mathfrak{b} = (e_2 + e_3)_r$, $\mathfrak{a} \cup \mathfrak{b} = (e_1 + e_2 + e_3)_r$. Hence

$$\begin{aligned}\bar{D}(\mathfrak{a}\cup\mathfrak{b})+\bar{D}(\mathfrak{a}\cap\mathfrak{b}) &= \bar{D}((e_1+e_2+e_3)_r)+\bar{D}((e_3)_r) = \bar{R}(e_1+e_2+e_3)+\bar{R}(e_3)\\ &= \bar{R}(e_1)+\bar{R}(e_2)+\bar{R}(e_3)+\bar{R}(e_3)\\ &\qquad\qquad\qquad\text{(by Definition 18.1 }(\bar{\mathrm{g}})\text{)}\\ &= (\bar{R}(e_1)+\bar{R}(e_3))+(\bar{R}(e_2)+\bar{R}(e_3))\\ &= \bar{R}(e_1+e_3)+\bar{R}(e_2+e_3) \quad \text{(by Definition 18.1 }(\bar{\mathrm{g}})\text{)}\\ &= \bar{D}((e_1+e_3)_r)+\bar{D}((e_2+e_3)_r) = \bar{D}(\mathfrak{a})+\bar{D}(\mathfrak{b}).\end{aligned}$$

To prove (d), let $\mathfrak{c}$ be inverse to $\mathfrak{a}$ in $\mathfrak{b}$. Then $\mathfrak{c} \neq (0)$; and if $\mathfrak{c} = (u)_r$, then $\bar{D}(\mathfrak{c}) = \bar{R}(u) > 0$, since if $\bar{R}(u) = 0$, then $u = 0$ by Definition 18.1 ($\bar{\mathrm{b}}$) contrary to $(u)_r \neq (0)$. Now

$$\begin{aligned}\bar{D}(\mathfrak{b}) &= \bar{D}(\mathfrak{a}\cup\mathfrak{c}) = \bar{D}(\mathfrak{a})+\bar{D}(\mathfrak{c})-\bar{D}(\mathfrak{a}\cap\mathfrak{c}) && \text{(by (c))}\\ &= \bar{D}(\mathfrak{a})+\bar{D}(\mathfrak{c})-0 && \text{(by (b))}\\ &> \bar{D}(\mathfrak{a}).\end{aligned}$$

LEMMA 18.3: Let $(\mathfrak{a}_i;\ i = 1, 2, \cdots)$, $\mathfrak{a}_i \in L$ be a sequence such that $i < j$ implies $\mathfrak{a}_i \geqq \mathfrak{a}_j$. Then $\Pi(\mathfrak{a}_i;\ i = 1, 2, \cdots)$ exists in L (i.e., the $\mathfrak{a}_i$ have a greatest lower bound in L), and

$$\text{(1)} \qquad \bar{D}(\Pi(\mathfrak{a}_i;\ i = 1, 2, \cdots)) = \lim_{i\to\infty} (\bar{D}(\mathfrak{a}_i)).$$

PROOF: We shall show first that there exists a sequence e_i of idempotents with $(e_i)_r = \mathfrak{a}_i$ and $e_{i+1} \in \mathfrak{R}(e_i)$ for $i = 1, 2, \cdots$. Let e_1 be any idempotent with $(e_1)_r = \mathfrak{a}_1$. It clearly suffices to show that if $i = 2, 3, \cdots$ and if e_{i-1} has already been defined so that $(e_{i-1})_r = \mathfrak{a}_{i-1}$, then e_i may be selected so as to satisfy $e_i \in \mathfrak{R}(e_{i-1})$, $(e_i)_r = \mathfrak{a}_i$. Since $\mathfrak{a}_i \geqq \mathfrak{a}_{i-1} = (e_{i-1})_r$, and since $(1 - e_{i-1})_r$ is inverse to $\mathfrak{a}_{i-1}$ by Theorem 2.1, we have $\mathfrak{a}_i \cap (1 - e_{i-1})_r = (0)$, and there exists an inverse $\mathfrak{b}_i$ of $\mathfrak{a}_i$ such that $\mathfrak{b}_i \geqq (1 - e_{i-1})_r$ by Part I, the Corollary to Theorem 1.4. Hence, by Theorem 2.1 there exists an idempotent e_i such that $\mathfrak{a}_i = (e_i)_r$, $(1 - e_i)_r = \mathfrak{b}_i$. Now $\mathfrak{a}_i \leqq \mathfrak{a}_{i-1}$ yields $e_i \in (e_{i-1})_r$, i.e., $e_i = e_{i-1}e_i$. Moreover, $(1 - e_{i-1})_r \leqq \mathfrak{b}_i = (1 - e_i)_r$, whence by Lemma 2.1(i), $(1 - e_{i-1})_r{}^l \geqq (1 - e_i)_r{}^l$, i.e., $(e_{i-1})_l \geqq (e_i)_l$, whence $e_i = e_i e_{i-1}$. Thus $e_i = e_i e_{i-1} = e_{i-1} e_i$, and $e_i \in \mathfrak{R}(e_{i-1})$. Consequently the desired sequence exists.

Now $x \in \mathfrak{R}(e_i)$ implies $xe_{i-1} = xe_ie_{i-1} = xe_i = x$, $e_{i-1}x = e_{i-1}e_ix = e_ix = x$, i.e., that $x \in \mathfrak{R}(e_{i-1})$. Thus $\mathfrak{R}(e_i) \subset \mathfrak{R}(e_{i-1})$, whence $i \geqq j$ implies $\mathfrak{R}(e_i) \subset \mathfrak{R}(e_j)$, and $i \geqq j$ implies $e_i \in \mathfrak{R}(e_j)$, i.e., $e_ie_j = e_je_i = e_i$. Now if $i \geqq j$, then $(e_j - e_i)^2 = e_j^2 - e_je_i - e_ie_j + e_i^2 = e_j - e_i - e_i + e_i = e_j - e_i$, and hence $e_j - e_i$ is idempotent. Moreover $e_i(e_j - e_i) = (e_i - e_j)e_i = 0$. Then by Definition 18.1 ($\bar{\mathrm{g}}$), $\bar{R}(e_j) = \bar{R}(e_i + (e_j - e_i)) = \bar{R}(e_i) + \bar{R}(e_j - e_i)$, whence

$$(2) \qquad \bar{R}(e_i - e_j) = \bar{R}(e_j - e_i) = \bar{R}(e_j) - \bar{R}(e_i),$$

since $\bar{R}(-a) = \bar{R}(a)$ for all $a \in \mathfrak{R}$.

If $i \leqq j$, then (2) can be written as:

$$(3) \qquad \bar{R}(e_j - e_i) = \bar{R}(e_i - e_j) = \bar{R}(e_i) - \bar{R}(e_j).$$

Therefore, for every i, j, we have by (2), (3),

$$(4) \qquad \bar{R}(e_i - e_j) = |\bar{R}(e_i) - \bar{R}(e_j)|.$$

Now by (2), $i \geqq j$ implies $\bar{R}(e_j) - \bar{R}(e_i) \geqq 0$, whence $\bar{R}(e_j) \geqq \bar{R}(e_i)$. But every $\bar{R}(e_i)$ is on the closed real interval $0 \leqq x \leqq 1$, whence the sequence $\bar{R}(e_i)$, being a bounded monotone decreasing sequence of real numbers, has a real limit on this interval. Thus $\lim_{i,j\to\infty} |\bar{R}(e_i) - \bar{R}(e_j)| = 0$, and by (4), $\lim_{i,j\to\infty} \bar{R}(e_i - e_j) = 0$. Since $\mathfrak{R}$ is complete, there exists $e \in \mathfrak{R}$ such that $\lim_{i\to\infty} \bar{R}(e_i - e) = 0$, i.e. e is the limit of (e_i). Now $e_i e_j = e_j e_i = e_i$ for $i \geqq j$, and for every $j = 1, 2, \cdots$ it is clear that $\lim^{(\bar{R})}_{i\to\infty, i \geqq j} e_i = \lim_{i\to\infty} e_i = e$, whence by the continuity of multiplication, $ee_j = e_j e = e$, and $e \in \mathfrak{R}(e_j)$. Now again by the continuity of multiplication,

$$e^2 = e \cdot e = (\lim^{(\bar{R})}_{j\to\infty} e_j)e = \lim^{(\bar{R})}_{j\to\infty} (e_j e) = \lim^{(\bar{R})}_{j\to\infty} e = e,$$

and e is idempotent. Clearly $e \in \mathfrak{R}(e_j) \leqq (e_j)_r = \mathfrak{a}_j$, and it follows that

$$(5) \qquad (e)_r \leqq \mathfrak{a}_j \qquad (j = 1, 2, \cdots).$$

Let $\mathfrak{a} \leqq \mathfrak{a}_j$ $(j = 1, 2, \cdots)$; then $\mathfrak{a} = (u)_r \leqq \mathfrak{a}_j$ with $u \in \mathfrak{R}$, whence $u \in \mathfrak{a}_j$ $(j = 1, 2, \cdots)$. Thus $u \in (e_j)_r$, i.e., $u = e_j u$, and by continuity of multiplication

$$u = \lim^{(\bar{R})}_{j\to\infty} e_j u = (\lim^{(\bar{R})}_{j\to\infty} e_j)u = eu;$$

therefore $u \in (e)_r$, and $\mathfrak{a} \leqq (e)_r$. Hence $(e)_r$ is effective as $\Pi(\mathfrak{a}_j; i = 1, 2, \cdots)$. Finally

$$\bar{D}(\Pi(\mathfrak{a}_j;\ j = 1, 2, \cdots)) = \bar{D}((e)_r) = \bar{R}(e),$$
$$\bar{D}(\mathfrak{a}_j) = (\bar{D}(e_j)_r) = \bar{R}(e_j),$$

and $\lim_{j\to\infty} \bar{R}(e_j - e) = 0$ implies $\lim_{j\to\infty} \bar{R}(e_j) = \bar{R}(e)$ by the continuity of $\bar{R}(a)$ (Lemma 18.1 (g)). Therefore

$$\bar{D}(\Pi(\mathfrak{a}_j;\ j = 1, 2, \cdots)) = \lim_{j\to\infty} (\bar{D}(\mathfrak{a}_j)),$$

and (1) is established.

Lemma 18.4: Let $(\mathfrak{a}_i;\ i = 1, 2, \cdots)$, $\mathfrak{a}_i \in L$, be a given sequence such that $i < j$ implies $\mathfrak{a}_i \geqq \mathfrak{a}_j$, and let $\mathfrak{b} \in L$. Then $\Pi(\mathfrak{a}_i;\ i = 1, 2, \cdots)$, $\Pi(\mathfrak{a}_i \cup \mathfrak{b};\ i = 1, 2, \cdots)$ exist, and

(6) $$(\Pi(\mathfrak{a}_i;\ i = 1, 2, \cdots)) \cup \mathfrak{b} = \Pi(\mathfrak{a}_i \cup \mathfrak{b};\ i = 1, 2, \cdots).$$

PROOF: Since $i<j$ implies $\mathfrak{a}_i \geqq \mathfrak{a}_j$, $\mathfrak{a}_i \cup \mathfrak{b} \geqq \mathfrak{a}_j \cup \mathfrak{b}$, both $\Pi(\mathfrak{a}_i; i=1, 2, \cdots)$, $\Pi(\mathfrak{a}_i \cup \mathfrak{b}; i=1, 2, \cdots)$ exist by Lemma 18.3. Now $(\Pi(\mathfrak{a}_i; i=1, 2, \cdots)) \cup \mathfrak{b} \leqq \mathfrak{a}_j \cup \mathfrak{b}$ for every $j = 1, 2, \cdots$, whence

(7) $$(\Pi(\mathfrak{a}_i;\ i = 1, 2, \cdots)) \cup \mathfrak{b} \leqq \Pi(\mathfrak{a}_i \cup \mathfrak{b};\ i = 1, 2, \cdots).$$

Applying (1) (Lemma 18.3) to the sequences $(\mathfrak{a}_i)$, $(\mathfrak{a}_i \cup \mathfrak{b})$, $(\mathfrak{a}_i \cap \mathfrak{b})$, which all satisfy the hypotheses of Lemma 18.3, we have

$$\bar{D}(\Pi(\mathfrak{a}_i;\ i = 1, 2, \cdots)) = \lim_{i\to\infty} (\bar{D}(\mathfrak{a}_i)),$$

$$\bar{D}(\Pi(\mathfrak{a}_i \cup \mathfrak{b};\ i = 1, 2, \cdots)) = \lim_{i\to\infty} (\bar{D}(\mathfrak{a}_i \cup \mathfrak{b})),$$

$$\bar{D}(\Pi(\mathfrak{a}_i \cap \mathfrak{b};\ i = 1, 2, \cdots)) = \lim_{i\to\infty} (\bar{D}(\mathfrak{a}_i \cap \mathfrak{b})).$$

Now by the Corollary to Definition 18.3 (c),

$$\begin{aligned}\bar{D}((\Pi(\mathfrak{a}_i;\ i = 1, 2, \cdots)) \cup \mathfrak{b}) &= \bar{D}(\Pi(\mathfrak{a}_i;\ i = 1, 2, \cdots)) + \bar{D}(\mathfrak{b}) \\ &\qquad - \bar{D}((\Pi(\mathfrak{a}_i;\ i = 1, 2, \cdots)) \cap \mathfrak{b}) \\ &= \bar{D}(\Pi(\mathfrak{a}_i;\ i = 1, 2, \cdots)) + \bar{D}(\mathfrak{b}) - \bar{D}(\Pi(\mathfrak{a}_i \cap \mathfrak{b};\ i = 1, 2, \cdots)) \\ &= \lim_{i\to\infty} (\bar{D}(\mathfrak{a}_i)) + \bar{D}(\mathfrak{b}) - \lim_{i\to\infty} (\bar{D}(\mathfrak{a}_i \cap \mathfrak{b})) = \lim_{i\to\infty} (\bar{D}(\mathfrak{a}_i) + D(\mathfrak{b}) - D(\mathfrak{a}_i \cap \mathfrak{b})) \\ &= \lim_{i\to\infty} (\bar{D}(\mathfrak{a}_i \cup \mathfrak{b})) \qquad \text{(by the Corollary to Definition 18.3 (c))} \\ &= \bar{D}(\Pi(\mathfrak{a}_i \cup \mathfrak{b};\ i = 1, 2, \cdots)),\end{aligned}$$

i.e.,

(8) $$\bar{D}((\Pi(\mathfrak{a}_i;\ i = 1, 2, \cdots)) \cup \mathfrak{b}) = \bar{D}(\Pi(\mathfrak{a}_i \cup \mathfrak{b};\ i = 1, 2, \cdots)).$$

By the contrapositive of the Corollary to Definition 18.3 (d), (7), (8) together yield (6).

LEMMA 18.5: Let Ω be any Cantor ordinal number, and let $S=(\mathfrak{a}_\alpha; \alpha<\Omega)$, $\mathfrak{a}_\alpha \in L$, be a given sequence such that $\alpha < \beta < \Omega$ implies $\mathfrak{a}_\alpha \geqq \mathfrak{a}_\beta$, and let $\mathfrak{b} \in L$. Then $\Pi(\mathfrak{a}_\alpha;\ \alpha < \Omega)$, $\Pi(\mathfrak{a}_\alpha \cup \mathfrak{b};\ \alpha < \Omega)$ exist, and

(9) $$(\Pi(\mathfrak{a}_\alpha;\ \alpha < \Omega)) \cup \mathfrak{b} = \Pi(\mathfrak{a}_\alpha \cup \mathfrak{b};\ \alpha < \Omega).$$

PROOF: Define η as the greatest lower bound of all $\bar{D}(\mathfrak{a}_\alpha)$, $\alpha < \Omega$. Suppose that there exists $\alpha_0 < \Omega$ such that $\bar{D}(\mathfrak{a}_{\alpha_0}) = \eta$. Then $\beta \geqq \alpha_0$ implies $\mathfrak{a}_\beta \leqq \mathfrak{a}_{\alpha_0}$; but $\bar{D}(\mathfrak{a}_\beta) \geqq \eta = \bar{D}(\mathfrak{a}_{\alpha_0})$, whence $\mathfrak{a}_\beta < \mathfrak{a}_{\alpha_0}$ yields $\bar{D}(\mathfrak{a}_\beta) < \bar{D}(\mathfrak{a}_{\alpha_0})$ (by the Corollary to Definition 18.3 (d)), contrary to $\bar{D}(\mathfrak{a}_\beta) \geqq D(\mathfrak{a}_{\alpha_0})$, and it follows that $\mathfrak{a}_\beta = \mathfrak{a}_{\alpha_0}$. For $\beta < \alpha_0$, $\mathfrak{a}_\beta \geqq \mathfrak{a}_{\alpha_0}$, whence for every

$\beta < \Omega$, $\mathfrak{a}_\beta \geqq \mathfrak{a}_{\alpha_0}$. Thus $\Pi(\mathfrak{a}_\alpha;\ \alpha < \Omega) \equiv \mathfrak{a}_{\alpha_0}$, $\Pi(\mathfrak{a}_\alpha \cup \mathfrak{b};\ \alpha < \Omega) \equiv \mathfrak{a}_{\alpha_0} \cup \mathfrak{b}$ are effective, and (9) holds.

Suppose now that no $\alpha < \Omega$ exists such that $\bar{D}(\mathfrak{a}_\alpha) = \eta$. We shall replace the set of all ordinals $\alpha < \Omega$ by a denumerable set K. For every $i = 1, 2, \cdots$, there exists $\alpha_i < \Omega$ such that $\bar{D}(\mathfrak{a}_{\alpha_i}) \leqq \eta + 1/i$; let (β_i) be any sequence $\beta_1 \leqq \beta_2 \leqq \cdots < \Omega$ such that $\beta_i \geqq \alpha_i$. Let $\alpha < \Omega$; then since $\bar{D}(\mathfrak{a}_\alpha) > \eta$, there exists $i = i_\alpha = 1, 2, \cdots$, such that $\bar{D}(\mathfrak{a}_\alpha) > \eta + 1/i$, whence $\bar{D}(\mathfrak{a}_\alpha) > \bar{D}(\mathfrak{a}_{\alpha_i}) \geqq \bar{D}(\mathfrak{a}_{\beta_i})$ (by the Corollary to Definition 18.3 (d)). Suppose now that $\alpha \geqq \beta_i$; then $\mathfrak{a}_\alpha \leqq \mathfrak{a}_{\beta_i}$, and $\bar{D}(\mathfrak{a}_\alpha) \leqq \bar{D}(\mathfrak{a}_{\beta_i})$, contrary to $\bar{D}(\mathfrak{a}_\alpha) > \bar{D}(\mathfrak{a}_{\beta_i})$. Hence $\alpha < \beta_i$, and $\mathfrak{a}_\alpha \geqq \mathfrak{a}_{\beta_i}$. The sequences $(\mathfrak{a}_{\beta_i};\ i = 1, 2, \cdots)$, $(\mathfrak{a}_{\beta_i} \cup \mathfrak{b};\ i = 1, 2, \cdots)$ satisfy the hypothesis of Lemma 18.4, whence $\Pi(\mathfrak{a}_{\beta_i};\ i = 1, 2, \cdots)$, $\Pi(\mathfrak{a}_{\beta_i} \cup \mathfrak{b};\ i = 1, 2, \cdots)$ exist, and

$$(10) \qquad (\Pi(\mathfrak{a}_{\beta_i};\ i = 1, 2, \cdots)) \cup \mathfrak{b} = \Pi(\mathfrak{a}_{\beta_i} \cup \mathfrak{b};\ i = 1, 2, \cdots).$$

Now for every $\alpha < \Omega$, and $i = i_\alpha$, $\Pi(\mathfrak{a}_{\beta_i};\ i = 1, 2, \cdots) \leqq \mathfrak{a}_{\beta_i} \leqq \mathfrak{a}_\alpha$; moreover, if $\mathfrak{c} \leqq \mathfrak{a}_\alpha$ $(\alpha < \Omega)$, then $\mathfrak{c} \leqq \mathfrak{a}_{\beta_i}$ $(i = 1, 2, \cdots)$, and $\mathfrak{c} \leqq \Pi(\mathfrak{a}_{\beta_i};\ i = 1, 2, \cdots)$. Thus $\Pi(\mathfrak{a}_{\beta_i};\ i = 1, 2, \cdots)$ is effective as $\Pi(\mathfrak{a}_\alpha;\ \alpha < \Omega)$. Similarly, $\Pi(\mathfrak{a}_{\beta_i} \cup \mathfrak{b};\ i = 1, 2, \cdots)$ is effective as $\Pi(\mathfrak{a}_\alpha \cup \mathfrak{b};\ \alpha < \Omega)$, whence these products exist; then (10) yields (9).

COROLLARY: Axiom III_1 is satisfied by L.

PROOF: This is a restatement of condition (10) of Lemma 18.5.

LEMMA 18.6: If S is any subset of L, then the greatest lower bound $\Pi(S)$ exists (in L).

PROOF: Suppose that there exists a set $S \subset L$ without a greatest lower bound. Then there exists a set $S_0 \subset L$ of minimum power such that $\Pi(S_0)$ does not exist. Now every finite set $S = (\mathfrak{a}_1, \cdots, \mathfrak{a}_n)$ possesses a greatest lower bound $\Pi(S) = \mathfrak{a}_1 \cap \cdots \cap \mathfrak{a}_n$. Hence S_0 is infinite. Let Ω be the first ordinal number of the power of S_0. Then Ω is a limit-ordinal, and S_0 may be replaced by a system $(\mathfrak{b}_\alpha;\ \alpha < \Omega)$. For every $\alpha < \Omega$ the set $S^\alpha \equiv (\mathfrak{b}_\gamma;\ \gamma < \alpha)$ has the power of α, which is less than the power of Ω, i.e., of S_0. Hence by our hypothesis, S^α possesses a greatest lower bound $\Pi(S^\alpha) \equiv \mathfrak{a}_\alpha$. Define the system $S' \equiv (\mathfrak{a}_\alpha;\ \alpha < \Omega)$. Since $\alpha < \beta < \Omega$ implies $S^\alpha \subset S^\beta$, $\mathfrak{a}_\alpha = \Pi(S^\alpha) \geqq \Pi(S^\beta) = \mathfrak{a}_\beta$, we may apply Lemma 18.5 and obtain the existence of $\Pi(S')$. But

$$\begin{aligned} \Pi(S') &= \Pi(\mathfrak{a}_\alpha;\ \alpha < \Omega) = \Pi(\Pi(\mathfrak{b}_\gamma;\ \gamma < \alpha);\ \alpha < \Omega) \\ &= \Pi(\mathfrak{b}_\gamma;\ \gamma < \Omega) \qquad \text{(since } \Omega \text{ is a limit-ordinal)} \\ &= \Pi(S_0); \end{aligned}$$

hence $\Pi(S_0)$ exists, contracting our hypothesis, whence it follows that for every set $S \subset L$, $\Pi(S)$ exists, and the proof is complete.

COROLLARY: Axiom II_2 is satisfied by L.

PROOF: This is a restatement of Lemma 18.6.

We now drop the assumptions that $\mathfrak{R}$ is a regular, irreducible, complete rank-ring.

THEOREM 18.1: If $\mathfrak{R}$ is a regular ring, then $\bar{R}_{\mathfrak{R}}$ is a continuous or finite-dimensional geometry (i.e., satisfies Axioms I—VI) if and only if $\mathfrak{R}$ is an irreducible , complete rank-ring; in this case, the rank-function $\bar{R}(a)$ is unique and is related to the dimension-function $D(\mathfrak{a})$ by $\bar{R}(a) = D((a)_r)$.

PROOF: Clearly if $\bar{R}_{\mathfrak{R}}$ is a continuous, or finite-dimensional, geometry, then $\mathfrak{R}$ is a rank-ring as was shown in Chapter XVII (cf. Definition 17.1, Theorem 17.1). Moreover, $\mathfrak{R}$ is complete by Theorem 17.4, and irreducible by Theorem 17.5. Conversely, let $\mathfrak{R}$ be a complete, irreducible rank-ring. It has been noted that Axioms I, II, IV, V follow from the regularity of $\mathfrak{R}$. Since $\mathfrak{R}$ is complete, we may apply the Corollary to Lemma 18.5 to obtain Axiom II_2. Since this implies Axiom II_1 (cf. the remark on page 3, Part I) the proof of Axiom II is complete. Now we may apply the Corollary to Lemma 18.5, to obtain Axiom III_1. The right-left dual of this states, because $L = \bar{R}_{\mathfrak{R}}$ and $L' = \bar{L}_{\mathfrak{R}}$ are anti-isomorphic, that Axiom III_2 holds also. This completes the proof of Axiom III. The Axiom VI follows from the irreducibility of $\mathfrak{R}$ by Theorem 2.9. Finally, by Theorem 17.2 the rank-function $R(a)$ is uniquely characterized by its properties $(\bar{\text{a}})$, $(\bar{\text{b}})$, $(\bar{\text{c}})$, $(\bar{\text{e}})$, $(\bar{\text{g}})$, (as $D((a)_r)$), when $L = \bar{R}_{\mathfrak{R}}$ satisfies I—VI.

PART III

PART III · CHAPTER I

Center of a Continuous Geometry

In Part I both the existence and uniqueness of a (normalized) dimension function for continuous geometries were proved with the help of Axioms I—VI. In this chapter we shall examine systems satisfying Axioms I—V and shall see that the existence of general dimension functions can be proved with the help of these axioms alone. We are clearly entitled to use, and shall presuppose, all the material of Part I dependent only on Axioms I—V, i.e., all the material preceding Theorem 5.10 (Part I, page 38).

Let us recall briefly the results of Part I, Chapter V, preceding Theorem 5.10. We defined the ternary, binary and unary relations $(x, y, z)D$, $(x, y)D$, $(x)D$ by the respective properties

$$(x + y)z = xz + yz,$$
$$(x, y, z)D \text{ for every } z \in L,$$
$$(x, y)D \text{ for every } y \in L,$$

and then proved them to be self dual and symmetric. The set of all elements $x \in L$ with $(x)D$ was shown to be a (complemented) Boolean algebra with $+, \cdot, <$ as in L; we shall refer to this set as the *center* of L and shall denote it by Z. (We shall use the term *distributive lattice* to describe a lattice in which the equivalent distributive laws $(a + b)c = ac + bc$, $ab + c = (a + c)(b + c)$ hold. If such a lattice is complemented, i.e., if each element possesses an inverse in the sense of Axiom V, then it is referred to as a *Boolean algebra*. Distributive lattices and Boolean algebras are discussed in detail in Appendix 1.) Thus

$$(1) \qquad (x + y)z = xz + yz, \quad xy + z = (x + z)(y + z)$$

if at least one of the elements x, y, z is in Z. The unique inverse of $x \in Z$, which then is also in Z, will be denoted by $-x$. It is clear that $0, 1 \in Z$, and that Axiom VI is equivalent to the statement that Z is the set consisting of 0, 1 alone, since by Part I, Theorem 5.4, Z is the set of all elements of L with unique inverses. If $a \in Z, b \in L$, then $ab = 0$ if and only if $b \leqq -a$. For, $b \leqq -a$ implies $ab \leqq a(-a) = 0$, $ab = 0$; conversely, $ab = 0$ implies

$$b = 1 \cdot b = (a + (-a))b = ab + (-a)b \qquad \text{(since } (a)D)$$
$$= 0 + (-a)b \leqq -a.$$

The properties of Z just stated will be used continually throughout the sequel.

THEOREM 1.1: If either $S \subset Z$ or $b \in Z$, then

$$\Sigma(S) \cdot b = \Sigma(ab;\ a \in S),\ \Pi(S) + b = \Pi(a + b;\ a \in S).$$

PROOF: By duality it suffices to prove the first formula. Let S be finite, $S = (a_1, \cdots, a_n)$. (If $n = 0$, $S = \Theta$, and the statement is evident. Hence we may suppose $n = 1, 2, \cdots$.) Then we must prove that

$$(a_1 + \cdots + a_n)b = a_1 b + \cdots + a_n b.$$

This follows by $(n-1)$-fold application of (1), since either $a_1, \cdots, a_n \in Z$, or $b \in Z$. Consider now an arbitrary set S. If the relation $\Sigma(S) \cdot b = \Sigma(ab;\ a \in S)$ does not hold for all sets S when either $S \subset Z$ or $b \in Z$, then there exists a set $\bar{S}$ of minimum power for which it fails to hold. Let Ω be the smallest ordinal corresponding to the power of $\bar{S}$, whence $\bar{S}$ may be replaced by a system $(a_\alpha;\ \alpha < \Omega)$. By our first considerations, $\bar{S}$ and Ω must both be infinite; moreover Ω has no immediate predecessor. Define $\bar{S}_\alpha \equiv (a_\gamma;\ \gamma < \alpha)$, and $b_\alpha \equiv \Sigma(\bar{S}_\alpha) = \Sigma(a_\gamma;\ \gamma < \alpha)$ for every $\alpha < \Omega$. The power of $\bar{S}_\alpha$ is the power of α and hence is less than the power of Ω, i.e., of $\bar{S}$. Thus the assumption concerning $\bar{S}$ implies

$$b_\alpha b = \Sigma(\bar{S}_\alpha) \cdot b = \Sigma(ab;\ a \in \bar{S}_\alpha) = \Sigma(a_\gamma b;\ \gamma < \alpha)$$

for every $\alpha < \Omega$. Now $\alpha < \beta < \Omega$ implies $\bar{S}_\alpha \subset \bar{S}_\beta$ whence $b_\alpha \leqq b_\beta$. Therefore

$$\begin{aligned}
\Sigma(\bar{S}) \cdot b &= \Sigma(a_\gamma;\ \gamma < \Omega) \cdot b \\
&= \Sigma(\Sigma(a_\gamma;\ \gamma < \alpha);\ \alpha < \Omega) \cdot b \\
&= \Sigma(b_\alpha;\ \alpha < \Omega) \cdot b = (\lim_{\alpha \to \Omega}{}^{*}\ b_\alpha) \cdot b \\
&= \lim_{\alpha \to \Omega}{}^{*}\ (b_\alpha b) \qquad\qquad \text{(by III}_2\text{)} \\
&= \Sigma(b_\alpha b;\ \alpha < \Omega) = \Sigma(\Sigma(a_\gamma b;\ \gamma < \alpha);\ \alpha < \Omega) \\
&= \Sigma(a_\gamma b;\ \gamma < \Omega) = \Sigma(ab;\ a \in \bar{S}),
\end{aligned}$$

contrary to the definition of $\bar{S}$. Hence the theorem is proved.

THEOREM 1.2: The center Z is an unrestrictedly additive and multiplicative Boolean subalgebra of L, i.e. for every set $S \subset Z$ its least upper bound $\Sigma(S)$ in L and its greatest lower bound $\Pi(S)$ in L are in Z.

PROOF: By duality it suffices to prove $\Sigma(S) \in Z$. For every $x, y \in L$ application of Theorem 1.1 to S and $b = x + y$, x, y yields

$$\begin{aligned}(x+y)\Sigma(S) &= \Sigma((x+y)a;\ a \in S)\\ &= \Sigma(xa+ya;\ a\in S) \qquad (\text{since } a \in S \subset Z)\\ &= \Sigma(xa;\ a\in S)+\Sigma(ya;\ a\in S)\\ &= x\Sigma(S)+y\Sigma(S).\end{aligned}$$

Therefore $\Sigma(S) \in Z$.

One might reasonably ask at this point whether those Boolean algebras which arise as centers of continuous geometries (i.e., systems satisfying Axioms I—V) have further properties distinguishing them abstractly from other unrestrictedly additive and multiplicative (i.e., *continuous*) Boolean algebras. It will be shown in Appendix I that this is not the case, i.e., that any continuous Boolean algebra Z is the center of a suitable continuous geometry L; in fact L may be taken equal to Z. Hence if Z is the center of a continuous geometry L, the properties of Z alone are precisely the properties of any continuous Boolean algebra; we shall therefore devote our attention to the relations between Z and L.

DEFINITION 1.1: Let $a \in L$, and let $Z(a, 1)$ be the set of all elements $b \in Z$ such that $a \leqq b$ (cf. Part I, Definition 1.5). Then we define the *central envelope* $e(a)$ as the product $\Pi(Z(a, 1))$, i.e., as the smallest element of Z which is $\geqq a$.

NOTE: $e(a) = 0$ if $a = 0$, and if Axiom VI holds, then $e(a) = 1$ if $a \neq 0$. We shall find the envelope $e(a)$ an important tool in our analysis of Z, when Axiom VI is not assumed.

THEOREM 1.3:

(a) $a \in L$ implies $e(a) \in Z$, and $a \leqq e(a)$;
(b) $e(a) = a$ if and only if $a \in Z$;
(c) $a < b$ implies $e(a) \leqq e(b)$;
(d) $e(\Sigma(S)) = \Sigma(e(a);\ a \in S)$ for every $S \subset L$;
(e) $e(\Pi(S)) \leqq \Pi(e(a);\ a \in S)$ for every $S \subset L$;
(f) $e(\bar{a}b) = \bar{a}e(b)$ if $\bar{a} \in Z$, $b \in L$.

PROOF: (a) Since $a \in L$ implies $Z(a, 1) \subset Z$, we have by Theorem 1.1 that $\Pi(Z(a, 1)) \in Z$, i.e., $e(a) \in Z$ for every $a \in L$. Obviously, $a \leqq e(a)$.

(b) Since $e(a) \in Z$, $a = e(a)$ implies $a \in Z$. Conversely, if $a \in Z$, then a belongs to $Z(a, 1)$, and is clearly the smallest element of $Z(a, 1)$. Hence $e(a) = \Pi(Z(a, 1)) = a$.

(c) If $a < b$, $a < b \leqq e(b)$. But $e(b) \in Z$, whence $e(a) \leqq e(b)$.

(d) If $a \in S$, then $\Sigma(S) \geqq a$, whence by (c), $e(\Sigma(S)) \geqq e(a)$. Therefore $e(\Sigma(S)) \geqq \Sigma(e(a);\ a \in S)$. Moreover, $e(a) \in Z$ for every $a \in S$, whence by Theorem 1.1, $\Sigma(e(a);\ a \in S) \in Z$. Since $e(a) \geqq a$ for every $a \in S$, we have

$\Sigma(e(a);\ a \in S) \geqq \Sigma(S)$. Consequently

$$\Sigma(e(a);\ a \in S) \geqq e(\Sigma(S)).$$

Thus it results that $\Sigma(e(a);\ a \in S) = e(\Sigma(S))$.

(e) If $a \in S$, then $\Pi(S) \leqq a$, whence by (c), $e(\Pi(S)) \leqq e(a)$. Therefore $e(\Pi(S)) \leqq \Pi(e(a);\ a \in S)$.

(f) We apply (e) to the case $S=(\bar{a}, b)$ and obtain $e(\bar{a}b) \leqq e(\bar{a})e(b) = \bar{a}e(b)$, since $\bar{a} \in Z$. Hence $e(\bar{a}b) \leqq \bar{a}e(b)$. Now $e(\bar{a}b) + (-\bar{a}) \geqq \bar{a}b + (-\bar{a}) \geqq \bar{a}b + (-\bar{a})b = (\bar{a} + (-\bar{a}))b = 1 \cdot b = b$. Since $e(\bar{a}b) + (-\bar{a}) \in Z$, it follows that $e(\bar{a}b) + (-\bar{a}) \geqq e(b)$. Now $\bar{a}e(b) \leqq \bar{a}(e(\bar{a}b) + (-\bar{a})) = \bar{a}e(\bar{a}b) + \bar{a}(-\bar{a}) = \bar{a}e(\bar{a}b) + 0 = \bar{a}e(\bar{a}b) \leqq e(\bar{a}b)$. Hence $\bar{a}e(b) \leqq e(\bar{a}b)$. This completes the proof that $\bar{a}e(b) = e(\bar{a}b)$.

THEOREM 1.4: If $\bar{a} \in Z$, then

(a) $a \sim b$ implies $\bar{a}a \sim \bar{a}b$,
(b) $a \prec b$ implies $\bar{a}a \precsim \bar{a}b$,
(c) $a \approx b$ implies $\bar{a}a \approx \bar{a}b$.

Moreover,

(d) $a \sim b$ implies $e(a) = e(b)$,
(e) $a \prec b$ implies $e(a) \leqq e(b)$.

PROOF: (a) Since $a \sim b$, there exists x such that $a + x = b + x$, $ax = bx$. Then since $\bar{a} \in Z$, $\bar{a}a + \bar{a}x = \bar{a}(a + x) = \bar{a}(b + x) = \bar{a}b + \bar{a}x$, $\bar{a}a \cdot \bar{a}x = \bar{a}ax = \bar{a}bx = \bar{a}b \cdot \bar{a}x$. Thus $\bar{a}a \sim \bar{a}b$.

(b) If $a \prec b$, there exists b' with $a \sim b' < b$. Then $\bar{a}a \sim \bar{a}b' \leqq \bar{a}b$ by (a), whence $\bar{a}a \precsim \bar{a}b$.

(c) Since $a \approx b$ means the existence of a positive integer n and a sequence $(a_i;\ i=0, \cdots, n)$ with $a=a_0$, $b=a_n$, $a_{i-1} \sim a_i$ $(i = 1, \cdots, n)$, we have $\bar{a}a = \bar{a}a_0$, $\bar{a}b = \bar{a}a_n$, and by (a) $\bar{a}a_{i-1} \sim \bar{a}a_i$ $(i = 1, \cdots, n)$. Hence $\bar{a}a \approx \bar{a}b$.

(d) Since $-e(a) \in Z$, we have by (a) that $a \sim b$ implies $(-e(a))\,a \sim (-e(a))\,b$. But $(-e(a))\,a \leqq (-e(a))\,e(a) = 0$, whence $(-e(a))\,b \sim 0$, i.e., $(-e(a))\,b = 0$. Thus $b \leqq -(-e(a)) = e(a)$, and since $e(a) \in Z$, this yields $e(b) \leqq e(a)$. Similarly $e(a) \leqq e(b)$, whence $e(a) = e(b)$.

(e) If $a \prec b$, there exists b' with $a \sim b' < b$. Then by (d), $e(a) = e(b') \leqq e(b)$, and hence $e(a) \leqq e(b)$.

LEMMA 1.1: If $ab = 0$, $(a, b)D$, then $e(a)e(b) = 0$.

PROOF: We apply Part I, Theorem 5.8. Since $(a, b)D$, $ab = 0$, there exists $a^* \in Z$ with $a \leqq a^*$, $a^*b = 0$. Thus $e(a) \leqq e(a^*) = a^*$, whence

$e(a)b \leqq a^*b = 0$, i.e., $e(a)b = 0$. Now by Theorem 1.3 (f) with $\bar{a} = e(a) \in Z$, $e(a)e(b) = e(e(a)b) = e(0) = 0$.

LEMMA 1.2: If $(a, b)D$ is false, then there exist $a_1 \leqq a$, $b_1 \leqq b$ with $a_1 \sim b_1$, $a_1 \neq 0$, $e(a_1) = e(b_1) \neq 0$.

PROOF: For some $x \in L$, $xa + xb \neq x(a + b)$. Hence, $xa + xb < x(a + b)$. Let y be an inverse to $xa + xb$ in $x(a+b)$. Then $y \neq 0$, $ya \leqq yax(a+b) \leqq yxa \leqq y(xa + xb) = 0$, i.e., $ya = 0$; similarly, $yb = 0$. But $y \leqq a + b$. Let $a_1 = a(b + y)$, $b_1 = b(a + y)$. Then $a_1 \sim b_1$ (mod y), since $a_1 y \leqq ay = 0$, $b_1 y \leqq by = 0$, $a_1 + y = (a + y)(b + y) = b_1 + y$. Finally, $a_1 \neq 0$, for $a_1 = 0$ implies $(a, b, y)\perp$, hence $y(a+b) = 0$, contradicting $0 \neq y \leqq a + b$; and $0 < a_1 \leqq e(a_1)$ by Lemma 1.4(d).

LEMMA 1.3: If $e(a) = e(b) \neq 0$, then there exist $a_1 \leqq a$, $b_1 \leqq b$ with $a_1 \sim b_1$, $a_1 \neq 0$, $e(a_1) = e(b_1) \neq 0$.

PROOF: If $ab \neq 0$, we can choose $a_1 = b_1 = ab$. If $ab = 0$, then Lemma 1.1 implies that $(a, b)D$ is false. Then Lemma 1.2 applies to establish existence of a_1, b_1 as required.

THEOREM 1.5: If $(a, b)D$, then $e(a)e(b) = e(ab)$.

PROOF: Let a' be an inverse of ab in a and b' an inverse of ab in b. By Part I, Theorem 5.6, $(a', b')D$, $a'b' = 0$. Hence by Lemma 1.1, $e(a')e(b') = 0$. Thus

$$e(a)e(b) = e(ab + a')e(ab + b') = (e(ab) + e(a'))\,(e(ab) + e(b'))$$

(by Theorem 1.3 (d))

$$= e(ab) + e(a')e(b') = e(ab),$$

the last equality but one holding by the dual of the distributive law.

THEOREM 1.6: If $a \in L$, then the center of the lattice $L(0, a)$ (which satisfies Axioms I—V by Part I, Corollary to Theorem 1.3) is the set $(\bar{a}a;\ \bar{a} \in Z)$.

PROOF: If $\bar{a} \in Z$, we have for every $b, c \in L(0, a)$, i.e., for $b, c \leqq a$ and hence for $b + c \leqq a$,

$$\bar{a}a(b + c) = \bar{a}(b + c) = \bar{a}b + \bar{a}c = \bar{a}a \cdot b + \bar{a}a \cdot c,$$

whence $(\bar{a}a, b, c)D$. Therefore $(\bar{a}a)D$ in $L(0, a)$ (not in L), i.e., $\bar{a}a$ belongs to the center of $L(0, a)$. Conversely, let a_0 belong to the center of $L(0, a)$, and let b be inverse to a_0 in a, and c any element of L. Then $a_0, b, ac \in L(0, a)$, and

$$(a_0 + b)c = (a_0 + b) \cdot ac = a_0 \cdot ac + b \cdot ac = a_0 c + bc,$$

whence $(a_0, b, c)D$. Thus $(a_0, b)D$ (in L). Moreover, $a_0 b = 0$, whence by Part I, Theorem 5.8, there exists $\bar{a} \in Z$ with $a_0 \leqq \bar{a}$, $\bar{a}b = 0$. Now $\bar{a}a = \bar{a}(a_0 + b) = \bar{a}a_0 + \bar{a}b = a_0 + 0 = a_0$, whence $a_0 \varepsilon (\bar{a}a;\ \bar{a} \in Z)$.

APPENDIX 1

This appendix is devoted to certain questions involving Boolean algebras, which throw light on various aspects of the theory of continuous geometries. It may, however, be omitted without endangering the understanding of the subsequent chapters. (For a comprehensive modern study of Boolean algebras, and their applications to other parts of mathematics, the reader is referred to M. H. Stone, Trans. American Mathematical Society, vol. 40 (1936), pages 37—111. Quotations of other pertinent literature are also given there.)

DEFINITION A.1: Let us consider a *partially ordered set* L, that is, a system $(L, <)$ satisfying Axiom I (Part I, Chapter I). If $\aleph$ is a (finite or infinite) cardinal number, then we shall say that L is an *$\aleph$-lattice* in case the following weakened form of Axiom II (Part I, Chapter I) holds:

$\mathrm{II}_1^{\aleph}$: For every set $S \subset L$ of power $< \aleph$ there is an element $\Sigma(S)$ in L which is *a least upper bound* of S, i.e.,

(a) $\Sigma(S) \geqq a$ for every $a \in S$,

(b) $x \geqq a$ for every $a \in S$ implies $x \geqq \Sigma(S)$.

$\mathrm{II}_2^{\aleph}$: For every set $S \subset L$ of power $< \aleph$ there is an element $\Pi(S)$ in L which is a *greatest lower bound* of S, i.e.,

(a) $\Pi(S) \leqq a$ for every $a \in S$,

(b) $x \leqq a$ for every $a \in S$ implies $x \leqq \Pi(S)$.

(The condition $< \aleph$ may seem at first less natural than $\leqq \aleph$. Thus in our terminology, if all finite sets $S \subset L$ have least upper bounds, then $\mathrm{II}_1^{\aleph_0}$ holds [$\aleph_0$ being the cardinal number of the natural numbers], although the property $\mathrm{II}_1^{\aleph_0}$ has nothing whatever to do with denumerable sets; or, if all denumerable sets have least upper bounds, then $\mathrm{II}_1^{\aleph_1}$ holds [$\aleph_1$ being the smallest non-denumerably infinite cardinal number], although, of course there is no reference to sets having power $\aleph_1$. However, we use the condition $< \aleph$ as a matter of convenience; the condition $\leqq \aleph$ is a special case of it, since the cardinal numbers are well-ordered, while the converse is clearly not true.)

NOTE (i): If existent, $\Sigma(S)$, $\Pi(S)$ are unique. (Cf. the Corollary after Axiom II in Part I, Chapter I.)

NOTE (ii): For $\aleph = 0$, conditions $\mathrm{II}_1^{\aleph}$, $\mathrm{II}_2^{\aleph}$ are vacuously satisfied by any partially ordered set L. Since for a set $S = (a)$ with one element, $\Sigma(S) \equiv \Pi(S) \equiv a$ is effective, $S = \Theta$ is the only significant case for $\aleph = 1, 2$. Clearly $a = \Sigma(\Theta)$ means that $a \leqq x$ for every $x \in L$, and $a = \Pi(\Theta)$ means that $a \geqq x$ for every $x \in L$. Hence $\mathrm{II}_1^{\aleph}$, $\mathrm{II}_2^{\aleph}$ state for $\aleph = 1, 2$ merely the existence of 0 and 1 in L (cf. Part I, Definition 1.2).

Note (iii): If $\aleph = 3$, the only implications (in addition to the existence of 0, 1) of $II_1^{\aleph}$, $II_2^{\aleph}$ are the existence of $\Sigma(S)$, $\Pi(S)$ for sets $S = (a, b)$ of two elements. As in Part I, Definition 1.3, we shall write $a + b \equiv \Sigma((a, b))$, $ab \equiv \Pi((a, b))$. Since $a + b$ is effective as a least upper bound for (b, a), we have $a + b = b + a$; since both $(a + b) + c$ and $a + (b + c)$ are effective as least upper bound for (a, b, c), we have $(a+b) + c = a + (b+c)$. Hence addition is commutative and associative, and by duality the same is true for multiplication. Finally, it is obvious that the three statements $a \leqq b$, $a + b = b$, $ab = a$ are equivalent. (Cf. Part I, Theorem 1.1.)

Note (iv): If $II_1^{\aleph}$ holds for $\aleph = 3$, then $a_1 + \cdots + a_n$ is effective as least upper bound for $S = (a_1, \cdots, a_n)$ for every $n = 1, 2, \cdots$, whence $II_1^{\aleph}$ holds for $\aleph = \aleph_0$, and hence also for every $\aleph = 4, 5, \cdots$. By duality the same is true for $II_2^{\aleph}$. An $\aleph$-lattice with $\aleph = 3, 4, \cdots, \aleph_0$ will be referred to simply as a *lattice.*

Note (v): If $\aleph >$ power of L, then properties $II_1^{\aleph}$, $II_2^{\aleph}$ coincide respectively with Axioms II_1, II_2, respectively in Part I, Chapter I (since then $S \subset L$ implies power of $S \leqq$ power of $L < \aleph$). (For a broader criterion cf. Theorem A.4.) Such a lattice is called a *continuous lattice.*

Note (vi): For each $\aleph$, $II_1^{\aleph}$ and $II_2^{\aleph}$ are dual to each other. If $\aleph >$ power of L, the two properties imply each other (by virtue of Note (v) and the discussion in Part I, page 3), but for $\aleph \leqq$ power of L, they have distinct meanings.

Definition A.2: A *complemented* $\aleph$-lattice ($\aleph \geqq 3$) is an $\aleph$-lattice L, satisfying Axiom V (Part I, Chapter I), i.e., one such that for every $a \in L$ there exists an *inverse* of a, that is, an element $x \in L$ with $a + x = 1$, $ax = 0$. (In the case of Note (iv), L is then a *complemented lattice,* and in the case of Note (v), L is a *complemented continuous lattice.*)

Lemma A.1: If L is an $\aleph$-lattice, the two dual *distributive laws,*

IV_1': $(a + b)c = ac + bc$ for every $a, b, c \in L$,

IV_2': $ab + c = (a + c)(b + c)$ for every $a, b, c \in L$,

imply each other.

Proof: By duality it suffices to prove that IV_1' implies IV_2'. We have, if IV_1' holds,

$$\begin{aligned}(a + c)(b + c) &= ab + ac + cb + cc && \text{(by } IV_1'\text{)}\\ &= ab + ac + cb + c = ab + c && \text{(since } ac,\ bc \leqq c\text{)},\end{aligned}$$

so that IV_2' holds.

Definition A.3: An $\aleph$-lattice L ($\aleph \geqq 3$) is a *distributive $\aleph$-lattice* if it satisfies the equivalent distributive laws IV_1', IV_2'. (In the case of Note (iv), L is then a *distributive lattice,* and in the case of Note (v), L is a *continuous distributive lattice.*) If L satisfies Axiom V, i.e., is complemented, then we call it an $\aleph$-*Boolean algebra* (merely *Boolean algebra* in the case of Note (iv) and *continuous Boolean algebra* in the case of Note (v)).

It should be observed that IV_1' (and hence IV_2' also) implies the modular Axiom IV (Part I, Chapter I), as one verifies by putting $a \leqq c$ in IV_1'.

In what follows, L is assumed to be a $\aleph$-Boolean algebra with $\aleph \geqq 3$.

THEOREM A.1:

(a) Every $a \in L$ has a unique inverse $x \in L$. This inverse will be denoted by $-a$.

(b) The correspondence $a \to -a$ is an involutoric dual-automorphism of L (i.e., anti-automorphism of L—cf. Part II, Definition 4.1—with $-(-a) = a$ for every $a \in L$).

(c) $ab = 0$ is equivalent to $a \leqq -b$.

PROOF: By V, a has at least one inverse x. If $b \leqq x$, then $ab \leqq ax = 0$, whence $ab = 0$; conversely, $ab = 0$ implies $b = 1 \cdot b = (a + x)b = ab + xb = 0 + xb = xb \leqq x$. Hence $ab = 0$ is equivalent to $b \leqq x$. Thus for two inverses x_1, x_2 of a, the statements: $b \leqq x_1$ and $b \leqq x_2$ are equivalent; whence $x_1 \leqq x_2$, $x_2 \leqq x_1$, i.e., $x_1 = x_2$. This proves (a).

If $-a$ denotes the unique inverse x of a, the above yields that $ab = 0$ and $b \leqq -a$ are equivalent, whence (c) is established.

Since a is an inverse of $-a$, we have $-(-a) = a$ by (a). Since $a \cdot (-b) = 0$ and $(-b) \cdot a = 0$ are equivalent, we have by (c) that $-b \leqq -a$ is equivalent to $a \leqq -(-b) = b$. But $a = b$ and $-a = -b$ are obviously equivalent; hence $a < b$ and $-a > -b$ are also equivalent. This proves that $a \to -a$ is a dual-automorphism of L, and (b) is established.

COROLLARY: For every $a, b \in L$, $-(ab) = (-a) + (-b)$, $-(a+b) = (-a)(-b)$.

PROOF: This is immediate from Theorem A.1 (b).

THEOREM A.2: For every set $S \subset L$ with power $< \aleph$ and every $b \in L$

$$\Sigma(S) \cdot b = \Sigma(ab;\ a \in S), \quad \Pi(S) + b = \Pi(a + b;\ a \in S).$$

PROOF: By duality, it suffices to prove the first formula. We consider first the case $ab = 0$ for every $a \in S$. Then $a \leqq -b$ for every $a \in S$, whence $\Sigma(S) \leqq -b$, $\Sigma(S) \cdot b = 0$ (by Theorem A.1 (b)). Now let us consider the general case. Clearly $a \in S$ implies $\Sigma(S) \geqq a$, $\Sigma(S) \cdot b \geqq ab$, whence $\Sigma(S) \cdot b \geqq \Sigma(ab;\ a \in S)$. Define $b_1 \equiv b(-\Sigma(a'b;\ a' \in S))$. Then for every $a \in S$

$$ab_1 = ab(-\Sigma(a'b;\ a' \in S)) \leqq \Sigma(a'b;\ a' \in S)\,(-\Sigma(a'b;\ a' \in S)) = 0,$$

and $ab_1 = 0$. The result proved above then yields $\Sigma(S)b_1 = 0$, i.e.,

$\Sigma(S)b(-\Sigma(a'b;\ a' \in S)) = 0$, and it follows that $\Sigma(S)b \leqq \Sigma(a'b;\ a' \in S)$ (by Theorem A.1 (b), (c)). Thus $\Sigma(S)b = \Sigma(ab;\ a \in S)$.

COROLLARY: For every $n = 1, 2, \cdots$, and any sets $S_1, \cdots, S_n \subset L$, all with powers $< \aleph$,

$$\Sigma(S_1) \cdot \ldots \cdot \Sigma(S_n) = \Sigma(a_1 \cdot \ldots \cdot a_n;\ a_i \in S_i\ (i = 1, \cdots, n)),$$
$$\Pi(S_1) + \cdots + \Pi(S_n) = \Pi(a_1 + \cdots + a_n;\ a_i \in S_i\ (i = 1, \cdots, n)).$$

PROOF: This results from n-fold application of the formulas in Theorem A.2.

THEOREM A.3:

(a) L satisfies Axioms I—V (Part I, Chapter I) if and only if L is continuous.

(b) L satisfies Axiom VI (Part I, Chapter I) if and only if L consists of the elements 0, 1 only.

(c) For a lattice L' satisfying Axioms I--V, the center Z (as defined at the beginning of Chapter I) is L' if and only if L' is a Boolean algebra.

PROOF: (a) The continuity of L is necessary by Axiom II. To prove its sufficiency, we note that Axioms I, V hold by the definition of a Boolean algebra, II coincides with continuity, IV is a consequence of IV'_1, and III follows from Theorem A.2.

(b) By virtue of Theorem A.1 (a), Axiom VI states that for every $a \in L$ either $a = 0$ or $a = 1$.

(c) Clearly L' is a Boolean algebra if and only if IV'_1 holds, i.e., $(x, y, z)D$ for every $x, y, z \in L'$, i.e., $(x)D$ for every $x \in L'$. This last condition means $Z = L'$.

An immediate consequence of this theorem is that the centers Z of lattices L' satisfying Axioms I—V are abstractly all continuous Boolean algebras (cf. the remark following Theorem 1.2). Among these lattices L', the irreducible lattices (viz., those satisfying VI) and the (continuous) Boolean algebras represent two opposite extremes, viz., $Z = (0, 1)$ and $Z = L'$.

Henceforth we shall assume $\aleph$ to be infinite. (This is not an essential restriction, since $\aleph = 3, 4, 5, \ldots$ may be replaced by $\aleph = \aleph_0$.) The following theorem gives a sufficient condition that a Boolean algebra be continuous.

THEOREM A.4: In order that an $\aleph$-Boolean algebra L be continuous, the following condition is sufficient:

(*) For every set $T \subset L$ such that $a, b \in T$, $a \neq b$ implies $ab = 0$, it is so that the power of T is $< \aleph$.

PROOF: Since II_2 follows from II_1, it suffices to prove that (*) implies II_1. Let us suppose that (*) holds and that II_1 fails to hold. Then there exists a set $S \subset L$ without a least upper bound, and therefore there exists a set $\bar{S}$ of minimum power having this property. Let Ω be the smallest ordinal corresponding to this power, and replace $\bar{S}$ by the system $(a_\alpha; \alpha < \Omega)$. Since L is an $\aleph$-Boolean algebra, the power of Ω is $\geqq \aleph$. Now define $\bar{S}_\alpha \equiv (a_\gamma; \gamma < \alpha)$ for every $\alpha < \Omega$. The power of $\bar{S}_\alpha$ is that of α and therefore less than that of Ω, i.e., of $\bar{S}$. Hence the definition of $\bar{S}$ yields the existence of $b_\alpha = \Sigma(\bar{S}_\alpha) = \Sigma(a_\gamma; \gamma < \alpha)$ for each $\alpha < \Omega$. Clearly if $\Sigma(b_\alpha; \alpha < \Omega)$ exists, i.e., if $\Sigma(\Sigma(a_\gamma; \gamma < \alpha); \alpha < \Omega)$ exists, it is effective as $\Sigma(a_\gamma; \gamma < \Omega) = \Sigma(\bar{S})$. Hence by our hypothesis $\Sigma(b_\alpha; \alpha < \Omega)$ does not exist.

Now for each $\alpha < \Omega$ there exists a smallest $\alpha' < \Omega$ such that $b_{\alpha'} = b_\alpha$; let us denote this α' by α^*. Thus $\alpha' < \alpha^*$ implies $b_{\alpha'} \neq b_{\alpha^*}$. Let K be the set of all $\alpha^* < \Omega$. Then each element in the family $(b_\alpha; \alpha < \Omega)$ is in the family $(b_{\alpha^*}; \alpha^* \in K)$ and conversely, and so $\Sigma(b_{\alpha^*}; \alpha^* \in K)$ does not exist. Therefore the power of K is $\geqq \aleph$. Now K is a well-ordered set. If $\alpha^*, \beta^* \in K$, $\alpha^* < \beta^*$, then since $S_{\alpha^*} \subset S_{\beta^*}$, we have $b_{\alpha^*} \leqq b_{\beta^*}$, and since $b_{\alpha^*} \neq b_{\beta^*}$ (because $\beta^* \in K$), $b_{\alpha^*} < b_{\beta^*}$. Thus if K has a last (greatest) element α^0, b_{α^0} is effective as $\Sigma(b_{\alpha^*}; \alpha^* \in K)$, which is impossible. Hence, every $\alpha^* \in K$ has an immediate successor $\overline{\alpha^*}$. Define $c_{\alpha^*} \equiv b_{\overline{\alpha^*}}(-b_{\alpha^*})$. If $\alpha^*, \beta^* \in K$, $\alpha^* \neq \beta^*$, say $\alpha^* < \beta^*$, then $\overline{\alpha^*} \leqq \beta^*$, $b_{\overline{\alpha^*}} \leqq b_{\beta^*}$, whence

$$c_{\alpha^*} c_{\beta^*} \leqq b_{\overline{\alpha^*}}(-b_{\beta^*}) \leqq b_{\beta^*}(-b_{\beta^*}) = 0, \text{ and } c_{\alpha^*} c_{\beta^*} = 0.$$

If $c_{\alpha^*} = 0$, then $b_{\overline{\alpha^*}}(-b_{\alpha^*}) = 0$, whence $b_{\overline{\alpha^*}} \leqq b_{\alpha^*}$, contrary to $b_{\alpha^*} < b_{\overline{\alpha^*}}$. Therefore $c_{\alpha^*} \neq 0$ $(\alpha^* \in K)$, and consequently $\alpha^* \neq \beta^*$ implies $c_{\alpha^*} \neq c_{\beta^*}$. Define the set $T \equiv (c_{\alpha^*}; \alpha^* \in K)$. Then T has the same power as K, and hence its power is $\geqq \aleph$. But T satisfies the hypotheses of condition (*), whence the power of T is $< \aleph$. This contradiction establishes the desired result, i.e., that L is continuous.

It should be observed that only the property $II_1^\aleph$ (and not $II_2^\aleph$) was used in the proof of Theorem A.4, so that the hypothesis that L is an $\aleph$-Boolean algebra, i.e., that both $II_1^\aleph$ and $II_2^\aleph$ hold is stronger than necessary. The theorem may, moreover, be extended without great difficulty to modular lattices. In this case the hypothesis on T in (*) should be replaced by the condition $T\perp$, i.e., that T be an independent set (cf. Part I, Definition 2.1). In fact, in a Boolean algebra the hypothesis in (*) is equivalent to independence as can be seen by an application of the Corollary to Theorem A.2 with $n = 2$, $S_1 = J$, $S_2 = K$ to Definition 2.1 in Part I. Condition (*) could be given in other forms; for $\aleph = \aleph_0$ it is equivalent to the chain condition. When applied to modular lattices, (*)

has similar implications for Axiom III to those established in Theorem A.4 for Axiom II, i.e., it permits the weakening of the statement of Axiom III to a statement of that axiom in which Ω has power less than $\aleph$ without weakening the axiom. These questions will not be pursued further.

DEFINITION A.4: A non-empty set $I \subset L$ is an $\aleph$-*ideal in* L in case I possesses the properties

(a) $S \subset I$, power of $S < \aleph$ implies $\Sigma(S) \in I$,

(b) $a \in I$, $b \in L$ implies $ab \in I$.

(Condition (b) is equivalent to $a \in I$, $c \in L$, $c \leqq a$ implies $c \in I$.)

DEFINITION A.5: If I is an $\aleph$-ideal in L, and if $a, b \in L$, then a and b are *equivalent modulo* I ($a \sim b \pmod{I}$) in case there exists $u \in I$ with $a + u = b + u$.

LEMMA A.2: If I is an $\aleph$-ideal in L, then $a \sim b \pmod{I}$ is equivalent to $(a + b)(-ab) \in I$.

PROOF: Evidently

$$\begin{aligned} a + (a + b)(-ab) &= (a + b)a + (a + b)(-ab) \\ &= (a + b)(a + (-ab)) \\ &= (a + b)(a + (-a) + (-b)) \\ &\qquad \text{(by the Corollary to Theorem A.1)} \\ &= (a + b) \cdot 1 = a + b, \end{aligned}$$

and similarly $b + (a + b)(-ab) = a + b$. If $u \equiv (a + b)(-ab) \in I$, then since $a + u = b + u$, $a \sim b \pmod{I}$. Conversely let $a \sim b \pmod{I}$; then there exists $v \in I$ with $a + v = b + v$. Hence $a + b + v = (a + v) + (b + v) = (a + v)(b + v) = ab + v$, whence $(a+b)(-ab) \leqq (ab+v)(-ab) = ab(-ab) + v(-ab) = 0 + v(-ab) = v(-ab) \leqq v$, and consequently $(a + b)(-ab) \in I$.

LEMMA A.3: The relation $a \sim b \pmod{I}$ (I an $\aleph$-ideal in L) is reflexive, symmetric and transitive.

PROOF: $a \sim a \pmod{I}$ since $a + u = a + u$ for any u in I and I is assumed not empty; and obviously $a \sim b \pmod{I}$ implies $b \sim a \pmod{I}$. To prove the transitivity, let $a \sim b \pmod{I}$, $b \sim c \pmod{I}$, i.e., $a + u = b + u$, $b + v = c + v$, $u, v \in I$. Then $a + (u + v) = (a + u) + v = (b + u) + v = (b + v) + u = (c + v) + u = c + (u + v)$, and $u + v \in I$. Therefore $a \sim c \pmod{I}$.

DEFINITION A.6: Let A_a be the set of all $x \in L$ with $x \sim a$, (mod. I) [I being an $\aleph$-ideal in L]. The system $(A_a;\ a \in L)$ is a mutually exclusive and exhaustive partition of L into equivalence classes. We denote the equivalence classes by $A, B, C, \cdots$ and the set of all equivalence classes by L/I.

DEFINITION A.7: $A \leqq B$ means the existence of $a \in A$, $b \in B$ with $a \leqq b$. $A < B$ means $A \leqq B$, $A \neq B$.

LEMMA A.4: Let $a \in A$, $b \in B$. Then $A \leqq B$ is equivalent to each of the following conditions:

(a) There exists $u \in I$ with $a \leqq b + u$.

(b) $a(-ab) \in I$.

PROOF: We shall prove $A \leqq B \to (a) \to (b) \to A \leqq B$. First, suppose $A \leqq B$. Then there exist $a' \in A$, $b' \in B$ with $a' \leqq b'$. Now $a \sim a'$, $b \sim b'$ (mod I), whence there exist $u', v' \in I$ such that $a + u' = a' + u'$, $b + v' = b' + v'$. Now define $u \equiv u' + v' \in I$. Then $b + u = b + u' + v' = (b + v') + u' = (b' + v') + u' \geqq (a' + v') + u' = (a' + u') + v' = (a + u') + v' \geqq a$, and (a) is established.

Next, suppose (a) holds. Then since $a \leqq b + u$, $a = a(b + u) \leqq ab + u$, whence

$$a(-ab) \leqq (ab+u)(-ab) = ab(-ab) + u(-ab) = 0 + u(-ab) = u(-ab) \leqq u.$$

Thus since $u \in I$, $a(-ab) \in I$, and (b) is proved.

Finally, let (b) hold. Then define $u \equiv a(-ab) \in I$. Since $b + u = (b + u) + u$, $b \sim b + u$ (mod I), and $b' \equiv b + u \in B$. Now $b' = b + u = b + a(-ab) \geqq a \cdot ab + a \cdot (-ab) = a(ab + (-ab)) = a \cdot 1 = a$, whence $b' \geqq a$, and $A \leqq B$. This completes the proof.

LEMMA A.5: The set L/I is partially ordered with respect to the relation $<$.

PROOF: Since $A < A$ is impossible by the definition of $<$, we need prove that the relation $<$ is transitive. Let $A < B$, $B < C$; then $A \leqq B$, $B \leqq C$. Hence for every $a \in A$, $b \in B$, $c \in C$ there exist $u, v \in I$ such that $a \leqq b + u$, $b \leqq c + v$ by Lemma A.4(a). Therefore $a \leqq c + (u + v)$, $u + v \in I$, and we have $A \leqq C$ by Lemma A.4(a). If $A = C$, then $A \leqq B$, $B \leqq A$, whence $a(-ab) = b(-ab) = 0$ by Lemma A.4(b). Hence $(a + b)(-ab) = a(-ab) + b(-ab) = 0 \in I$, and $a \sim b$ (mod I) by Lemma A.2, i.e., $A = B$. But this is impossible since $A < B$, whence $A < C$.

THEOREM A.5: L/I is an $\aleph$-Boolean algebra if I is an $\aleph$-ideal.

PROOF: We shall verify Axioms $\text{II}^{\aleph}$, IV′, V, since Axiom I has been established in Lemma A.5. In order to prove $\text{II}^{\aleph}$, consider a set $\mathscr{S} \subset L/I$ with power $< \aleph$. For each $A \in \mathscr{S}$ select an element $a \in A$ and denote the set of these elements by S. Hence $S \subset L$, power of $S < \aleph$, and $\mathscr{S} = (A_a; a \in S)$. Thus $\Sigma(S)$ and $\Pi(S)$ exist and for every $a \in S$, $\Pi(S) \leqq a \leqq \Sigma(S)$,

$A_{\Pi(S)} \leqq A_a \leqq A_{\Sigma(S)}$ i.e., $A_{\Pi(S)} \leqq B \leqq A_{\Sigma(S)}$ for every $B \in \mathscr{S}$. Suppose now that $C \geqq B$ for every $B \in \mathscr{S}$. Fix $a_0 \in C$; then $A_{a_0} = C \geqq A_a$ for every $a \in S$, whence there exists for each $a \in S$ an element $u_a \in I$ with $a_0 + u_a \geqq a$. Then $a_0 + \Sigma(u_a;\ a \in S) \geqq \Sigma(S)$, and since I is an $\aleph$-ideal, it follows that $\Sigma(u_a;\ a \in S) \in I$. Thus $C = A_{a_0} \geqq A_{\Sigma(S)}$. Suppose that $C \leqq B$ for every $B \in \mathscr{S}$. Then fix $a_0 \in C$; $A_{a_0} = C \leqq A_a$ for every $a \in S$, whence there exists for each $a \in S$ an element $u_a \in I$ with $a_0 \leqq a + u_a$. Thus $a_0 \leqq a + \Sigma(u_{a'};\ a' \in S)$ for every $a \in S$, and consequently $a_0 \leqq \Pi(a + \Sigma(u_{a'};\ a' \in S);\ a \in S) = \Pi(S) + \Sigma(u_{a'};\ a' \in S)$. As before, $\Sigma(u_{a'};\ a' \in S) \in I$, whence $C = A_{a_0} \leqq A_{\Pi(S)}$. Thus $A_{\Sigma(S)}$, $A_{\Pi(S)}$ are effective as $\Sigma(\mathscr{S})$, $\Pi(\mathscr{S})$, respectively, in L/I. Thus $\mathrm{II}_1^{\aleph}$, $\mathrm{II}_2^{\aleph}$ are established. In particular, if $\mathscr{S} = (A_a,\ A_b)$, then this discussion shows that

$$(\dagger') \qquad A_a + A_b = A_{a+b},\ A_a A_b = A_{ab}.$$

Thus Properties IV′, V hold for L/I since they hold for L, by virtue of (†′).

Corollary: If $S \subset L$, power of $S < \aleph$, then

$$(\dagger) \qquad \Sigma(A_a;\ a \in S) = A_{\Sigma(S)},\ \Pi(A_a;\ a \in S) = A_{\Pi(S)}.$$

Proof: This follows from the proof of the theorem. Note that (†′) is a special case of (†) with $S = (a,\ b)$.

Note: If power of $S \geqq \aleph$, then formula (†) need not be true, even if $\Sigma(S)$ exists in L for this set S. For, consider Example 2 or 3 at the end of this appendix, P being the interval $0 \leqq x \leqq 1$, S the class of all one-element subsets of P. Then $A_a = 0$ for each $a \in S$, whence $\Sigma(A_a;\ a \in S) = 0$, but $\Sigma(S)$ exists and is equal to $P = 1$.

Theorem A.6: In order that L/I (I being an $\aleph$-ideal) be a continuous Boolean algebra, the following condition is sufficient:

(‡) For every set $T \subset L$ such that $a,\ b \in T$, $a \neq b$ implies $ab = 0$, and such that $a \in T$ implies $a \notin I$, it is so that the power of T is $< \aleph$.

Proof: By virtue of Theorems A.5, A.4 it suffices to establish condition (*) of Theorem A.4 for L/I. Let $\mathfrak{J}' \subset L/I$, and let $AB = 0$ for every $A,\ B \in \mathfrak{J}'$ such that $A \neq B$. Now if power of $\mathfrak{J}' \not< \aleph$ then power of $\mathfrak{J}' \geqq \aleph$, and there is a subset $\mathfrak{J}''$ of $\mathfrak{J}'$ with power $\aleph$. If $0 \equiv A_0 \in \mathfrak{J}''$, define $\mathfrak{J} \equiv \mathfrak{J}'' - (0)$; otherwise define $\mathfrak{J} \equiv \mathfrak{J}''$. Then $\mathfrak{J}$ has power $\aleph$, since $\aleph$ is infinite, and $0 \notin \mathfrak{J}$. Let Ω be the smallest ordinal of power $\aleph$ and write $\mathfrak{J} = (A^\alpha;\ \alpha < \Omega)$; for every $\alpha < \Omega$ select $a^\alpha \in A^\alpha$. For each $\alpha < \Omega$ the set $(a^\gamma;\ \gamma < \alpha)$ has the power of α, which is less than the power of Ω, i.e. less than $\aleph$. Hence $\Sigma(a^\gamma;\ \gamma < \alpha)$ exists $(\alpha < \Omega)$; define for each $\alpha < \Omega$: $\bar{a}^\alpha \equiv a^\alpha(-\Sigma(a^\gamma;\ \gamma < \alpha))$. If $\beta < \alpha < \Omega$, then

$$\bar{a}^\beta \bar{a}^\alpha \leqq a^\beta(-\Sigma(a^\gamma;\ \gamma < \alpha)) \leqq \Sigma(a^\gamma;\gamma < \alpha)\,(-\Sigma(a^\gamma;\ \gamma < \alpha)) = 0,$$

whence $\bar{a}^\beta \bar{a}^\alpha = 0$. Hence by symmetry $\bar{a}^\beta \bar{a}^\alpha = 0$ if $\beta \neq \alpha$ and $\alpha, \beta < \Omega$. Suppose now that $\bar{a}^\alpha \notin I$ for every $\alpha < \Omega$. Then $\bar{a}^\alpha \neq 0$ $(\alpha < \Omega)$, and $\alpha \neq \beta$ implies $\bar{a}^\alpha \neq \bar{a}^\beta$. Thus $T = (\bar{a}^\alpha;\ a < \Omega)$ has the power of Ω, i.e., has power $\aleph$. For $a, b \in T$, $a \neq b$ (i.e., $a = \bar{a}^\beta$, $b = \bar{a}^\alpha$, $\beta \neq \alpha$), we have $ab = 0$, and $a \in T$ implies $a \in I$; therefore by (‡) T has power $< \aleph$, which is impossible. Consequently, there exists $\alpha < \Omega$ with $\bar{a}^\alpha \in I$. Now if $\gamma < \alpha$, then $A^\gamma A^\alpha = 0$, i.e., $a^\gamma a^\alpha \in I$. Since the power of the set of all $\gamma < \alpha$ is the power of α and is hence less than the power of Ω, i.e., less than $\aleph$, we have that $\Sigma(a^\gamma a^\alpha;\ \gamma < \alpha) \in I$, whence by Theorem A.2, $\Sigma(a^\gamma;\ \gamma < \alpha) \cdot a^\alpha \in I$. Now $a^\alpha = a^\alpha \cdot 1 = a^\alpha \big((\Sigma(a^\gamma; \gamma < \alpha) + (-\Sigma(a^\gamma; \gamma < \alpha)) \big) = a^\alpha \Sigma(a^\gamma;\ \gamma < \alpha) + a^\alpha(-\Sigma(a^\gamma;\ \gamma < \alpha)) = a^\alpha \Sigma(a^\gamma;\ \gamma < \alpha) + \bar{a}^\alpha$. Therefore $a^\alpha \in I$, since $\bar{a}^\alpha$, $a^\alpha \Sigma(a^\gamma; \gamma < \alpha) \in I$. Consequently $a^\alpha \sim 0 \pmod{I}$, whence $A^\alpha = A_{a^\alpha} = 0$, $0 \in \mathfrak{J}$, contrary to $0 \notin \mathfrak{J}$. Thus our indirect proof shows that (*) holds for L/I, and the proof is complete.

This theorem is important in that it enables us to construct continuous Boolean algebras L/I from $\aleph$-Boolean algebras which are not continuous.

We examine now three typical examples of continuous Boolean algebras.

Example 1: Let P be an arbitrary set, and L the class of all subsets of P. For $a, b \in L$, define $a < b$ to mean a is a proper subset of b, i.e., $a \subset b$, $a \neq b$. Then L is a partially ordered set; in fact L is a continuous lattice, since for $S \subset L$ the set-theoretical sum $\mathfrak{S}(S)$ of all sets of S is effective as $\Sigma(S)$, and the intersection $\mathfrak{P}(S)$ of all sets of S is effective as $\Pi(S)$. Clearly 0 is the empty set Θ and 1 the entire set P. The distributive law (IV′) is now immediately verified, and every $a \in L$ is seen to possess an inverse $x \in L$, viz., the complement of the set a in P, whence V holds.

Example 2: Let P be a space in which a measure function $\mu^*(a)$, with the usual properties of Lebesgue (-Stieltjes-Radon-) outer measure, is defined. The class L_μ of all measurable subsets of P is a subclass of the class L in Example 1, such that 0, 1 (i.e., Θ, P) $\in L_\mu$, $a \in L_\mu$ implies $(-a) \in L_\mu$, and $\Sigma(S)$, $\Pi(S) \in L_\mu$ if S is a finite or denumerably infinite subset of L. Hence L_μ is an $\aleph_1$-Boolean algebra.

Let I_μ be the set of all subsets a of P with $\mu^*(a) = 0$. It is well known that every such subset a is measurable, that $a \in I_\mu$ implies $b \in I_\mu$ for every $b \leqq a$, and that $\Sigma(S) \in I_\mu$ if S is a finite or denumerably infinite subset of L_μ. Hence I_μ is an $\aleph_1$-ideal in L_μ. Therefore by Theorem A.5, L_μ/I_μ is an $\aleph_1$-Boolean algebra. We show now that L_μ, I_μ satisfy condition (‡) in Theorem A.6. Let $T \subset L_\mu$, and let $a, b \in T$ imply $ab = 0$, and $a \in T$ imply $a \notin I_\mu$. Let $(\tilde{a}_i;\ i = 1, 2, \cdots)$ be a sequence of measurable sets with $\Sigma(\tilde{a}_i;\ i = 1, 2, \cdots) = P$ and all $\mu(\tilde{a}_i)$ finite. (The existence of such a sequence is an essential property of Lebesgue measure.) For each $a \epsilon T$ we have $a \notin I_\mu$, whence $\mu(a) > 0$, and since $\mu(a) = \lim_{j\to\infty} \mu(a\Sigma(\tilde{a}_i;\ i = 1, \cdots, j))$, there exists j such that $\mu(a\Sigma(\tilde{a}_i;\ i = 1, \cdots, j)) \neq 0$. Therefore there is a positive integer ν with $\mu(a\Sigma(\tilde{a}_i;\ i = 1, \cdots, j)) > 1/\nu$. For each j, ν let $T_{j\nu}$ be the set of elements $a \epsilon T$ with this last property. Then $T = \mathfrak{S}(T_{j\nu};\ j, \nu = 1, 2, \cdots)$. Now let $b^{(1)}, \cdots, b^{(k)}$ be k different elements of $T_{j\nu}$; then $b^{(\rho)}b^{(\sigma)} = 0$ for $\rho, \sigma = 1, \cdots, k$, $\rho \neq \sigma$, and so

$$\frac{k}{\nu} \leqq \sum_{\rho=1}^{k} \mu(b^{(\rho)}\Sigma(\tilde{a}_i;\ i = 1, \cdots, j))$$

$$= \mu\big(\Sigma(b^{(\rho)}\Sigma(\tilde{a}_i;\ i = 1, \cdots, j);\ \rho = 1, \cdots, k)\big)$$

$$\leqq \mu(\Sigma(\tilde{a}_i;\ i = 1, \cdots, j)) \leqq \sum_{i=1}^{j} \mu(\tilde{a}_i),$$

whence $k \leqq \nu \sum_{i=1}^{j} \mu(\tilde{a}_i) < +\infty$. Thus $T_{j\nu}$ is finite, and therefore T, which is the set-theoretical sum of all the $T_{j\nu}$, is finite or denumerably infinite. Thus the power of T is less than $\aleph_1$, and Theorem A.6 yields that L_μ/I_μ is a continuous Boolean algebra.

NOTE: The class L_μ/I_μ of "all measurable sets up to sets of measure zero" is important, since it may be considered as the basic system for the logical treatment of classical mechanics — with better reasons than L or L_μ. (Cf. J. v. Neumann, *Annals of Mathematics*, vol. 33 (1932), pages 595—598, and G. Birkhoff and J. v. Neumann, *Annals of Mathematics*, vol. 37 (1936), page 825.)

If P is a topological space of the usual structure, then for every $a \epsilon L_\mu$ there exists $u \epsilon I_\mu$ such that $a' = a + u$ is a Borel set in P. Since $a' \sim a$ (mod I_μ), this means that every element of L_μ/I_μ, i.e., every equivalence class, contains an element of L_b, the class of all Borel subsets of L_μ (cf. Example 3 below; L_b is also an $\aleph_1$-Boolean algebra). Let $I_{\mu b}$ be the intersection of L_b and I_μ. Then $I_{\mu b}$ is clearly an $\aleph_1$ ideal in L_b, and by the above note each element of L_μ/I_μ contains exactly one element of $L_b/I_{\mu b}$, and each element of $L_b/I_{\mu b}$ is contained in exactly one element of L_μ/I_μ. Therefore L_μ/I_μ and $L_b/I_{\mu b}$ are lattice-isomorphic. Thus we might consider $L_b/I_{\mu b}$ instead of L_μ/I_μ.

EXAMPLE 3: Let P be a metric, separable and complete space. (Cf. for example F. Hausdorff, *Mengenlehre*, Berlin, 1927, particularly pages 94, 124, 129, where all topological notions which we use are explained in detail.) The class L_b of all Borel subsets of P is a subsystem of L which possesses the properties stated at the beginning of Example 2 for L_μ i.e., L_b is an $\aleph_1$-Boolean algebra.

A subset of P is *nowhere dense*, in case its closure possesses no inner points; it is of *first category* in case it is the set-theoretical sum of a finite or denumerably infinite class of nowhere dense sets. (Cf. Hausdorff, loc. cit., pages 138—145.) P itself is not of first category since it is supposed to be non-empty; cf. Hausdorff, loc. cit., page 142. The class I_{cb} of all Borel sets of first category is clearly an $\aleph_1$-ideal in L_b. Therefore by Theorem A.5, L_b/I_{cb} is defined and is an $\aleph_1$-Boolean algebra. Moreover, L_b, I_{cb} satisfy the condition (‡) in Theorem A.6, as we shall now prove in three steps.

(i) For every closed set $a \in L_b$ there exists an open set $a' \in L_b$ with $a \sim a'$ (mod I_{cb}).

Proof: Let a' be the set of inner points of a, u the complement of a' in a. Then $a = a' + u$, a is closed, a' is open, and therefore u is closed; u is clearly nowhere dense, whence $u \in I_{cb}$, $a \sim a'$ (mod I_{cb}).

(ii) For each $a \in L_b$ there exists an open set $a^* \in L_b$ with $a \sim a^*$ (mod I_{cb}).

Proof: Denote the set of all $a \in L_b$ such that there exists an open set $a^* \in L_b$ for which $a \sim a^*$ (mod I_{cb}) by L'. Clearly every open set is in L'. If $a_1, a_2, \cdots \in L'$, then $\Sigma(a_1, a_2, \cdots) \in L'$. For, if a_i^* corresponds to a_i, then $a^* \equiv \Sigma(a_1^*, a_2^*, \cdots)$ is open too, and will be effective as $(\Sigma(a_1, a_2, \cdots))^*$. If $a \in L'$, then the complement $-a$ of a is in L' also. For, if a^* is open, $a \sim a^*$ (mod I_{cb}), then $-a \sim -a^*$ (mod I_{cb}), and since $-a^*$ is closed, we have by (i) the existence of an open set $\overline{a^*}$ with $-a^* \sim \overline{a^*}$ (mod (I_{cb}), whence $-a \sim \overline{a^*}$ (mod I_{cb}). If $a_1, a_2, \cdots \in L'$, then $\Pi(a_1, a_2, \cdots) \in L'$, since $\Pi(a_1, a_2, \cdots) = -\Sigma(-a_1, -a_2, \cdots)$. Thus all Borel sets are in L', i.e., $L' = L_b$, which is the desired result.

(iii) Condition (‡) is satisfied by L_b, I_{cb}.

Proof: Let $T \subset L_b$, and suppose that $a, b \in T$, $a \neq b$ implies $ab = 0$, and that $a \in T$ implies $a \notin I_{cb}$. For each $a \in T$ there is an open set a^* with $a \sim a^*$ (mod I_{cb}) by (ii). Define $T^* \equiv (a^*;\ a \in T)$. Since $a^* \sim a \nsim 0$ (mod I_{cb}), $a^* \neq 0$. If $a^*, b^* \in T^*$, $a^* \neq b^*$, then $a^* \sim a$, $b^* \sim b$ (mod I_{cb}) with $a, b \in T$, $a \neq b$, and hence $a^*b^* \sim ab = 0$ (mod I_{cb}), $a^*b^* \in I_{cb}$. But a^*b^* is an open set, whence it must be empty if it is of first category (cf. loc. cit., page 142); hence $a^*b^* = 0$. Thus T^* is a class of mutually disjoint, non-empty, open sets in the separable space P, and consequently T^* is finite or denumerably infinite. Therefore power of T = power of $T^* < \aleph_1$, and (‡) is established.

Theorem A.6 now applies and yields the result that L_b/I_{cb} is a continuous Boolean algebra.

The continuous Boolean algebras described in Example 1 are called *atomistic* (cf. A. Tarski, Fundamenta Mathematica, vol. 24 (1935), pages 191—198. In Appendix 2 we shall define the term "atomistic" differently, but we shall prove the equivalence of the two definitions). In a certain sense they may be considered the simplest kind of continuous Boolean algebras, because all finite Boolean algebras are such, as will be proved. We shall see that Examples 2, 3 are not instances of Example 1, and we may conclude from this that infinite Boolean algebras may be essentially more complicated than finite Boolean algebras. The importance of atomistic Boolean algebras for the "reduction theory" of lattices in general, and of continuous geometries in particular, consists in that they characterize the case of "direct decomposability into irreducible components". This will be discussed in Appendix 2.

Characteristic differences between the continuous Boolean algebras of our three examples can be best formulated in terms of certain generalized forms of the distributive law. We shall state and discuss six generalized distributive laws (A)—(F).

First, Boolean algebras were defined by the common distributive law (IV_1):

(A) $$(a + b)c = ac + bc.$$

We found that the continuous Boolean algebras satisfy automatically the more general distributive laws (Theorem A.2 and its Corollary)

(B) $$\Sigma(S_1) \cdot \ldots \cdot \Sigma(S_2) = \Sigma(a_1 \cdot \ldots \cdot a_n;\ a_i \in S_i\ (i = 1, \cdots, n)),$$

(C) $$\Pi(a_1 + \cdots + a_n;\ a_i \in S_i\ (i = 1, \cdots, n)) = \Pi(S_1) + \cdots + \Pi(S_n).$$

The most general "distributive law" is the following (cf. A. Tarski, loc. cit., page 195, "Postulate $\mathscr{D}_3$"):

(D) $$\Pi(\Sigma(a_\alpha^\rho;\ \alpha \in K_\rho);\ \rho \in H) = \Sigma(\Pi(a_{\alpha(\rho)}^\rho;\ \rho \in H);\ \alpha(\rho) \in K_\rho(\rho \in H)).$$

A special case of (D) of some importance arises by putting all $K_\rho \equiv (1, 2)$:

(E) $$\Pi(a_1^\rho + a_2^\rho;\ \rho \in H) = \Sigma(\Pi(a_{\alpha(\rho)}^\rho;\ \rho \in H);\ \alpha(\rho) = 1,\ 2\ (\rho \in H)).$$

Obviously (B) is a special case of (D). That (C) is also a "distributive law", i.e. is a special case of (D), can be seen in the following manner. According to (D) the right member of (C) should be

$$\Sigma(\Pi(a_{l(a_1,\ldots,a_n)};\ a_i \in S_i\ (i = 1, \cdots, n));\ l(a_1, \cdots, a_n) = 1, \cdots, n$$
$$\text{for every } (a_1, \cdots, a_n) \text{ with } a_i \in S_i\ (i = 1, \cdots, n)).$$

Among these terms $\Pi(a_{l(a_1,\ldots,a_n)};\ a_i \in S_i \quad (i = 1, \cdots, n))$ the $\Pi(S_i)$ $(i = 1, \cdots, n)$ occur, as is seen by putting $l(a_1, \cdots, a_n) \equiv i$ identically. Consider now an arbitrary term $\Pi(a_{l(a_1,\ldots,a_n)};\ a_i \in S_i \quad (i = 1, \cdots, n))$, and an integer $i = 1, \cdots, n$. If for each $a_i \in S_i$ there exist $a_j \in S_j$ $(j \neq i)$ such that $l(a_1, \cdots, a_n) = i$, then this term is $\leqq \Pi(a_i;\ a_i \in S_i) = \Pi(S_i) \leqq \Pi(S) + \cdots + \Pi(S_n)$. Hence if our term is not $\leqq \Pi(S_1) + \cdots + \Pi(S_n)$, then for each $i = 1, \cdots, n$ there exists $a_i^0 \in S_i$ such that for every $a_j \in S_j$ $(j \neq i)$ we have $l(a_1, \cdots, a_n) \neq i$ if $a_i = a_i^0$. Hence $l(a_1^0, \cdots, a_n^0)$ is different from all numbers $1, \cdots, n$, which is impossible. Thus each of our terms is $\leqq \Pi(S_1) + \cdots + \Pi(S_n)$, and hence their sum, i.e. the right member of (C), is $\Pi(S_1) + \cdots + \Pi(S_n)$.

The continuous Boolean algebras of Example 1 satisfy (D), and hence also (E), as is immediately verified. Conversely, every continuous Boolean algebra which satisfies (E) is atomistic, that is, one of the type of Example 1 (cf. A. Tarski,loc. cit., pages 195—197, "Postulate $\mathscr{D}_3$", and "Theorems 5, 6"; also Theorem B.3 in our Appendix 2). Hence (E) (or equally well, (D)) is characteristic of the atomistic among all continuous Boolean algebras.

If L is finite, then each K_ρ, and hense H, may be assumed finite in (D). Then (D) becomes a special case of (B) (in fact, an iteration of (A)), and hence it necessarily holds. Thus the finite Boolean algebras are always atomistic; this fact may be easily established by direct methods, of course.

As we shall now show, the continuous Boolean algebras of Examples 2 and 3 violate (E) (and so *a fortiori* (D)) and are therefore non-atomistic. Define P to be the set of all real numbers $x \geqq 0, < 1$, μ^* to be the common Lebesgue outer measure, $H \equiv (1, 2, \cdots)$, a_α^ρ to be the equivalence class of the set of all numbers $x \geqq 0, < 1$ with ρ^{th} dyadic digit $= \alpha$ $(= 0,1)$. Since $a_0^\rho + a_1^\rho$ is the equivalence class of $P = 1$, the left side of (E) is 1. But each $\Pi(a_{\alpha(\rho)}^\rho;\ \rho = 1, 2, \cdots)$ is the equivalence class of a set with only one point, which is in I_μ and I_{cb} in the two respective examples, and therefore $\Pi(a_{\alpha(\rho)}^\rho;\ \rho = 1, 2, \cdots) = 0$. Hence the right side of (E) is 0.

A special case of (D) which can be used to distinguish Examples 2 and 3 is the following:

(F) Put in (D) $H \equiv K_1 = K_2 = \cdots = (1, 2, \cdots)$, and assume $a_\alpha^\rho \leqq a_\beta^\rho$ if $\alpha \leqq \beta$.

Indeed, Example 2 satisfies (F): Since the left member of (D) is clearly $\geqq$ the right member, it suffices to prove that the sum of a suitable denumerable number of terms on the right side is equal to the left side. Hence we may consider elements of L_μ, and not of L_μ/I_μ for the a_α^ρ. (For non-denumerably infinite sums this procedure, which is based on the note after Theorem A.5, would not be legitimate, since I_μ is only an $\aleph_1$-ideal.) We must show that $\Pi(\Sigma(a^\rho_\alpha;\ \alpha=1,2,\cdots);\ \rho=1,2,\cdots)$ differs from $\Sigma'(\Pi(a^\rho_{\alpha(\rho)};\ \rho=1,2,\cdots))$ only by a set of measure zero, Σ' being extended over a suitable denumerable set of terms. It will do just as well to compare $\tilde{a}_i\Pi(\Sigma(a^\rho_\alpha;\ \alpha=1,2,\cdots);\ \rho=1,2,\cdots)$ and $\tilde{a}_i\Sigma'(\Pi(a^\rho_{\alpha(\rho)};\ \rho=1,2,\cdots)$, where Σ' may depend on $i=1,2,\cdots$, since $\Sigma(\tilde{a}_i;\ i=1,2,\cdots)=P$ (the $\tilde{a}_i$, $i=1,2,\cdots$ being defined as in Example 2). Since $\tilde{a}_i\Pi(\Sigma(a^\rho_\alpha;\ \alpha=1,2,\cdots);\ \rho=1,2,\cdots)\geqq\tilde{a}_i\Pi(a^\rho_{\alpha(\rho)};\ \rho=1,2,\cdots)$, and since $\mu(\tilde{a}_i)$ is finite, it suffices to find for each $\varepsilon>0$ a term $\tilde{a}_i\Pi(a^\rho_{\alpha(\rho)};\ \rho=1,2,\cdots)$ with measure $\geqq\mu(\tilde{a}_i\Pi(\Sigma(a^\rho_\alpha;\ \alpha=1,2,\cdots);\ \rho=1,2,\cdots))-\varepsilon$, for then these terms for $\varepsilon=1,\frac{1}{2},\frac{1}{3},\cdots$ give together desired Σ'.

Assume that for an integer $\sigma=1,2,\cdots$ the $\alpha(\rho)$, $\rho<\sigma$ have been chosen and that

$$\mu(\tilde{a}_i\cdot a^1_{\alpha(1)}\cdot\ldots\cdot a^{\sigma-1}_{\alpha(\sigma-1)}\cdot\Pi(\Sigma(a^\rho_\alpha;\ \alpha=1,2,\cdots);\ \rho=\sigma,\sigma+1,\cdots)$$
$$>\mu(\tilde{a}_i\cdot\Pi(\Sigma(a^\rho_\alpha;\ \alpha=1,2,\cdots);\ \rho=1,2,\cdots))-\varepsilon.$$

Then since

$$\mu(\tilde{a}_i a^1_{\alpha(1)}\cdot\ldots\cdot a^{\sigma-1}_{\alpha(\sigma-1)}\cdot\Pi(\Sigma(a^\rho_\alpha;\ \alpha=1,2,\cdots);\ \rho=\sigma,\sigma+1,\cdots)$$
$$=\lim_{\alpha(\sigma)\to\infty}\mu(\tilde{a}_i\cdot a^1_{\alpha(1)}\cdot\ldots\cdot a^{\sigma-1}_{\alpha(\sigma-1)}\cdot a^\sigma_{\alpha(\sigma)}\cdot\Pi(\Sigma(a^\rho_\alpha;\alpha=1,2,\cdots);\rho=\sigma+1,\cdots)),$$

there exists $\alpha(\sigma)$ with

$$\mu(\tilde{a}_i\cdot a^1_{\alpha(1)}\cdot\ldots\cdot a^{\sigma-1}_{\alpha(\sigma-1)}\cdot a^\sigma_{\alpha(\sigma)}\cdot\Pi(\Sigma(a^\rho_\alpha;\ \alpha=1,2,\cdots);\ \rho=\sigma+1,\cdots))$$
$$>\mu(\tilde{a}_i\cdot\Pi(\Sigma(a^\rho_\alpha;\ \alpha=1,2,\cdots);\ \rho=1,2,\cdots))-\varepsilon.$$

Thus there is an infinite sequence $\alpha(\rho)$, $\rho=1,2,\cdots$ such that for all $\rho=1,2,\cdots$

$$\mu(\tilde{a}_i\cdot a^1_{\alpha(1)}\cdot\ldots\cdot a^{\sigma-1}_{\alpha(\sigma-1)})>\mu(\tilde{a}_i\cdot\Pi(\Sigma(a^\rho_\alpha;\alpha=1,2,\cdots);\rho=1,2,\cdots))-\varepsilon.$$

Now if σ tends to ∞, we have

$$\mu(\tilde{a}_i\cdot\Pi(a^\rho_{\alpha(\rho)};\ \rho=1,2,\cdots))=\lim_{\sigma\to\infty}\mu(\tilde{a}_i\cdot a^1_{\alpha(1)}\cdot\ldots\cdot a^{\sigma-1}_{\alpha(\sigma-1)})$$
$$\geqq\mu(\tilde{a}_i\cdot\Pi(\Sigma(a^\rho_\alpha;\ \alpha=1,2,\cdots);\ \rho=1,2,\cdots))-\varepsilon.$$

This completes the proof.

We see that Example 3 violates (F) thus: Let $(\xi_\rho;\ \rho=1,2,\cdots)$ be a sequence of elements of P which is everywhere dense in P. Let $S_{\alpha,\rho}$, α, $\rho=1,2,\cdots$, be the open sphere of radius $1/\alpha$ and center ξ_ρ, and let $C_{\alpha,\rho}$ be its complement. Let a^α_ρ be the equivalence class of $C_{\alpha,\rho}$ (in L_b/I_{cb}). Clearly $S_{1,\rho}\supset S_{2,\rho}\supset\cdots$, $C_{1,\rho}\subset C_{2,\rho}\subset\cdots$, $a^1_\rho\leqq a^2_\rho\leqq\cdots$

Now $\Sigma(a_\alpha^\rho;\ \rho = 1, 2, \cdots)$ is the equivalence class of the sum of all $C_{1,\rho}$, $C_{2,\rho}, \cdots$, that is, of the complement of the intersection of all $S_{1,\rho}, S_{2,\rho}, \cdots$, which is ξ_ρ. Hence it is equal to the equivalence class of P, which is 1. Therefore the left side of (F) is 1. Every $\Pi(a_{\alpha(\rho)}^\rho;\ \rho=1, 2, \cdots)$ is the equivalence class of the intersection of $C_{\alpha(1)}^{(1)}, C_{\alpha(2)}^{(2)}, \cdots$, which is a closed set, and contains no $\xi_1, \xi_2, \cdots$; hence it is nowhere dense. Therefore its equivalence class is 0, and so the right member of (F) is 0, whence (F) does not hold.

APPENDIX 2

In this appendix we propose to discuss the "decomposability" of continuous geometries which are not necessarily irreducible, i.e., of systems L satisfying Axioms I—V, but not necessarily VI.

DEFINITION B.1: If I is an arbitrary set (of indices α), a system $(\bar{a}_\alpha;\ \alpha \in I)$, where $\bar{a}_\alpha \in L$, $\bar{a}_\alpha \neq 0$ $(\alpha \in I)$, is a *direct sum decomposition*, or more briefly, a *decomposition, of* L if it possesses the following property: For every $x \in L$ there exists for each $\alpha \in I$ one and only one $u_\alpha \in L(0, \bar{a}_\alpha)$ (i.e., $u_\alpha \leqq \bar{a}_\alpha$) such that $x = \Sigma(u_\alpha;\ \alpha \in I)$. We then call $(u_\alpha;\ \alpha \in I)$ the *decomposition* of x.

NOTE: One readily verifies that if I has two elements, then this definition coincides with the definition of direct sum decomposition in Part I, Definition 5.2.

LEMMA B.1: Let $\bar{a}_\alpha$, $\alpha \in I$ be a decomposition of L. Then
(a) if x, y have the respective decompositions u_α, $\alpha \in I$ and v_α, $\alpha \in I$, then $x \leqq y$ is equivalent to $u_\alpha \leqq v_\alpha$ for every $\alpha \in I$;
(b) if for each $\beta \in J$ there is defined an element $x^\beta \in L$, and if x^β has the decomposition u_α^β, $\alpha \in I$, then $\Sigma(x^\beta;\ \beta \in J)$, $\Pi(x^\beta;\ \beta \in J)$ have the respective decompositions $\Sigma(u_\alpha^\beta;\ \beta \in J)$, $\alpha \in I$, and $\Pi(u_\alpha^\beta;\ \beta \in J)$, $\alpha \in I$;
(c) $u_\alpha = \bar{a}_\alpha x$ for every $\alpha \in I$ if x has decomposition u_α, $\alpha \in I$.

PROOF: (a) Since $x = \Sigma(u_\alpha; \alpha \in I)$, $y = \Sigma(v_\alpha;\ \alpha \in I)$ implies $x + y = \Sigma(u_\alpha + v_\alpha;\ \alpha \in I)$, and u_α, $v_\alpha \leqq \bar{a}_\alpha$ implies $u_\alpha + v_\alpha \leqq \bar{a}_\alpha$, it follows that $u_\alpha + v_\alpha$, $\alpha \in I$ is the decomposition of $x + y$. Now $x \leqq y$ means $x + y = y$, and hence is equivalent to $u_\alpha + v_\alpha = v_\alpha$ for every $\alpha \in I$, i.e., to $u_\alpha \leqq v_\alpha$ for every $\alpha \in I$.

(b) This follows from (a) by virtue of the definitions of Σ, Π in terms of $\leqq$, and the fact that the u_α vary over all $L(0, \bar{a}_\alpha)$ while x varies over all L.

(c) We have for any given $\beta \in I$,

$$x = x + \bar{a}_\beta x = \Sigma(u_\alpha; \alpha \in I) + \bar{a}_\beta x = \Sigma(u_\alpha;\ \alpha \in I, \alpha \neq \beta) + u_\beta + \bar{a}_\beta x.$$

Hence if we define

$$u'_\alpha \equiv \begin{cases} u_\beta + \bar{a}_\beta x & (\alpha = \beta) \\ u_\alpha & (\alpha \neq \beta), \end{cases}$$

then $x = \Sigma(u'_\alpha;\ \alpha \in I)$, and $u'_\alpha \leqq \bar{a}_\alpha$ $(\alpha \in I)$. Therefore $u_\alpha = u'_\alpha$ and in particular $u_\beta = u'_\beta \geqq \bar{a}_\beta x$. On the other hand $u_\beta \leqq \bar{a}_\beta$, x, whence $u_\beta \leqq \bar{a}_\beta x$. Therefore $u_\beta = \bar{a}_\beta x$.

THEOREM B.1: The system $\bar{a}_\alpha$, $\alpha \in I$ is a decomposition of L if and only if α) $(\bar{a}_\alpha;\ \alpha \in I) \perp$, β) $\Sigma(\bar{a}_\alpha;\ \alpha \in I) = 1$, γ) $\bar{a}_\alpha \in Z$, $\bar{a}_\alpha \neq 0$ for $\alpha \in I$.

PROOF: If α), β), γ) hold, then Theorem 1.1 in Part III yields that for every $x \in L$

$$x = 1 \cdot x = \Sigma(\bar{a}_\alpha;\ \alpha \in I) \cdot x = \Sigma(\bar{a}_\alpha x;\ \alpha \in I) = \Sigma(u_\alpha;\ \alpha \in I),$$

with $u_\alpha \equiv \bar{a}_\alpha x \leqq \bar{a}_\alpha$. Moreover, if $x = \Sigma(u'_\alpha;\ \alpha \in I)$, $u'_\alpha \leqq \bar{a}_\alpha$, then Theorem 1.1 in Part III and α) yield

$$\begin{aligned} \bar{a}_\beta x &= \bar{a}_\beta \Sigma(u'_\alpha;\ \alpha \in I) \\ &= \Sigma(u'_\alpha \bar{a}_\beta;\ \alpha \in I) \\ &= u'_\beta, \end{aligned}$$

whence $u'_\beta = a_\beta x$. This proves the uniqueness of the u_α, whence $\bar{a}_\alpha$, $\alpha \in I$ is a decomposition.

Conversely, let $\bar{a}_\alpha$, $\alpha \in I$ be a decomposition of L. Then there exist (for $x = 1$) $\bar{u}_\alpha \leqq \bar{a}_\alpha$ such that

$$1 = \Sigma(\bar{u}_\alpha;\ \alpha \in I) \leqq \Sigma(\bar{a}_\alpha;\ \alpha \in I),$$

whence $\Sigma(\bar{a}_\alpha;\ \alpha \in I) = 1$, and β) is established. For $J \subset I$ define $x \equiv \Sigma(\bar{a}_\beta;\ \beta \in J)$. Then if we define

$$u_\alpha \equiv \begin{cases} \bar{a}_\alpha & (\alpha \in J) \\ 0 & (\alpha \notin J), \end{cases}$$

u_α, $\alpha \in I$ is the decomposition of x, whence Lemma B.1 yields $\Sigma(\bar{a}_\beta;\ \beta \in J) \cdot \bar{a}_\alpha = 0$ if $\alpha \notin J$. Hence if $\alpha_1, \cdots, \alpha_k$ are k different elements of I, then for each $i = 1, \cdots, k-1$ we may define $J \equiv (\alpha_1, \cdots, \alpha_{i-1})$, $\alpha \equiv \alpha_i$, and we have $(\bar{a}_{\alpha_1} + \cdots + \bar{a}_{\alpha_{i-1}})\,\bar{a}_{\alpha_i} = 0$. Therefore $(\bar{a}_{\alpha_1}, \cdots, \bar{a}_{\alpha_k}) \perp$ by Part I, Theorem 2.2, and hence $(\bar{a}_\alpha;\ \alpha \in I) \perp$ by Part I, Theorem 2.3. This proves α). Finally, let $x \in L$. By Lemma B.1,

$$x = \Sigma(u_\alpha;\ \alpha \in I) = \Sigma(\bar{a}_\alpha x;\ \alpha \in I).$$

Replacing x by $\Sigma(\bar{a}_\alpha;\ \alpha \in I,\ \alpha \neq \beta)x$, we have by α)

$$\Sigma(\bar{a}_\alpha;\ \alpha \in I,\ \alpha \neq \beta)x = \Sigma(\bar{a}_\alpha x;\ \alpha \in I,\ \alpha \neq \beta).$$

By combining these results, we find

$$x = \Sigma(\bar{a}_\alpha;\ \alpha \in I,\ \alpha \neq \beta)x + \bar{a}_\beta x.$$

Now $\bar{a}_\beta$ and $\Sigma(\bar{a}_\alpha;\ \alpha \in I, \alpha \neq \beta)$ are inverses by α), β), whence the equation just established states that $(\bar{a}_\beta,\ \Sigma(\bar{a}_\alpha;\ \alpha \in I,\ \alpha \neq \beta))D$, and by Part I, Theorem 5.3 (a) that $(\bar{a}_\beta)D$. Hence $\bar{a}_\beta \in Z(\beta \in I)$, and γ) is proved.

Definition B.2: An element $\bar{a} \neq 0$ of a continuous Boolean algebra M is *atomistic* in case $\bar{b} \in M$, $\bar{b} \leqq \bar{a}$ implies $\bar{b} = 0$ or $\bar{b} = \bar{a}$. The set M_{at} is the set of all atomistic elements $\bar{a} \in M$; M is *atomistic* in case $\Sigma(M_{\mathrm{at}}) = 1$. (Cf. A. Tarski, Fundamenta Mathematica vol. **24** (1935), pages 191—192, "Definition C" and "Postulate D_1".)

Lemma B.2: If M is a continuous Boolean algebra, $M_{\mathrm{at}} \perp$.

Proof: We must prove that $J,\ K \subset M_{\mathrm{at}}$, $\mathfrak{P}(J,\ K) = \Theta$ implies $\Sigma(J) \cdot \Sigma(K) = 0$. By the Corollary to Theorem A.2 (in Appendix 1), $\Sigma(J)\Sigma(K) = \Sigma(\bar{a}\bar{b};\ \bar{a} \in J,\ \bar{b} \in K)$, whence it suffices to show that $\bar{a},\ \bar{b} \in M_{\mathrm{at}}$, $\bar{a} \neq \bar{b}$ implies $\bar{a}\bar{b} = 0$. If $\bar{a}\bar{b} \neq 0$, then $\bar{a}\bar{b} \leqq \bar{a},\ \bar{b}$ implies $\bar{a} = \bar{a}\bar{b} = \bar{b}$ contrary to $\bar{a} \neq \bar{b}$. Hence $\bar{a} \neq \bar{b}$ implies $\bar{a}\bar{b} = 0$.

Lemma B.3: Let M be the center Z of L. Then $L(0,\ \bar{a})$, $\bar{a} \in Z$, is irreducible (satisfies Axiom VI) if and only if $\bar{a}$ is atomistic.

Proof: We know that $L(0,\ \bar{a})$ is irreducible if and only if its center consists of 0, $\bar{a}$ only. By Theorem 1.6 this means that $\bar{a}\bar{c} = 0$ or $\bar{a}$ for every $\bar{c} \in Z$, i.e., $\bar{b} = 0$ or $\bar{a}$ for every $\bar{b} \leqq \bar{a}$, $\bar{b} \in Z$, i.e. that $\bar{a}$ is atomistic.

Theorem B.2: L possesses a decomposition $\bar{a}_\alpha$, $\alpha \in I$ for which all $L(0,\ \bar{a}_\alpha)$, $\alpha \in I$, are irreducible if and only if Z is atomistic. In this case the set $(\bar{a}_\alpha;\ \alpha \in I)$ is Z_{at}.

Proof: A decomposition of the desired sort is characterized by conditions: α)—γ) in Theorem B.1, together with $\bar{a}_\alpha \in Z_{\mathrm{at}}$ for every $\alpha \in I$, by Lemma B.3. If $(\bar{a}_\alpha; \alpha \in I) = Z_{\mathrm{at}}$, then the last condition is automatically satisfied as is also γ). Moreover, α) follows from Lemma B.2, and β) means that $\Sigma(Z_{\mathrm{at}}) = 1$, i.e., that Z is atomistic. Hence the condition is sufficient. Conversely, suppose that the $\bar{a}_\alpha$, $\alpha \in I$ give a decomposition of the desired sort. Then $\Sigma(Z_{\mathrm{at}}) \geqq \Sigma(\bar{a}_\alpha;\ \alpha \in I) = 1$, whence $\Sigma(Z_{\mathrm{at}}) = 1$, and Z is atomistic. This proves the necessity of the condition. Finally, $(\bar{a}_\alpha;\ \alpha \in I) \subset Z_{\mathrm{at}}$. If $(\bar{a}_\alpha;\ \alpha \in I) \neq Z_{\mathrm{at}}$ then there exists $\bar{b} \in Z_{\mathrm{at}}$ such that $\bar{b} \neq \bar{a}_\alpha$ $(\alpha \in I)$. Thus by Lemma B.2, $\Sigma(\bar{a}_\alpha;\ \alpha \in I) \cdot \bar{b} = 0$. But the left side is $1 \cdot \bar{b} = \bar{b} \neq 0$, and we have reached a contradiction. Hence $(\bar{a}_\alpha;\ \alpha \in I) = Z_{\mathrm{at}}$.

The decomposition of L into irreducible parts $L(0,\ \bar{a})$ is very desirable since it permits us to describe L in an elementary manner (by means of Lemma B.1) by irreducible continuous geometries. Theorem B.2 shows that if the decomposition is to be accomplished at all, it can be done in one way only, and that a necessary and sufficient condition for this is the atomistic character of Z. We shall now prove a double criterion for the atomistic character of Z, which establishes connections with the analysis of Appendix 1. (Cf. the remarks made after the "generalized distributive law" (E) in Appendix 1.)

THEOREM B.3: Either of the two following conditions is necessary and sufficient that a continuous Boolean algebra M be atomistic:

(a) M is isomorphic to a continuous Boolean algebra of Example 1 (i.e., there exists a set P such that M is isomorphic to the algebra of all subsets of P).

(b) The general distributive law (E) (in Appendix 1) holds in M.

PROOF: We shall prove that M atomistic implies (a), (a) implies (b), (b) implies M atomistic.

Suppose M is atomistic, i.e., $\Sigma(M_{\mathrm{at}}) = 1$. For every $\bar{a} \in M$ we have

$$\bar{a} = \bar{a} \cdot 1 = \bar{a} \cdot \Sigma(M_{\mathrm{at}}) = \Sigma(\bar{a}\bar{b};\ \bar{b} \in M_{\mathrm{at}}) = \Sigma(\bar{a}\bar{b};\ \bar{b} \in M_{\mathrm{at}},\ \bar{a}\bar{b} \neq 0).$$

Now if $\bar{b} \in M_{\mathrm{at}}$, $\bar{a}\bar{b} \neq 0$, then since $\bar{a}\bar{b} \leqq \bar{b}$, we have $\bar{a}\bar{b} = \bar{b}$, $\bar{b} \leqq \bar{a}$. Conversely, $\bar{b} \in M_{\mathrm{at}}$, $\bar{b} \leqq \bar{a}$ yields $\bar{a}\bar{b} = \bar{b} \neq 0$. Hence $\bar{a} = \Sigma(\bar{b};\ \bar{b} \in M_{\mathrm{at}}, \bar{b} \leqq \bar{a})$. Thus the $\Sigma(S)$, $S \subset M_{\mathrm{at}}$ exhaust M. If S, $T \subset M_{\mathrm{at}}$, then $S \subset T$ implies $\Sigma(S) \leqq \Sigma(T)$. Assume conversely $\Sigma(S) \leqq \Sigma(T)$. If $\bar{a} \in S$ but $\bar{a} \notin T$, then by Lemma B.2, $\Sigma(T) \cdot \bar{a} = 0$. But $\Sigma(T) \cdot \bar{a} \geqq \Sigma(S)\bar{a} = \bar{a}$ (since $\bar{a} \leqq \Sigma(S)$) which implies $\bar{a} = 0$, contrary to $\bar{a} \in M_{\mathrm{at}}$. Hence $\bar{a} \in S$ implies $\bar{a} \in T$, i.e., $S \subset T$. Thus $S \subset T$ if and only if $\Sigma(S) \leqq \Sigma(T)$. Interchanging S and T, and combining the two statements, we have $S = T$ if and only if $\Sigma(S) = \Sigma(T)$. Thus the correspondence $S \rightleftarrows \Sigma(S)$ of all $S \subset M_{\mathrm{at}}$ with all $\bar{a} \in M$ is one-to-one. Since $S \subset T$ is equivalent to $\Sigma(S) \leqq \Sigma(T)$, this correspondence is a lattice-isomorphism. Hence the continuous Boolean algebra of Example 1 in Appendix 1 is lattice-isomorphic to M, if we put $P = M_{\mathrm{at}}$.

The statement: (a) implies (b), is obvious.

Suppose now that (b) holds. Apply (E) with $H = M$, $a_1^a = a$, $a_2^a = -a$ for $a \in H$. Then $a_1^a + a_2^a = a + (-a) = 1$, whence the left member of (E) is 1. Thus (E) becomes

$$\begin{aligned} 1 &= \Sigma(\Pi(a^a_{\alpha(a)};\ a \in M);\ \alpha(a) = 1,\ 2\ \ (a \in M)) \\ &= \Sigma\big(\Pi(b(a);\ a \in M);\ b(a) = a \text{ or } -a\ \ (a \in M)\big) \\ &= \Sigma\big(\Pi(b(a);\ a \in M);\ b(a) = a \text{ or } -a\ \ (a \in M),\ \Pi(b(a);\ a \in M) \neq 0\big). \end{aligned}$$

Now each term $\Pi(b(a);\ a \in M)$ (in the last expression) is necessarily atomistic. For, if $c \epsilon M$, we have $\Pi(b(a);\ a \epsilon M) \leqq b(c) = c$ or $-c$; hence if $c \leqq \Pi(b(a);\ a \epsilon M)$, then either $c = \Pi(b(a);\ a \epsilon M)$ or $c \leqq -c$, $c = c(-c) = 0$. Therefore $1 \leqq \Sigma(M_{\mathrm{at}})$, $\Sigma(M_{\mathrm{at}}) = 1$, and M is atomistic.

Since in all cases except those in which the center Z of L is atomistic there is no hope of obtaining a decomposition of L, it is necessary to find other methods of analyzing L. This will be done in the following chapters.

PART III · CHAPTER II

Transitivity of Perspectivity and Properties of Equivalence Classes

We are now in a position to apply the method of Theorem 5.14 in Part I to obtain a result which, while weaker than Theorem 5.14, is sufficient to establish the equivalence of the relations $\sim$ and $\approx$. The proof of Theorem 5.14 is essentially dependent upon Axiom VI, and the theorem is not true when Axiom VI is not assumed.

LEMMA 2.1: If $ab = 0$, there exist a', a'', b', b'' such that: a', a'' are inverses in a; b', b'' are inverses in b; $a' \sim b'$, $(a'', b'')D$, $a''b'' = 0$.

PROOF: Let Ω be the first ordinal corresponding to the first power greater than that of L. We shall proceed to construct two independent transfinite systems $(a_\alpha;\ \alpha < \alpha_0)$, $(b_\alpha;\ \alpha < \alpha_0)$, such that $a_\alpha \leqq a$, $b_\alpha \leqq b$ $(\alpha < \alpha_0 < \Omega)$ and having other properties which we shall give later.

Suppose a_β, b_β $(\beta < \alpha < \Omega)$ have been defined, and suppose $a_\beta \leqq a$, $b_\beta \leqq b$ $(\beta < \alpha)$. It is desired to define a_α, b_α if possible. Since $\Sigma(a_\beta;\ \beta < \alpha) \leqq a$, $\Sigma(b_\beta;\ \beta < \alpha) \leqq b$, there exist inverses $\bar{a}_\alpha$, $\bar{b}_\alpha$ of $\Sigma(a_\beta;\ \beta < \alpha)$ in a and of $\Sigma(b_\beta;\ \beta < \alpha)$ in b, respectively. If there exist $\bar{a}$, $\bar{b}$ with $\bar{a} \neq 0$, $\bar{b} \neq 0$, $\bar{a} \leqq \bar{a}_\alpha$, $\bar{b} \leqq \bar{b}_\alpha$, $\bar{a} \sim \bar{b}$, define $a_\alpha \equiv \bar{a}$, $b_\alpha \equiv \bar{b}$; otherwise we leave a_α, b_α undefined. If $\alpha' \leqq \Omega$ is such that all a_α, b_α are defined for $\alpha < \alpha'$, then clearly $\Sigma(a_\beta;\ \beta < \alpha) \cdot a_\alpha = 0$ for every $\alpha < \alpha'$. Let J_0 be a finite set of ordinals $\alpha < \alpha'$, and let the elements of J_0 (ordered increasingly) be denoted by $\alpha_1, \cdots, \alpha_n$. Then $(a_{\alpha_1} + \cdots + a_{\alpha_{i-1}})a_{\alpha_i} \leqq \Sigma(a_\beta;\ \beta < \alpha_i)a_{\alpha_i} = 0$ for $i = 2, \cdots, n$, so that $(a_{\alpha_1}, \cdots, a_{\alpha_n})\perp$, i.e., $(a_\alpha;\ \alpha \in J_0)\perp$. Thus Part I, Theorem 2.3 yields $(a_\alpha;\ \alpha < \alpha')\perp$. Similarly, $(b_\alpha;\ \alpha < \alpha')\perp$. Since $a_\alpha \neq 0$ $(\alpha < \alpha')$, all the a_α are distinct. Now it is impossible that a_α should be defined for every $\alpha < \Omega$, for otherwise the set of all the a_α would have greater power than L. Hence there exists a smallest α_0 for which a_{α_0}, b_{α_0} are undefined. Thus there exist no elements a_{α_0}, b_{α_0} such that $a_{\alpha_0} \neq 0$, $b_{\alpha_0} \neq 0$, $a_{\alpha_0} \leqq \bar{a}_{\alpha_0}$, $b_{\alpha_0} \leqq \bar{b}_{\alpha_0}$, $a_{\alpha_0} \sim b_{\alpha_0}$. By Part I, Theorem 5.7, we have then, $(\bar{a}_{\alpha_0}, \bar{b}_{\alpha_0})D$, $\bar{a}_{\alpha_0} \cdot \bar{b}_{\alpha_0} = 0$. Since $(a_\alpha;\ \alpha < \alpha_0)\perp$, $(b_\alpha;\ \alpha < \alpha_0)\perp$, $\Sigma(a_\alpha;\ \alpha < \alpha_0) \cdot \Sigma(b_\alpha;\ \alpha < \alpha_0) \leqq ab = 0$, we have $(a_\alpha, b_\alpha;\ \alpha < \alpha_0)\perp$ by

Part I, Theorem 2.1. But $a_\alpha \sim b_\alpha$ $(\alpha < \alpha_0)$, whence $\Sigma(a_\alpha;\ \alpha < \alpha_0) \sim \Sigma(b_\alpha;\ \alpha < \alpha_0)$ by Part I, Theorem 3.6. Now define

$$a' \equiv \Sigma(a_\alpha;\ \alpha < \alpha_0),\ a'' \equiv \bar{a}_{\alpha_0},\ b' \equiv \Sigma(b_\alpha;\ \alpha < \alpha_0),\ b'' \equiv \bar{b}_{\alpha_0}.$$

Then a', a'' are inverses in a; b', b'' are inverses in b; $a' \sim b'$, $(a'', b'')D$, $a''b'' = 0$.

THEOREM 2.1: For every a, $b \in L$, there exist a', a'', b', b'' such that: a', a'' are inverses in a; b', b'' are inverses in b; $a' \sim b'$; $(a'', b'')D$, $a''b'' = 0$.

PROOF: Let a_1 be an inverse of ab in a, and b_1 an inverse of ab in b. We have

$$ab \cdot a_1 = 0,\ (ab + a_1)b_1 = ab_1 = a \cdot bb_1 = ab \cdot b_1 = 0,$$

whence $(ab,\ a_1,\ b_1)\perp$ (Part I, Theorem 2.2). Since $a_1 b_1 = 0$, we may apply Lemma 2.1 to a_1, b_1; thus we have a_1', a_1'', b_1', b_1'', related to a_1, b_1 in the manner described in Lemma 2.1. Since $(ab,\ a_1,\ b_1)\perp$, $a_1' \leqq a_1$, $b_1' \leqq b_1$, we have $(ab,\ a_1',\ b_1')\perp$. Define $a' \equiv ab + a_1'$, $b' \equiv ab + b_1'$, $a'' \equiv a_1''$, $b'' \equiv b_1''$. Then by Part I, Theorem 3.5, $a' \sim b'$; moreover, $(a'', b'')D$, $a''b'' = 0$. Since

$$a_1' a'' = 0,\ ab(a_1' + a'') = aba_1 = 0,$$

it results that $(ab,\ a_1',\ a'')\perp$, whence $a'a'' = (ab + a_1')a'' = 0$. Moreover,

$$a' + a'' = (ab + a_1') + a'' = ab + (a' + a'') = ab + a_1 = a.$$

Therefore a', a'' are inverses in a. Similarly b', b'' are inverses in b. Hence a', a'', b', b'' have the desired properties.

THEOREM 2.2: $a \approx b$ is equivalent to $a \sim b$.

PROOF: Obviously $a \sim b$ implies $a \approx b$. Conversely, assume $a \approx b$. By Theorem 2.1 there exist a' a'', b', b'' related to a, b as described in Theorem 2.1. Since $(a'', b'')D$, $a''b'' = 0$, we have by Lemma 1.1 (or Theorem 1.5) that $e(a'')e(b'') = 0$. By Theorem 1.4 (c), $a \approx b$ yields $e(a'')a \approx e(a'')b$. Now

$$e(a'')b = e(a'')(b' + b'') = e(a'')b' + e(a'')b'' = e(a'')b' + 0 = e(a'')b',$$

and consequently $e(a'')a \approx e(a'')b'$. On the other hand, $a' \sim b'$, whence by Theorem 1.4(c), $e(a'')a' \sim e(a'')b'$. Thus $e(a'')a' \approx e(a'')a$. Since $a' \leqq a$, $e(a'')a' \leqq e(a'')a$, it follows that $e(a'')a' = e(a'')a$ by Part I, Theorem 4.4 Hence

$$a' \geqq e(a'')a' = e(a'')a \geqq e(a'')a'' = a'',$$

and so $a'' = a'a'' = 0$. Similarly $b'' = 0$. Hence $a = a'$, $b = b'$, and $a \sim b$, since $a' \sim b'$.

THEOREM 2.3: The relation $\sim$ is transitive.

PROOF: This is trivial by Theorem 2.2, since the relation $\approx$ is transitive.

We are now in a position to carry over several essential results from Part I, Chapter VI, without changing the proofs given there.

THEOREM 2.4:

(a) If $a \leqq c$, $b \leqq d$, $a \sim b$, $c \sim d$, and if a', b' are inverses of a, b in c, d respectively, then $a' \sim b'$.

(b) If $ab = 0$, $ef = 0$, $a \sim e$, $b \sim f$, then $a + b \sim e + f$.

PROOF: The proofs of (a) and (b) are literally those of Lemmas 6.3 and 6.5 in Part I, since those proofs depend only on the transitivity of the relation $\sim$, which we established here in Theorem 2.3.

COROLLARY: If $(a_i;\ i = 1, \cdots, n)\perp$, $(b_i;\ i = 1, \cdots, n)\perp$ are such that $a_i \sim b_i$ for $i = 1, \cdots, n$ then $\Sigma(a_i; i = 1, \cdots, n) \sim \Sigma(b_i; i = 1, \cdots, n)$.

PROOF: This follows by induction on $n = 1, 2, \cdots$ from Theorem 2.4 (b).

We continue the discussion, which is to a certain extent parallel to that one in the latter part of Part I, Chapter VI. The relation $\ll$, which will be defined presently (Definition 2.3) is however new and quite characteristic for the present general situation.

DEFINITION 2.1:

(a) $a \prec b$ means that there exists b^* such that $a \sim b^* < b$;

(b) $a \succ b$ means that there exists b^* such that $a \sim b^* > b$.

COROLLARY: The relations $\prec$, $\succ$ are dual to each other.

THEOREM 2.5:

(a) $a \succ b$ if and only if $b \prec a$.

(b) If $a \sim a'$, $b \sim b'$, then $a \prec b$ is equivalent to $a' \prec b'$.

(c) $a \prec b$, $b \prec c$ implies $a \prec c$.

PROOF: See Part I, Theorem 6.1. Note that (d) of that theorem is omitted here, since it does not hold.

DEFINITION 2.2: Let A_a denote the class of all elements x such that $x \sim a$, and let $\mathscr{L}$ denote the class of all A_a, $a \in L$.

COROLLARY: The system $(A_a;\ a \in L)$ is a mutually exclusive and exhaustive partition of L into subclasses. (Cf. Part I, Definition 6.2 and its Corollary.)

The elements of $\mathscr{L}$ will be denoted by A, B, C, $\cdots$.

DEFINITION 2.3: $a \ll b$ (*a strongly less than b*) means that for every $\bar{a} \in Z$ either $\bar{a}a \prec \bar{a}b$ or $\bar{a}a = \bar{a}b = 0$; $a \gg b$ means $b \ll a$.

COROLLARY: If $a \ll b$, then $a \prec b$ or $a = b = 0$.

PROOF: This results by putting $\bar{a} = 1$ in Definition 2.3.

The relation $\ll$ is introduced here as a convenience, chiefly because multiplication by elements of Z leaves it invariant, as will be seen (Theorem 2.6 (a)). Except for the case $a = b = 0$, the statement $a \prec\!\!\prec b$ is considerably stronger than the statement $a \prec b$, since the former says that $\bar{a}a \prec \bar{a}b$ for *every* $\bar{a} \in Z$ for which $\bar{a}a \neq 0$ or $\bar{a}b \neq 0$, and the latter says that $\bar{a}a \prec \bar{a}b$ for *one* such element $\bar{a}$, viz., $\bar{a} = 1 \in Z$.

THEOREM 2.6:

(a) $a \prec\!\!\prec b$, $\bar{c} \in Z$ implies $\bar{c}a \prec\!\!\prec \bar{c}b$;

(b) $a \prec\!\!\prec b$ implies $e(a) \leqq e(b)$;

(c) $a \precsim b$, $b \prec\!\!\prec c$, $c \precsim d$ implies $a \prec\!\!\prec d$;

(d) $a \prec\!\!\prec b$, $a' \sim a$, $b' \sim b$ implies $a' \prec\!\!\prec b'$;

(e) $a \prec\!\!\prec b$ if and only if there exists $a' \sim a$ with $a' \leqq b$, $a' \prec\!\!\prec b$;

(f) if a', a'' are inverses in a, then $a' \prec\!\!\prec a$ is equivalent to $e(a'') = e(a)$.

PROOF: (a) Let $a \prec\!\!\prec b$, $\bar{c} \in Z$. Then if $\bar{a}$ is any element of Z, $\bar{a}\bar{c} \in Z$, and $\bar{a}\bar{c} \cdot a \prec \bar{a}\bar{c} \cdot b$ or $\bar{a}\bar{c} \cdot a = \bar{a}\bar{c} \cdot b = 0$. But this states that for every $\bar{a} \in Z$, $\bar{a} \cdot \bar{c}a \prec \bar{a} \cdot \bar{c}b$ or $\bar{a} \cdot \bar{c}a = \bar{a} \cdot \bar{c}b = 0$, i.e., that $\bar{c}a \prec\!\!\prec \bar{c}b$.

(b) Let $a \prec\!\!\prec b$. Then either $a \prec b$ or $a = b = 0$. In the latter case $e(a) = e(b)$; in case $a \prec b$, $e(a) \leqq e(b)$ by Theorem 1.4 (e).

(c) Consider $\bar{a} \in Z$. By Theorem 1.3 (a), (b): $\bar{a}a \precsim \bar{a}b$, $\bar{a}c \precsim \bar{a}d$; but $\bar{a}b \prec \bar{a}c$ or $\bar{a}b = \bar{a}c = 0$. Hence either $\bar{a}a \prec \bar{a}d$ or $\bar{a}a \precsim 0 \precsim \bar{a}d$. In the latter case $\bar{a}a = 0 \precsim \bar{a}d$; if $\bar{a}d \sim 0$, then $\bar{a}d = 0$, and $\bar{a}a = \bar{a}d = 0$. If $\bar{a}d \succ 0$, then $0 = \bar{a}a \prec \bar{a}d$, whence in all cases $\bar{a}a \prec \bar{a}d$ or $\bar{a}a = \bar{a}d = 0$. Therefore $a \prec\!\!\prec d$.

(d) This is a special case of (c) in which a, b, c, d there are replaced by a', a, b, b'.

(e) If $a \prec\!\!\prec b$, then $a = b = 0$ or $a \prec b$. In the former case $a' = a = 0$ has the properties $a' \sim a$, $a' \leqq b$, $a' \prec\!\!\prec b$. In the latter case, there exists a' with $a \sim a' < b$, whence by (d), $a' \prec\!\!\prec b$. Conversely, if $a \sim a' \leqq b$, $a' \prec\!\!\prec b$, then $a \prec\!\!\prec b$ by (d).

(f) Now $a' \prec\!\!\prec a$ means that for each $\bar{a} \in Z$ either $\bar{a}a' \prec \bar{a}a$ or $\bar{a}a' = \bar{a}a = 0$. Since $a' \leqq a$, we have $\bar{a}a' \leqq \bar{a}a$ for every $\bar{a} \in Z$. Hence $\bar{a}a' \prec \bar{a}a$ is equivalent to $\bar{a}a' \neq \bar{a}a$ (by Part I, Theorem 4.4), and $\bar{a}a' = \bar{a}a = 0$ is equivalent to $\bar{a}a = 0$. Now $\bar{a} \in Z$ and $a' + a'' = a$ yield $\bar{a}a' + \bar{a}a'' = \bar{a}(a' + a'') = \bar{a}a$; moreover, $\bar{a}a' \cdot \bar{a}a'' = \bar{a} \cdot a'a'' = 0$, whence $\bar{a}a'$, $\bar{a}a''$ are inverses in $\bar{a}a$. Therefore $\bar{a}a' \neq \bar{a}a$ is equivalent to $\bar{a}a'' \neq 0$. Thus we see that $a' \prec\!\!\prec a$ means that for each $\bar{a} \in Z$ either $\bar{a}a'' \neq 0$ or $\bar{a}a = 0$; in other words, $a' \prec\!\!\prec a$ means that $\bar{a} \in Z$, $\bar{a}a'' = 0$ implies $\bar{a}a = 0$, i.e., $\bar{a} \in Z$, $a'' \leqq -\bar{a}$ implies $a \leqq -\bar{a}$. Replacing $-\bar{a}$ by $\bar{b}$, we obtain the equivalent condition, $\bar{b} \in Z$, $a'' \leqq \bar{b}$ implies $a \leqq \bar{b}$. This clearly implies $(\bar{b} = e(a''))$ and is implied by the statement $e(a) \leqq e(a'')$. But since $a'' \leqq a$, $e(a'') \leqq e(a)$. Consequently $a' \ll a$ if and only if $e(a) = e(a'')$.

At this point we give a reformulation of Theorem 2.1 together with some consequences.

THEOREM 2.7: For every $a, b \in L$, there exist e_1, e_2, e_3 such that

(a) $e_1, e_2, e_3 \in Z$, (b) $(e_1, e_2, e_3)\perp$, (c) $e_1 + e_2 + e_3 = 1$,
(d) $e_1 a \gg e_1 b$, (e) $e_2 a \ll e_2 b$, (f) $e_3 a \sim e_3 b$.

The elements e_1, e_2, e_3 may be defined as $e(a''), e(b''), -(e(a'') + e(b''))$ respectively, where a'', b'' are elements introduced in Theorem 2.1.

PROOF: Let e_1, e_2, e_3 be defined as stated in the theorem. We shall verify properties (a)—(f).

(a) Obvious.

(b) Since by Theorem 2.1 $(a'', b'')D$, $a''b'' = 0$, we have by Lemma 1.1 (or Theorem 1.5), $e_1 e_2 = e(a'')e(b'') = 0$. By definition, $e_3 = -(e_1+e_2)$, and so $(e_1 + e_2)e_3 = 0$. Hence $(e_1, e_2, e_3)\perp$.

(c) Obvious.

(d) Clearly

$$e_1 a = e(a'')a = e(a'')(a' + a'') = e(a'')a' + e(a'')a'' \quad (\text{since } e(a'') \in Z)$$
$$= e(a'')a' + a'' \quad (\text{since } a'' \leqq e(a''));$$

since $e(a'')b'' \leqq e(a'')e(b'') = 0$, we have

$$e_1 b = e(a'')b = e(a'')(b' + b'') = e(a'')b' + e(a'')b'' \quad (\text{since } e(a'') \in Z)$$
$$= e(a'')b' + 0 = e(a'')b'.$$

Hence, if $\bar{a} \in Z$,

$$\bar{a} \cdot e_1 a = \bar{a}e(a'') \cdot a' + \bar{a}a'', \quad \bar{a} \cdot e_1 b = \bar{a}e(a'') \cdot b'.$$

Since $a' \sim b'$, $\bar{a}e(a'') \cdot a' \sim \bar{a}e(a'') \cdot b'$ by Theorem 1.4 (a). Thus $\bar{a} \cdot e_1 b \prec \bar{a} \cdot e_1 a$ except when $\bar{a}e(a'') \cdot a' + \bar{a}a'' = \bar{a}e(a'') \cdot a'$, i.e. except when $\bar{a}a'' \leqq \bar{a}e(a'') \cdot a'$. The exceptional case is equivalent to

$$\bar{a}a'' = \bar{a}a'' \cdot \bar{a}e(a'') \cdot a' = \bar{a}e(a'') \cdot a'a'' = 0,$$

i.e., to $\bar{a}e_1 = \bar{a}e(a'') = 0$. But the condition $\bar{a}e_1 = 0$ implies $\bar{a}e_1 b = \bar{a}e_1 a = 0$. Therefore we have for every $\bar{a} \in Z$ either $\bar{a}e_1 b \prec \bar{a}e_1 a$ or $\bar{a}e_1 b = \bar{a}e_1 a = 0$, i.e., $e_1 b \ll e_1 a$.

(e) This is similar to (d).

(f) Evidently $e_3 a = e_3(a' + a'') = e_3 a' + e_3 a'' = e_3 a'$, since $e_3 a'' \leqq e_3 e(a'') = e_3 e_1 = 0$. Similarly, $e_3 b = e_3 b'$. But $a' \sim b'$ yields by Theorem 1.4 (a) that $e_3 a' \sim e_3 b'$, whence $e_3 a \sim e_3 b$.

THEOREM 2.8: If $a, b \in L$, there exists an element $q(a, b) \in Z$ such that for each $\bar{a} \in Z$, $\bar{a}a \lesssim \bar{a}b$ is equivalent to $\bar{a} \leqq q(a, b)$ (such an element $q(a, b)$ is clearly unique). If e_1, e_2, e_3 are the elements introduced in Theorem 2.7, then

$$q(a, b) = e_2 + e_3, \quad -q(b, a) = e_2, \quad -q(a, b) = e_1, \quad q(a, b)q(b, a) = e_3,$$

(of course, $q(a, 1) = 1$ — Ed.).

PROOF: We define $q(a, b) \equiv e_2 + e_3 \in Z$. By Theorem 2.7 we have $e_3a \sim e_3b$, $e_2a \prec\!\prec e_2b$, whence $e_2a \lesssim e_2b$. Thus there exists c with $e_2a \sim c \leqq e_2b$. Now $e_2a \cdot e_3a \leqq e_2e_3 = 0$, $c \cdot e_3b \leqq e_2e_3 = 0$. Hence Theorem 2.4 (b) applies, and $e_2a + e_3a \sim c + e_3b$. Since $e_2a + e_3a = (e_2 + e_3)a = q(a, b)a$, $c + e_3b \leqq e_2b + e_3b = (e_2 + e_3)b = q(a, b)b$, we have $q(a, b)a \lesssim q(a, b)b$.

Now suppose $\bar{a} \in Z$; then $\bar{a}q(a, b)a \lesssim \bar{a}q(a, b)b$ by Theorem 1.4 (a), (b). If $\bar{a} \leqq q(a, b)$, then $\bar{a}q(a, b) = \bar{a}$, and $\bar{a}a \lesssim \bar{a}b$.

Conversely, assume $\bar{a}a \lesssim \bar{a}b$. Then by Theorem 1.4 (a), (b), $e_1 \cdot \bar{a}a \lesssim e_1 \cdot \bar{a}b$, i.e., $\bar{a} \cdot e_1a \lesssim \bar{a} \cdot e_1b$. Since by Theorem 2.6 (d), $e_1a \succ\!\succ e_1b$, we have $\bar{a} \cdot e_1a = \bar{a} \cdot e_1b = 0$. But $\bar{a}e_1a = 0$ implies $\bar{a}e_1e(a) = 0$ by Theorem 1.3 (f). Since $e(a) \geqq e(a'') = e_1$, $e_1 \cdot e(a) = e_1$, this means $e_1\bar{a} = 0$, i.e., $\bar{a} \leqq -e_1$. But $q(a, b) = e_2 + e_3 = -e_1$, whence $\bar{a} \leqq q(a, b)$. This establishes the first part of the theorem.

By definition $q(a, b) = e_2 + e_3$. We have seen also that $q(a, b) = -e_1$, whence $e_1 = -q(a, b)$. By interchanging a, b (and hence e_1, e_2), we find $e_2 = -q(b, a)$. Finally $e_3 = -(e_1 + e_2) = -\big(-q(a, b) + (-q(b, a))\big) = q(a, b) \cdot q(b, a)$ (cf. the Corollary to Theorem A.1 in Appendix 1 to Chapter I).

DEFINITION 2.4:

(a) $A < B$ means that there exist $a \in A$, $b \in B$ with $a \prec b$.
(b) $A > B$ means that there exist $a \in A$, $b \in B$ with $a \succ b$.
(c) $A \ll B$ means that there exist $a \in A$, $b \in B$ with $a \prec\!\prec b$.
(d) $A \gg B$ means that there exist $a \in A$, $b \in B$ with $a \succ\!\succ b$.

COROLLARY:

(a) $A < B$ if and only if $B < A$;
(b) $A \ll B$ if and only if $B \gg A$;
(c) $A < B$ if and only if $a \prec b$ for every a, b such that $a \in A$, $b \in B$;
(d) $A < B$ if and only if for each $a \in A$ there exists $b \in B$ with $a < b$;
(e) $A < B$ if and only if there exist $a \in A$, $b \in B$ with $a < b$;
(f) $A \ll B$ if and only if $a \prec\!\prec b$ for every a, b such that $a \in A$, $b \in B$.

PROOF: Properties (a), (c), (d), (e) are essentially those in Corollaries 1—4 to Definition 6.3 of Part I; (b) is trivial, and (f) follows from Theorem

2.6 (d). Thus we see that the relations $<$, $\ll$ are defined for classes through representative elements and the respective relations $\prec$, $\prec\prec$ and are independent of the choice of representative elements.

Note that Corollary 6 to Definition 6.3 of Part I does not hold here; the set $\mathscr{L}$ is, however, partially ordered by the relations $<$, $>$.

DEFINITION 2.5: We define $\theta \equiv A_0$, $\dagger \equiv A_1$.

COROLLARY: $\theta \leqq A \leqq \dagger$ for every $A \in \mathscr{L}$.

DEFINITION 2.6: If $\bar{a} \in Z$, $A \in \mathscr{L}$, we define $\bar{a} \cdot A \equiv \bar{a}A$ as the unique common value of all $A_{\bar{a}a}$, $a \in A$ (cf. Theorem 1.4 (a)).

COROLLARY:

(a) $A \leqq B$ implies $\bar{a}A \leqq \bar{a}B$ for every $\bar{a} \in Z$;

(b) $A \ll B$ if and only if for every $\bar{a} \in Z$, $\bar{a}A < \bar{a}B$ or $\bar{a}A = \bar{a}B = \theta$;

(c) $A \ll B$, $\bar{a} \in Z$ implies $\bar{a}A \ll \bar{a}B$;

(d) $A \leqq B$, $B \ll C$, $C \leqq D$ implies $A \ll D$.

PROOF: (a) Select $a \in A$, $b \in B$. Then $a \precsim b$, whence by Theorem 1.4 (a), (b): $\bar{a}a \precsim \bar{a}b$, and $\bar{a}A \leqq \bar{a}B$.

(b) Select $a \in A$, $b \in B$. Then $A \ll B$ if and if only for every $\bar{a} \in Z$: $\bar{a}a \prec \bar{a}b$ or $\bar{a}a = \bar{a}b = 0$. Since $\bar{a}a \in \bar{a}A$, $\bar{a}b \in \bar{a}B$, this is equivalent to $\bar{a}A < \bar{a}B$ or $\bar{a}A = \bar{a}B = \theta$ for every $\bar{a} \in Z$.

(c) This is immediate from Theorem 2.6 (a).

(d) This follows from Theorem 2.6 (c).

THEOREM 2.9: A class $A \in \mathscr{L}$ contains exactly one element if and only if A is of the form $A_{\bar{a}}$, $\bar{a} \in Z$.

PROOF: Let $A = A_{\bar{a}}$ contain only the element $\bar{a}$. Suppose x is an inverse of $\bar{a}$, and let a be any inverse of x. Then $a \sim \bar{a}$, whence $a = \bar{a}$, and x has but one inverse, $\bar{a}$. Thus $x \in Z$, whence also $\bar{a} = -x \in Z$. Conversely, let $A = A_{\bar{a}}$, $\bar{a} \in Z$, and suppose $a \sim \bar{a}$. Then there exists x such that x is inverse to a, $\bar{a}$. Since $\bar{a} \in Z$, $x = -\bar{a} \in Z$, whence x has but one inverse $\bar{a}$, and therefore $a = \bar{a}$.

NOTE: It should be observed that the elements of Z are in one-to-one correspondence with the classes $A \in \mathscr{L}$ containing but one element. When it is convenient to do so, we shall replace Z by the set of all classes $A \in \mathscr{L}$ containing exactly one element. When this is done, we have, of course, $Z \subset \mathscr{L}$.

We define $A + B$ and $A - B$ for certain pairs A, $B \in \mathscr{L}$ in the same way as in Part I, Definitions 6.3, 6.4:

DEFINITION 2.7: We shall say that $A + B$ *exists* in case there exist $a \in A$, $b \in B$ such that $ab = 0$. When $A + B$ exists it is defined as the unique class C which is equal to A_{a+b} for every a, b such that $a \in A$, $b \in B$,

$ab = 0$ (the existence and uniqueness following from Theorem 2.4 (b)). Thus $A + B$ depends on A, B only.

DEFINITION 2.8: We shall say that $A - B$ *exists* in case there exist $a \in A$, $b \in B$ such that $a \geqq b$, i.e., in case $A \geqq B$. When $A - B$ exists it is defined as the unique class C which is equal to $A_{b'}$ for every b' such that there exist $a \in A$, $b \in B$, $b \leqq a$, for which b' is inverse to b in a (the existence and uniqueness of C following from Theorem 2.4 (a)). Thus $A - B$ depends on A, B only.

We adopt here, as in Part I, the convention that the assertion that $A + B(A - B)$ has a property means first the assertion that $A + B$ $(A - B)$ exists and further that the class $A + B$ $(A - B)$ has the specified property. We now restate the essential properties of $A + B$, $A - B$.

THEOREM 2.10:

(i) If $A + B$ exists, then $A + B = B + A$.

(ii) If $A + B$, $(A + B) + C$ exist, then $(A + B) + C = A + (B+C)$.

(iii) $A + \theta = A$.

(iv) $\dagger - A$ exists for every $A \in \mathscr{L}$.

(v) $A + B$ exists if and only if $A \leqq \dagger - B$.

(vi) $A - B$ exists if and only if $A \geqq B$.

(vii) If $A + B$ exists, then $(A + B) - B = B$.

(viii) If $A - B$ exists, then $(A - B) + B = A$.

(ix) The equation $A + X = B$ has a solution X if and only if $A \leqq B$, in which case the solution is unique and equal to $B - A$.

(x) If $X \leqq Y$, and if $A + Y$ exists, then $A + X$ exists and $A + X \leqq A + Y$.

(xi) If $A + X$, $A + Y$ exist, then $X \lesseqgtr Y$ are equivalent to $A + X \lesseqgtr A + Y$, respectively.

(xii) $X \lesseqgtr Y$ are equivalent to $\dagger - X \gtreqless \dagger - Y$, respectively.

PROOF: Properties (i)—(x) are stated and proved in Part I (cf. Theorems 6.2, 6.3), and the proofs given there are valid here. Properties (xi), (xii) were also proved in Part I, but the proofs given there make use of the complete ordering of the set $\mathscr{L}$, i.e. of Axiom VI. We shall give proofs of (xi), (xii), independent of Axiom VI.

(xi) Clearly $X = Y$ implies $A + X = A + Y$. Now by (x): $X < Y$ implies $A + X < A + Y$; interchanging X and Y, we see that $X > Y$ implies $A + X > A + Y$.

Conversely, suppose first that $A + X = A + Y$. Then by (vii): $X = (A + X) - A = (A + Y) - A = Y$. If $A + X < A + Y$, there exists by (ix), $U > \theta$ such that $(A + X) + U = A + Y$. Then (ii) gives

$A + (X + U) = A + Y$, whence by the result just proved, $X + U = Y$, and by (ix), $X < Y$. Interchanging X and Y, we see that $A + X > A + Y$ implies $X > Y$.

(xii) If $X < Y$, then (xi), together with (viii), (x) yields $X + (\dagger - X) = Y + (\dagger - Y) > X + (\dagger - Y)$, $\dagger - X > \dagger - Y$. Interchanging X and Y, we see that $X > Y$ implies $\dagger - X < \dagger - Y$. Clearly $X = Y$ implies $\dagger - X = \dagger - Y$. This proves the forward implications. Since $U = \dagger - (\dagger - U)$, the converse statements result if we replace X, Y by $\dagger - X$, $\dagger - Y$, respectively.

DEFINITION 2.9: For every $A \in \mathcal{L}$ we define $e(A)$ as the unique common value of all $e(a)$, $a \in A$ (cf. Theorem 1.4 (d)).

COROLLARY:

(a) $A \ll B$ implies $e(A) \leqq e(B)$,
(b) $A \ll B$ is equivalent to $e(B - A) = e(B)$.

PROOF: These statements are immediate by Theorem 2.6 (b), (f).

THEOREM 2.11:

(a) If $\bar{a} \in Z$ and $A + B$ exists, then $\bar{a}A + \bar{a}B$ exists and is equal to $\bar{a}(A + B)$.
(b) If $\bar{a} \in Z$ and if $A - B$ exists, then $\bar{a}A - \bar{a}B$ exists and is equal to $\bar{a}(A - B)$.
(c) If $\bar{a}$, $\bar{b} \in Z$, $\bar{a}\bar{b} = 0$, then $\bar{a}A + \bar{b}A$ exists and is equal to $(\bar{a} + \bar{b})A$.
(d) If $\bar{a}$, $\bar{b} \in Z$, $A \in \mathcal{L}$, then $\bar{a}(\bar{b}A) = (\bar{a}\bar{b})A$.

PROOF: (a) Select $a \in A$, $b \in B$ with $ab = 0$; then $\bar{a}a \in \bar{a}A$, $\bar{a}b \in \bar{a}B$, $\bar{a}a \cdot \bar{a}b = \bar{a} \cdot ab = 0$. Since $\bar{a} \in Z$, $\bar{a}(a + b) = \bar{a}a + \bar{a}b$. Hence $\bar{a}(A + B) = \bar{a}A_{a+b} = A_{\bar{a}(a+b)} = A_{\bar{a}a+\bar{a}b} = \bar{a}A + \bar{a}B$.

(b) Since $A = B + (A - B)$, we have by (a), $\bar{a}A = \bar{a}(B + (A - B)) = \bar{a}B + \bar{a}(A - B)$. Hence $\bar{a}A - \bar{a}B$ exists and is equal to $\bar{a}(A - B)$ by Theorem 2.10 (ix).

c) Select $a \in A$. Then $\bar{a}a \in \bar{a}A$, $\bar{b}a \in \bar{b}A$, $\bar{a}a \cdot \bar{b}a = \bar{a}\bar{b} \cdot a = 0$. Since $\bar{a}$, $\bar{b} \in Z$, we have $(\bar{a} + \bar{b})a = \bar{a}a + \bar{b}a$. Hence $(\bar{a} + \bar{b})A = A_{(\bar{a}+\bar{b})a} = A_{\bar{a}a+\bar{b}a} = \bar{a}A + \bar{b}A$.

(d) Select $a \in A$. Then $\bar{a}(\bar{b}A) = \bar{a}A_{\bar{b}a} = A_{\bar{a}(\bar{b}a)} = A_{(\bar{a}\bar{b})a} = (\bar{a}\bar{b})A$.

DEFINITION 2.10: Define $0A \equiv \theta$. If $n = 1, 2, \cdots$, and $(n - 1)A$ has been defined, then put $nA \equiv (n - 1)A + A$ if $(n - 1)A + A$ exists. Otherwise nA is undefined. (Cf. Part I, Definition 6.6.) We say that nA *exists* in case nA is defined.

THEOREM 2.12: Let A, $B \in \mathcal{L}$ and $n, m = 0, 1, 2, \cdots$.

(a) If either $(n+m)A$ exists or nA, mA, $nA + mA$ exist, then $(n+m)A = nA + mA$.
(b) If mA, $n(mA)$ exist, then $(nm)A = n(mA)$.
(c) If $n \geqq m$, and if nA exists, then $(n-m)A = nA - mA$.
(d) If $A + B$, $n(A+B)$ exist, then $n(A+B) = nA + nB$.
(e) If $A - B$, nA exist, then $n(A-B) = nA - nB$.
(f) If nA exists, then $\bar{a}(nA) = n(\bar{a}A)$ for every $\bar{a} \in Z$.

PROOF: (a) This is immediate by iteration of Definition 2.8.

(b) For $n = 0$, (b) is obvious. Otherwise, (b) is merely an iteration of (a) (with m and n in (a) equal to our m).

(c) By (a), $nA = (n-m)A + mA$, whence $nA - mA = (n-m)A$ by Theorem 2.10 (ix).

(d) For $n = 0, 1$ this is evident. Suppose it holds for $n - 1$. Then

$$\begin{aligned} n(A+B) &= (n-1)(A+B) + (A+B) && \text{(by (a))} \\ &= ((n-1)A + (n-1)B) + (A+B) \\ &= ((n-1)A + A) + ((n-1)B + B) \\ & && \text{(by Theorem 2.10 (ii))} \\ &= nA + nB && \text{(by (a)).} \end{aligned}$$

(e) By (d), $nA = n(A-B) + nB$, whence $nA - nB = n(A-B)$ by Theorem 2.10 (ix).

(f) This is obvious for $n = 0, 1$. Suppose it holds for $n - 1$. Then $\bar{a}(nA) = \bar{a}((n-1)A + A) = \bar{a}(n-1)A + \bar{a}A = (n-1)(\bar{a}A) + \bar{a}A = n\bar{a}A$.

THEOREM 2.13: nA is defined for every $n = 0, 1, 2, \cdots$ if and only if $A = \theta$.

PROOF: See Part I, Theorem 6.4; the proof given there is valid here.

THEOREM 2.14: Let $A, B \in \mathscr{L}$ and $n = 0, 1, 2, \cdots$. Then there exists a unique element $q_n(A, B) \in Z$ such that for every $\bar{a} \in Z$ the existence of $n \cdot \bar{a}A$ together with the relation $n\bar{a}A \leqq \bar{a}B$ is equivalent to $\bar{a} \leqq q_n(A, B)$ (of course, $q_1(A, B)$ will coincide with $q(a, b)$ for $a \in A$, $b \in B$ — Ed.).

PROOF: It is obvious that if $q_n(A, B)$ exists, it is unique. We prove the existence by induction. For $n = 0$, every $\bar{a} \in Z$ has the properties that $n\bar{a}A$ exists and $n\bar{a}A \leqq \bar{a}B$, whence $q_0(A, B) \equiv 1$ is effective. Assume $n = 1, 2, \cdots$, and that $q_{n-1}(A, B)$ exists. For brevity, denote $q_{n-1}(A, B)$ by f'. Thus $(n-1)f'A$ exists, and $(n-1)f'A \leqq f'B$. Thus there exists X with $(n-1)f'A + X = f'B$ by Theorem 2.10 (ix). Moreover, for every $\bar{a} \in Z$ such that $\bar{a} \leqq f'$, we have $(n-1)\bar{a}f'A + \bar{a}X = \bar{a}f'B$, i.e.,

$(n-1)\bar{a}A+\bar{a}X=\bar{a}B$. Now select $a_0 \in A$, $x_0 \in X$ and define $f \equiv f'q(a_0, x_0)$. We shall prove that f is effective as $q_n(A, B)$.

Let $\bar{a}$ be any element of Z. Suppose first that $\bar{a} \leqq f$. Then $\bar{a} \leqq f'$, $\bar{a} \leqq q(a_0, x_0)$, whence by the latter condition $\bar{a}a_0 \precsim \bar{a}x_0$ (cf. Theorem 2.8), i.e., $\bar{a}A \leqq \bar{a}X$. Thus $n\bar{a}A = (n-1)\bar{a}A+\bar{a}A$ exists and is $\leqq (n-1)\bar{a}A+\bar{a}X = \bar{a}B$ by Theorem 2.10 (x). Conversely, let $n\bar{a}A \leqq \bar{a}B$. Then $(n-1)\bar{a}A \leqq \bar{a}B$, whence $\bar{a} \leqq f'$. Therefore $(n-1)\bar{a}A+\bar{a}A \leqq \bar{a}B = (n-1)\bar{a}A+\bar{a}X$, from which we may conclude that $\bar{a}A \leqq \bar{a}X$ by Theorem 2.10 (xi), i.e., that $\bar{a}a_0 \precsim \bar{a}x_0$. But this means $\bar{a} \leqq q(a_0, x_0)$ by Theorem 2.8; since we had $\bar{a} \leqq f'$, it results that $\bar{a} \leqq f'q(a_0, x_0) = f$. Consequently, if we define $q_n(A, B) \equiv f$, we see that for $\bar{a} \in Z$ the existence of $n\bar{a}A$ together with $n\bar{a}A \leqq \bar{a}B$ is equivalent to $\bar{a} \leqq q_n(A, B)$, and the proof is complete.

Corollary: For every $n = 1, 2, \cdots$, $q_n(A, B) \leqq q_{n-1}(A, B)$, and $q_0(A, B) = 1$; i.e.,

$$1 = q_0(A, B) \geqq q_1(A, B) \geqq q_2(A, B) \geqq \cdots.$$

Theorem 2.15: If $A, B \in \mathscr{L}$, then $\Pi(q_n(A, B); n=0, 1, 2, \cdots) = -e(A)$.

Proof: Select $a_0 \in A$. Then $(-e(A))a_0 = (-e(a_0))a_0 \leqq (-e(a_0))e(a_0) = 0$, whence $(-e(A))A = (-e(A))A_{a_0} = A_{(-e(A))a_0} = \theta$. Hence for each $n=0, 1, 2, \cdots$, $n(-e(A))A = n\theta$ exists and is equal to $\theta \leqq (-e(A))B$. Thus $-e(A) \leqq q_n(A, B)$, and it follows that $\bar{a} \equiv \Pi(q_n(A, B); n=0, 1, 2, \cdots) \geqq -e(A)$. But $\bar{a} \in Z$, $\bar{a} \leqq q_n(A, B)$ for every $n=0, 1, 2, \cdots$, whence $n\bar{a}A$ exists for every n; therefore $\bar{a}A = \theta$ by Theorem 2.13. Then $\theta = \bar{a}A = \bar{a}A_{a_0} = A_{\bar{a}a_0}$, whence $\bar{a}a_0 \sim 0$, i.e., $\bar{a}a_0 = 0$. Therefore by Theorem 1.3(f), $\bar{a}e(a_0) = 0$, and $\bar{a} \leqq -e(a_0)$. But $e(a_0) = e(A)$, and we have proved $\bar{a} \leqq -e(A)$. This, together with $\bar{a} \geqq -e(A)$ yields $\Pi(q_n(A, B); n = 0, 1, 2, \cdots) = \bar{a} = -e(A)$.

Definition 2.11: For every $A, B \in \mathscr{L}$ we define

$$r_n(A, B) \equiv q_n(A, B) \ (-q_{n+1}(A, B))$$

for $n = 0, 1, 2, \cdots$, and $r_\infty(A, B) \equiv -e(A)$.

Corollary: For every $n = 0, 1, 2, \cdots, \infty$,

$$r_n(A, B) \in Z \ (r_0(A, B) = -q_1(A, B)$$

and $r_0(A, \dagger) = -\dagger = 0$ — Ed.).

Theorem 2.16: For every $A, B \in \mathscr{L}$ it is so that

(a) $(r_n(A, B); n = 0, 1, 2, \cdots, \infty) \perp$,

(b) $\Sigma(r_n(A, B); n = 0, 1, 2, \cdots, \infty) = 1$,

(c) $r_n(A, B) \cdot B - n\,(r_n(A, B) \cdot A)$ exists and is $\ll r_n(A, B) \cdot A$ for $n = 0, 1, 2, \cdots$,

(d) $r_\infty(A, B) \cdot A = \theta$.

PROOF: (a), (b) We note first that for $n = 0, 1, 2, \cdots$,

$$\begin{aligned}(1)\quad q_{n+1}(A, B)+r_n(A, B) &= q_{n+1}(A, B)+q_n(A, B)(-q_{n+1}(A, B))\\ &= q_n(A, B)q_{n+1}(A, B)+q_n(A, B)(-q_{n+1}(A, B))\\ &\qquad\text{(by the Corollary to Theorem 2.14)}\\ &= q_n(A, B)(q_{n+1}(A, B)+(-q_{n+1}(A, B))\\ &= q_n(A, B)\cdot 1 = q_n(A, B).\end{aligned}$$

Applying this result to $n = 0, 1, 2, \cdots, m$, we find

$$r_0(A, B) + r_1(A, B) + \cdots + r_m(A, B) + q_{m+1}(A, B) = q_0(A, B) = 1.$$

Hence $\Sigma(r_n(A, B); n = 0, 1, 2, \cdots) + q_{m+1}(A, B) \geqq 1$, whence $\Sigma(r_n(A, B); n = 0, 1, 2, \cdots) + q_{m+1}(A, B) = 1$. Thus $\Sigma(r_n(A, B); n = 0, 1, 2, \cdots) + q_p(A, B) = 1$ for $p = 1, 2, \cdots$. Therefore

$$\begin{aligned}1 &= \Pi(\Sigma(r_n(A, B); n = 0, 1, 2, \cdots) + q_p(A, B); p = 1, 2, \cdots)\\ &= \Sigma(r_n(A, B); n=0, 1, 2, \cdots) + \Pi(q_p(A, B); p=1, 2, \cdots) \quad \text{(by III}_1\text{)}\\ &= \Sigma(r_n(A, B); n = 0, 1, 2, \cdots) + (-e(A)) \quad \text{(by Theorem 2.15)}\\ &= \Sigma(r_n(A, B); n=0, 1, 2, \cdots) + r_\infty(A, B) = \Sigma(r_n(A, B); n=0, 1, \cdots, \infty),\end{aligned}$$

and (b) is established.

Now if $m \leqq n$, then $r_n(A, B) \leqq q_{n+1}(A, B) + r_n(A, B) = q_n(A, B) \leqq q_m(A, B)$ by (1). Thus $r_n(A, B) + \cdots + r_{n-k+1}(A, B) \leqq q_{n-k+1}(A, B)$ for $k = 1, 2, \cdots, n$. But $r_{n-k}(A, B) = q_{n-k}(A, B)(-q_{n-k+1}(A, B)) \leqq -q_{n-k+1}(A, B)$, whence

$$(r_n(A, B) + \cdots + r_{n-k+1}(A, B))r_{n-k}(A, B) = 0 \quad (k = 1, \cdots, n).$$

Thus by Part I, Theorem 2.2, $(r_n(A, B), \cdots, r_0(A, B)) \perp$ for $n = 0, 1, \cdots$. Hence by Part I, Theorem 2.3, $(r_n(A, B); n = 0, 1, 2, \cdots) \perp$. Now

$$\begin{aligned}r_n(A, B) &= q_n(A, B)(-q_{n+1}(A, B)) \leqq -q_{n+1}(A, B)\\ &\leqq -\Pi(q_m(A, B); m = 0, 1, 2, \cdots) = -(-e(A))\\ &\qquad\text{(by Theorem 2.15)}\\ &= -r_\infty(A, B),\end{aligned}$$

whence $\Sigma(r_n(A, B); n = 0, 1, 2, \cdots) \leqq -r_\infty(A, B)$, i.e.,

$$\Sigma(r_n(A, B); n = 0, 1, 2, \cdots)r_\infty(A, B) = 0.$$

Therefore by Part I, Theorem 2.1, $(r_n(A, B); n = 0, 1, \cdots, \infty) \perp$ and (a) is established.

(c) We note first that $r_n(A, B) \leqq q_n(A, B)$, whence by Theorem 2.14, $nr_n(A, B) \cdot A$ exists and is $\leqq r_n(A, B) \cdot B$. Hence by Theorem 2.10 (vi), $A' \equiv r_n(A, B) \cdot B - nr_n(A, B) \cdot A$ exists. Define $A'' \equiv r_n(A, B)A$, and select $b_0 \in A'$, $a_0 \in A''$; define $f \equiv q(a_0, b_0)$. By Theorem 2.7, $fa_0 \precsim fb_0$, i.e.

$$fr_n(A, B)A \leqq fr_n(A, B)B - nfr_n(A, B)A. \tag{2}$$

Hence $fr_n(A, B)A \leqq f - nfr_n(A, B)$, and $nfr_n(A, B) + fr_n(A, B)$ exists by Theorem 2.10 (v). Thus $(n+1)fr_n(A, B)$ exists and (2) yields by Theorem 2.10 (xi), (viii), that $(n+1)fr_n(A, B) \cdot A \leqq fr_n(A,B) \cdot B$. Hence by Theorem 2.14, $fr_n(A, B) \leqq q_{n+1}(A, B)$. But we have also $fr_n(A, B) \leqq r_n(A, B) \leqq -q_{n+1}(A, B)$ whence

$$fr_n(A, B) \leqq q_{n+1}(A, B)\ (-q_{n+1}(A, B)) = 0.$$

Now

$$\begin{aligned} r_n(A, B) &= 1 \cdot r_n(A, B) = (f + (-f)) \cdot r_n(A, B) = fr_n(A, B) + (-f)r_n(A, B) \\ &= 0 + (-f)r_n(A, B) = (-f)r_n(A, B). \end{aligned}$$

But Theorems 2.7, 2.6 yield $(-f)b_0 \prec\prec (-f)a_0$, whence $r_n(A, B)(-f)b_0 \prec\prec r_n(A, B)(-f)a_0$, and therefore $r_n(A, B)b_0 \prec\prec r_n(A, B)a_0$. This states that $r_n(A, B)(r_n(A, B) \cdot B - nr_n(A, B) \cdot A) \ll r_n(A, B) \cdot r_n(A, B) \cdot A$, i.e., that $r_n(A, B) \cdot B - nr_n(A, B) \cdot A \ll r_n(A, B) \cdot A$.

(d) As was shown at the beginning of the proof of Theorem 2.15, $r_\infty(A, B) \cdot A = (-e(A))A = \theta$.

PART III · CHAPTER III

Minimal Elements

DEFINITION 3.1: An element $a \in L$ is *minimal* in case $x \ll a$ implies $x = 0$. An element $A \in \mathscr{L}$ is *minimal* in case $X \in \mathscr{L}$, $X \ll A$ implies $X = \theta$.

COROLLARY: If $a \in L$ is any element in $A \in \mathscr{L}$, then a is minimal if and only if A is minimal.

LEMMA 3.1: If a is minimal, and if $b \precsim a$, then b is minimal.

PROOF: By Theorem 2.6(c), $x \ll b$ implies $x \ll a$ and therefore $x = 0$.

COROLLARY: If A is minimal, and if $B \leqq A$, then B is minimal.

LEMMA 3.2: An element $a \in L$ is minimal if and only if $c < a$ implies $e(c) < e(a)$.

PROOF: We shall prove an equivalent statement, viz., that a is not minimal if and only if there exists $c < a$ such that $e(c) = e(a)$. Suppose first that a is not minimal. Then there exists $x \ll a$ such that $x \neq 0$. By Theorem 2.6(e), there exists x' such that $x' \sim x$, $x' \leqq a$, $x' \ll a$. Define c as any inverse of x' in a; $x \neq 0$ implies $x' \neq 0$, and hence $c \neq a$. Then by Theorem 2.6(f), $e(c) = e(a)$. Conversely let there exist $c < a$ with $e(c) = e(a)$, and define x' as any inverse of c in a. By Theorem 2.6 (f), $x' \ll a$. But $x' \neq 0$, since $c \neq a$; therefore a is not minimal.

LEMMA 3.3: If I is any set (of indices σ) and $(a_\sigma;\ \sigma \in I)$ is any system of elements of L, then there exists a system $(a'_\sigma;\ \sigma \in I)$ such that $a'_\sigma \leqq a_\sigma$ $(\sigma \in I)$, $\Sigma(a'_\sigma;\ \sigma \in I) = \Sigma(a_\sigma;\ \sigma \in I)$, $(a'_\sigma;\ \sigma \in I)\perp$.

PROOF: Since we may assume that I is well ordered, there is no loss in generality in replacing I by the set of all ordinals $\alpha < \Omega$ where Ω is an existent fixed ordinal. We define the a'_α inductively as follows. For each $\alpha < \Omega$ let a'_α be an inverse of $a_\alpha \Sigma(a_\beta;\ \beta < \alpha)$ in a_α. Then $a'_\alpha \cdot \Sigma(a'_\beta;\ \beta < \alpha) \leqq a'_\alpha a_\alpha \cdot \Sigma(a'_\beta;\ \beta < \alpha) \leqq a'_\alpha a_\alpha \Sigma(a_\beta;\ \beta < \alpha) = 0$. If $(\alpha_1, \cdots, \alpha_n)$ is any finite set of ordinals $< \Omega$ (ordered increasingly), we have $(a'_{\alpha_1} + \cdots + a'_{\alpha_{k-1}})a'_{\alpha_k} = 0$ $(k = 2, \cdots, n)$, whence by Part I, Theorem 2.2, $(a'_{\alpha_i};\ i = 1, \cdots, n)\perp$. Therefore by Part I, Theorem 2.3, $(a'_\alpha;\ \alpha < \Omega)\perp$. We shall prove now that

$$(1) \qquad \Sigma(a'_\beta;\ \beta < \alpha) = \Sigma(a_\beta;\ \beta < \alpha)$$

for every $\alpha \leqq \Omega$. If (1) fails to hold for some $\alpha \leqq \Omega$, there exists a smallest ordinal α_0 for which (1) does not hold. Now $\alpha_0 \neq 0$, since for $\alpha = 0$ both members of (1) are zero, and therefore (1) holds. Suppose that α_0 has an immediate predecessor α_1. Then (1) holds for $\alpha = \alpha_1$.

$$\Sigma(a_\beta;\ \beta < \alpha_1) = \Sigma(a'_\beta;\ \beta < \alpha_1).$$

Hence

$$\begin{aligned}\Sigma(a_\beta;\ \beta < \alpha_0) &= \Sigma(a_\beta;\ \beta < \alpha_1) + a_{\alpha_1}\\ &= \Sigma(a_\beta;\ \beta < \alpha_1) + a'_{\alpha_1} + a_{\alpha_1}\Sigma(a_\beta;\ \beta < \alpha_1)\\ &= \Sigma(a_\beta;\ \beta < \alpha_1) + a'_{\alpha_1} \text{ (since the third term is} \leqq \text{the first)}\\ &= \Sigma(a'_\beta;\ \beta < \alpha_1) + a'_{\alpha_1} \text{ (since (1) holds for } \alpha = \alpha_1)\\ &= \Sigma(a'_\beta;\ \beta < \alpha_0),\end{aligned}$$

and (1) is seen to hold for $\alpha = \alpha_0$, contrary to the definition of α_0. Thus α_0 has no immediate predecessor. Now for every $\gamma < \alpha_0$,

$$\Sigma(a'_\beta;\ \beta < \gamma) = \Sigma(a_\beta;\ \beta < \gamma),$$

whence

$$\Sigma(\Sigma(a'_\beta;\ \beta < \gamma);\ \gamma < \alpha_0) = \Sigma(\Sigma(a_\beta;\ \beta < \gamma);\ \gamma < \alpha_0),$$

i.e.,

$$\Sigma(a'_\beta;\ \beta < \alpha_0) = \Sigma(a_\beta;\ \beta < \alpha_0),$$

since α_0 has no immediate predecessor. This again contradicts the definition of α_0, and therefore we may conclude that (1) holds for every $\alpha \leqq \Omega$. Thus all the desired properties are established for the system $(a'_\alpha;\ \alpha < \Omega)$, and the proof is complete.

LEMMA 3.4: If $S \subset L$, and if S has the two properties

(2) $$a \in S,\ b \leqq a \text{ implies } b \in S,$$

(3) $$T \subset S,\ T \perp \text{ implies } \Sigma(T) \in S,$$

then there exists an element $a_0 \in L$ such that S is the set of all elements $a \in L$ such that $a \leqq a_0$.

PROOF: Let us replace S by the system of elements $(a_\sigma;\ \sigma \in I)$ where $I = S$, $a_a \equiv a$, and apply Lemma 3.3. Thus we obtain a system $(a'_\sigma;\ \sigma \in I)$, with $a'_\sigma \leqq a_\sigma$ $(\sigma \in I)$, $\Sigma(a'_\sigma;\ \sigma \in I) = \Sigma(a_\sigma;\ \sigma \in I) = \Sigma(S)$, $(a'_\sigma;\ \sigma \in I)\perp$. By (2), every $a'_\sigma \in S$, whence if T is the set $(a'_\sigma;\ \sigma \in I)$, then $T \subset S$. Moreover, $T\perp$, and by (3), $\Sigma(T) \in S$. But this means $a_0 \equiv \Sigma(S) \in S$; consequently $a \leqq a_0$ implies $a \in S$ by (2). On the other hand $a \in S$ evidently implies $a \leqq a_0$. Hence $S = (a;\ a \leqq a_0)$.

DEFINITION 3.2: An element $\bar{a} \in Z$ is *discrete* in case there exists a minimal element $a \in L$ such that $\bar{a} = e(a)$.

COROLLARY: If $\bar{a}$ is discrete, $\bar{b} \in Z$, $\bar{b} \leqq \bar{a}$, then $\bar{b}$ is discrete.

PROOF: Since $\bar{a}$ is discrete, $\bar{a}$ is of the form $\bar{a} = e(a)$ with a minimal. Now $\bar{b}a \leqq a$, whence $\bar{b}a \precsim a$, and $\bar{b}a$ is minimal by Lemma 3.1. Then by Theorem 1.3 (f), $e(\bar{b}a) = \bar{b}e(a) = \bar{b}\bar{a} = \bar{b}$, and $\bar{b}$ is discrete.

LEMMA 3.5: If S is the set of all $\bar{a} \in Z$ such that $\bar{a}$ is discrete, then there exists $a_0 \in Z$ such that $S = (\bar{a};\ \bar{a} \leqq a_0,\ \bar{a} \in Z)$.

PROOF: We apply Lemma 3.4 — with L replaced by Z — by establishing conditions (2), (3) for S. Suppose $\bar{a} \in S$, $\bar{b} \in Z$, $\bar{b} \leqq \bar{a}$. Then by the Corollary to Definition 3.2, $\bar{b}$ is discrete, and it follows that $\bar{b} \in S$. Thus (2) holds. To prove (3), consider $T \subset S$, $T \perp$. We may evidently replace T by an independent system $(e(a_\alpha);\ \alpha \in I)$, where the a_α are minimal. Now by Theorem 1.3 (d),

$$\Sigma(e(a_\alpha);\ \alpha \in I) = e(\Sigma(a_\alpha;\ \alpha \in I)). \tag{4}$$

We shall prove that $\Sigma(a_\alpha;\ \alpha \in I)$ is minimal. For this purpose we consider $a \ll \Sigma(a_\alpha;\ \alpha \in I)$ and we need only prove that $a = 0$.

For each $\beta \in I$

$$e(a_\beta)a \ll e(a_\beta)\Sigma(a_\alpha;\ \alpha \in I) = \Sigma(e(a_\beta)a_\alpha;\ \alpha \in I),$$

the last equality holding by Theorem 1.1. But if $\alpha \neq \beta$, $e(a_\beta)a_\alpha \leqq e(a_\beta)e(a_\alpha) = 0$, since $(e(a_\alpha);\ \alpha \in I)\perp$. Hence for every $\beta \in I$, $e(a_\beta)a \ll e(a_\beta)a_\beta = a_\beta$, and $e(a_\beta)a = 0$, since a_β is minimal. Thus $\Sigma(e(a_\alpha)a;\ \alpha \in I) = 0$, whence by Theorem 1.1, $a\Sigma(e(a_\alpha);\ \alpha \in I) = 0$, and therefore also $ae(\Sigma(a_\alpha;\ \alpha \in I)) = 0$. But we had $a \ll \Sigma(a_\alpha;\ \alpha \in I)$, whence it follows from Theorem 1.4 (d), (e), that $e(a) \leqq e(\Sigma(a_\alpha;\ \alpha \in I))$. Consequently $a = ae(a) \leqq ae(\Sigma(a_\alpha;\ \alpha \in I)) = 0$, i.e., $a = 0$. This proves that $\Sigma(a_\alpha; \alpha \in I)$ is minimal and therefore that $\Sigma(e(a_\alpha);\ \alpha \in I) = \Sigma(T)$ is discrete, i.e., is in S. Thus (3) is established, and we obtain by Lemma 3.4 the existence of a_0 such that $S = (\bar{a};\ \bar{a} \leqq a_0,\ \bar{a} \in Z)$.

THEOREM 3.1: There exists a unique element $\Delta \in Z$ such that $\bar{a} \in Z$ is discrete if and only if $\bar{a} \leqq \Delta$.

PROOF: Define Δ as the element a_0 introduced in Lemma 3.5. Then Δ obviously has the desired properties. Its uniqueness is trivial.

COROLLARY: For $\bar{a} \in Z$, $\bar{a}$ is discrete and $\leqq -\Delta$ if and only if $\bar{a} = 0$; for $a \in L$, a is minimal and $\leqq -\Delta$ if and only if $a = 0$.

PROOF: Let $\bar{a}$ be discrete and $\leqq -\Delta$. By Theorem 3.1, $\bar{a} \leqq \Delta$, whence $\bar{a} \leqq \Delta(-\Delta) = 0$, $\bar{a} = 0$. Let a be minimal and $\leqq -\Delta$. Then $e(a)$ is discrete, $e(a) \leqq -\Delta$, whence by the result just proved $e(a) = 0$, and so $a = 0$. The converse implications are trivial.

LEMMA 3.6:

(a) For every discrete $\bar{a} \in Z$ all minimal elements a with $e(a) = \bar{a}$ give rise to the same equivalence class A_a. We denote this class by $A^{\bar{a}}$.
(b) If $\bar{a} \in Z$ is discrete, $e(A^{\bar{a}}) = \bar{a}$.
(c) If $\bar{a} \in Z$ is discrete, $A^{\bar{a}} = \bar{a}A^{\Delta}$.

PROOF: (a) We shall show that if a, b are minimal with $e(a) = e(b)$, then $a \sim b$. Define $f \equiv q(a, b)$ (cf. Theorem 2.8). Then by Theorems 2.8, 2.7, $(-f)b \ll (-f)a$; but $(-f)a \leqq a$, whence $(-f)a \lesssim a$, and therefore $(-f)b \ll a$ by Theorem 2.6(c). Since a is minimal, $(-f)b = 0$. Thus $f \geqq b$, $f \geqq e(b) = e(a) \geqq a$, whence a, $b \leqq f$. Hence, since $fa \lesssim fb$, we have $a \lesssim b$. Similarly, $b \lesssim a$, and consequently $a \sim b$.

(b) If a is minimal, $\bar{a} = e(a) = e(A_a) = e(A^{\bar{a}})$.

(c) Since $\bar{a}$ is discrete, $\bar{a} \leqq \Delta$. Let a be minimal with $e(a) = \Delta$. Then $\bar{a}a \leqq a$ is minimal by Lemma 3.1 and by Theorem 1.3 (f), $e(\bar{a}a) = \bar{a}e(a) = \bar{a}\Delta = \bar{a}$. Hence $A^{\bar{a}} = A_{\bar{a}a} = \bar{a}A_a = \bar{a}A^{\Delta}$.

DEFINITION 3.3: For $n = 0, 1, 2, \cdots, \infty$, define $r_n^{\Delta} \equiv r_n(A^{\Delta}, \dagger)$.

THEOREM 3.2:

(a) $r_0^{\Delta} = 0$,
(b) $r_n^{\Delta} \in Z$ $(n = 1, 2, \cdots, \infty)$,
(c) $(r_n^{\Delta};\ n = 1, 2, \cdots, \infty)\perp$,
(d) $\Sigma(r_n^{\Delta};\ n = 1, 2, \cdots, \infty) = 1$,
(e) $\Sigma(r_n^{\Delta};\ n = 1, 2, \cdots) = \Delta$,
(f) $r_\infty^{\Delta} = -\Delta$,
(g) $nr_n^{\Delta}A^{\Delta} = r_n^{\Delta} \cdot \dagger$ $(n = 0, 1, 2, \cdots)$.

PROOF: (g) Let $n = 0, 1, 2, \cdots$. Then by Theorem 2.16 (c), $r_n^{\Delta} \cdot \dagger - nr_n^{\Delta}A^{\Delta} \ll r_n^{\Delta}A^{\Delta}$. Let $b \in r_n^{\Delta} \cdot \dagger - nr_n^{\Delta}A^{\Delta}$, and let a be a minimal element of A^{Δ}. Then $r_n^{\Delta}a \in r_n^{\Delta}A^{\Delta}$ (cf. Definition 2.6), and since $r_n^{\Delta}a \lesssim a$ (because $r_n^{\Delta}a \leqq a$), $r_n^{\Delta}a$ is minimal by Lemma 3.1. Now $b \ll r_n^{\Delta}a$, whence $b = 0$, and therefore $r_n^{\Delta} \cdot \dagger - nr_n^{\Delta}A^{\Delta} = \theta$, i.e., $r_n^{\Delta} \cdot \dagger = nr_n^{\Delta}A^{\Delta}$.

(a) For $n = 0$, (g) yields $\theta = 0 \cdot r_0^{\Delta}A^{\Delta} = r_0^{\Delta} \cdot \dagger = A_{r_0^{\Delta}}$ whence $r_0^{\Delta} \sim 0$, i.e. $r_0^{\Delta} = 0$ (more generally, for any equivalence class A, $r_0(A, \dagger) = 0$, by the Corollary to Definition 2.11 — Ed.).

(b) This is trivial by definition.

For, $\cdots$ (c), (d), $\cdots$ are precisely (a), (b) of Theorem 2.16 as applied to the special case at hand, since $r_0^{\Delta} = 0$.

(f) We have $r_\infty^{\Delta} = r_\infty(A^{\Delta}, \dagger) = -e(A^{\Delta}) = -\Delta$ by Theorem 2.16 (d), Lemma 3.6 (b).

(e) Since $\Sigma(r_n^{\Delta};\ n = 1, 2, \cdots) + r_\infty^{\Delta} = 1$, it follows by (c) that $\Sigma(r_n^{\Delta};\ n = 1, 2, \cdots) \in Z$ is the unique inverse of $r_\infty^{\Delta} = -\Delta$. But Δ is also such an inverse; therefore (e) holds.

This completes for the time being the study of the element Δ and its subelements. We turn now to a discussion of the non-minimal elements and the subelements of $-\Delta$.

LEMMA 3.7: If $a \in L$ is not minimal, then there exist $b, c \leqq a$ such that $bc = 0$, $e(b) = e(c) \neq 0$, $b \sim c$.

PROOF: Since a is not minimal, there exists by Lemma 3.2 an element $b_1' < a$ such that $e(b_1') = e(a)$. Let c_1 be inverse to b_1' in a, whence $c_1 \neq 0$. Thus $0 \neq e(c_1) \leqq e(a)$. Define $b_1 \equiv e(c_1)b_1'$. Thus $b_1, c_1 \leqq a$, $b_1c_1 = 0$, and $e(b_1) = e(c_1)e(b_1') = e(c_1) \neq 0$. Now Lemma 1.3 applies to b_1, c_1 and hence there exist $b \leqq b_1$, $c \leqq c_1$ with $b \sim c$, $e(b) = e(c) \neq 0$. Also, $bc \leqq b_1c_1 = 0$ and $b, c \leqq a$, since $b \leqq b_1 \leqq a$, $c \leqq c_1 \leqq a$.

THEOREM 3.3: Let $a \in L$. If $e(a) \leqq -\Delta$ (i.e., if $a \leqq -\Delta$), and if S is the set of all $\bar{a} \in Z$ such that there exist $b, c \leqq a$ with $bc = 0$, $b \sim c$, $e(b) = e(c) = \bar{a}$, then $S = (\bar{a};\ \bar{a} \leqq e(a),\ \bar{a} \in Z)$.

PROOF: We prove first that S satisfies conditions (2), (3) (for Z instead of L) of Lemma 3.4. Suppose $\bar{a} \in S$, $\bar{b} \leqq \bar{a}$. Since $\bar{a} \in S$, there exist b, $c \leqq a$ with $bc = 0$, $b \sim c$, $e(b) = e(c) = \bar{a}$. If we define $b' \equiv \bar{b}b$, $c' \equiv \bar{b}c$, then $b', c' \leqq a$, $b'c' = \bar{b}b \cdot \bar{b}c = \bar{b} \cdot bc = 0$, $b' \sim c'$ (since $b \sim c$ yields $\bar{b}b \sim \bar{b}c$ by Theorem 1.4 (a)), and

$$\begin{aligned} e(b') = e(\bar{b}b) &= \bar{b}e(b) && \text{(by Theorem 1.3(f))} \\ &= \bar{b}\bar{a}\ (=\bar{b}) \\ &= \bar{b}e(c) = e(\bar{b}c) && \text{(by Theorem 1.3(f))} \\ &= e(c'). \end{aligned}$$

Thus $\bar{b} \in S$, and (2) holds.

In order to prove (3), let $T \subset S$, $T \perp$. Replace T by a system $(\bar{a}_\alpha;\ \alpha \in I) \perp$, and let $\bar{a} \equiv \Sigma(\bar{a}_\alpha;\ \alpha \in I) = \Sigma(T)$. For each $\alpha \in I$ there exist b_α, $c_\alpha \leqq a$ such that $b_\alpha c_\alpha = 0$, $b_\alpha \sim c_\alpha$, $e(b_\alpha) = e(c_\alpha) = \bar{a}_\alpha$. Now since $(\bar{a}_\alpha;\ \alpha \in I) \perp$, $b_\alpha + c_\alpha \leqq \bar{a}_\alpha$ $(\alpha \in I)$ (since b_α, $c_\alpha \leqq e(b_\alpha) = e(c_\alpha) = \bar{a}_\alpha$), we have also $(b_\alpha + c_\alpha;\ \alpha \in I) \perp$. But since $b_\alpha c_\alpha = 0$ $(\alpha \in I)$, this implies that the combined system $(b_\alpha, c_\alpha;\ \alpha \in I)$ is independent by Part I, Theorem 2.4. Now define $b \equiv \Sigma(b_\alpha;\ \alpha \in I)$, $c \equiv \Sigma(c_\alpha;\ \alpha \in I)$. Then $bc = 0$; $b, c \leqq a$, and by Part I, Theorem 3.6, $b \sim c$. Finally,

$$\begin{aligned} e(b) = e(\Sigma(b_\alpha;\ \alpha \in I)) &= \Sigma(e(b_\alpha);\ \alpha \in I) && \text{(by Theorem 1.3(d))} \\ &= \Sigma(\bar{a}_\alpha;\ \alpha \in I) = \bar{a}, \end{aligned}$$

and similarly $e(c) = \bar{a}$. Hence $\bar{a} \in S$, i.e., $\Sigma(T) \in S$, and (3) is established.

By Lemma 3.4 then, there exists $\bar{a}_0 \in Z$ such that $S = (\bar{a};\ \bar{a} \leqq \bar{a}_0,\ \bar{a} \in Z)$. Now since $\bar{a}_0 \in S$, there exist b_0, $c_0 \leqq a$ with $e(b_0) = \bar{a}_0$; hence $\bar{a}_0 = e(b_0) \leqq e(a)$. We shall prove now that $\bar{a}_0 = e(a)$.

Suppose $\bar{a}_0 \neq e(a)$. Then $\bar{a}_0 < e(a)$ and there exists an inverse $\bar{a}_1 \neq 0$ (in Z) of $\bar{a}_0$ in $e(a)$. Now $a\bar{a}_1 \leqq e(a) \leqq -\Delta$. Moreover, since $\bar{a}_1 \leqq e(a)$, we have $e(a\bar{a}_1) = \bar{a}_1 e(a) = \bar{a}_1 \neq 0$, whence $a\bar{a}_1 \neq 0$. Consequently $a\bar{a}_1$ is not minimal, since by the Corollary to Theorem 3.1, $a\bar{a}_1 \leqq -\Delta$, $a\bar{a}_1$ minimal implies $a\bar{a}_1 = 0$. Hence by Lemma 3.7 there exist $b, c \leqq a\bar{a}_1 \leqq a$ with $bc = 0$, $b \sim c$, $e(b) = e(c) \neq 0$. Now, clearly, $e(b) \in S$, whence $e(b) \leqq \bar{a}_0$. But $b \leqq \bar{a}_1$ implies $e(b) \leqq e(\bar{a}_1)$, so $e(b) \leqq \bar{a}_1\bar{a}_0 = 0$, $e(b) = 0$, contrary to $e(b) \neq 0$. Hence our original assumption $\bar{a}_0 < e(a)$ is false, and it results that $\bar{a}_0 = e(a)$. This completes the proof.

COROLLARY: If $e(a) \leqq -\Delta$, then for every $\bar{a} \leqq e(a)$ there exist $b, c \leqq a$ with $bc = 0$, $b \sim c$, $e(b) = e(c) = \bar{a}$; in particular there exist $b, c \leqq a$ with $bc = 0$, $b \sim c$, $e(b) = e(c) = e(a)$.

PROOF: This is trivial, being weaker than the statement proved in Theorem 3.3.

THEOREM 3.4: If $e(A) \leqq -\Delta$, then there exists $B \leqq A$ such that $2B$ exists, $2B \leqq A$, and $e(B) = e(A)$.

PROOF: Select $a \in A$. Then $e(a) = e(A) \leqq -\Delta$, and by the Corollary to Theorem 3.3 there exist $b, c \leqq a$ with $bc = 0$, $b \sim c$, $e(b) = e(c) = e(a)$. Define $B \equiv A_b$, $C \equiv A_c$. Then $B = C \leqq A$, $e(B) = e(C) = e(A)$. Since $bc = 0$, $B + C$ exists and is equal to A_{b+c}, and since $b + c \leqq a$, we have $B + C = A_{b+c} \leqq A_a = A$. But $B = C$, whence $2B = B + B = B + C \leqq A$. Thus $2B \leqq A$, $e(B) = e(A)$.

Changes from the 1935—37 Edition and Comments on the Text

BY ISRAEL HALPERIN

Page 1, Chapter I of Part I: In private conversation in 1937 von Neumann emphasized the need for a discussion of the inter-relationships of axioms I—VI. A remarkable result in this direction is due to Irving Kaplansky (cf. *Any orthocomplemented complete modular lattice is a continuous geometry*, Annals of Mathematics, vol. 61 (1955), pages 524—541), namely: Axiom III is a consequence of I, II, IV, and V′. Here, V′ denotes *orthocomplementation*; it postulates: For each element a there exists a particular complement a' such that $a \leqq b$ implies $a' \geqq b'$, and for all a, $a'' = a$. For other theorems of this type, cf. also Ichiro Amemiya and Israel Halperin, *Complemented modular lattices*, Canadian Journal of Mathematics, vol. 11 (1959), pages 481—520.

Careful reading of this book shows that III_1 is required only for countable sequences of lattice elements. Hence, this restricted III_1, together with I, II, III_2, IV and V imply III for arbitrary sets of lattice elements.

Page 1, line 9: Axiom II (Continuity) is called *completeness* by many authors.

Page 3, line 1: An inverse of a is called a *complement* of a by many authors.

Page 3, line 24: The Dedekind-MacNeille completion of L' may fail to satisfy the modularity axiom, even though L' itself satisfies this axiom (cf. N. Funayama, *On the completion by cuts of distributive lattices*, Proceedings of the Japan Academy, vol. 20 (1944), pages 1—2. Let L' denote the lattice of all finite-dimensional subspaces of a Hilbert space together with their orthogonal complements, ordered by set inclusion. Let L denote the lattice of *all closed* subspaces. Then L is the completion of L' (clearly); L is *not* modular (cf. Birkhoff and von Neumann, Annals of Mathematics, vol. 37 (1936), particularly page 832), yet L' is modular and orthocomplemented.

Page 4, line 23: The proof in the original Notes has been altered slightly.

Page 5, line 21: That $L(a, b)$ also satisfies VI if L also satisfies VI is shown in Theorem 5.12 of Part I (cf. page 39).

Page 5, line 23: An inverse of a in b is called a *relative complement of a in b* by many authors; a lattice with zero element is called *relatively complemented* if there exists at least one relative complement of a in b whenever $a \leqq b$.

Page 8, Chapter II of Part I: This theory of independence requires Axiom III_2 but does not require III_1, V, or VI. The use of III_1 in the proof of Theorem 2.6 can be avoided as follows: let b, c denote the left and right sides respectively, of (6) on page 12. Then clearly, $b \geqq c$. Hence by IV, $b = c + bd$, where d denotes $\Sigma(a_\sigma; \sigma \notin \mathfrak{P}(I_\rho; \rho \in J))$. But Theorem 2.5 shows that $bd = 0$. Hence $b = c$, as required.

For an independence theory which does not postulate III_1 *or* III_2, cf. Amemiya and Halperin, Canadian Journal of Mathematics, vol. 11 (1959), pages 481—520, particularly § 3.

Page 16, Chapter III and IV: The main purpose of these chapters is to prove transitivity of perspectivity in a restricted case, that is: $a \approx b$, $ab = 0$ imply $a \sim b$ (Theorem 4.2 on page 30). The proof of this fact does not use Axiom VI, and II and III are required only for countable sequences of lattice elements. For a shorter proof of this restricted transitivity of perspectivity, cf. John von Neumann and Israel Halperin, *On the transitivity of perspective mappings*, Annals of Mathematics, vol. 41 (1940), pages 87—93. The shorter proof uses Axiom II and III_2 for *arbitrary* sets of lattice elements but does not require III_1 or VI.

Page 32, Chapter V: This chapter (to and including Theorem 5.3) does not require any of the Axioms III_1, V, VI. Irving Kaplansky (cf. Annals of Mathematics, vol. 61 (1955), pages 524—541, especially pages 537, 538) observed that even III_2 can be dropped if V (complementation) is postulated, as follows. In any modular lattice with zero element, call x, y *completely disjoint* and write $(x, y)P$ if: $x_1 \leqq x$, $y_1 \leqq y$, $x_1 \sim y_1$ imply $x_1 = y_1 = 0$. Then $(x, y)P$ and $x \approx y$ together imply $x = y = 0$ (cf. von Neumann and Halperin, Annals of Mathematics, vol. 41 (1940) page 91, Lemma 9). Consequently, if $c \geqq x + y$ and the lattice is relatively complemented, then $(x, y)P$ is equivalent to: $x \leqq$ every relative complement of y in c (cf. Amemiya and Halperin, Canadian Journal of Mathematics, vol. 11 (1959), page 485, § 2.4); hence in such a lattice, if $(x, a_\alpha)P$ holds for all α and Σa_α exists, then $(x, \Sigma a_\alpha)P$ holds. Now Theorem 5.7 shows that Lemma 5.2 holds if II, V hold (without postulating III, VI).

Page 37, line 12: The proof in the original Notes has been altered slightly.

Page 41, line 3: This proof of the transitivity of perspectivity for irreducible geometries and the proof for reducible geometries in Part III (cf. page 265) use Axioms II and III (for arbitrary sets of lattice elements). For a more direct proof, in which II and III are required only for countable sequences of lattice elements, cf. Israel Halperin, *On the transitivity of perspectivity in continuous geometries*, Transactions of the American Mathematical Society, vol. **44** (1938), pages 537—562.

Page 48, line 69: Minimal sequences were used previously in the construction of a dimension function in the paper *On rings of operators* by F. J. Murray and J. von Neumann, Annals of Mathematics, vol. **37** (1936), pages 116—229.

Page 66, line 17: The inferences (A), (B), (C), (D), (E), and a restricted form of (F) are valid when $\mathscr{D}$ is replaced by a more general *regular* ring (cf. pages 90—92).

Page 69, Chapter II of Part II: Important parts of this chapter are valid for systems which are regular rings, except that the associative law for multiplication is required to hold only in a weak form (cf. Ichiro Amemiya and Israel Halperin, *Complemented modular lattices derived from non-associative rings*, Acta Szeged, vol. 20 (1959), pages 181—201, particularly §§ 2, 3).

More general rings called Baer rings which include regular rings have been defined and analysed by Irving Kaplansky, *Rings of operators*, mimeographed notes (1955), University of Chicago.

Page 70, line 9: If (γ) holds, then a one-sided unit in $\mathfrak{R}$ is necessarily a two-sided unit in $\mathfrak{R}$.

Page 72, line 3 from the bottom: This theorem requires that the regular ring possesses a left unit 1 (this means: $1x = x$ for all x). If a left unit is not postulated, the proofs of Lemma 2.2, 2.3, and 2.4, and that of Theorem 2.3 are easily modified to show that $\bar{R}_{\mathfrak{R}}$ is a modular lattice with zero element which is relatively complemented (cf. also K. D. Fryer and Israel Halperin, *The von Neumann coordinatization theorem for complemented modular lattices*, Acta Szeged, vol. **17** (1956), pages 203—249, particularly page 209.

Page 73, line 12 from the bottom: The proof in the original Notes has been shortened slightly.

Page 74, line 10: The proof in the original Notes has been shortened slightly.

Page 76, line 4: The proof in the original Notes has been shortened slightly.

Page 81, line 19: The proof in the original Notes has been shortened slightly.

The proof that $\mathfrak{R}_n$ is regular if $\mathfrak{R}$ is regular seems to require that $\mathfrak{R}$ should have a unit (i.e. an identity element). But the proof is easily modified to avoid this assumption.

Theorem 2.13 is deduced here from Lemmas 2.11, 2.12, and 2.13, but the technique used to prove these lemmas can be applied directly to yield a short inductive proof of Theorem 2.13. A different inductive proof which relies on Theorem 2.11 is given by B. Brown and N. H. McCoy, Proceedings of the American Mathematical Society, vol. 1 (1950), pages 165—171, particularly pages 167—169.

Page 81, lines 5, 6 from the bottom: The proof in the original Notes has been shortened slightly.

Page 85, line 13: The condition in the original Notes that $\mathfrak{M}$ contain the operator 1 has been dropped.

Page 90, lines 13, 14: Theorem XIV in M. and N. shows how to imbed a (II_1) factor $\mathfrak{M}$ in a ring $\mathcal{U}(\mathfrak{M})$ by adding certain unbounded operators to $\mathfrak{M}$; the discussion here has shown that $\mathcal{U}(\mathfrak{M})$ is a regular ring. But, in fact, if $\mathfrak{M}$ is a (II_1) factor, then the projection geometry of $\mathfrak{M}$ is a continuous geometry and coincides with $\bar{R}_{\mathcal{U}(\mathfrak{M})}$, and $\mathcal{U}(\mathfrak{M})$ is a *continuous* ring (cf. Part II, Chapter XVII). The coordinatization procedure of Theorem 14.1 on page 208, when applied to the projection geometry of $\mathfrak{M}$, yields precisely $\mathcal{U}(\mathfrak{M})$. Thus $\mathcal{U}(\mathfrak{M})$ can be obtained in two ways: by adding unbounded operators to $\mathfrak{M}$, and by coordinatization.

If $\mathfrak{M}$ is a non-finite factor, its projection geometry is a complemented, continuous (i.e. complete) lattice, but it is not modular and does not satisfy Axiom III for continuous geometry. Thus, it cannot be the principal right-ideal lattice of a regular ring. Hence, $\mathfrak{M}$ cannot be imbedded in a *regular* ring which has a lattice of all principal right ideals isomorphic to the projection geometry of $\mathfrak{M}$.

For an extension of the construction of $\mathcal{U}(\mathfrak{M})$, cf. S. K. Berberian, *The regular ring of a finite AW*-algebra*, Annals of Mathematics, vol. 65 (1957), pages 224—240.

Page 90, lines 19—21: These lines require interpretation. Von Neumann applied an imbedding procedure to vector spaces $\mathscr{D}^n$ over an arbitrary division ring $\mathscr{D}$ to construct a continuous geometry $L_\infty(\mathscr{D})$, say (cf. the Foreword to this book). But he explained in his manuscripts (still unpublished) how to apply a corresponding imbedding procedure to n-th order matrix rings $\mathscr{D}_n$ to obtain a *continuous* ring $\mathscr{D}_\infty$, say. Then $L_\infty(\mathscr{D})$ is isomorphic to $\bar{R}_{\mathfrak{R}}$ where $\mathfrak{R} = \mathscr{D}_\infty$.

Thus the $\mathscr{D}_\infty$ are a wide generalization of the ring $\mathscr{C}_\infty$ (where $\mathscr{C}$ denotes the division ring of all complex numbers), the role of complex numbers being played by an arbitrary division ring (division algebra in von Neumann's terminology). But $\mathscr{U}(\mathfrak{M})$ is not isomorphic to $\mathscr{C}_\infty$ (cf. the Foreword to this book).

Page 105, line 20: The proof shows that if $\bar{\mathfrak{a}}_i$ $(i = 1, \cdots, n)$ is a basis for L (*homogeneous* basis need not be postulated), then every element $\mathfrak{u}$ in L can be expressed in the form $\mathfrak{u} = \mathfrak{u}_1 \cup \cdots \cup \mathfrak{u}_n$ with each $\mathfrak{u}_m$ of the form

$$\mathfrak{u}_m = (\bar{\mathfrak{a}}_1 \cup \cdots \cup \bar{\mathfrak{a}}_{m-1} \cup \mathfrak{b}_m) \cap \prod_{i=1}^{m-1} (\mathfrak{c}_{mi} \cup \sum_{\substack{j=1 \\ j\neq i}}^{m-1} \bar{\mathfrak{a}}_i)$$

with $\mathfrak{b}_m \leqq \bar{\mathfrak{a}}_m$ and $\mathfrak{c}_{mi}$ in L_{mi} for all i, m. This representation for arbitrary $\mathfrak{u}$ in L, plays a key role in the proof of the coordinatization theorem (Theorem 14.1 on page 208).

Page 109, line 9 from the bottom: The words "and $I \underset{(1,h)}{\rightleftarrows} I'$" have been added to display (9) in the original Notes and the proof has been shortened accordingly.

Page 115, line 1: These equivalent properties mean that the dual-automorphism is an *orthocomplementation.*

Page 117, Chapters V to XIV of Part II: These Chapters have for their main result the coordinatization theorem (Theorem 14.1 on page 208). The procedure here is the following: define a system $\mathfrak{S}$ of L-numbers (Chapter VI); define addition and multiplication in $\mathfrak{S}$ so that $\mathfrak{S}$ becomes a regular associative ring with identity element (Chapters VI, VII, VIII); lastly, prove the coordinatization theorem for $L(0, \sum_{j=1}^{m} \bar{\mathfrak{a}}_j)$ for all m by extension from $m-1$ to m (Chapters IX—XIV). The most difficult step by far is the extension procedure; this is done by two methods: one applies when $m < n$, the other applies when $m > 3$.

For a direct coordinatization procedure which avoids all extensions, cf. K. D. Fryer and Israel Halperin, *On the coordinatization theorem of J. von Neumann*, Canadian Journal of Mathematics, vol. 7 (1955), pages 432—444, and *The von Neumann coordinatization theorem for complemented modular lattices*, Acta Szeged, vol. 17 (1956), pages 203—249. The second of these papers extends the coordinatization theorem to a class of lattices of order $n = 3$, the lattice analogues of Desarguesian plane geometries. An extension of the coordinatization theorem to lattices of order 3 which are analogues of Moufang projective planes and a corresponding theory of regular *non-associative* rings is given in three papers: (1) K. D. Fryer and Israel Halperin, *On the construction of coordinates for non-Desarguesian complemented modular lattices*, I, II, Proceedings of the Royal Netherlands Academy (Amsterdam), vol. 61 (1958), pages 142—161; (2) Ichiro Amemiya and Israel Halperin, *On the coordinatization of complemented modular lattices*, Proceedings of the Royal Netherlands Academy (Amsterdam), vol. 62 (1959), pages 70—78; (3) Ichiro Amemiya and Israel Halperin. *Complemented modular lattices derived from non-associative rings*, Acta Szeged, vol. 20 (1959), pages 181—201.

An extension of von Neumann's coordinatization theorem has been announced recently by B. Jonsson; this requires only that the unit in the complemented modular lattice L be a union of n independent elements $\mathfrak{a}_i$ with $n \geqq 3$, with each $\mathfrak{a}_i$ perspective to a subelement of $\mathfrak{a}_1$ for $i > 3$, with $\mathfrak{a}_i$ perspective to $\mathfrak{a}_1$ for $i = 2, 3$, and with $L(0, \sum_{i=1}^{3} \mathfrak{a}_i)$ satisfying Desarguesian-type conditions.

Page 118, line 1: The proof in the original Notes has been shortened.

Page 118, line 4 from the bottom: The proof in the original Notes has been shortened slightly.

Page 118, line 11: The definition of *frame* has been added to the definition of *normalized frame* in the original Notes.

Page 119, line 22: The rest of this Chapter is devoted to the proof of Theorem 5.1 on page 127, concerning the projective isomorphisms P and their properties. But the proof of the coordinatization theorem (Theorem 14.1 on page 208) can be modified so as to avoid the use of Theorem 5.1 (cf. Fryer and Halperin, Acta Szeged, vol. 12 (1956), pages 203—249, particularly pages 230—234). Then Theorem 5.1 can be deduced (immediately) from the coordinatization theorem.

Page 123, line 4: The proof of Lemma 5.9 can be shortened considerably, as follows. Choose a Δ of shortest possible length (in (9) on page 123) such that $\Phi \neq$ Identity (assuming such Δ exist). Then Δ must begin:

$$\begin{pmatrix} i & j \\ t_1 & j \end{pmatrix} \begin{pmatrix} t_1 & j \\ t_1 & t_2 \end{pmatrix} \begin{pmatrix} t_1 & t_2 \\ t_3 & t_2 \end{pmatrix} \begin{pmatrix} t_3 & t_2 \\ t_3 & t_4 \end{pmatrix} \cdots$$

with $t_1 \neq i, j$; $t_2 \neq t_1, j$. Suppose $t_2 \neq i$, if possible; then without changing Φ we could replace the first two factors by $\begin{pmatrix} i & j \\ i & t_2 \end{pmatrix} \begin{pmatrix} i & t_2 \\ t_1 & t_2 \end{pmatrix}$; then we could replace $\begin{pmatrix} i & t_2 \\ t_1 & t_2 \end{pmatrix} \begin{pmatrix} t_1 & t_3 \\ t_3 & t_2 \end{pmatrix}$ by $\begin{pmatrix} i & t_2 \\ t_3 & t_2 \end{pmatrix}$. Since such a shortening of Δ is not possible we must have $t_2 = i$. The same argument shows not only that $t_3 = j$, but that Δ begins

$$\begin{pmatrix} i & j \\ k & j \end{pmatrix} \begin{pmatrix} k & j \\ k & i \end{pmatrix} \begin{pmatrix} k & i \\ j & i \end{pmatrix} \begin{pmatrix} j & i \\ j & k \end{pmatrix} \begin{pmatrix} j & k \\ i & k \end{pmatrix} \begin{pmatrix} i & k \\ i & j \end{pmatrix} \cdots.$$

Then Lemma 5.8 shows that Δ was *not* of shortest length, a contradiction; so Lemma 5.9 must hold.

Page 132, line 8: If this δ were chosen as the definition of $\beta\gamma$ (which seems more natural) the coordinatization theorem would come out in terms of *left* principal ideals and *left* submodules.

Page 136, Chapters VII, VIII of Part II: The discussion of addition of L-numbers and the proof of distributivity can be shortened considerably (cf. K. D. Fryer and Israel Halperin, *Coordinates in geometry*, Transactions of the Royal Society of Canada, vol. **48** (1954), pages 11—26, particularly pages 18—24).

Page 136, line 2 from the bottom: The Corollary and its proof have been added to the original Notes.

Page 138, line 19: The original proofs of (2) and (3) have been simplified.

Page 144, line 23: The proof of Theorem 7.2 can be shortened to a few lines (without using special elements $\mathfrak{g}_1, \mathfrak{f}_1$) by using the fact that $n \geqq 4$ and applying the technique used to prove the preceding Theorem 7.1.

Page 145, line 7 from the bottom: The proof in the original Notes has been simplified slightly.

Page 147, line 9 from the bottom: The proof in the original Notes has been simplified slightly.

Page 153, line 9: The proof in the original Notes has been simplified slightly.

Page 158, line 8: Theorem 8.1 (distributivity) can be proved by direct calculation in a few lines without semi-isomorphisms and without Lemmas 8.1—8.4 (cf. Fryer and Halperin, Transactions of the Royal Society of Canada, vol. 48 (1954), particularly pages 23—24)). However, the use of *two* expressions for addition is essential, one to verify right distributivity, the other to verify left distributivity.

Page 169, line 1: The element β in the expression $E = (\beta; \gamma^1, \cdots, \gamma^{m-1})$ can, in practice, be restricted to be *idempotent*. It would be better to write E as $[\gamma^1, \cdots, \gamma^{m-1}, \beta]$. In fact, if β is idempotent, then $[\gamma^1, \cdots, \gamma^{m-1}, \beta] = [\gamma^1\beta, \cdots, \gamma^{m-1}\beta, \beta]$, as is easily verified, and the "coordinatization" of Theorem 14.1 on page 208 will assign to the lattice element E precisely the right submodule of all vectors $(\gamma^1\beta, \cdots, \gamma^{m-1}\beta, \beta, 0, \cdots, 0)\xi$ with ξ arbitrary in $\mathfrak{S}$.

Page 170, line 5: The proof of Lemma 10.3 can be shortened considerably, as follows. Let $\mathfrak{w}$ be defined as in the proof in the text. Then $\sum_{j=1}^{m-1} \bar{\mathfrak{a}}_j$ is an axis of perspectivity between $L(0, \bar{\mathfrak{a}}_m)$ and $L(0, \mathfrak{w})$ since each of $\bar{\mathfrak{a}}_m$, $\mathfrak{w}$ is an inverse of $\sum_{j=1}^{m-1} \bar{\mathfrak{a}}_j$ in $\sum_{j=1}^{m} \bar{\mathfrak{a}}_j$. But $\bar{\mathfrak{a}}_m$ is the join of $(\beta)_m$ and $(1-\beta)_m$. Hence the perspective map of $\bar{\mathfrak{a}}_m$ is the join of the perspective maps of $(\beta)_m$ and $(1-\beta)_m$. This is precisely (4) on page 170.

Page 186, line 8: (**) may be put in the following equivalent form:

$$(\sum_{j=1}^{m-1} \bar{\mathfrak{a}}_j) \cap \big((1; \gamma^1, \cdots, \gamma^{m-1}) \cup (\beta; \delta^1, \cdots, \delta^{m-1})\big)$$

is unchanged if for any ε in $\mathfrak{S}$, and any $i = 1, \cdots, m-1$ the elements γ^i and δ^i are simultaneously replaced by $\gamma^i + \varepsilon$, $\delta^i + \varepsilon$ respectively; in other words, this simultaneous alteration of γ^i and δ^i may change the element $E = (1; \gamma^1, \cdots, \gamma^{m-1}) \cup (\beta; \gamma^1, \cdots, \gamma^{m-1})$ but it does not change $(\sum_{j=1}^{m-1} \bar{\mathfrak{a}}_j) \cap E$.

The proof of (**) for the case $m < n$, given in Chapter XII of Part II uses a certain projective automorphism W (cf. page 191, line 3 from below). This automorphism W is constructed explicitly as a product of three perspective mappings PQR. The restriction $m < n$ is needed precisely because "extra space" is needed to define these perspective mappings although W is an *automorphism* of $L(0, \sum_{j=1}^{m} \bar{\mathfrak{a}}_j)$ and does not itself involve "extra space". It has been observed by Ichiro Amemiya (cf. *On the representation of complemented modular lattices*, Journal of the Mathematical Society of Japan, vol. 9 (1957), pages 263—279) that W can be defined in another way, namely, in a *piece-wise* fashion which does not require "extra-space" (just as we define, in piece-wise fashion, an automorphism W_0 of a projective plane, with the properties: (i) W_0 leaves fixed each point in a given line in the plane, (ii) $pW_0 = q$ for two given different points p, q neither of which is on the given line, and (iii) $aW_0 = a$ for a given point a which is not on the given line and is collinear (but different from each of) p, q. Because of Amemiya's observation, *the restriction* $m < n$ *can be dropped* and Chapter XII of Part II, suitably modified, completes the proof of the coordinatization theorem. This means: Chapters XIII and XIV of Part II (to, but not including the final Theorem 14.1 on page 208) are not needed.

The proof of Theorem 14.1 given by Fryer and Halperin (cf. Canadian Journal of Mathematics, vol. 7 (1955), pages 432—444) is, on the other hand, somewhat related to these Chapters XIII and XIV of Part II and *avoids* the use of automorphisms W.

Page 189, line 9: In the original Notes a slightly different Lemma 11.6 was derived from Lemma 11.5 by use of the projective mappings P (cf. Theorem 5.1 on page 127). H. Löwig (in a letter to von Neumann in 1942) pointed out that the use of these mappings in that proof was not justified.

Page 198, lines 1—2: H. Löwig pointed out (in a letter to von Neumann in 1942) that Lemma 13.1 is a special case of Lemma 12.7.

Page 198, line 4: Lemma 13.2 was found by me in 1937, in the form required for its use in Theorem 13.1 (that is, with $\mathfrak{b} = \bar{\mathfrak{a}}_1 \cup \bar{\mathfrak{a}}_2$ and $\mathfrak{a} = \bar{\mathfrak{a}}_2$) and communicated orally to von Neumann. It permitted the replacement of the original 6-page proof of Theorem 13.1 of Chapter II by the proof given here. Von Neumann then remarked that Lemma 13.2 could be put in the general form given in the text. Later, H. Löwig (in a letter to von Neumann in 1942) pointed out that the original 6-page proof of Theorem 13.1 was faulty.

Von Neumann himself authorized the substitution of the proof given here for Theorem 13.1 in place of the original one. This is the only serious change in this book from the 1935—37 Edition of the Notes.

Page 201, line 5: The proof in the original Notes has been changed slightly.

Page 208, line 10 from the bottom: The proof of this theorem actually shows that if L has a normalized frame $\mathscr{B} = (\bar{\mathfrak{a}}_i, \bar{\mathfrak{c}}_{ij}; i \neq j, i, j = 1, \cdots, n)$ then L is isomorphic to $\bar{R}_{\mathfrak{R}}$ where $\mathfrak{R} = \mathfrak{S}_n$ with $\mathfrak{S}$ a regular ring. Now $\mathfrak{R} = \mathfrak{S}_n$ has been shown to be unique but $\mathfrak{S}$ itself may depend (apparently) on the normalized frame $\mathfrak{R}$. It is clear that $\mathfrak{S}$ depends on $\bar{\mathfrak{a}}_1, \bar{\mathfrak{a}}_2, \bar{\mathfrak{a}}_3, \bar{\mathfrak{c}}_{12}, \bar{\mathfrak{c}}_{13}$ alone, without knowledge of the value of n, since $\mathfrak{S}$ can be constructed "inside" $L(0, \sum_{i=1}^{3} \bar{\mathfrak{a}}_i)$. It is not difficult to show $\mathfrak{S}$ is actually determined by $\bar{\mathfrak{a}}_1$ alone; this means, $\mathfrak{S}(\mathscr{B})$ is the same for all homogeneous bases $\mathscr{B}$ for L which contain a given $\bar{\mathfrak{a}}_1$ as member. Write $\mathfrak{S}(\bar{\mathfrak{a}}_1)$ for this $\mathfrak{S}(\mathscr{B})$.

Then it follows that $\mathfrak{S}(\mathfrak{a}) = \mathfrak{S}(\mathfrak{b})$ if each of $\mathfrak{a}$, $\mathfrak{b}$ is a member of some homogeneous basis (not necessarily of the same order) provided that $\mathfrak{a} \approx \mathfrak{b}$ or even if $\mathfrak{a} = \sum_{i=1}^{N} \mathfrak{a}_i$, $\mathfrak{b} = \sum_{i=1}^{N} \mathfrak{b}_i$ hold with independent joins and finite N and $\mathfrak{a}_i \approx \mathfrak{b}_i$ for each i.

$\mathfrak{S}(\mathfrak{a})$ is a division ring $\mathscr{D}$ for some $\mathfrak{a}$ if and only if L is a finite dimensional projective geometry; in this case the division ring $\mathscr{D}$ is unique, $\mathfrak{a}$ is an atom, and the order n of the basis containing such an $\mathfrak{a}$ is unique; the order m of every homogeneous basis is then a divisor of n and the corresponding ring $\mathfrak{S}$ is precisely the matrix ring $\mathscr{D}_{n/m}$.

If L permits a suitable (unique) normalized dimension function, for example if L is a continuous geometry, then $\mathfrak{a}$ is a member of a homogeneous basis of order m if and only if $\mathfrak{a}$ has (normalized) dimension $1/m$. In this case again $\mathfrak{S}(\mathfrak{a})$ is determined by the value of m alone, $= \mathfrak{S}(m)$ say, and $\mathfrak{S}(m_1)$ is precisely the matrix ring $\mathfrak{R}_{m_2}$ where $\mathfrak{R} = \mathfrak{S}(m_1 m_2)$.

The preceding paragraphs can be expressed in the following way: If n is an integer $\geqq 2$, and $\mathfrak{S}_1$, $\mathfrak{S}_2$ are two regular rings such that the matrix rings $(\mathfrak{S}_1)_n$ and $(\mathfrak{S}_2)_n$ are isomorphic, then it is not known whether in general, $\mathfrak{S}_1$ and $\mathfrak{S}_2$ need to be isomorphic; it can be shown that this is so if $\mathfrak{S}_1$ is a division ring, or a matrix ring over a division ring, or if $\mathfrak{S}_1$ is (more generally) a rank-ring (cf. Chapter XVIII of Part II).

Suppose $\mathfrak{R} = \mathfrak{S}_n$ with $n \geqq 3$. Then Theorem 4.5 of Part II (cf. page 114) asserts that the lattice $\bar{R}_{\mathfrak{R}}$ possesses an orthocomplementation if and only if $\mathfrak{R}$ possesses a corresponding involutoric anti-automorphism x^* such that $x^*x = 0$ implies $x = 0$. Now the elements of $\bar{R}_{\mathfrak{R}}$ can be represented as right-linear sets of vectors $x = (\alpha_i;\ i = 1, \cdots, n)$ with all α_i in $\mathfrak{S}$) (cf. Appendix 3 on page 90). It can be shown that there exists an involutoric anti-automorphism $\alpha \to \alpha^*$ of $\mathfrak{S}$ and a form $(x, y) = \sum_{i=1}^{n} \alpha_i^* \varphi_i \beta_i$ in $\mathfrak{S}$ with the properties: (i)$\varphi_i^* = \varphi_i$ and (ii) $(x, x) = 0$ implies $x = 0$, such that x and $y = (\beta_i)$ are orthogonal in $\bar{R}_{\mathfrak{R}}$ if and only if $(x, y) = 0$. This was shown first for the case that $\mathfrak{S}$ is a division ring by G. Birkhoff and J. von Neumann, *The logic of quantum mechanics*, Annals of Mathematics, vol. **37** (1936), pages 823—843; the result was extended to the general regular $\mathfrak{S}$ by F. Maeda, *Representations of orthocomplemented modular lattices*, Journal of Science of Hiroshima University, vol. **14** (1950), pages 1—4

(Maeda's restriction that $\bar{R}_{\mathfrak{R}}$ should possess an *orthogonal* homogeneous basis is superfluous).

Page 211, lines 1, 2 from the bottom: The proof in the original Notes has been shortened slightly.

Page 221, line 5: Other theorems on the automorphisms of a complemented modular lattice are given in the hand-written manuscripts (still unpublished) of von Neumann.

Page 222, line 4: The symbol L_∞ is not used explicitly on pages 57—58.

Page 225, line 10: The proof of (d) in the original Notes has been transposed to follow that of (e) and has been simplified slightly.

Page 233, lines 3, 4 from the bottom: These lines have been added to the proof in the original Notes.

Page 241, line 3: The use of Axiom III in the proof of Theorem 1.1 can be avoided by using Axioms IV and V (cf. the previous comment relative to page 8, Chapter II of Part I) as follows. Let $x = \sum(S) \cdot b$, $y = \sum(ab;\, a \in S)$. Clearly $x \geqq y$. Let z be an inverse of y in x. Then for each $a \in S$, $az = axz \leqq abz \leqq yz = 0$.

If $S \subset Z$, let z' be a complement of z. Then for each $a \in S$, we have $(a)D$, hence $a = a(z + z') = az + az' = 0 + az' \leqq z'$. Therefore, from the definition of $\sum(S)$, we have $\sum(S) \leqq z'$. Hence $z = zx \leqq z \cdot \sum(S) \leqq zz' = 0$, that is $z = 0$ and so $x = y$.

If $b \in Z$, let b' be a complement of b. Then for every $a \in S$, $a = a(b + b') = ab + ab'$ and $\Sigma(S) = \Sigma(ab;\, a \in S) + \Sigma(ab';\, a \in S)$. Hence $\sum(S) \cdot b = (\sum(ab;\, a \in S)) \cdot b + (\sum(ab';\, a \in S)) \cdot b = \sum(ab;\, a \in S)$, since $(\sum(ab';\, a \in S)) \cdot b \leqq b'b = 0$.

In this proof even (IV) need not be postulated if Z consists of all x for which $(u, v, w)D$ holds whenever any of u, v, w, coincides with x.

Page 244, line 3: Lemma 1.2, 1.3 and their proofs have been here added to the original Notes; they are used to simplify the proof of Lemma 3.7 given in the original Notes.

Page 246, line 18: A continuous lattice is called a *complete* lattice by many authors.

Page 250, line 6: The restriction "non-empty" has been added to the original Notes.

Page 260, line 13: The proof in the original Notes has been altered very slightly.

Page 281, line 1: The condition $b \lesssim c$ in the statement of Lemma 3.7 in the original Notes has been changed to $b \sim c$ and the proof has been simplified slightly.

Page 281, line 9: The condition $b \lesssim c$ in the statement of Theorem 3.3 in the original Notes has been changed to $b \sim c$.

Page 282: The original Notes break off abruptly at this point, although the manuscripts of von Neumann (still unpublished) may go further.

It is clear how to define a minimal sequence A_n in $L(0, -\Delta)$, that is, a sequence A_n such that for all n:

$$2A_{n+1} \leqq A_n \quad \text{and} \quad e(A_n) = -\Delta.$$

Then whenever $A \leqq e \leqq -\Delta$, with e a central element and $0 \leqq \lambda \leqq 1$, write $A \leqq \lambda e$ to mean: for every rational real number $p/q \geqq \lambda$ there exists an integer m_0 such that $A \leqq pA_m$, $e \geqq qA_m$ hold for all $m \geqq m_0$.

It follows that for each $A \leqq -\Delta$ there exists a "resolution of $-\Delta$"; that is, a family of central elements $e_\lambda(A)$ $(0 \leqq \lambda \leqq 1)$ such that $e_\lambda(A) \leqq e_\mu(A)$ whenever $\lambda \leqq \mu$, $e_1(A) = -\Delta$ and the $e_\lambda(A)$ satisfy:
(i) $Ae_\lambda(A) \leqq \lambda e_\lambda(A)$ for all $0 \leqq \lambda \leqq 1$,
(ii) $e \in Z$, $Ae \leqq \lambda e$, $e \leqq (-\Delta)(1 - e_\lambda(A))$ together imply $Ae = 0$.
This leads to a numerical dimension function $D(A) = \int_0^1 \lambda \, dwe_\lambda(A)$ for every continuous non-negative measure w on the Boolean sub-algebra $Z(0, -\Delta)$. (This means: $w(e)$ is defined and $0 \leqq w(e) \leqq 1$ for all central elements $e \leqq -\Delta$; $w(e_1 + e_2) = w(e_1) + w(e_2)$ if $e_1 \cap e_2 = 0$ and $w(\sum_{n=1}^{\infty} e_n) = \lim_{n\to\infty} w(e_n)$ if $e_1 \leqq e_2 \leqq \cdots$ is an increasing sequence in $Z(0, -\Delta)$.) Cf. the Foreword to this book.

A numerical dimension function (which, when normalized, is unique) can be constructed on reducible continuous geometries which are *irreducible relative to an equivalence* induced by a given suitable family of lattice automorphisms (cf. Israel Halperin, *Dimensionality in reducible geometries*, Annals of Mathematics, vol. 40 (1939), pages 581—599; cf. also F. Maeda, *Dimensionsverbände reduzibler Geometrien* (in Japanese), Journal of Science of Hiroshima University, vol. A 13 (1944), pages 11—40; an earlier paper by F. Maeda in Japanese, on this topic, appeared in 1939).

A completion of von Neumann's dimension discussion which leads to a single *vector-valued* dimension and to a sub-direct decomposition of the lattice itself, has been given by Tsurane Iwamura, *On continuous geometries* I, Japanese Journal of Mathematics, vol. 19 (1944), pages 57—71. In a second paper *On continuous geometries* II, Journal of the

Mathematical Society of Japan, vol. 2 (1950), pages 149—164, Iwamura gave a dimension function for a reducible continuous geometry which is irreducible relative to an *abstract* congruence relation. Later, a lattice-theoretic dimensionality theory that applies to certain orthocomplemented (but not necessarily modular) lattices was given by L. H. Loomis (cf. *The lattice theoretic background of the dimension theory*, Memoirs of the American Mathematical Society (1955)). Loomis' theory applies to the projection geometry of every factor of *M.* and *N.* (including the non-finite ones). A discussion which includes these cases and also all continuous geometries (not all continuous geometries are orthocomplemented) was given by S. Maeda, *Dimension functions on certain general lattices*, Journal of Science of Hiroshima University, vol. 19 (1955), pages 211—237 (cf. also S. Maeda, *On the lattice of projections of a Baer *-ring*, Journal of Science of Hiroshima University, vol. 22 (1958), pages 75—88).

A satisfactory description of all reducible geometries in terms of irreducible ones, that is, a satisfactory *direct decomposition* theory has not yet been published. T. Iwamura's description (cf. above) and a very general description due to O. Frink, *Complemented modular lattices and projective spaces of infinite dimension*, Transactions of the American Mathematical Society, vol. 60 (1946), pages 452—467 both yield subdirect decompositions. Von Neumann gave a direct decomposition theory for reducible continuous geometries which are projection geometries of rings of operators in Hilbert space (cf. the Foreword to this book and his paper *On rings of operators. Reduction theory*, Annals of Mathematics, vol. 50 (1949), pages 401—485). A wide class of reducible continuous geometries which permit such a direct decomposition (into "integrals" of irreducible geometries) was given in the paper by Israel Halperin, *Dimensionality in reducible geometries*, Annals of Mathematics, vol. 40 (1939), pages 581—599, but the problem in general remains open.

INDEX